Lecture Notes in Computer Science　　5034

Commenced Publication in 1973
Founding and Former Series Editors:
Gerhard Goos, Juris Hartmanis, and Jan van Leeuwen

T0224473

Rudolf Fleischer   Jinhui Xu (Eds.)

Algorithmic Aspects in Information and Management

4th International Conference, AAIM 2008
Shanghai, China, June 23-25, 2008
Proceedings

 Springer

Volume Editors

Rudolf Fleischer
Fudan University
Department of Computer Science and Engineering
Shanghai 200433, China
E-mail: rudolf@fudan.edu.cn

Jinhui Xu
State University of New York at Buffalo
Department of Computer Science and Engineering
Buffalo, NY 14260, USA
E-mail: jinhui@cse.buffalo.edu

Library of Congress Control Number: 2008927528

CR Subject Classification (1998): F.2.1-2, E.1, G.1-3, J.1

LNCS Sublibrary: SL 3 – Information Systems and Application, incl. Internet/Web
and HCI

ISSN   0302-9743
ISBN-10   3-540-68865-X Springer Berlin Heidelberg New York
ISBN-13   978-3-540-68865-5 Springer Berlin Heidelberg New York

Springer is a part of Springer Science+Business Media

springer.com

© Springer-Verlag Berlin Heidelberg 2008
Printed in Germany

Typesetting: Camera-ready by author, data conversion by Scientific Publishing Services, Chennai, India
Printed on acid-free paper   SPIN: 12278879   06/3180   5 4 3 2 1 0

Preface

This volume contains the proceedings of the 4th International Conference on Algorithmic Aspects in Information and Management (AAIM 2008), held June 23–25, 2008, at Fudan University, Shanghai, China. This conference is intended for original algorithmic research on immediate applications and/or fundamental problems pertinent to information management and management science, broadly construed. The first three conferences were held in Xi'an (2005), Hong Kong (2006), and Portland (2007).

Submissions to this year's conference were conducted electronically. A total of 53 papers were submitted from Algeria, Australia, Bangladesh, Canada, China, France, Germany, Iran, Ireland, Italy, Japan, Latvia, Mexico, Netherlands, UK, Ukraine, and the USA. Each paper was evaluated by three Program Committee members from an international Program Committee (listed on the following pages), assisted in some cases by external reviews and comments. Eventually, 31 papers were accepted, but one paper was later withdrawn. In addition to these selected papers, the conference also included two invited keynote talks by Ding-Zhu Du (University of Texas at Dallas and Xi'an Jiaotong University, China) and Vijay V. Vazirani (Georgia Institute of Technology), and a special session on Fixed Parameter Tractable (FPT) algorithms by Mike Fellows, Rolf Niedermeier, Jianer Chen, and Michael Langston.

We would like to thank all the people who made this meeting possible: the authors for submitting their papers to AAIM 2008, the Program Committee members and external reviewers (listed on the following pages) for their excellent work, the speakers of the FPT special session, and the two invited keynote speakers. Finally, we would like to thank Fudan University, the Software School of Fudan University, and the National Science Foundation of China (NSFC) for their support, and the local organizers and our colleagues for their assistance.

June 2008

Rudolf Fleischer
Jinhui Xu

Organization

AAIM 2008 was organized by the Software School of Fudan University, Shanghai, China.

Program Committee

Tetsuo Asano, JAIST, Japan
Jianer Chen, Texas A&M University, USA
Wei Chen, Microsoft Research Asia, Beijing, China
Francis Chin, The University of Hong Kong, Hong Kong
Ovidiu Daescu, University of Texas at Dallas, USA
Xiaotie Deng, City University of Hong Kong, Hong Kong
Leah Epstein, University of Haifa, Israel
Rolf Fagerberg, University of Southern Denmark, Denmark
Rudolf Fleischer, Fudan University, Shanghai, China (Co-chair)
Xin He, University at Buffalo, The State University of New York, USA
Tsan-Sheng Hsu, Academia Sinica Taipei, Taiwan
Ming-Yang Kao, Northwestern University, USA
Naoki Katoh, Kyoto University, Japan
Guohui Lin, University of Alberta, Canada
Alberto Marchetti-Spaccamela, Sapienza University of Rome, Italy
Ulrich Meyer, Universität Frankfurt, Germany
Matthias Müller-Hannemann, MLU Halle, Germany
M. Muthukrishnan, Google, USA
Giri Narasimhan, Florida International University, USA
Shang-Hua Teng, University of Boston, USA
Xiaodong Wu, University of Iowa, USA
Chen Xi, IAS, Princeton, USA
Jinhui Xu, State University of New York at Buffalo, USA (Co-chair)
Christos Zaroliagis, CTI and University of Patras, Greece
Norbert Zeh, Dalhousie University, Canada
Huaming Zhang, University of Alabama in Huntsville, USA
Yunhong Zhou, HP Labs, USA

External Reviewers

Vincent Conitzer
Refael Hassin
Asaf Levin

Naoyuki Kamiyama
Christos H. Makris
Peter Noel

Rob Stee   Lei Xu
Zhiyi Tan   Li Zhang
Fang Wu   Yongding Zhu

AAIM Steering Committee

Tetsuo Asano, JAIST, Japan
Franz Aurenhammer, TU Graz, Austria
Bernard Chazelle, Princeton University, USA
Danny Z. Chen, University of Notre Dame, USA
Siu-Wing Cheng, HKUST, Hong Kong
Otfried Cheong, KAIST, Korea
Rudolf Fleischer, Fudan University, China
Wen-Lian Hsu, Academia Sinica Taipei, Taiwan
Tao Jiang, University of California at Riverside, USA
Naoki Katoh, Kyoto University, Japan
Ming-Yang Kao, Northwestern University, USA
Xiang-Yang Li, Illinois Institute of Technology, Chicago, USA
C.K. Poon, City University of Hong Kong, Hong Kong
Rajeev Raman, Leicester University, UK
Peter Widmayer, ETH Zürich, Switzerland
Yinfeng Xu, Xi'an Jiaotong University, China
Binhai Zhu, Montana State University, USA

Table of Contents

Double Partition: $(6 + \varepsilon)$-Approximation for Minimum Weight Dominating Set in Unit Disk Graphs

Ding-Zhu Du

Science College, Xi'an Jiaotong University, Xi'an, China
and
Department of Computer Science and Engineering,
University of Texas at Dallas, Richardson, TX 75083
dzdu@utdallas.edu

Abstract. We introduce a new technique on partition, called double partition. With this new type of partition, we obtain a polynomial time $(6 + \varepsilon)$-approximation ($\varepsilon > 0$) for the minimum weight dominating set problem in unit disk graphs, which improves a recent result of a 72-approximation given by Ambühl et al. for solving a long-standing open problem. As a corollary, we obtain a $(9.875 + \varepsilon)$-approximation for the minimum weight connected dominating set problem in unit disk graphs.

R. Fleischer and J. Xu (Eds.): AAIM 2008, LNCS 5034, p. 1, 2008.

Nash Bargaining Via Flexible Budget Markets

Vijay V. Vazirani

College of Computing

Abstract. In his seminal 1950 paper, John Nash defined the bargaining problem; the ensuing theory of bargaining lies today at the heart of game theory. In this work, we initiate an algorithmic study of Nash bargaining problems.

We consider a class of Nash bargaining problems whose solution can be stated as a convex program. For these problems, we show that there corresponds a market whose equilibrium allocations yield the solution to the convex program and hence the bargaining problem. For several of these markets, we give combinatorial, polynomial time algorithms, using the primal-dual paradigm.

Unlike the traditional Fisher market model, in which buyers spend a fixed amount of money, in these markets, each buyer declares a lower bound on the amount of utility she wishes to derive. The amount of money she actually spends is a specific function of this bound and the announced prices of goods.

Over the years, a fascinating theory has started forming around a convex program given by Eisenberg and Gale in 1959. Besides market equilibria, this theory touches on such disparate topics as TCP congestion control and efficient solvability of nonlinear programs by combinatorial means. Our work shows that the Nash bargaining problem fits harmoniously in this collage of ideas.

R. Fleischer and J. Xu (Eds.): AAIM 2008, LNCS 5034, p. 2, 2008.
© Springer-Verlag Berlin Heidelberg 2008

On the Minimum Hitting Set of Bundles Problem

Eric Angel[1], Evripidis Bampis[1], and Laurent Gourvès[2]

[1] IBISC CNRS, Université d'Evry, France
{angel,bampis}@ibisc.fr
[2] CNRS LAMSADE, Université de Paris-Dauphine, France
laurent.gourves@lamsade.dauphine.fr

Abstract. We consider a natural generalization of the classical MIN-IMUM HITTING SET problem, the MINIMUM HITTING SET OF BUNDLES problem (MHSB) which is defined as follows. We are given a set $\mathcal{E} = \{e_1, e_2, \ldots, e_n\}$ of n elements. Each element e_i $(i = 1, \ldots, n)$ has a non negative cost c_i. A *bundle* b is a subset of $\mathcal{E}$. We are also given a collection $\mathcal{S} = \{S_1, S_2, \ldots, S_m\}$ of m sets of bundles. More precisely, each set S_j $(j = 1, \ldots, m)$ is composed of $g(j)$ distinct bundles $b_j^1, b_j^2, \ldots, b_j^{g(j)}$. A solution to MHSB is a subset $\mathcal{E}' \subseteq \mathcal{E}$ such that for every $S_j \in \mathcal{S}$ at least one bundle is covered, i.e. $b_j^l \subseteq \mathcal{E}'$ for some $l \in \{1, 2, \cdots, g(j)\}$. The *total cost* of the solution, denoted by $C(\mathcal{E}')$, is $\sum_{\{i|e_i \in \mathcal{E}'\}} c_i$. The goal is to find a solution with *minimum* total cost.

We give a deterministic $N(1 - (1 - \frac{1}{N})^M)$-approximation algorithm, where N is the maximum number of bundles per set and M is the maximum number of sets an element can appear in. This is roughly speaking the best approximation ratio that we can obtain since, by reducing MHSB to the vertex cover problem, it implies that MHSB cannot be approximated within 1.36 when $N = 2$ and $N - 1 - \epsilon$ when $N \geq 3$. It has to be noticed that the application of our algorithm in the case of the MIN k−SAT problem matches the best known approximation ratio.

Keywords: MINIMUM HITTING SET, MIN k−SAT, approximation algorithm.

1   Introduction

The minimum HITTING SET OF BUNDLES problem (MHSB) is defined as follows. We are given a set $\mathcal{E} = \{e_1, e_2, \ldots, e_n\}$ of n elements. Each element e_i $(i = 1, \ldots, n)$ has a non negative cost c_i. A *bundle* b is a subset of $\mathcal{E}$. We are also given a collection $\mathcal{S} = \{S_1, S_2, \ldots, S_m\}$ of m sets of bundles. More precisely, each set S_j $(j = 1, \ldots, m)$ is composed of $g(j)$ distinct bundles $b_j^1, b_j^2, \ldots, b_j^{g(j)}$. A solution to MHSB is a subset $\mathcal{E}' \subseteq \mathcal{E}$ such that for every $S_j \in \mathcal{S}$ at least one bundle is covered, i.e. $b_j^l \subseteq \mathcal{E}'$ for some $l \in \{1, 2, \ldots, g(j)\}$. The *total cost* of the solution, denoted by $C(\mathcal{E}')$, is $\sum_{\{i|e_i \in \mathcal{E}'\}} c_i$. Notice that, the cost of an element appearing in several bundles is counted once. The objective is to find a solution with minimum total cost.

R. Fleischer and J. Xu (Eds.): AAIM 2008, LNCS 5034, pp. 3–14, 2008.
© Springer-Verlag Berlin Heidelberg 2008

The special case of the MHSB problem, in which a bundle is only an element of $\mathcal{E}$ is the classical MINIMUM HITTING SET problem[1]. It is known to be equivalent to the classical MINIMUM SET COVER problem: positive and negative approximability results for the MINIMUM HITTING SET can be directly derived from the classical MINIMUM SET COVER problem [1]. [2]

1.1 Applications of the MHSB Problem

Our motivation to study the MHSB problem comes not only from its own theoretical interest, but also from the fact that it models many other combinatorial optimization problems of the literature. We illustrate this fact with the MULTIPLE-QUERY OPTIMIZATION problem (MQO) in database systems [10] and the MIN k-SAT problem [3].

In an instance of the MQO problem, we are given a set $Q = \{q_1, q_2, \ldots, q_k\}$ of k database queries and a set $T = \{t_1, t_2, \ldots, t_r\}$ of r tasks. A plan p is a subset of T and a query q_i can be solved by $n(i)$ distinct plans $P_i = \{p_i^1, p_i^2, \ldots, p_i^{n(i)}\}$. Each plan is a set of elementary tasks, and each task t_j has a cost (processing time) $c_j \in \mathbb{Q}^+$. Solving the problem consists in selecting one plan per query, and the cost of a solution is the sum of the costs of the tasks involved in the selected plans (the cost of a task which belongs to at least one selected plan is counted once). Clearly, a query of the MQO problem corresponds to a subset of $\mathcal{S}$ in the MHSB problem, a plan to a bundle, and a task to an element of $\mathcal{E}$. In this context, N is the maximum number of plans per query and M, is the maximum number of queries a task can appear in.

MQO was shown to be NP-hard in [10], and different solution methods have been proposed, including heuristics, branch and bound algorithms [10] and dynamic programming [9]. Up to now, no approximation algorithms with guaranteed performance were known for MQO.

As another application, we consider the MIN $k-$SAT problem. The input consists of a set $\mathcal{X} = \{x_1, \ldots, x_t\}$ of t variables and a collection $\mathcal{C} = \{C_1, \ldots, C_z\}$ of z disjunctive clauses of at most k literals (a constant ≥ 2). A literal is a variable or a negated variable in $\mathcal{X}$. A solution is a truth assignment for $\mathcal{X}$ with cost equal to the number of satisfied clauses. The objective is to find a truth assignment minimizing the number of satisfied clauses. (See Section 4 for the reduction of MIN $k-$SAT to the MHSB problem.) Kohli et al [7] showed that the problem is NP-hard and gave a k-approximation algorithm. Marathe and Ravi [8] improved this ratio to 2, while Bertsimas et al [3] showed that the problem is approximable within $2(1 - \frac{1}{2^k})$. Recently, Avidor and Zwick [2] improved the result for $k = 2$ (ratio 1.1037) and $k = 3$ (ratio 1.2136).

---

[1] Given a collection $\mathcal{S}$ of subsets of a finite set $\mathcal{E}$, and nonnegative costs for every element of $\mathcal{E}$, a *minimal hitting set* for $\mathcal{S}$ is a subset $\mathcal{E}' \subseteq \mathcal{E}$ such that $\mathcal{E}'$ contains at least one element from each subset in $\mathcal{S}$ and the total cost of $\mathcal{E}'$ is minimal.

[2] Recall that in the MINIMUM SET COVER problem, given a universe set $\mathcal{U}$, and nonnegative costs for every element of $\mathcal{U}$, a collection $\mathcal{T}$ of subsets of $\mathcal{U}$, we look for a subcollection $\mathcal{T}' \subseteq \mathcal{T}$, such that the union of the sets in $\mathcal{T}'$ is equal to $\mathcal{U}$, and $\mathcal{T}'$ is of minimal cost.

1.2   Contribution

We give a deterministic $N(1-(1-\frac{1}{N})^M)$-approximation algorithm for the MHSB problem, where N is the maximum number of bundles per set and M is the maximum number of sets an element can appear in. Our algorithm follows a rather classical scheme in the area of approximation algorithms: LP formulation, randomized rounding, derandomization. However, the analysis of the performance guarantee is quite involved. The approximation ratio is, roughly speaking, the best that we can expect for the MHSB problem since, by reducing MHSB to the vertex cover problem, it implies that MHSB cannot be approximated within 1.36 when $N = 2$ and $N - 1 - \epsilon$ when $N \geq 3$.

Our algorithm matches the best approximation ratio for the MIN $k-$SAT problem (for general k) obtained by the algorithm of Bertsimas $et\ al.$ [3] and it can also be applied in the case of the MQO problem.

1.3   Organization of the Paper

We consider the inapproximability of MHSB in Section 2 while Section 3 is devoted to its approximability. We first consider greedy strategies (proofs are omitted) yielding to an M-approximation algorithm, followed by LP-based approximation algorithms. We first give a simple N-approximation algorithm and a randomized $N(1-(1-1/N)^M)$-expected approximation algorithm. In Subsection 3.3, we apply a derandomization technique to derive a deterministic $N(1-(1-1/N)^M)$-approximation algorithm. An analysis of the integrality gap conducted in Subsection 3.4 shows that the approximation result is the best we can expect. Section 4 emphasizes the link between MHSB and MIN$k-$SAT. We finally conclude in Section 5.

2   Inapproximability

We exploit the fact that the MINIMUM HITTING SET problem can be formulated as a MIN VERTEX COVER in hypergraphs. In the latter problem, we are given a hypergraph H and the goal is to find the smallest subset of the vertex set with non empty intersection with each hyperedge of H. Here, we are interested in the particular case of this problem where each hyperedge is composed of exactly k vertices (meaning that for the hitting set instance, each subset $S \in \mathcal{S}$ is such that $|S| = k$). We denote this case by MIN-HYPER $k-$VERTEX COVER. When $k = 2$, we get the classical MIN VERTEX COVER problem on graphs. MIN-HYPER $k-$VERTEX COVER admits a k-approximation algorithm. This result is essentially tight when $k \geq 3$ since Dinur $et\ al$ [4] recently proved that for every $\epsilon > 0$, MIN-HYPER $k-$VERTEX COVER cannot be approximated within ratio $k - 1 - \epsilon$. When $k = 2$, the MIN VERTEX COVER problem cannot be approximated within $10\sqrt{5} - 21 \approx 1.36$ [5] while there is a $2 - \frac{2\ln\ln|V|}{\ln|V|}(1 - o(1))$-approximation algorithm [6].

From the above discussion we can deduce the following result.

Theorem 1. *If there is a ρ-approximation algorithm for the* MHSB *problem, then there is an approximation algorithm with the same ratio ρ for the* MIN-HYPER $k-$VERTEX COVER *problem.*

As a corollary of Theorem 1, MHSB cannot be approximated within $10\sqrt{5} - 21 - \epsilon$ when $N = 2$ and $N - 1 - \epsilon$ when $N \geq 3$.

3   Approximation Algorithms

3.1   Greedy Algorithms

We first consider the greedy algorithm (denoted by GREEDY 1) which selects $\text{argmin}_{1 \leq l \leq g(j)}\{C(b_j^l)\}$ for every j in $\{1, \ldots, m\}$. Actually, GREEDY 1 takes the cheapest bundle of a set without considering what is chosen for the others.

Proposition 1. GREEDY 1 *is an M-approximation algorithm for* MHSB *and the result is tight.*

We turn to a more evolved greedy algorithm which, unlike GREEDY 1, takes into account the bundles selected for other sets. The algorithm, denoted by GREEDY 2, is based on the one that was originally used for SET COVER (see [11]). Given $\mathcal{E}' \subseteq \mathcal{E}$, let $B(\mathcal{E}') = |\{S_j \in \mathcal{S} \mid \exists b_j^l \subseteq \mathcal{E}'\}|$. Actually, $B(\mathcal{E}')$ is the number of sets in $\mathcal{S}$ hit by $\mathcal{E}'$. Let $Eff(b_j^l)$ be the *effective cost* of a bundle defined as

$$Eff(b_j^l) = \begin{cases} \frac{C(b_j^l \setminus \mathcal{E}')}{B(\mathcal{E}' \cup b_j^l) - B(\mathcal{E}')} & \text{if } B(\mathcal{E}' \cup b_j^l) > B(\mathcal{E}') \\ +\infty & \text{otherwise} \end{cases}$$

The algorithm uses a set $\mathcal{E}'$ which is empty at the beginning. While $B(\mathcal{E}') < m$, GREEDY 2 computes the effective cost of each bundle and add to $\mathcal{E}'$ the one which minimizes this function. Unfortunately, we can show that GREEDY 2 does not improve the performance guarantee of GREEDY 1.

Proposition 2. GREEDY 2 *is a ρ-approximation algorithm for* MHSB *such that $\rho \geq M$.*

3.2   LP-Based Algorithms

Solving MHSB may also consist in choosing a bundle for each set of $\mathcal{S}$. This helps to formulate the problem as an integer linear program (ILP).

$$\begin{array}{llr} \text{minimize} & \sum_{1 \leq i \leq n} x_i\, c_i & (1) \\ \text{subject to} & \sum_{l=1}^{g(j)} x_{j,l} \geq 1 & j = 1 \ldots m \quad (2) \\ & \sum_{\{l \mid e_i \in b_j^l\}} x_{j,l} \leq x_i & \forall(i,j) \text{ s.t. } e_i \text{ appears in a bundle} \quad (3) \\ & & \text{of } S_j \\ & x_{j,l} \in \{0,1\} & j = 1 \ldots m \text{ and } l = 1 \ldots g(j) \quad (4) \\ & x_i \in \{0,1\} & i = 1 \ldots n \quad (5) \end{array}$$

Each bundle b_j^l is represented by a variable $x_{j,l}$ ($x_{j,l} = 1$ means b_j^l is a subset of the solution, $x_{j,l} = 0$ otherwise). Each element e_i is represented by a variable x_i ($x_i = 1$ means e_i belongs to the solution, otherwise $x_i = 0$). Among all bundles of a subset S_j, at least one is selected because of the first constraint $\sum_{l=1}^{g(j)} x_{j,l} \geq 1$. The second constraint ensures that all elements of a selected bundle appear in the solution. Since the objective function $\sum_{1 \leq j \leq r} x_j \, c_j$ has to be minimized, an element which does not belong to any selected bundle will not belong to the solution. Let LP be the linear relaxation of the ILP, i.e. replace (4) and (5) by

$$x_{j,l} \geq 0 \qquad j = 1 \ldots m \text{ and } l = 1 \ldots g(j) \tag{6}$$
$$x_i \geq 0 \qquad i = 1 \ldots n \tag{7}$$

In the sequel, OPT and OPT_f are respectively the cost of a solution of ILP and LP (f stands for fractional). We will also use the following fact.

Remark 1. If $\{x\}$ is an optimal solution of the LP defined by (1),(2),(3),(6),(7), we can easily prove that $\sum_{l=1}^{g(j)} x_{j,l} = 1$ for $j = 1, 2, \ldots, m$ and that $x_{j,l} \leq 1$ for every j and l.

Let us now consider the first simple algorithm that we call D-ROUNDING: Solve LP and for $j = 1$ to m, select $b_j^{h_j}$ where $h_j = argmax_{1 \leq l \leq g(j)}\{x_{j,l}\}$ (ties are broken arbitrarily).

Theorem 2. D-ROUNDING *is N-approximate.*

Proof. Let $\{x^*\}$ (*resp.* $\{x\}$), be an optimal assignment for ILP (*resp.* LP). One has:

$$\sum_{1 \leq i \leq n} x_i \, c_i \leq \sum_{1 \leq i \leq n} x_i^* \, c_i$$

Let $\{\tilde{x}\}$ be the solution returned by D-ROUNDING ($\tilde{x}_i = 1$ if e_i belongs to the solution and $\tilde{x}_i = 0$ otherwise). For any fixed i, if $\tilde{x}_i = 1$ then $x_i \geq 1/N$. Indeed, we take the variable whose value is the greatest (at least $1/N$ since $N = \max_j\{g(j)\}$). Then, we have $\tilde{x}_i \leq N \, x_i$ and

$$\sum_{i=1}^n \tilde{x}_i \, c_i \leq N \sum_{i=1}^n x_i \, c_i \leq N \sum_{i=1}^n x_i^* \, c_i \qquad \square$$

Now, we consider a randomized algorithm (called R-ROUNDING) which exploits a natural idea for rounding an optimal fractional solution. It consists in interpreting fractional values of 0-1 variables as probabilities. Formally, the algorithm is as follows: Solve LP and for $j = 1$ to m, select randomly a bundle of S_j with a probability distribution $\{x_{j,1}, \ldots, x_{j,g(j)}\}$.

We prove that R-ROUNDING has a better approximation ratio than D-ROUNDING but, for the sake of readability, we first state two propositions and a lemma whose proofs will be given later.

Proposition 3. *Given two integers $M \geq 2$, $N \geq 2$ and a real $x \in [0,1]$, the function $f(M, N, x) = \frac{(1-x)^M - 1 + Mx}{M - N(1-(1-1/N)^M)}$ is nonnegative, increasing and convex.*

Proposition 4. *Let N, M and P be three positive integers such that $P \leq N$. Let r_1, r_2, ..., r_P be a set of non negative reals such that $\sum_{i=1}^{P} r_i \leq 1$. The following inequality holds*

$$\sum_{i=1}^{P} f(M, N, r_i) \leq f(M, N, \sum_{i=1}^{P} r_i)$$

Lemma 1. *Given an instance of the MHSB problem where $M = \max_i |\{S_j : \exists l \ s.t. \ e_i \in b_j^l\}|$, $N = \max_j \{g(j)\}$ and $\{x\}$ is an optimal assignment for LP, there exists a feasible assignment $\{\tilde{x}\}$ for LP which satifies*

$$\sum_{i=1}^{n} \tilde{x}_i \, c_i \leq \sum_{i=1}^{n} f(M, N, x_i) c_i \tag{8}$$

We now state the main result about R-ROUNDING.

Theorem 3. R-ROUNDING *is $N\left(1 - (1 - \frac{1}{N})^M\right)$-approximate (in expectation).*

Proof. Let u_i be the probability of the event "e_i belongs to the solution returned by R-ROUNDING". Notice that $1 - u_i \geq (1 - x_i)^M$. Indeed, one has $1 - u_i = \prod_{\{j|e_i \in \text{bundle of } s_j\}} \sum_{\{l'|e_i \notin b_j^{l'}\}} x_{j,l'} = \prod_{\{j|e_i \in \text{bundle of } s_j\}} (1 - \sum_{\{l|e_i \in b_j^l\}} x_{j,l})$ $\geq \prod_{\{j|e_i \in \text{bundle of } s_j\}} (1 - x_i) \geq (1 - x_i)^M$. The last but one inequality comes from inequality (3), and the last inequality comes from the definition of M, which is the maximum number of sets an element can appear in. Since $1 - u_i \geq (1 - x_i)^M$, one has $u_i \leq 1 - (1 - x_i)^M$. The expected cost of the solution is then bounded as follows:

$$\mathbf{E}[C(\mathcal{E}')] = \sum_{i=1}^{n} u_i \, c_i \leq \sum_{i=1}^{n} \left(1 - (1 - x_i)^M\right) c_i \tag{9}$$

Using Lemma 1, we know that $\sum_{i=1}^{n} x_i \, c_i \leq \sum_{i=1}^{n} \tilde{x}_i \, c_i$ since $\{\tilde{x}\}$ is feasible while $\{x\}$ is optimal. Using Lemma 1 again we obtain

$$\sum_{i=1}^{n} x_i \, c_i \leq \sum_{i=1}^{n} f(M, N, x_i) c_i = \sum_{i=1}^{n} \frac{(1 - x_i)^M - 1 + Mx_i}{M - N(1 - (1 - 1/N)^M)} c_i$$

This inequality becomes

$$\left(M - N(1 - (1 - 1/N)^M)\right) \sum_{i=1}^{n} x_i \, c_i \leq \sum_{i=1}^{n} \left((1 - x_i)^M - 1 + Mx_i\right) c_i$$

$$\left(N(1 - (1 - 1/N)^M) - M\right) \sum_{i=1}^{n} x_i \, c_i \geq \sum_{i=1}^{n} \left(1 - (1 - x_i)^M - Mx_i\right) c_i$$

$$N(1 - (1 - 1/N)^M) \sum_{i=1}^{n} x_i \, c_i \geq \sum_{i=1}^{n} \left(1 - (1 - x_i)^M\right) c_i$$

Using this last inequality, inequality (9) and $\sum_{i=1}^{n} x_i c_i = OPT_f \leq OPT$ we get the expected result:

$$N(1 - (1 - 1/N)^M)OPT \geq \mathbf{E}[C(\mathcal{E}')] \qquad \square$$

Retrospectively, it was not possible to give a more direct proof of Theorem 3 using $N(1 - (1 - 1/N)^M)x \geq 1 - (1 - x)^M$, because it does not hold when $x \in (0, \frac{1}{N})$.

Proof of Proposition 3

Proof. The function $f(M, N, x)$ is increasing between 0 and 1 since $f'(M, N, x) = (M - N(1 - (1 - 1/N)^M))^{-1}(M - M(1 - x)^{M-1}) \geq 0$. Indeed, we know that $M - N(1 - (1 - 1/N)^M) \geq 0$.

$$(1 - 1/N)^M \geq 1 - M/N$$
$$1 - (1 - 1/N)^M \leq M/N$$
$$N(1 - (1 - 1/N)^M) \leq M \qquad (10)$$
$$0 \leq M - N(1 - (1 - 1/N)^M)$$

Furthermore, $M - M(1 - x)^{M-1} \geq 0$ because $M \geq 1$ and $0 \leq x \leq 1$. As a consequence, $f(M, N, x) \geq 0$ when $0 \leq x \leq 1$ because $f(M, N, 0) = 0$ and $f(M, N, x)$ increases.

The function $f(M, N, x)$ is convex when $0 \leq x \leq 1$ since $f''(M, N, x) = (M - N(1 - (1 - 1/N)^M))^{-1}(M(M - 1)(1 - x)^{M-2}) \geq 0$. $\qquad \square$

Proof of Proposition 4

Proof. Let E be an experiment with N disjoint outcomes $\Omega = \{O_1, O_2, \ldots, O_N\}$. Every outcome occurs with a probability r_i. Then, $\sum_{i=1}^{N} r_i = 1$. Let O'_i be the event "O_i occurs a least once when E is conducted M times". Its probability $\mathbf{Pr}[O'_i]$ is $1 - (1 - r_i)^M$. Given a nonnegative integer $P \leq N$, we clearly have $\sum_{i=1}^{P} r_i \leq 1$. Furthermore, the probability of the event "at least one event in $\{O_1, \ldots, O_P\}$ occurs when E is conducted M times" is $1 - (1 - \sum_{i=1}^{P} r_i)^M$. We clearly have

$$\sum_{i=1}^{P} (1 - (1 - r_i)^M) \geq 1 - (1 - \sum_{i=1}^{P} r_i)^M, \qquad (11)$$

since $\sum_{i=1}^{P} \mathbf{Pr}[O'_i] \geq \mathbf{Pr}[\bigcup_{i=1}^{P} O'_i]$. Inequality (11) gives

$$\sum_{i=1}^{P} ((1 - r_i)^M - 1 + Mr_i) \leq (1 - \sum_{i=1}^{P} r_i)^M - 1 + M \sum_{i=1}^{P} r_i \qquad (12)$$

if we multiply by -1 and add $\sum_{i=1}^{P} Mr_i$ on both sides. Finally, we saw in the proof of Proposition 3 that $M - N(1 - (1 - \frac{1}{N})^M) \geq 0$. Then, one can divide both parts of inequality (12) by $M - N(1 - (1 - \frac{1}{N})^M)$ to get the result. $\qquad \square$

Proof of Lemma 1

Proof. Let $\{x\}$ be the values assigned to the variables when LP is solved. Consider the following algorithm which, given $\{x\}$, computes new values $\{\tilde{x}\}$.

1   **For** $j = 1$ to m **Do**
1.1   **For** $l = 1$ to $g(j)$ **Do**
                 $\tilde{x}_{j,l} := f(M, N, x_{j,l})$
             End For
         End For
2   **For** $i = 1$ to n **Do**
             $\tilde{x}_i = \max_j \{\sum_{\{l|e_i \in b_j^l\}} \tilde{x}_{j,l}\}$
         End For

Actually, the algorithm gives to the variable representing bundle b_j^l the value $f(M, N, x_{j,l})$ which is always nonnegative by Proposition 3 and Remark 1. Then, $\{\tilde{x}\}$ fulfills constraints (6) of LP.

We now show that for all j in $\{1, \ldots, m\}$, we have $\sum_{l=1}^{g(j)} \tilde{x}_{j,l} \geq 1$. Let $P = g(j')$ for a fixed j' belonging to $\{1, \ldots, m\}$. Since $x_{j,l}$ is optimal by Remark 1, we know that

$$\sum_{l=1}^{P} x_{j',l} = 1 \tag{13}$$

By the convexity of f (see Proposition 3), we have

$$\frac{1}{P} \sum_{l=1}^{P} f(M, N, x_{j',l}) \geq f(M, N, \frac{1}{P} \sum_{l=1}^{P} x_{j',l})$$

Using inequality (13), we get

$$f(M, N, \frac{1}{P} \sum_{l=1}^{P} x_{j',l}) = f(M, N, \frac{1}{P})$$

from which we deduce that

$$\frac{1}{P} \sum_{l=1}^{P} f(M, N, x_{j',l}) \geq f(M, N, \frac{1}{P}) = \frac{(1 - \frac{1}{P})^M - 1 + M/P}{M - N(1 - (1 - \frac{1}{N})^M)}$$

$$\sum_{l=1}^{P} f(M, N, x_{j',l}) \geq \frac{M - P(1 - (1 - \frac{1}{P})^M)}{M - N(1 - (1 - \frac{1}{N})^M)} \tag{14}$$

Since $P \leq N$, $N \geq 2$ and $M \geq 2$ we can prove the following inequality.

$$M - P(1 - (1 - \frac{1}{P})^M) \geq M - N(1 - (1 - \frac{1}{N})^M) \tag{15}$$

Using (14) and (15) we deduce

$$\sum_{l=1}^{P} f(M, N, x_{j',l}) \geq 1 \tag{16}$$

Since no particular hypothesis was made for j', we deduce that $\{\tilde{x}\}$ fulfills constraints (2) of LP. Each variable $\tilde{x}_i$ receives the value $\max_j \{\sum_{\{l|e_i \in b_j^l\}} \tilde{x}_{j,l}\}$ at step 2 of the algorithm. Thus, $\{\tilde{x}\}$ fulfills constraints (3) of LP. Since every $\tilde{x}_{j,l}$ is nonnegative, we know that $\tilde{x}_i$ is also nonnegative and $\{\tilde{x}\}$ fulfills constraints (7) of LP. We can conclude that $\{\tilde{x}\}$ is a feasible assignment for LP. The remaining part of the proof concerns inequality (8).

Take an element $e_i \in \mathcal{E}$. We know from step 2 that there is a q in $\{1, \ldots, m\}$ such that

$$\tilde{x}_i = \sum_{\{l|e_i \in b_q^l\}} \tilde{x}_{q,l} = \sum_{\{l|e_i \in b_q^l\}} f(M, N, x_{q,l}) \tag{17}$$

Using Proposition 4, we know that

$$\sum_{\{l|e_i \in b_q^l\}} f(M, N, x_{q,l}) \leq f(M, N, \sum_{\{l|e_i \in b_q^l\}} x_{q,l}) \tag{18}$$

Constraint (3) of the LP says $\sum_{\{l|e_i \in b_q^l\}} x_{q,l} \leq x_i$. Since f is increasing between 0 and 1, we deduce

$$f(M, N, \sum_{\{l|e_i \in b_q^l\}} x_{q,l}) \leq f(M, N, x_i) \tag{19}$$

Using inequalities (17), (18) and (19) we know that $\tilde{x}_i \leq f(M, N, x_i)$ holds for every element e_i. We sum this inequality over all elements and obtain

$$\sum_{i=1}^{n} \tilde{x}_i \leq \sum_{i=1}^{n} f(M, N, x_i)$$

which is the expected result.   □

3.3   Derandomization

The derandomization of R-ROUNDING is done via the method of *conditional expectation* (see for example [11]). We get a deterministic algorithm called D2-ROUNDING.

> Solve LP
> $\mathbf{Pr}[h_j = l] = x_{j,l}$ where $j = 1 \ldots m$ and $l = 1 \ldots g(j)$
> **For** $j = 1$ **to** m **Do**
>   Let $l^* = \operatorname{argmin}_{1 \leq l \leq g(j)} \mathbf{E}[C(h) \mid h_1 = l_1, \ldots, h_{j-1} = l_{j-1}, h_j = l]$
>   Set $h_j = l^*$
> **End For**

Here $\mathbf{E}[C(h)]$ is the expected cost of a solution constructed by randomly choosing for each subset S_j a bundle (and therefore the elements inside) according to the distribution probability given by the values $x_{j,l}$ for $l = 1, \ldots, g(j)$. This expected cost can be computed in polynomial time: If we note u_i the probability that element e_i belongs to the solution, recall that one has $u_i = 1 - \prod_{\{j | e_i \in \text{bundle of } S_j\}} \sum_{\{l' | e_i \notin b_j^{l'}\}} x_{j,l'}$, and we have $\mathbf{E}[C(h)] = \sum_{i=1}^n u_i c_i$. In the same way, $\mathbf{E}[C(h) \mid h_1 = l_1, \ldots, h_{j-1} = l_{j-1}, h_j = l]$ denotes the conditional expectation of $C(h)$ *provided* that we have chosen the bundle $b_{j'}^{l_{j'}}$ for the set $S_{j'}$ (for $1 \le j' \le j - 1$), and bundle b_j^l for the set S_j. In the same way than before, this conditional expectation can be exactly computed in polynomial time.

Theorem 4. D2-ROUNDING *is a deterministic* $N(1 - (1 - \frac{1}{N})^M)$*-approximation algorithm.*

Proof. In the following, we show that the expected cost never exceeds the original one.

Suppose we are given $l = (l_1 \ldots l_{j'})$, a partial solution of the problem such that $l_1 \in \{1, \ldots, g(1)\}$, $l_2 \in \{1, \ldots, g(2)\}$, $\ldots$, $l_{j'} \in \{1, \ldots, g(j')\}$ and $j' \in \{1, \ldots, m-1\}$.

$$\mathbf{E}[C(h) \mid h_1 = l_1, \ldots, h_{j'} = l_{j'}] = \sum_{l=1}^{g(j'+1)} \mathbf{E}[C(h) \mid h_1 = l_1, \ldots, h_{j'} = l_j, h_{j'+1} = l]$$
$$. \mathbf{Pr}[h_{j'+1} = l \mid h_1 = l_1, \ldots, h_{j'} = l_{j'}]$$
$$= \sum_{l=1}^{g(j'+1)} \mathbf{E}[C(h) \mid h_1 = l_1, \ldots, h_{j'} = l_{j'}, h_{j'+1} = l] \, x_{j'+1,l}$$

If $l' = argmin_{1 \le l \le g(j'+1)} \mathbf{E}[C(h) \mid h_1 = l_1, \ldots, h_{j'} = l_{j'}, h_{j'+1} = l]$ then

$$\mathbf{E}[C(h) \mid h_1 = l_1, \ldots, h_{j'} = l_{j'}, h_{j'+1} = l'] \le \mathbf{E}[C(h) \mid h_1 = l_1, \ldots, h_{j'} = l_{j'}]$$

At each step, the algorithm chooses a bundle (fixes its probability to 1) and the new expected cost does not exceed the previous one. Since $\mathbf{E}[C(h)] \le N(1 - (1 - \frac{1}{N})^M) OPT$ at the beginning of the algorithm, D2-ROUNDING converges to a solution whose total cost is $N(1 - (1 - \frac{1}{N})^M)$-approximate. $\square$

3.4   Integrality Gap

Theorem 5. *The integrality gap of the LP is* $N(1 - (1 - \frac{1}{N})^M)$.

Proof. Given N and m, we can build an instance as follows.

- $\mathcal{S} = \{S_0, \ldots, S_{m-1}\}$
- $S_j = \{b_j^0, \ldots, b_j^{N-1}\}$, $j = 0, \ldots, m-1$
- $\mathcal{E} = \{e_0, \ldots, e_{N^m-1}\}$
- $c_i = 1 \ \forall e_i \in \mathcal{E}$

- Take $i \in \{0, \ldots, N^m - 1\}$ and let α be the representation of i with the numeral N-base system, i.e. $i = \sum_{j=0}^{m-1} \alpha(i,j) N^j$ where $\alpha(i,j) \in \{0, \ldots, N-1\}$. We set $e_i \in b_j^l$ if $\alpha(i,j) = l$.

We view solutions as vectors whose jth coordinate indicates which bundle of S_j is selected. Given a solution h, an element e_i is not selected if, for $j = 0, \ldots, N-1$, we have $\alpha_i^j \neq h_j$. Then, exactly $(N-1)^m$ elements are not selected. The total cost is always $N^m - (N-1)^m$. Now consider LP. If the variable $x_{j,l}$ of each bundle b_j^l is equal to $1/N$ then the fractional cost of the solution is N^{m-1}. Indeed, an element e_i appears in exactly one bundle per S_j and the value of its variable x_i in LP is also $1/N$. As a consequence, we have $OPT_f = N^{m-1}$. Since $M = m$ in the instance, we get the following ratio

$$\frac{OPT}{OPT_f} = \frac{N^M - (N-1)^M}{N^{M-1}} = N(1 - (1 - \frac{1}{N})^M) \qquad \square$$

4 About MIN k−SAT

Theorem 6. *If there is a ρ-approximation algorithm for MHSB then there is an approximation algorithm with the same ratio ρ for MIN k−SAT.*

Proof. Let A be a ρ-approximation algorithm for MHSB. Take an arbitrary instance of MIN k−SAT and build a corresponding instance of MHSB as follows. The collection $\mathcal{S}$ is made of t sets $S_1, \ldots, S_t$, one for each variable of $\mathcal{X}$. Each set S_j is composed of two bundles b_j^T and b_j^F. The set $\mathcal{E}$ contains z elements $e_1, \ldots, e_z$, one for each clause. Each element e_i has a cost $c_i = 1$. Finally, $b_j^T = \{e_i \mid C_i$ contains the unnegated variable $x_j\}$ and $b_j^F = \{e_i \mid C_i$ contains the negated variable $x_j\}$. The resulting instance of MHSB is such that $N = 2$ and $M = k$.

Let τ be a truth assignment for the instance of MIN k−SAT with cost $C(\tau)$. One can easily derive from τ a solution h for the corresponding instance of MHSB with cost $C(h) = C(\tau)$. Indeed, let h_j be T if x_j is assigned the value in τ, otherwise $h_j = F$. Conversely, let h be a solution for the MHSB instance (with $N = 2$ and $M = k$). One can easily derive a truth assignment τ for the corresponding instance of MIN k−SAT with cost $C(h) = C(\tau)$. Indeed, x_j gets the value *true* if $h_j = T$, otherwise x_j is assigned the value *false*. $\qquad \square$

As a corollary of Theorem 6, MIN k−SAT admits a $2(1 - \frac{1}{2^k})$-approximation algorithm because D2-ROUNDING is a $N(1 - (1 - 1/N)^M)$-approximation algorithm and the reduction is such that $N = 2$ and $M = k$. This result is equivalent to the one proposed by Bertsimas et al. [3].

5 Concluding Remarks

Among the deterministic approximation algorithms that we considered, D2-ROUNDING is clearly the best in terms of performance guarantee since N

$(1 - (1 - 1/N)^M) < \min\{N, M\}$ (see inequality (10)). Because of the integrality gap, improving this ratio with an LP-based approximation algorithm requires the use of a different (improved) formulation. An interesting direction would be to use semidefinite programming and an appropriate rounding technique as used by Halperin [6] for vertex cover in hypergraphs.

References

1. Ausiello, G., D'Atri, A., Protasi, M.: Structure preserving reductions among convex optimization problems. Journal of Computer and System Sciences 21(1), 136–153 (1980)
2. Avidor, A., Zwick, U.: Approximating MIN 2-SAT and MIN 3-SAT. Theory of Computer Systems 38(3), 329–345 (2005)
3. Bertsimas, D., Teo, C.-P., Vohra, R.: On dependent randomized rounding algorithms. Operation Research Letters 24(3), 105–114 (1999)
4. Dinur, I., Guruswami, V., Khot, S., Regev, O.: A new multilayered PCP and the hardness of hypergraph vertex cover. In: Proceedings of STOC 2003, pp. 595–601 (2003)
5. Dinur, I., Safra, S.: The importance of being biased. In: Proceedings of STOC 2002, pp. 33–42 (2002)
6. Halperin, E.: Improved Approximation Algorithms for the Vertex Cover Problem in Graphs and Hypergraphs. SIAM J. Comput. 31(5), 1608–1623 (2002)
7. Kohli, R., Krishnamurty, R., Mirchandani, P.: The minimum satisfiability problem. SIAM Journal on Discrete Mathematics 7, 275–283 (1994)
8. Marathhe, M.V., Ravi, S.S.: On approximation algorithms for the minimum satisfiability problem. Information Processing Letters 58, 23–29 (1996)
9. Toroslu, I.H., Cosar, A.: Dynamic programming solution for multiple query optimization problem. Information Processing Letters 92(3), 149–155 (2004)
10. Sellis, T.K.: Multiple-Query Optimization. Transactions on Database Systems 13(1), 23–52 (1988)
11. Vazirani, V.V.: Approximation Algorithms. Springer, Heidelberg (2001)

Speed Scaling with a Solar Cell

Nikhil Bansal[1], Ho-Leung Chan[2], and Kirk Pruhs[2,*]

[1] IBM T.J. Watson Research, P.O. Box 218, Yorktown Heights, NY
nikhil@us.ibm.com
[2] Computer Science Department, University of Pittsburgh
{hlchan,kirk}@cs.pitt.edu

Abstract. We consider the speed scaling problem of scheduling a collection of tasks with release times, deadlines, and sizes so as to minimize the energy recharge rate. This is the first theoretical investigation of speed scaling for devices with a regenerative energy source. We show that the problem can be expressed as a polynomial sized convex program. We that using the KKT conditions, one can obtain an efficient algorithm to verify the optimality of a schedule. We show that the energy optimal YDS schedule, is 2-approximate with respect to the recharge rate. We show that the online algorithm BKP is $O(1)$-competitive with respect to recharge rate.

1 Introduction

Chip manufacturers such as Intel, AMD and IBM have made it a priority to redesign their chips to consume less power and to provide various hardware and software capabilities for power management. All of these chip manufacturers make chips that use dynamic speed scaling as a power management technique. Typically, the power consumed varies as the cube of the processor speed, and hence this can yield significant energy/temperature reductions.

The first theoretical investigation of speed scaling algorithms was in the seminal paper by Yao, Demers and Shenker [16]. They considered the problem of feasibly scheduling a collection of jobs with arbitrary release times, deadlines and sizes to minimize the energy consumed, and gave both offline and online algorithms for it. Subsequently, there has been a lot of work on improving these results and extending them to optimize various other objectives such as, flow time, throughput and so on [2,3,4,5,7,8,11,13,14,15,17].

All of these theoretical investigations of speed scaling as an energy management technique involve problems where the goal is to minimize the total energy used. This would be the appropriate objective if the energy source was a battery, and the goal was to extend the battery's lifetime. But some devices, most notably some sensors, also contain technologies that allow then to harvest energy from their environment. The most common energy harvesting technology is solar cells. Some sensors also contain technology that allows then to scavenge

* Supported in part by NSF grants CNS-0325353, CCF-0514058 and IIS-0534531.

R. Fleischer and J. Xu (Eds.): AAIM 2008, LNCS 5034, pp. 15–26, 2008.

energy from ambient vibrations. To give some feel for state of technology, batteries can store about 1 Joule of energy per cubic millimeter, while solar cells can provide approximately 100 micro-Watt per square centimeter in bright sunlight, and vibration devices can provide nano-Watts per cubic millimeter [1].

In this paper we initiate a study of speed scaling for energy management in devices that contain both a battery and an energy harvesting device, which we will henceforth assume for simplicity is a solar cell. Our goal is understand how the presence of a solar cell will affect the resulting speed scaling problems. For simplicity, we assume that the solar cell generates energy at a fixed rate (although many of a our results apply to a more general setting). We consider the deadline feasibility problem introduced in [16] because it is the most studied, and probably the best understood, speed scaling problem in the literature. We consider the objective minimizing the recharge rate, subject to the deadline feasibility constraint.

1.1   Related Results

Before explaining our results, let us recap what is known about speed scaling with a deadline feasibility constraint on battery only devices. The standard assumption is that when the processor is run at speed s, then the power consumption is $P(s) = s^\alpha$ for some constant $\alpha > 1$ [10,4,12]. For CMOS based devices, which will likely remain the dominant technology for the near term future, the well known cube-root rule is that the speed s is roughly proportional to the cube-root of the power P, or equivalently, $P(s) = s^3$, i.e., [16] gave an optimum greedy algorithm YDS. For the online version, they gave an algorithm AVR and showed that it is $2^{\alpha-1}\alpha^\alpha$ competitive. It was recently shown that AVR is in fact $(2 - \epsilon)\alpha^\alpha$ competitive, where ϵ is a vanishingly small function of α. [16] also proposed another algorithm OA, which was shown by [4] to be α^α competitive. [4] gave another online algorithm BKP and showed that is was $2(\alpha/(\alpha - 1))^\alpha e^\alpha$ competitive(this is the best known competitive ratio for large α. It is also known that any algorithm must have competitive ratio of at least $e^{\alpha-1}/\alpha$ [6] and hence the result cannot be improved substantially. Improved results for the practically interesting cases of $\alpha = 2$ and $\alpha = 3$have also been obtained recently [6].

1.2   Our Results

We consider both the offline and online versions of the minimum recharge rate problem. In Section 3 we show that the offline problem can be expressed as a convex program and hence can be solved to any desired accuracy by using standard techniques such as the Ellipsoid Method. We then explore this convex program further in Section 4. We analyze the consequences of the KKT conditions for optimality for this program and obtain an equivalent set of combinatorial properties that are both necessary and sufficient for a solution to be optimal. This gives us several insights into the structure of an optimum solution and also allows us to obtain an efficient test to determine whether a solution is optimum or not.

In Section 5, we show that the YDS algorithm, that is optimal for the no-recharge case, is in fact a 2-approximation for the minimum recharge rate problem. We also show that this bound is tight. In the special case when the release times and deadlines of jobs are similarly ordered, we show that YDS is optimal. Finally, in Section 6, we consider the online setting, and show that we show that the BKP algorithm [4] is $O(1)$ competitive for the problem. In particular, BKP achieves a competitive ratio of $4(\frac{\alpha}{\alpha-1})^{\alpha}e^{\alpha}$.

In summary, our results seem to suggest that the inclusion of a solar cell in some sense changes the resulting speed scaling problem by a constant factor. As evidence of this, the optimal energy schedule YDS, is an $O(1)$-approximate schedule for recharge rate, and the algorithm BKP is $O(1)$-competitive for both energy and recharge rate. The KKT conditions reveal that cutting the recharge-rate optimal schedule at the points where the battery is empty, partitions the schedule into energy optimal YDS schedules. So there is some relationship between energy optimal and recharge-rate optimal schedules. However, computing the recharge-rate optimal schedule is still seemingly much harder than computing an energy optimal schedule because it is not clear how to partition the work amongst these YDS subschedules of the recharge-rate optimal schedule.

1.3   Formal Problem Statement

We consider a system that consists of a battery that can be charged at a rate of R, i.e. the energy reserve of the battery increases by R units per unit time, from an external source such as a solar cell. The battery is used to run a processor.

The input is a collection of jobs, where each job i has a release time r_i when it arrives into the system, work w_i that must be performed to complete the job, and a deadline d_i by which this work must be completed. In the online version of the problem, the scheduler learn about job i at time r_i. At this point it also learns w_i and d_i. A schedule specifies at each time which job is run, and at what speed. Note that if the processor runs at speed s, the power is consumed at rate $R - s^{\alpha}$. Thus the energy level at any time t' is $\int_{t=0}^{t'}(R - s(t)^{\alpha})dt$. We say that a schedule is feasible if the system never runs out of power. That is, at any time the energy level of the battery is non-negative. In the minimum recharge rate problem, that we consider in this paper, the goal is to construct a feasible schedule that minimizes the recharge rate R required.

We assume that the energy level of the battery at $t = 0$ is 0. This is without loss of generality; given an instance I with battery level E_0 at $t = 0$, we can construct another instance I' with battery level 0 at $t = 0$ and with all the jobs in I shifted forward in time by E_0/R units. We also assume that there is no upper bound on the amount of energy that the battery can hold. This assumption is justified as the problem is most interesting when there is not much "slack" in the instance, and works in the regime where the battery is never saturated.

2 Preliminaries

We begin by reviewing the algorithms YDS and BKP for energy efficient scheduling, as well as the KKT conditions for convex programming.

The Algorithm YDS. Given an instance I, let the density of any time interval $[t, t']$ be defined as $\text{den}(t, t') = w(t, t')/(t' - t)$, where $w(t, t')$ is the total size of jobs with release time at least t and deadline at most t'. Intuitively, $\text{den}(t, t')$ is the minimum average speed at which any feasible algorithm must work during the interval $[t, t']$. YDS applies the following steps until all jobs are scheduled: Let $[t, t']$ be the highest density interval. The speed is set to $\text{den}(t, t')$ during $[t, t']$. Then the instance is modified such that times $[t, t']$ did not exist. That is, all deadlines $d_i > t$ are modified $d'_i = \max\{t, d_i - (t' - t)\}$, and all release times $r_i > t$ are modified to $r'_i = \max\{t, r_i - (t' - t)\}$, and the process is repeated. The jobs are scheduled in the earliest deadline first order.

We note that each job is run at fixed speed in the YDS schedule. This speed is fixed with respect to time, but may be different for different jobs. Moreover, if job i runs at speed s, then the speed at any time during $[r_i, d_i]$ is at least s.

Another useful (but non-algorithmic) view of YDS is the following. Start with an arbitrary schedule, and keep improving the schedule as follows until no longer possible: If some job i runs at time t when the processor speed is s, and there is some other time t' where job i can run (i.e. $r_i \leq t' \leq d_i$) but the speed at t' is less than s, then move infinitesimally small work of job i from t to t'.

The Algorithm BKP. At any time t and $t_1 < t \leq t_2$, let $w(t, t_1, t_2)$ be the total size of jobs that have release time at least t_1, deadline at most t_2 and have been released by time t. Intuitively, it is an estimation for the density of the interval $[t_1, t_2]$ based on the jobs released by t. Let $v(t)$ be defined by

$$v(t) = \max_{t' > t} \frac{w(t, t - (e - 1)(t' - t), t')}{e(t' - t)}$$

Then, at any time t, BKP runs at speed $e \cdot v(t)$ and processes the unfinished job with earliest deadline. BKP is known to be $2(\frac{\alpha}{\alpha-1})^{\alpha} e^{\alpha}$-competitive [4].

The KKT Conditions. Consider a convex program

$$\min f_0(x) \quad s.t.$$
$$f_i(x) \leq 0 \quad i = 1, \ldots, n$$

Assume the functions f_i are all differentiable. Let λ_i, $i = 1, \ldots, n$ be a variable (Lagrangian multiplier) associated with f_i. Then the necessary KKT conditions for solutions x and λ to be primal and dual optimal are:

$$f_i(x) \leq 0 \quad i = 1, \ldots, n \quad (1)$$
$$\lambda_i \geq 0 \quad i = 1, \ldots, n \quad (2)$$
$$\lambda_i f_i(x) = 0 \quad i = 1, \ldots, n \quad (3)$$

$$\nabla f_0(x) + \sum_{i=1}^{n} \lambda_i \nabla f_i(x) = 0 \qquad (4)$$

We refer to the above four equations as Condition 1, 2, 3 and 4 of the KKT conditions, respectively. Condition 3 is commonly known as complementary slackness. If the program is strictly feasible, i.e., there is some point x where $f_i(x) < 0$ for $i = 1, \ldots, n$, then these conditions are also sufficient [9].

3   Convex Programming Formulation

In this section, we give a convex program to find the minimum recharge rate, which implies that the problem can solved optimally in polynomial time. For simplicity of description, we give a pseudo-polynomial sized time indexed formulation, but as we show later the size can be made polynomial.

Let I be any job sequence. Recall that the release time, size and deadline of a job i are denoted as r_i, w_i and d_i, respectively. Without loss of generality, we assume that the release time and deadline of each job are integers. Let $w_{i,j}$ be the amount of work done on job i during time $[j-1, j]$. Then, minimizing the recharge rate R can be written as the following program CP.

$$\min R \quad s.t.$$

$$w_i - \sum_{t:r_i < t \le d_i} w_{i,t} \le 0 \qquad \forall i = 1, 2, \ldots \qquad (5)$$

$$\sum_{t:t \le j} \left(\sum_{x:r_x < t \le d_x} w_{x,t} \right)^\alpha - Rj \le 0 \qquad \forall j = 1, 2, \ldots \qquad (6)$$

$$-w_{i,j} \le 0 \qquad \forall i, j = 1, 2, \ldots \qquad (7)$$

The constraints (5) enforce that each job is completed. Constraints (6) enforce that the battery is non-negative at any integral time. We need to show that the optimal solution for CP gives the minimum recharge rate. This is not completely obvious since CP does not explicitly enforce that the battery does not run out of energy at some non-integral time.

Lemma 1. *The optimal solution R for CP is the minimum recharge rate to complete all jobs in I.*

Proof. Constraints (5) guarantee that each job is completed. Consider any time j and let E_{j-1} and E_j be the energy in the battery at time $j-1$ and j, respectively. Let $s = \sum_{i:r_i < j \le d_i} w_{i,j}$ be the speed during $[j-1, j]$ and let $R' = s^\alpha$. Then for any $w \in [0, 1]$, at time $j-1+w$ the energy in the battery is $E_{j-1} + w(R - R')$. If $R - R' \ge 0$, then $E_{j-1} + w(R - R') \ge E_{j-1} \ge 0$; else if $R - R' < 0$, then $E_{j-1} + w(R - R') \ge E_{j-1} + (R - R') = E_j \ge 0$. Hence, if $E_{j-1} \ge 0$ and $E_j \ge 0$, then the battery is not depleted at any time during $[j-1, j]$. This implies that the schedule returned by CP is feasible. Conversely, every feasible schedule must satisfy the constraints stated in CP. Hence R is the minimum recharge rate. $\square$

Since CP is convex, we can apply the standard methods to determine R to any desired accuracy. We remark that CP has pseudo-polynomial size as the number of variables and equations are depend upon the time horizon. However, given the insight provided by the KKT conditions in the next section, we can reduce the size to polynomial by considering only those time points that are the release time or deadline of a job. We can redefine $w_{i,j}$ to be the work done on job i between the $(j-1)$-th and j-th time points. We also need to modify the left size of (6) such that the speed during that interval is $\sum_i w_{i,j}$ divided by the length of the interval. The resulting convex program gives the minimum recharge rate.

4   Recognizing an Optimal Schedule

We now study the consequences of the KKT conditions when applied to CP and the structural properties they impose on an optimal solution. This will lead to a simple algorithm to recognize an optimal schedule.

For our convex program CP, the constraints are differentiable and strictly feasible, so the KKT conditions are both necessary and sufficient for a solution to be optimal. Associate a dual variable α_i for the equation for job i in constraints (5) of CP. Associate a dual variable β_j for the equation for time j in constraints (6). Associate a dual variable $\gamma_{i,j}$ for the equation for job i and time j in (7). Now consider the four KKT conditions 1-4. Condition 1 states that the optimal solution satisfies the constraints of CP (and hence is feasible). Condition 2 states that α_i, β_j and $\gamma_{i,j}$ are non-negative. For Condition 3, the equations become

$$\alpha_i \left(w_i - \sum_{t:r_i < t \leq d_i} w_{i,t} \right) = 0 \qquad \forall i = 1, 2, \ldots \tag{8}$$

$$\beta_j \left(\sum_{t:t \leq j} \left(\sum_{x:r_x < t \leq d_x} w_{x,t} \right)^\alpha - Rj \right) = 0 \qquad \forall j = 1, 2, \ldots \tag{9}$$

$$\gamma_{i,j} w_{i,j} = 0 \qquad \forall i, j = 1, 2, \ldots \tag{10}$$

Equation (9) implies that β_j is positive only if the battery is empty at time j. Equation (10) implies that $\gamma_{i,j}$ is zero if job i is processed during $[j-1, j]$.

We now consider Condition 4. We list out separately the terms corresponding to each partial derivative in the gradient. When the derivative is taken w.r.t. R, we obtain that

$$1 - \sum_j j\beta_j = 0 \tag{11}$$

When the derivative is taken w.r.t. variable $w_{i,j}$, we obtain that

$$\alpha_i + \gamma_{i,j} = \alpha \left(\sum_{x:r_x < j \leq d_x} w_{x,j} \right)^{\alpha-1} \left(\sum_{t:j \leq t} \beta_t \right) \qquad \forall i, j = 1, 2, \ldots \tag{12}$$

Note that $\sum_{x:r_x < j \leq d_x} w_{x,j}$ is the speed of the schedule during $[j-1, j]$. Hence, the above equation gives a relationship of how the speed depends on α, β and γ.

As we now show, these KKT conditions are equivalent to the following combinatorial properties that a schedule must satisfy.

Lemma 2. *Let I be any job sequence and S be a schedule for I. Then, S is optimal if and only if it satisfies the following 4 properties.*

1. *S completes all jobs and the battery is not depleted at any integral time.*
2. *There exists time $T > 0$ such that the battery has zero energy at T and no job with deadline after T is processed by T.*
3. *Let T be the smallest time satisfying Property 2. Let $0 = t_0 < t_1 < \ldots < t_k = T$ be times up to T such that the battery has zero energy. Then, for each interval $[t_{y-1}, t_y]$, $y = 1, \ldots, k$, the work processed during $[t_{y-1}, t_y]$ is scheduled using the YDS schedule.*
4. *There exists multipliers $m_1, m_2, \ldots, m_{k-1} \geq 1$ for $t_1, t_2, \ldots, t_{k-1}$ with the following property. Let $s_{i,y}$ denote the speed that job i is processed during $[t_{y-1}, t_y]$. Then, if i is processed during both $[t_{y-1}, t_y]$ and $[t_{y'-1}, t_{y'}]$, $y < y'$, we have $s_{i,y'}/s_{i,y} = m_y m_{y+1} \cdots m_{y'-1}$.*

Remark: We note that the multipliers $m_1, \ldots, m_{k-1}$ are independent of the jobs, and hence the ratios $s_{iy'}/s_{iy}$ are identical for each job i.

Proof. We first show sufficiency, that is, if S satisfies the 4 properties stated in Lemma 2, then S also satisfies the KKT conditions for CP and hence S is optimal. We then show that these properties are also necessary. In particular, we show that if S does not satisfy these properties, then there is another feasible schedule with a smaller recharge rate, implying that S is not optimal. We now give the details.

Consider the values of $w_{i,j}$ and R implied by S. The first property above implies that $w_{i,j}$ and R satisfy the constraints of CP and hence Condition 1 of the KKT conditions. The remaining three properties allow us to determine the values of α, β and γ satisfying Condition 2, 3 and 4 of the KKT conditions. We first give some intuition. Assume job i is processed during $[j-1, j]$ for some time j. By (10) it follows that $\gamma_{i,j} = 0$ and by (12) it follows that if $\sum_{t:j \leq t} \beta_t > 0$, then the speed during $[j-1, j]$ is

$$\sum_{x:r_x < j \leq d_x} w_{x,j} = \left(\frac{\alpha_i}{\alpha \sum_{t:j \leq t} \beta_t} \right)^{1/(\alpha-1)} \tag{13}$$

Note that α_i is a constant for job i. Hence, if job i is processed during $[j-1, j]$ with speed s and is processed during $[j'-1, j']$ with speed s', then we have that $s'/s = (\sum_{t:j \leq t} \beta_t)^{1/(\alpha-1)}/(\sum_{t:j' \leq t} \beta_t)^{1/(\alpha-1)}$, or equivalently $\sum_{t:j \leq t} \beta_t = (s'/s)^{\alpha-1} \sum_{t:j' \leq t} \beta_t$. It means that in any optimum schedule, when a job is processed during two different time intervals, the ratio of speeds should be determined by the values of β. Note that this is exactly what property 4 in Lemma 2 also guarantees. This allows us set the values β can be set consistently. We now give the calculation to derive α, β and γ from the properties of Lemma 2.

Consider $t_1 < \ldots < t_k = T$ and $m_1, \ldots, m_{k-1}$ as defined by the third and fourth property of Lemma 2. We set β_j to zero for all $j \notin \{t_1, \ldots, t_k\}$. Note that it satisfies requirement (9) of the KKT conditions. For $j \in \{t_1, \ldots, t_k\}$, we set β_j such that they satisfy the following system of linear equations.

$$\sum_{t:t_y \leq t} \beta_t = (m_y)^{\alpha-1} \sum_{t:t_{y+1} \leq t} \beta_t \qquad y = 1, \ldots, k-1 \qquad (14)$$

$$1 - \sum_{y=1}^{k} t_y \beta_{t_y} = 0 \qquad (15)$$

This system has a unique non-negative solution, as (14) can be written as $\beta_{t_y} = ((m_y)^{\alpha-1} - 1) \sum_{t:t_{y+1} \leq t} \beta_t$. Hence, by considering the equation from $y = k - 1$ down to $y = 1$, we can express each of $\beta_{t_{k-1}}, \ldots, \beta_{t_1}$ in terms of β_{t_k}. Substituting these expressions into (15), we obtain a unique solution for β_{t_k}, as well as β_{t_y} for $y = k - 1, \ldots, 1$. Note that $\beta_{t_k} > 0$ and $\beta_{t_y} \geq 0$ for $y = k - 1, \ldots, 1$. This completely specifies β. Note that by (15), the values of β satisfy requirement (11) of the KKT conditions.

To calculate the values of α, we consider each job i. Let $[j_i - 1, j_i]$ be the earliest time interval during which i is processed. Then, α_i is set to

$$\alpha_i = \alpha \left(\sum_{x:r_x < j_i \leq d_x} w_{x,j_i} \right)^{\alpha-1} \left(\sum_{t:j_i \leq t} \beta_t \right). \qquad (16)$$

Note that $\alpha_i \geq 0$. As all jobs are completed by S, the KKT condition given by (8) is satisfied for any value of α. Finally, to calculate the values of γ, we consider any job i and any time j, $r_i < j \leq d_i$. We set $\gamma_{i,j}$ to

$$\gamma_{i,j} = \alpha \left(\sum_{x:r_x < j \leq d_x} w_{x,j} \right)^{\alpha-1} \left(\sum_{t:j \leq t} \beta_t \right) - \alpha_i. \qquad (17)$$

This guarantees that the KKT conditions specified by (12) are satisfied. It remains to show that $\gamma_{i,j} \geq 0$ and (10) is satisfied. This is trivially true if i has not been processed until time t_k, because $\gamma_{i,j} = 0$ in that case. If i has been processed by time t_k, recall that $[j_i - 1, j_i]$ is the first interval that i is processed. Consider any time $[j - 1, j]$ such that $r_i < j \leq d_i$. Let y and y' be values that $t_{y-1} < j_i \leq t_y$ and $t_{y'-1} < j \leq t_{y'}$. Then, by (17) and (14) we have that

$$\gamma_{i,j} = \alpha \left(\sum_{x:r_x < j \leq d_x} w_{x,j} \right)^{\alpha-1} \left(\sum_{t:j \leq t} \beta_t \right) - \alpha_i$$

$$= \alpha \left(\frac{\sum_{x:r_x < j \leq d_x} w_{x,j}}{m_y m_{y+1} \cdots m_{y'-1}} \right)^{\alpha-1} (m_y m_{y+1} \cdots m_{y'-1})^{\alpha-1} \left(\sum_{t:t_{y'} \leq t} \beta_t \right) - \alpha_i$$

$$= \alpha \left(\frac{\sum_{x:r_x < j \leq d_x} w_{x,j}}{m_y m_{y+1} \cdots m_{y'-1}} \right)^{\alpha-1} \left(\sum_{t:t_y \leq t} \beta_t \right) - \alpha_i \qquad (18)$$

If i is processed during $[j - 1, j]$, then by property 4 in Lemma 2, the speed $\sum_{x:r_x < j \leq d_x} w_{x,j}$ equals $m_y m_{y+1} \cdots m_{y'-1}$ times that during $[j_i - 1, j_i]$.

Hence, (18) implies that

$$\gamma_{i,j} = \alpha \Big(\sum_{x:r_x < j_i \leq d_x} w_{x,j_i} \Big)^{\alpha-1} \Big(\sum_{t:j_i \leq t} \beta_t \Big) - \alpha_i$$

which is identically equal to 0 by (16). Thus the KKT conditions given by (10) are satisfied in this case. Finally consider the case when i is not processed during $[j-1, j]$. By property 3 in Lemma 2, the schedule during $[t_{y-1}, t_y]$ is a YDS schedule, hence it must be that the speed $\sum_{x:r_x < j \leq d_x} w_{x,j}$ is at least as large as $m_y m_{y+1} \ldots m_{y'-1}$ times the speed during $[j_i - 1, j_i]$. By (18) and (16), this implies that $\gamma_{i,j} \geq 0$ if i is not processed during $[j-1, j]$. This completes the proof that the 4 properties above implies the KKT conditions.

It is relatively straightforward to see that the properties in Lemma 2 are necessary. □

Hence, to determine whether a schedule S minimizes the recharge rate, we can simply check for the above 4 properties. This gives our main result for this section.

Theorem 1. *Let I be any job sequence. Given a schedule S, we can determine in polynomial time whether S minimizes the recharge rate.*

Proof. Properties 1, 2 and 3 of Lemma 2 can be checked easily in polynomial time. To check Property 4 we can write as system of linear equations as follows. Let i be a job that is processed in both $[t_{y-1}, t_y]$ and $[t_{y'-1}, t_{y'}]$ for some $y < y'$ with speed $s_{i,y}$ and $s_{i,y'}$ respectively. We include an equation $\ln m_y + \ln m_{y+1} + \ldots + \ln m_{y'-1} = \ln(s_{i,y'}/s_{i,y})$. By considering all jobs and time intervals, we obtain a set of linear equations with variables of the type $\ln m_y$. There is a solution to these equations if and only if Property 4 is satisfied. □

5 Performance of YDS

In this section, we analyze the YDS schedule and show that it requires a recharge rate at most 2 times that of the optimum schedule, and that this bound is the best possible. Later we show that YDS is optimum for instances where the job deadlines and release times are ordered similarly.

Let I be any job sequence, and let OPT denote some optimum schedule. We first state a simple observation used to lower bound the energy usage of OPT.

Lemma 3. *Consider the YDS schedule for I. Let s be any speed and t be any time such that YDS has a speed at least s at t and has a speed strictly less than s immediately after t. Let J_a be the set of all jobs YDS has processed using a speed at least s until time t. Then, the energy usage of OPT for processing jobs in J_a by time t is at least that of YDS.*

Proof. We first notice that all jobs in J_a have deadlines at most t and are actually completed by YDS by time t. Furthermore, in the YDS schedule for I, jobs in J_a are processed identically as they would be in the YDS schedule for the instance

J_a, i.e. instance I with jobs in $I \setminus J_a$ removed. Therefore, YDS completes J_a using the minimum amount of energy. OPT needs to complete J_a by time t and must use at least the same amount of energy. □

We are now ready to prove the main result of this section.

Theorem 2. *YDS is a 2-approximation for minimizing the recharge rate.*

Proof. For any schedule if $E(t)$ denotes the energy usage until time t, then by definition, the recharge rate required is $\max_t E(t)/t$. Consider some instance where OPT has recharge rate r, but YDS is infeasible even with recharge rate $2r$. Let t' be the earliest time when YDS runs out of energy, and let t be the earliest time after t' when the speed of YDS falls below $r^{1/\alpha}$. Consider the times during $[0, t']$ where speed of YDS is $\geq r^{1/\alpha}$, and let E be the total energy used during these times. Since YDS is working at speed strictly less than $r^{1/\alpha}$ during other times in $[0, t']$, it follows that the total energy used by YDS during $[0, t']$ is strictly less than $E + rt'$.

We now apply Lemma 3 at time t with $s = r^{1/\alpha}$, and define J_a accordingly. As the energy used by YDS for jobs in J_a is at least $E + (t - t')r$, it follows that the energy usage of OPT on jobs in J_a is at least $E + (t - t')r$. However, as OPT has recharge rate r, it follows that $rt \geq E + (t - t')r$ and hence $E \leq rt'$. However as the total energy used by the YDS during the interval $[0, t']$ is strictly less than $E + rt'$, this implies that the total energy used by YDS is strictly less than $2rt'$ which contradicts the assumption that YDS ran out of energy at t' with recharge rate $2r$. □

We remark that the YDS schedule can be computed in $O(n^2 \log n)$ time [15], where n is the number of jobs. Therefore, this gives a polynomial time constant factor approximation algorithm for the recharge rate minimization problem. We also note that the above bound for YDS cannot be improved.

Observation 1. *The approximation ratio of YDS is at least 2 for the minimum recharge rate problem.*

Proof. Let ϵ be an arbitrarily small parameter such that $1/\epsilon$ is an integer. Consider the instance with two jobs, where job 1 has size $1/\epsilon^{1/\alpha}$, release time $1/\epsilon - 1$ and deadline $1/\epsilon$ and job 2 has size $1/\epsilon^2 - 1/\epsilon$, release time 0 and deadline $1/\epsilon^2$. Consider the schedule that stays idle during 0 to $1/\epsilon - 1$, finishes job 1 during $[1/\epsilon - 1, 1/\epsilon]$ consuming energy $1/\epsilon$, and then works at speed 1 during $[1/\epsilon, 1/\epsilon^2]$ on job 2. It is easily verified that it is feasible with a recharge rate of 1. YDS on the other hand, works at speed $(1/\epsilon^2 - 1/\epsilon)/(1/\epsilon^2 - 1) \approx 1 - \epsilon$ during $[0, 1/\epsilon - 1]$ on job 1. As it needs at least $1/\epsilon$ energy during $[1/\epsilon - 1, 1/\epsilon]$ for job 2, it is easily verified that a recharge rate of $2 - O(\epsilon)$ is necessary. □

Well-ordered Jobs: We also consider the special case where the jobs are *well-ordered*, i.e., for every jobs i_1, i_2, if the release time of i_1 is no later than the release time of i_2, then the deadline of i_1 is no later than the deadline of i_2. We can show that YDS is optimal for job sequences that are well-ordered. We omit the proof due to the page limit.

Theorem 3. *For well-ordered job sequences, YDS minimizes the recharge rate.*

6 An Online Algorithm

We now show that the BKP algorithm is constant competitive in the online setting. For any job sequence I, it is known that BKP uses no more than $2(\frac{\alpha}{\alpha-1})^\alpha e^\alpha$ times the total energy used by YDS [4]. In the following lemma, we show that in fact at any intermediate time t, the energy usage of BKP up to t is at most $2(\frac{\alpha}{\alpha-1})^\alpha e^\alpha$ times that of YDS.

Lemma 4. *Consider any job sequence I. Let $E(t)$ be the energy usage of YDS up to time t, and $E'(t)$ be that of BKP. Then, at any time t, $E'(t) \le 2(\frac{\alpha}{\alpha-1})^\alpha e^\alpha E(t)$.*

Proof. For the proof we define another algorithm ALG, that any time t runs at speed $p(t) = e \cdot \max_{t_1,t_2} w(t,t_1,t_2)/(t_2 - t_1)$, where $t_1 < t \le t_2$ and $w(t,t_1,t_2)$ denotes the amount of work that has release time at least t_1 and has deadline at most t_2. Recall that the speed of BKP at any time t is no greater than that of ALG. We will show that at any time t, the energy usage of ALG up to t is no greater than $2(\frac{\alpha}{\alpha-1})^\alpha e^\alpha E(t)$, which implies the lemma.

It is shown in [4] that ALG is $2(\frac{\alpha}{\alpha-1})^\alpha e^\alpha$-competitive in total energy usage. To show that the same guarantee holds for any intermediate time, consider any job sequence I and any time t. Without loss of generality, we assume t and the release time and deadline of every job are integers. Let I' be a job sequence constructed based on the YDS schedule for I: At any time $j < t$, a job is released with deadline $j + 1$ and size equal to the speed of YDS during $[j, j + 1]$, and the last job is released at time $t - 1$. As ALG is $2(\frac{\alpha}{\alpha-1})^\alpha e^\alpha$-competitive for total energy, the energy usage of ALG up to t with input I' is at most $2(\frac{\alpha}{\alpha-1})^\alpha e^\alpha$ times that of YDS for I'. To argue back about the job sequence I, we note that at any time up to t, YDS has the same speed for input I and I'. For ALG, we note that at any time $i < t$, the quantity $p(i)$ for input I is at most that for input I', and hence the speed of ALG for I is at most that for I'. This implies that for I, the energy usage of ALG up to time t is at most $2(\frac{\alpha}{\alpha-1})^\alpha e^\alpha$ times that of YDS. Since I and t are arbitrary, the lemma follows.   □

By Theorem 2 and Lemma 4 we obtain that

Theorem 4. *The BKP algorithm is $4(\frac{\alpha}{\alpha-1})^\alpha e^\alpha$-competitive for minimizing the recharge rate.*

References

1. http://www.eetimes.com/story/OEG20020405S0015
2. Albers, S., Fujiwara, H.: Energy-efficient algorithms for flow time minimization. In: Durand, B., Thomas, W. (eds.) STACS 2006. LNCS, vol. 3884, pp. 621–623. Springer, Heidelberg (2006)

3. Albers, S., Müller, F., Schmelzer, S.: Speed scaling on parallel processors. In: Proc. ACM Symposium on Parallel Algorithms and Architectures (SPAA), pp. 289–298 (2007)
4. Bansal, N., Kimbrel, T., Pruhs, K.: Speed scaling to manage energy and temperature. J. ACM 54(1) (2007)
5. Bansal, N., Bunde, D., Chan, H.-L., Pruhs, K.: Average rate speed scaling. In: LATIN 2008 (to appear, 2008)
6. Bansal, N., Chan, H.-L., Pruhs, K., Rogozhnikov-Katz, D.: Improved bounds for speed scaling in devices obeying the cube-root rule. In: IBM Research Technical Report
7. Bansal, N., Pruhs, K.: Speed scaling to manage temperature. In: Diekert, V., Durand, B. (eds.) STACS 2005. LNCS, vol. 3404, pp. 460–471. Springer, Heidelberg (2005)
8. Bansal, N., Pruhs, K., Stein, C.: Speed scaling for weighted flow time. In: SODA 2007: Proceedings of the eighteenth annual ACM-SIAM symposium on Discrete algorithms, pp. 805–813 (2007)
9. Boyd, S., Vandenberghe, L.: Convex Optimization. Cambridge University Press, Cambridge (2004)
10. Brooks, D.M., Bose, P., Schuster, S.E., Jacobson, H., Kudva, P.N., Buyuktosunoglu, A., Wellman, J.-D., Zyuban, V., Gupta, M., Cook, P.W.: Power-aware microarchitecture: Design and modeling challenges for next-generation microprocessors. IEEE Micro. 20(6), 26–44 (2000)
11. Chan, H.-L., Chan, W.-T., Lam, T.-W., Lee, L.-K., Mak, K.-S., Wong, P.W.H.: Energy efficient online deadline scheduling. In: SODA 2007: Proceedings of the eighteenth annual ACM-SIAM symposium on Discrete algorithms, pp. 795–804 (2007)
12. Irani, S., Shukla, S., Gupta, R.: Online strategies for dynamic power management in systems with multiple power saving states. Trans. on Embedded Computing Sys (2003); Special Issue on Power Aware Embedded Computing
13. Kwon, W.-C., Kim, T.: Optimal voltage allocation techniques for dynamically variable voltage processors. In: Proc. ACM-IEEE Design Automation Conf., pp. 125–130 (2003)
14. Li, M., Liu, B.J., Yao, F.F.: Min-energy voltage allocation for tree-structured tasks. Journal of Combinatorial Optimization 11(3), 305–319 (2006)
15. Li, M., Yao, F.F.: An efficient algorithm for computing optimal discrete voltage schedules. SIAM J. on Computing 35, 658–671 (2005)
16. Yao, F., Demers, A., Shenker, S.: A scheduling model for reduced CPU energy. In: Proc. IEEE Symp. Foundations of Computer Science, pp. 374–382 (1995)
17. Yun, H.S., Kim, J.: On energy-optimal voltage scheduling for fixed priority hard real-time systems. ACM Trans. on Embedded Computing Systems 2(3), 393–430 (2003)

Engineering Label-Constrained Shortest-Path Algorithms

Chris Barrett, Keith Bisset, Martin Holzer,
Goran Konjevod, Madhav Marathe, and Dorothea Wagner

Dept. of Computer Science and Virginia Bioinformatics Institute, Virginia Tech
{cbarrett, kbisset, mmarathe}@vbi.vt.edu
Fakultät für Informatik, Universität Karlsruhe (TH)
{mholzer, wagner}@ira.uka.de
Dept. of Computer Science and Engineering, Arizona State University
goran@asu.edu

Abstract. We consider a generalization of the shortest-path problem: given an alphabet Σ, a graph G whose edges are weighted and Σ-labeled, and a regular language $L \subseteq \Sigma^*$, the *L-constrained shortest-path problem* consists of finding a shortest path p in G such that the concatenated labels along p form a word of L. This definition allows to model, e. g., many traffic-planning problems. We present extensions of well-known speed-up techniques for the standard shortest-path problem, and conduct an extensive experimental study of their performance with various networks and language constraints. Our results show that depending on the network type, both goal-directed and bidirectional search speed up the search considerably, while combinations of these do not.

1 Introduction

Consider a *multimodal* road network, with roads differentiated by categories (highways, local streets etc.) and find in that network a shortest path from a given start to a destination point that uses highway at most once (thus avoiding on- and off-ramps). Another example is the *k-similar-path* problem, where we want to compute two shortest paths between the same pair of vertices such that the second path reuses at most k edges of the first (this can be useful to avoid traffic jams in vehicle routing).

To formalize such problems, we augment the network edges with appropriate labels and model the given restriction as a formal language. The labels of the edges on a shortest path concatenated must then form an element of the language. A detailed theoretical study of this *(formal-) language-constrained shortest-path problem* (LCSP) was undertaken in [4], where also a generalization of Dijkstra's algorithm to solve this problem is given, and in [3] an implementation of this algorithm was tested for the special case of linear regular languages (LINLCSP).

Building on this earlier work, we now consider the LCSP with *arbitrary* regular expressions (REGLCSP): we propose a concrete implementation of an algorithm solving this extended problem, and present adaptations of speed-up

R. Fleischer and J. Xu (Eds.): AAIM 2008, LNCS 5034, pp. 27–37, 2008.
© Springer-Verlag Berlin Heidelberg 2008

techniques designed for the standard, or *unimodal*, shortest-path problem to our setting. In a systematic experimental study on realistic transportation networks we investigate the applicability of *goal-directed search, Sedgewick-Vitter heuristic*, and *bidirectional search* as well as combinations of these. We also explore the scalability of our implementation by applying them to instances of increasing size as well as involving languages of varying complexity (both linear and general regular expressions).

Our experiments show that goal-directed search and the Sedgewick-Vitter heuristic yield substantial speed-up for REGLCSP on all European and some US road networks, while bidirectional search performs well especially on railway networks. Unlike in the unimodal case, combinations of bidirectional search with one of the other techniques do not perform any better than each variant applied separately. Experiments with k-similar paths confirm growing speed-up factors with increasing NFA sizes.

1.1   Related Work

Research on generalizations of the standard shortest-path problem has traditionally focused on the extension of Dijkstra's algorithm [5] to time-dependent cost functions (cf. [8]), while comparatively little work has been done on constraints restricting the set of feasible paths. There are reports on studies of *multimodal*, or *intermodal*, shortest paths in transportation science literature, however, generally of limited applicability. Regular languages as a model for constrained-shortest-path problems were first suggested in [9], and applications to database queries described in [13,7]. Our initial motivation for studying LCSP problems comes from the TRANSIMS project [1,2]. A theoretical study on algorithmic and complexity-related issues can be found in [4].

In [3], an algorithm for the REGLCSP problem with time-dependent edge weights obeying the FIFO property is described, where an *implicit* representation (cf. Section 3.1) is used; experimental results are presented only for LINLCSP. Also for time-dependent REGLCSP, [11] gives an implicit algorithm, running in linear time for FIFO weight functions. This algorithm is extended in [12] to allow for turn penalties; some experimental results are reported in both these papers. The present work deals with REGLCSP, where the focus is on an extensive experimental evaluation of speed-up techniques with diverse classes of larger networks.

2   Foundation

In this section we formally define the regular-language-constrained shortest-path problem, and describe two out of many algorithmic problems, *multimodal plans* and *k-similar paths*, that can be tackled using our problem formulation.

2.1   Problem Statement

The *regular-language-constrained shortest-path problem* (REGLCSP) is defined as follows. Given a finite alphabet Σ, a graph—also referred to by network—$G = (V, E, c, \ell)$ with cost function $c : E \to \mathbb{R}_+$ and label function $\ell : E \to \Sigma$, and a regular language $L \subseteq \Sigma^*$. For a *query* $(s, d) \in V \times V$ find a shortest s-d-path $p = (e_1, e_2, \ldots, e_k)$ in G such that $\ell(p) \in L$, where $\ell(p)$ is the concatenation $\ell(e_1) \cdot \ell(e_2) \cdots \ell(e_k)$ of the labels of p's edges. The cost—or length—of p is the sum of costs of p's edges. By Kleene's theorem, a regular language can be represented through a nondeterministic finite automaton (NFA), allowing for a concise description.

2.2 Applications

We describe two applications of the REGLCSP problem, which are also respected in our experiments (other examples include turn complexity, counting constraints, and trip chaining; cf. [3] for an overview).

Multimodal Plans. Consider a traveler who wants to take a bus from a start s to a destination point d and suppose transfers are undesirable, while walks from s to a bus stop and from a bus stop to d be allowed. To solve such a task, add to the given road network a vertex for every bus stop and an edge between each consecutive pair of stops. Label the edges according to the modes of travel allowed (e. g., c for car travel; w for walking on sidewalks and pedestrian bridges; b for bus transit). Now the traveler's restriction can be modeled as $w^* b^* w^*$, if we make sure that the network contains a zero-length w-edge for each change of bus.

k-Similar Paths. We want to consecutively route two (or more) vehicles from s to d such that the second uses at most k of the edges passed by the first one. This can be useful, e. g., to plan for a travel group different transfers between two fixed points. Note that the second path thus found may, depending on the network and the choice of k, be of greater length. To do this, find a shortest s-d-path p in the given network, label p's edges by t (for *taken*), the remaining ones by f (for *free*), and solve the s-d-query again for the expression $f^*(t \cup f^*)^k f^*$.

3 Algorithms

We now show how REGLCSP can be solved through a product network constructed from given network and NFA, and present some adaptations of unimodal speed-up techniques to our multimodal algorithm.

3.1 Product Network

Consider the direct product of a weighted, Σ-labeled network $G = (V, E, c, \ell_G)$ and an NFA $A = (Q, \Sigma, \delta, q_0, F)$ with set Q of vertices/states, alphabet Σ, transition function δ, start vertex q_0, and set F of final vertices; let T be the set of state transitions $t = (q_1, q_2)$ with $\delta(q_1) = q_2$ and labels $\ell_A(t) \in \Sigma$. The product network $P = G \times A$ is defined to have vertex set $\{(v, q) \mid v \in V, q \in Q\}$

and edge set $\{(e,t) \mid e \in E, t \in T, \ell_G(e) = \ell_A(t)\}$. The cost of an edge $(e,t) \in P$ corresponds to $c(e)$.

Theorem 1. *Finding an L-constrained shortest path for some $L \subseteq \Sigma^*$ and $(s,d) \in V \times V$ is equivalent to finding a shortest path in the product network $P = G \times A$ from vertex (s,q_0) to (d,f) for some $f \in F$.*

For the proof, one need only observe that there is a one-to-one correspondence between paths in P starting at (s,q_0) and ending at some vertex (d,f) and s-d-paths in G whose labeling belongs to L (for details please refer to [4]). Theorem 1 immediately yields the **RegLCSP Algorithm**, which performs an s-d-shortest-path search in P.

Implementation. Obviously, a direct implementation of this algorithm would require $\Theta(|G| \cdot |A|)$ space. We therefore propose a more practical way without having to compute an explicit representation of P: the algorithm considers pairs $(v,q) \in P$ but to iterate over (v,q)'s outgoing edges, simultaneously accesses the adjacency lists of v and q. To do this efficiently, we store the outgoing edges of both G and A bundled by their labels and keep pointers to the first edge of each bundle. Now we need only iterate over all labels $l \in \Sigma$ and consider each combination of vertices v' reachable from v via an edge labeled l and q' reachable from q via an edge labeled l. This *implicit representation* reduces storage space to $\Theta(|G| + |A|)$, while time complexity does not increase by more than a constant factor.

3.2   Speed-Up Techniques

In order to improve the above-described REGLCSP Algorithm, we adopt several approaches designed to speed up unimodal Dijkstra's algorithm: the key idea is to apply the genuine technique to the product network.

Goal-Directed Search (go). For given source and destination vertices s and d, goal-directed search, or A^* *search*, modifies the edge costs so that during the search, edges pointing roughly towards d are preferred to those pointing away from it. The effect is that potentially fewer vertices (and edges) have to be *touched* before d is found. With networks featuring *distance metric*, this modification, $\bar{c}$, can typically be achieved through Euclidean distances $\underline{\operatorname{dist}}(v,w)$ between vertices v and w:

$$\bar{c}(v,w) = c(v,w) - \underline{\operatorname{dist}}(v,d) + \underline{\operatorname{dist}}(w,d)$$

(in this case, c accounts for curves, bridges, etc., so that $\underline{\operatorname{dist}}$ provides a lower bound). When using *travel*, or *time*, *metric*, letting $\underline{\operatorname{dist}}'(\cdot,d) = \underline{\operatorname{dist}}(\cdot,d)/v_{\max}$ with $v_{\max} = \max_{\{v,w\} \in E} \underline{\operatorname{dist}}(v,w)/c(v,w)$ yields a feasible lower bound.

Sedgewick-Vitter Heuristic (sv). If we do not insist on exact shortest paths, a canonical extension of **go** is to bias the search to d even further: the Sedgewick-Vitter heuristic [10] uses as modified cost function $\bar{c}(v,w) = c(v,w) - \alpha \cdot \underline{\operatorname{dist}}(v,d) + \alpha \cdot \underline{\operatorname{dist}}(w,d)$ for some $\alpha \geq 1$, influencing the trade-off between gain in running time and path length increase. For previous work exploring this technique, cf. [6].

Bidirectional Search (bi). To reduce the search space, we run two simultaneous searches, *forward* and *backward*, starting from s and d, respectively; the expected improvement is a halving of the number of touched vertices[1]. A shortest s-d-path has been found when a vertex u is about to be scanned that has already been settled (i. e., its distance from the search's origin has become permanent) by the search in the other direction (note that the shortest path such found need not pass by u).

Combinations (bi+). We provide two variants for the combination of bi with either go or sv: *(1)* the cost function used for the backward search corresponds to that for the forward search; *or (2)* the cost function for both searches is the average of modified cost functions with respect to s and d, respectively:

$$\bar{c}(v,w) = c(v,w) + 1/2 \cdot (-\underline{\mathrm{dist}}(v,d) + \underline{\mathrm{dist}}(w,d) - \underline{\mathrm{dist}}(w,s) + \underline{\mathrm{dist}}(v,s)).$$

4 Experimental Study

The empiric part of this work systematically investigates our implementation of the REGLCSP Algorithm and the speed-up techniques described in the previous section. It can be seen as both a continuation and extension of [3] in that the restriction to linear expressions is waived and besides goal-directed search and the Sedgewick-Vitter heuristic, bidirectional search and combinations of techniques are employed. Our main focus is on the suitability of each technique for several networks and NFAs and the speed-ups attainable. Moreover, two different problems that can be tackled by REGLCSP (cf. Section 2.2) are considered.

4.1 Setup

Our experiments are conducted using realistic networks, representing various US and European road as well as European railway networks[2] (cf. Table 1). The road networks are weighted with actual distances (not necessarily Euclidean lengths) and labeled with values reflecting road category (from 1 to 4 for US and from 1 to 15 for EU networks, ranking from fast highways to local/rural streets). The railway network represents trains and other means of public transportation, where vertices mark railway stations/bus stops and edges denote non-stop connections between two embarking points, weighted with average travel times and labeled from 0 to 9 for rapid Intercity Express trains to slower local buses. One important difference between the US and the European road data collections is that the former come undirected, while the latter are directed.

We apply several specific language constraints of varying complexity, listed in Table 2: for the US networks, we distinguish between enforced use of highway

---

[1] The forward search will explore roughly r^2 vertices to find an r-edge shortest path, while the searches are likely to meet when each has explored roughly $(r/2)^2$ vertices.

[2] The US networks are taken from the TIGER/LINE collection, available at http://www.dis.uniroma1.it/~challenge9/data/tiger/. The European road and railway data were provided courtesy of PTV AG, Karlsruhe, and HaCon, Hannover.

Table 1. Network sizes. Left: US road networks; right: European road and railway networks. For each network, a short key, its type (road/rail), and the numbers n and m of vertices and edges, respectively, are indicated.

key	network	type	n	m	key	network	type	n	m
AZ	Arizona	road	545111	665827	LUX	Luxembourg	road	30087	70240
DC	Distr/Col	road	9559	14909	CHE	Switzerland	road	586025	1344496
GA	Georgia	road	738879	869890	DEU	Germany	rail	6900	24223

Table 2. Left: language constraints used with different networks (from top to bottom: US road, European road and railway). For each NFA, a short key, regular expression recognized by the NFA, informal description, and the numbers n and m of vertices and edges, respectively, are indicated. Right: complex NFA representing for a decomposition of the alphabet $\Sigma = \Sigma_1 \cup \Sigma_2$ the *unrestricted* expression, Σ^*.

	key	expression	description	n	m
US rd.	$\mathcal{H}$	$(3\cup4)^*(1\cup2)^+ \cdot$ $\cdot(3\cup4)^*$	highway usage	3	10
	$\mathcal{R}$	$(2\cup3\cup4)^*$	regional transfer	1	3
	$\mathcal{L}$	4^*	local streets	1	1
EU rd.	$\mathcal{S}$	$(1\cup\ldots\cup15)^*$	unrestr. (simple)	1	15
	$\mathcal{C}$	$(1\cup\ldots\cup15)^*$	unrestr. (complex)	5	72
	$\mathcal{L}$	$(10\cup11\cup12)^*$	local streets	1	3
EU r/w	$\mathcal{S}$	$(0\cup\ldots\cup9)^*$	unrestr. (simple)	1	10
	$\mathcal{C}$	$(0\cup\ldots\cup9)^*$	unrestr. (complex)	5	50
	$\mathcal{N}$	$(1\cup\ldots\cup9)^*$	normal-speed trains	1	9

(interstate or national), regional transfer (all categories but interstates), and use of local/rural streets only; for the European networks, we employ two different NFAs imposing no restriction at all (besides the 'canonical', $\mathcal{S}$, also an 'artificially made-complex' one, $\mathcal{C}$) as well as such restricting to local streets and avoiding high-speed trains (the latter usually being a little more expensive), respectively.

To measure the performance of each speed-up technique T, we compute the ratio tv_{pl}/tv_T of *touched vertices* (product vertices added to the priority queue), where tv_{pl} and tv_T stand for the number of touched vertices with *plain Dijkstra* (i.e., pure REGLCSP Algorithm) and with T, respectively. This definition of *speed-up* both is machine independent and proved to reflect actual running times quite precisely. Our code was compiled with the GCC (version 3.4) and executed on several 2- or 4-core AMD Opteron machines with between 8 and 32 GB of main memory. Unless otherwise noted, each series consists of 1000 queries.

4.2   Multimodal Routing

The term *multimodal* here is used in an extended sense since depending on the network type, it may refer either to multiple road categories or train classes. For comparability reasons, we explore the exact algorithms and the Sedgewick-Vitter heuristic separately.

Exact Algorithms. Assessment of our results is done in two steps, where we first provide a synopsis of the overall outcome and then detail on a few networks under the aspect of path lengths.

Synopsis. Figure 1 shows for each combination of network, NFA, and algorithm the average speed-up in terms of touched vertices in the product graph; the algorithms are distinguished along the x-axis (using the abbreviations introduced in Section 3.2, where pl stands for plain Dijkstra) and the NFAs are marked by their short keys (cf. Table 2). As a general result, it can be stated that both variants of the bidirectional/goal-directed combination (lumped together under bi+) always seem to be dominated by bi: there are just tiny differences in the number of vertices touched by bi and either one of the bigo variants (or even bisv). This is astounding insofar as in the unimodal case such combinations usually outperform both go and bi. Moreover, NFA size (mostly the number of vertices) has a direct impact on the number of touched vertices: the NFAs $\mathcal{H}$ and $\mathcal{C}$ incur considerably higher numbers of touched vertices than the others do.

One striking difference between the various US networks is that for the AZ and GA graphs, go search does not yield any improvement over pl at all, however, a speed-up factor of up to 2 (i.e., a reduction of 50 % of touched vertices) can be achieved for DC. Similar improvement of a factor of 2 is reached with European road networks, while go accelerates the DEU railway graph only marginally. On the other hand, bi gives good speed-up of around 2 for the railway network; little improvement in general for the US networks; and no speed-up, or even a slow-down (especially with NFA $\mathcal{L}$), for the European road networks.

Overall, the performance of each algorithm is strongly dependent on the network properties, such as density or the metric used. It is also noteworthy that some NFAs are so much restrictive that a larger number of queries cannot be answered: e.g., with $\mathcal{L}$, no feasible path is found for 34 and 53 % of the queries in the LUX and CHE networks, respectively.

Dijkstra Rank. To get a finer picture, we now consider, exemplarily for the LUX network, the speed-up values categorized by the lengths of the belonging shortest paths found, also called *Dijkstra rank*. Figure 2 shows in the form of standard box plots the average speed-up with the algorithms go and bi and NFA $\mathcal{C}$. The best factors are obtained when the Dijkstra rank lies somewhere in the middle of the complete range: a certain minimal distance between start and destination seems to be required for the speed-up technique to kick in; with higher ranks (both vertices are located near opposite borders of the network), however, the pl search is naturally bounded already, so that the speed-up factors decrease again.

Sedgewick-Vitter. Performance of the sv heuristic can be measured in terms of both reduction in the number of touched vertices and path length increase: the bigger the choice of α, i.e., the greater the distortion towards the target, the smaller gets the search space; however, with increasing α, accuracy of the found paths drops. For the LUX network, we observed that an α of 1.2 reduces the number of touched vertices by well over 20 % on average while the path

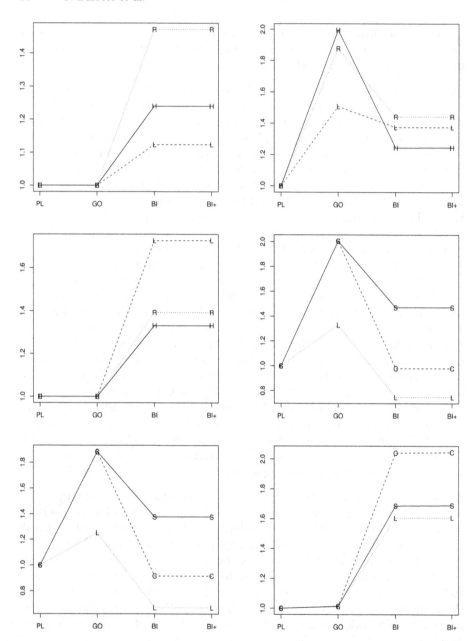

Fig. 1. Average speed-up in terms of touched vertices with each of the algorithms plain (`pl`), goal-directed (`go`), bidirectional (`bi`), and bidirectional/goal-directed combinations (`bi+`), applied to different networks (from top to bottom and left to right: AZ, DC; GA, LUX; CHE, DEU); the NFAs used are indicated by the characters on the lines (cf. Table 2).

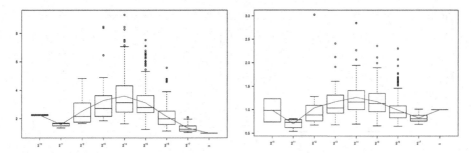

Fig. 2. Speed-up with `go` (left) and `bi` (right) applied to the LUX network and $\mathcal{C}$ NFA, categorized by Dijkstra rank. The x-axis denotes the (approximate) length of a path (∞ comprises infeasible queries); in each plot, the curve joins the mean values.

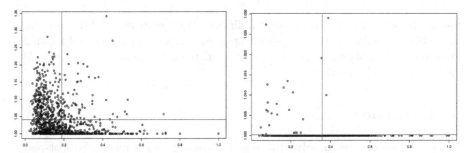

Fig. 3. Reduction in the number of touched vertices and path length increase with `sv`, applied to the LUX (left) and DEU (right) networks and the NFA $\mathcal{S}$ with α-parameters of 2 and 50, respectively. The x-axis denotes the share of touched vertices with `sv` in the number of vertices touched by `go`, while the y-axis denotes the path length increase. The horizontal and vertical lines mark the respective mean values.

lengths remain exact for all but a few queries. When raising α to 2, we save just over 80 % of touched vertices on average, but path lengths increase by around 4 % (cf. Figure 3). The picture for the DEU network looks similar, although much higher factors of α are needed to cause some effect: a reasonable choice seems to be 50 (the number of touched vertices diminishes to roughly a third with path quality almost unaffected).

4.3 k-Similar Paths

Besides exploring yet another practical application, the k-similar-path problem, as defined in Section 2.2, allows to construct NFAs of virtually arbitrary sizes in a natural way: the NFA restricting the second path found to $k \geq 0$ edges shared with the first one consists of $k + 1$ vertices and $2k + 1$ edges. Figure 4 shows for the DC network and increasing values of k both the number of vertices touched and speed-ups achieved with each algorithm. As can be predicted from theory, the curves joining the numbers of touched vertices exhibit to be almost

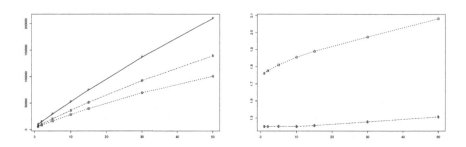

Fig. 4. k-similar-path computation: number of touched vertices (left) and speed-up (right) with the DC network and the algorithms `pl`, `go`, and `bi` for different choices of k (denoted along the x-axis)

linear (in fact, they appear slightly sublinear). With increasing NFA sizes (and hence bigger product networks), speed-ups also rise: with `bi` search, only a small growth is noticeable while with the `go` variant, the increase ranges between 1.75 (with $k = 1$) and 2 (with $k = 50$).

5   Conclusion

We have shown how to solve the regular-language-constrained shortest-path problem using product networks composed of a graph and an NFA, and proposed techniques to speed up point-to-point queries. In a practical implementation, all variants were empirically tested with a variety of real-world networks and constraints. Goal-directed and bidirectional search are found to give good speed-ups also for the multimodal setting. However, their performance greatly depends on the network properties. The Sedgewick-Vitter heuristic can, especially for railway graphs, take high α-parameters, while the paths found remain near-optimal. Surprisingly, combinations with bidirectional search do not perform better than the individual techniques. Investigating the application of k-shortest-path problems, it could be shown that the speed-up attainable increases with the size of the NFA.

In our opinion, the most interesting questions for future research include: adaptation and implementation of further speed-up techniques and heuristics; comparison between implicit and explicit product network representations; and integration of time-dependent cost functions.

Acknowledgments

The author Martin Holzer wants to give special thanks to Valentin Mihaylov for his assistance with parts of the implementation and execution of the experiments.

References

1. Barrett, C., Birkbigler, K., Smith, L., Loose, V., Beckman, R., Davis, J., Roberts, D., Williams, M.: An operational description of TRANSIMS. Technical report, Los Alamos National Laboratory (1995)
2. Barrett, C.L., Bisset, K., Holzer, M., Konjevod, G., Marathe, M.V., Wagner, D.: Engineering the label-constrained shortest-path algorithm. Technical report, NDSSL, Virginia Tech. (2007)
3. Barrett, C.L., Bisset, K., Jacob, R., Konjevod, G., Marathe, M.V.: Classical and contemporary shortest path problems in road networks: Implementation and experimental analysis of the TRANSIMS router. In: Möhring, R.H., Raman, R. (eds.) ESA 2002. LNCS, vol. 2461, pp. 126–138. Springer, Heidelberg (2002)
4. Barrett, C.L., Jacob, R., Marathe, M.V.: Formal-language-constrained path problems. SIAM J. Comput. 30(3), 809–837 (2000)
5. Dijkstra, E.W.: A note on two problems in connexion with graphs. Numerische Mathematik 1, 269–271 (1959)
6. Jacob, R., Marathe, M.V., Nagel, K.: A computational study of routing algorithms for realistic transportation networks. ACM Journal of Experimental Algorithms 4(6) (1999)
7. Mendelzon, A.O., Wood, P.T.: Finding regular simple paths in graph databases. SIAM J. Comput. 24(6), 1235–1258 (1995)
8. Orda, A., Rom, R.: Shortest-path and minimum-delay algorithms in networks with time-dependent edge-length. J. ACM 37(3), 607–625 (1990)
9. Romeuf, J.-F.: Shortest path under rational constraint. Information Processing Letters 28, 245–248 (1988)
10. Sedgewick, R., Vitter, J.S.: Shortest paths in euclidean graphs. Algorithmica 1(1), 31–48 (1986)
11. Sherali, H.D., Jeenanunta, C., Hobeika, A.G.: Time-dependent, label-constrained shortest path problems with applications. Transportation Science 37(3), 278–293 (2003)
12. Sherali, H.D., Jeenanunta, C., Hobeika, A.G.: The approach-dependent, time-dependent, label-constrained shortest path problems. Networks 48(2), 57–67 (2006)
13. Yannakakis, M.: Graph-theoretic methods in database theory. In: PODS, pp. 230–242 (1990)

New Upper Bounds on Continuous Tree Edge-Partition Problem

Robert Benkoczi[1], Binay Bhattacharya[2,*], and Qiaosheng Shi[2]

[1]Mathematics and Computer Science, University of Lethbridge,
Lethbridge, Alberta, Canada T1K 3M4
robert.benkoczi@uleth.ca
[2]School of Computing Science, Simon Fraser University,
Burnaby B.C., Canada. V5A 1S6
{binay,qshi1}@cs.sfu.ca

Abstract. We consider continuous tree edge-partition problem on a edge-weighted tree network. A continuous p-edge-partition of a tree is to divide it into p subtrees by selecting $p - 1$ cut points along the edges of the underlying tree. The objective is to maximize (minimize) the minimum (maximum) length of the subtrees. We present an $O(n \log^2 n)$-time algorithm for the max-min problem which is based on parametric search technique [7] and an efficient solution to the ratio search problem. Similar algorithmic technique, when applied to the min-max problem, results in an $O(n h_T \log n)$-time algorithm where h_T is the height of the underlying tree network. The previous results for both max-min and min-max problems are $O(n^2)$ [5].

1 Introduction

In this paper we consider continuous edge-partition (CEP, for short) problem on an edge-weighted tree network $T = (V(T), E(T), l)$, where $V(T)$ is the vertex set and $E(T)$ is the edge set, and each edge $e \in E(T)$ is associated with a positive length $l(e)$. We refer to interior points on an edge by their distances, along the edge, from the two endpoints of the edge. Let $\pi(x, y)$ be the unique simple path between a pair of points x and y in T and let $d(x, y)$ denote the length of $\pi(x, y)$.

For any two different points x, y on an edge e, if x and y are not the two endpoints of e, then we call the simple path from x to y a *partial edge* of e. We define the *length* of a subtree T' of T, denoted by $l(T')$, to be the total length of its edges and partial edges.

Let x be a point on an edge $e : \overline{uv} \in E(T)$. A *cut* at x is a splitting of e into two closed partial edges: one is from vertex u to x and the other is from x to vertex v. A cut at an interior point of an edge is called an *interior-cut*. A cut at an endpoint of an edge is called a *vertex-cut*. Note that a vertex cut is uniquely defined by a vertex and an edge incident to this vertex.

Let $p \geq 2$ be an integer. A *p-edge-partition* of T is a set of p subtrees induced by $p - 1$ cuts. The *max-min CEP problem* is to find a p-edge-partition of T that

---

* Research was partially supported by MITACS and NSERC.

R. Fleischer and J. Xu (Eds.): AAIM 2008, LNCS 5034, pp. 38–49, 2008.

maximizes the smallest length of a subtree; and the *min-max CEP problem* is to find a p-edge-partition of T that minimizes the largest length of a subtree.

Recently, Lin et. al. [5] proposed $O(n^2)$-time algorithms for the two problems, which improve the $O(n^2 \log(\min\{p, n\}))$ result of Halman and Tamir [4]. The proposed algorithms of Lin et. al. [5] are based on efficiently solving a problem, called *ratio search problem*. In this paper we consider a more general version of the ratio search problem that is defined as follows.

Definition 1 (Ratio search problem). *Given a positive integer q, a real number t, a non-increasing function $\mathcal{F} : \mathbf{R} \to \mathbf{R}$, a set of k non-negative real numbers $\Delta = \{b_i, i = 1, \ldots, k\}$, and each number b_i in Δ is associated with a non-negative integer number $g_i, 1 \le i \le k$, compute the largest real number z in $\{b_i/a_i, i = 1, \cdots, k \mid a_i \in [g_i + 1, g_i + q], b_i \in \Delta\}$ such that $\mathcal{F}(z) \ge t$.*

Lin et. al. [5] proposed an algorithm for the ratio search problem with uniform value of $g_i = 0, i = 1, \cdots, k$. We present an approach that solves the ratio search problem with the same time complexity, for non-uniform values of $g_i, i = 1, \cdots, k$. Using our efficient algorithm for the ratio search problem, we are able to solve the max-min CEP problem in $O(n \log^2 n)$ time, which is a substantial improvement of the previous results. For the min-max CEP problem, the proposed algorithm runs in time $O(n h_T \log n)$, where h_T is the height of the underlying tree.

The rest of the paper is organized as follows. Spine decomposition of a tree T is reviewed in Section 1.1. Section 2 provides a sub-quadratic algorithm for the max-min CEP problem on a tree network. The min-max problem is discussed in Section 3. An algorithm for the ratio search problem is presented in Section 4. Section 5 gives a brief conclusion.

1.1 Spine Decomposition of T

We consider a rooted binary tree T whose root vertex r_T is of degree one. We denote by $p(v)$ the parent of v in T. Let T_v be the subtree of T that is rooted at a vertex v. Let $N_l(v)$ be the number of leaves that are descendants of v in T.

A path $\pi(r_T, v')$ from the root r_T to a leaf v' of T is first identified such that for any two consecutive vertices v_i and v_{i+1} on $\pi(r_T, v')$ ($v_0 = r_T$, $p(v_{i+1}) = v_i$, and $v_m = v'$), the following condition is satisfied: for any child u of v_i other than v_{i+1}, $N_l(u) \le N_l(v_{i+1})$. That is, the path follows vertices from the root to a leaf such that the next vertex chosen is always the child of the current vertex with the most number of leaf descendants. The path $\pi(r_T, v')$ is called a *spine* and r_T is the root of this spine. We label the spine as the 1st-level spine.

The procedure is then applied recursively on each T'_u where u is a child of vertex v_i and T'_u is composed of T_u and the edge $\overline{uv_i}$, $i = 1, \cdots, m-1$. We label a spine, that hangs from another j^{th}-level spine, a $(j+1)^{th}$-level spine.

We have the following property about this spine decomposition of T.

Lemma 1. *[1] For any vertex $v \in V(T)$, the simple path $\pi(r_T, v)$ goes through at most $O(\log n)$ spines.*

In other words, the maximum level of spines in a spine decomposition of T is $O(\log n)$, denoted by τ.

2   The Max-Min Continuous Edge-Partition Problem

In this section, an $O(n \log^2 n)$-time algorithm for the max-min CEP problem is presented.

Let $l^*_{p(1)}$ be the length of the smallest subtree in an optimal solution to the max-min CEP problem. For any positive real number $l \leq l(T)$, define $Z_1(l)$ to be the largest number such that there exist an $Z_1(l)$-edge-partition in which the length of each subtree is at least l. A length l is *feasible* in the max-min model if $Z_1(l) \geq p$, and *infeasible*, otherwise.

Lemma 2. *In the max-min model,* $l^*_{p(1)} \leq l(T)/p$.

In [4], Halman and Tamir presented an $O(n)$-time algorithm to compute $Z_1(l)$ for a given positive length l. Here we only present its main steps.

A linear-time Feasibility Test A vertex v is called a *cluster vertex* of a rooted tree, if all its children are leaves of this tree.

Initially, $Z_1(l) = 0$. In the algorithm of Halman and Tamir [4] to compute $Z_1(l)$, each time a cluster vertex of the current tree, say v, is selected and the following steps are performed. Let $\{v_1, \cdots, v_t\}$ be the set of children of v. For each $k = 1, \cdots, t$, add n_k to $Z_1(l)$ where $n_k = \lfloor l(\overline{vv_k})/l \rfloor$ (corresponding to n_k interior-cuts on edge $\overline{vv_k}$), and let $a_k = l(\overline{v_i v_{i(k)}}) - n_k \times l$. If $\sum_{k=1}^{t} a_k \geq l$ then increase $Z_1(l)$ by 1 (corresponding to a vertex-cut on v and $\overline{vp(v)}$). Otherwise, increase the length of the edge $\overline{vp(v)}$ by $\sum_{k=1}^{t} a_k$. Delete all edges $\overline{vv_k}$ from the current tree. The process is repeated with a cluster of the updated tree. It is not hard to see that the running time of the approach is $O(n)$.

Lemma 3. *[4] Whether a given positive length l is feasible in the max-min CEP model can be determined in $O(n)$ time.*

Definition of $q(v), v \in V(T) \setminus \{r_T\}$ For each $v \in V(T) \setminus \{r_T\}$, let $q(v)$ be the smallest positive integer *s.t.* $l(\overline{vp(v)})/q(v)$ is feasible. In other words, $q(v) = \lceil l(\overline{vp(v)})/l^*_{p(1)} \rceil$. We first show, in Section 2.1, that all $q(v), v \in V(T) \setminus \{r_T\}$ can be computed in $O(n \log n)$ time.

2.1   Computing $q(v), v \in V(T) \setminus \{r_T\}$

Lemma 4. $p \leq \sum_{v \in V(T) \setminus \{r_T\}} q(v) \leq (p + n - 1)$.

Let z_1^* be the largest feasible real number in $Z_1 = \{l(\overline{vp(v)})/a \mid a \in [1, p], v \in V(T) \setminus \{r_T\}\}$. Lin et. al. [5] showed that $q(v) = \lceil l(\overline{vp(v)})/z_1^* \rceil$.

Lemma 5. *[5] For each $v \in V(T) \setminus \{r_T\}$, $q(v) = \lceil l(\overline{vp(v)})/z_1^* \rceil$.*

Therefore, it is suffice to present an $O(n \log n)$-time algorithm to compute z_1^*.

Compute z_1^* It is not hard to obtain the following inequalities.

$$q(v) = \lceil \frac{l(\overline{vp(v)})}{l_{p(1)}^*} \rceil \geq \lceil \frac{p \times l(\overline{vp(v)})}{l(T)} \rceil, v \in V \setminus \{r_T\} \text{ (Lemma 2)}, \qquad (1)$$

$$p \leq \sum_{v \in V \setminus \{r_T\}} \lceil \frac{p \times l(\overline{vp(v)})}{l(T)} \rceil < p + n - 1. \qquad (2)$$

We have the following result: $\sum_{v \in V \setminus \{r_T\}} (q(v) - \lceil \frac{p \times l(\overline{vp(v)})}{l(T)} \rceil) \leq n - 1$, by combining Equation 2 and Lemma 4. It implies that $0 \leq q(v) - \lceil \frac{p \times l(\overline{vp(v)})}{l(T)} \rceil \leq n - 1, v \in V \setminus \{r_T\}$ (Equation 1).

In other words, z_1^* is the largest feasible real number in $Z_1' = \{l(\overline{vp(v)})/a | a \in [\lceil \frac{p \times l(\overline{vp(v)})}{l(T)} \rceil, \lceil \frac{p \times l(\overline{vp(v)})}{l(T)} \rceil + n - 1], v \in V \setminus \{r_T\}\}$. The algorithm for the ratio search problem presented in Section 4 can be used to compute z_1^*. By using the linear time feasibility test (Lemma 3) and the result in Theorem 3, z_1^* can be computed in $O(n \log n)$. Therefore, we have the following lemma.

Lemma 6. *All $q(v), v \in V(T) \setminus \{r_T\}$, can be computed in $O(n \log n)$ time.*

2.2   Main Idea and the Overall Approach

Our approach for the solution of the max-min tree edge-partition model is to parametrically apply the linear-time feasibility test proposed by Halman and Tamir [4], using l as the single parameter, to compute $Z_1(l_{p(1)}^*)$ without specifying the value of $l_{p(1)}^*$ a priori. The approaches in [4,5] are also based on the general parametric approach of Megiddo [7]. The main difference between our approach and the two previous approaches [4,5] is described as follows. In their approaches [4,5], one feasibility test is needed at each vertex. But we plan to find all edge-partitions at all j^{th}-level spines in an optimal solution by solving $O(\log m_j)$ feasibility tests where m_j is the number of vertices in j^{th}-level spines, $j = 1, \cdots, \tau$. Its details are presented in Section 2.3. Basically, we show that the edge-partitions at j^{th}-level spines in an optimal solution can be computed in time $O(n \log n + m_j \log^2 m_j)$. Therefore, based upon Lemma 1, i.e., $\tau = O(\log n)$, the max-min CEP can be solved in $O(n \log^2 n)$ time since $\sum_{j=1}^{\tau} m_j = n$.

Theorem 1. *The max-min CEP problem can be solved in $O(n \log^2 n)$ time.*

2.3   Computing Edge-Partitions at all j^{th}-Level Spines, $1 \leq j \leq \tau$

We assume that, for any $k \in (j, \tau]$, the number of cuts at edges in k^{th}-level spines are known, and that the remainder of each $(j+1)^{th}$-level spine is known. In Fig. 1, an example is demonstrated. The bold pathes are 2nd-level spines and the dashed parts represent the remainders contributed from 3rd-level spines. We note that the remainder from each 3rd-level spine is a 1-degree polynomial of

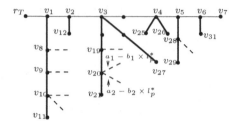

Fig. 1. 2nd-level spines and remainders of 3rd-level spines

$l_{p(1)}^*$ in the form of $\eta - \kappa \times l_{p(1)}^*$ where η is a positive real number and κ is a nonnegative integer number, and that each remainder is less than $l_{p(1)}^*$.

We first merge the remainders that are attached to the same vertex. For example, in Fig. 1, two remainders $\eta_1 - \kappa_1 \times l_{p(1)}^*$ and $\eta_2 - \kappa_2 \times l_{p(1)}^*$ from 3rd-level spines are attached to vertex v_{20}, then we remove the two remainders and attach v_{20} with a new remainder $\sum_{s=1}^{2} \eta_s - \sum_{s=1}^{2} \kappa_s \times l_{p(1)}^*$. We note that now it is possible to have new remainders that are $\geq l_{p(1)}^*$.

Let λ be the number of jth-level spines and m_j be the total number of vertices on these j^{th}-level spines. For each jth-level spine, i.e., $\Phi : \{v_0, \cdots, v_t\}$ (v_t is the root of Φ) in Fig. 2, we create a balanced binary tree structure over it. Before we present the balanced binary tree structure over each jth-level spine Φ, more notations and definitions are introduced as follows.

The edge connecting v_{i-1} and v_i is denoted by e_i, $i = 1, \cdots, t$. We define the remainder x_i of each edge e_i on Φ to be $l(e_i) - (q(v_{i-1}) - 1) \times l_{p(1)}^*$, $i = 1, \cdots, t$. Note that v_i is the parent of v_{i-1} in T and $q(v_{i-1}) = \lceil l(e_i)/l_{p(1)}^* \rceil$. Therefore, $0 < x_i \leq l_{p(1)}^*$. We continue with the hypothesis that $0 < x_i < l_{p(1)}^*$. For each $i, 1 \leq i < t$, we denote by y_i the remainder attached to vertex v_i (after merging remainders of $(j+1)^{th}$-level spines).

All x_i's and y_i's are in the form of $\eta - \kappa \times l_{p(1)}^*$ where η is a positive real number and κ is a nonnegative integer number. Let $x_i = \eta_i - \kappa_i \times l_{p(1)}^*$ and $y_k = \eta_k' - \kappa_k' \times l_{p(1)}^*$ where $i = 1, \cdots, t$ and $k = 1, \cdots, t-1$ (y_t is undefined).

Fig. 2 demonstrates a balanced binary tree structure over a spine Φ, denoted by T_Φ, where the vertices on Φ are leaves of T_Φ. We denote by T_u the subtree

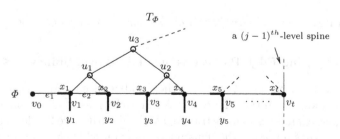

Fig. 2. A balanced binary tree structure over a jth-level spine Φ

of T_Φ rooted at a leaf or an internal node u of T_Φ. Let $V(T_u)$ be the set of leaf vertices in T_u. For example, in Fig. 2, $V(T_{v_2}) = \{v_2\}$ and $V(T_{u_2}) = \{v_3, v_4\}$. The term I_u is defined to be the smallest index of vertices contained in the subtree T_u of T_Φ, i.e., $I_{u_1} = I_{u_3} = 1$ and $I_{u_2} = 3$ in Fig. 2.

Preprocessing Step. In this step, we compute three sets of 1-degree polynomials with unknown $l^*_{p(1)}$, i.e., Z_r, Z'_r, Z''_r, and sort elements in these sets respectively. The three sets are described below.

For each vertex v_i in $V(T_u)$, let $z(u,i) = \sum_{k=I_u}^{i} x_k + \sum_{k=I_u}^{i-1} y_k$ and $z'(u,i) = \sum_{k=I_u}^{i} x_k + \sum_{k=I_u}^{i} y_k$. Clearly, there are $O(m_j \log m_j)$ such values in j^{th}-level spines since each vertex belongs to $O(\log m_j)$ rooted subtrees. All these $z(\cdot, \cdot)$ and $z'(\cdot, \cdot)$ values are in the form of $\eta - \kappa \times l^*_{p(1)}$ where η is a positive real number and κ is a nonnegative integer number. Also, we have the following property about these values.

Lemma 7. *For each vertex v_i in $V(T_u)$, $z(u,i) < n \times l^*_{p(1)}$ and $z'(u,i) < n \times l^*_{p(1)}$.*

Let A be the set of constant parts of all these $z(\cdot, \cdot)$ 1-degree polynomials. As we know, each η_i in A is associated with a nonnegative integer κ_i. According to Lemma 7, $\kappa_i \times l^*_{p(1)} < \eta_i < (\kappa_i + n) \times l^*_{p(1)}$ for each $\eta_i \in A$. For each $\eta \in A$, let $q(\eta)$ be the smallest positive integer *s.t.* $\eta/q(\eta)$ is feasible. In other words, $q(\eta) = \lceil \eta/l^*_{p(1)} \rceil$.

Let z^*_2 be the largest feasible real number in $Z_2 = \{\eta_i/f_i | \eta_i \in A, f_i \in [\kappa_i + 1, \kappa_i + n]\}$. We obtain that $q(\eta) = \lceil \eta/z^*_2 \rceil$, for each $\eta \in A$ [5]. According to Theorem 3 and Lemma 3, z^*_2 can be computed in $O(n \log n)$ since $|A| = O(m_j \log m_j)$. Therefore, all $q(\eta)$ ($\eta \in A$) can be computed in $O(n \log n)$ time.

Definition of Z_r and Z'_r: For each vertex v_i in $V(T_u)$, we let $z_r(u,i)$ be the remainder of $z(u,i)$ and let $z'_r(u,i)$ be $y_i + z_r(u,i)$, i.e., if $z(u,i) = \eta - \kappa \times l^*_{p(1)}$ then $z_r(u,i) = z(u,i) - (q(\eta) - \kappa - 1) \times l^*_{p(1)}$. Clearly, $0 < z_r(u,i) \le l^*_{p(1)}$. We continue with the hypothesis that $0 < z_r(u,i) < l^*_{p(1)}$. We define Z_r (resp. Z'_r) to be the set of these $z_r(\cdot, \cdot)$ (resp. $z'_r(\cdot, \cdot)$) 1-degree polynomials with unknown $l^*_{p(1)}$.

Definitions of Z''_r: Let v_i be a vertex lying in a jth-level spine $\Phi : \{v_0, \cdots, v_t\}$. In the balanced binary tree structure T_Φ, there are $O(\log m_j)$ subtrees containing all the vertices in path $\pi(v_{i+1}, v_{t-1})$, say $T_{u_1}, \cdots, T_{u_h}$ (where $I_{u_1} = i+1 < I_{u_2} < \cdots < I_{u_h} < t$). For each subtree $T_{u_s}, 1 \le s \le h$, let $z''(u_s, i) = \sum_{k=i+1}^{I_{u_s}-1}(x_k + y_k)$ (see Fig. 3). Clearly, these $z''(\cdot, \cdot)$ values are 1-degree polynomials with unknown l^*_p. Let $z''_r(u_s, i)$ be the remainder of $z''(u_s, i)$. Similar to the computation of $z_r(\cdot, \cdot)$, all these $z''_r(\cdot, \cdot)$ 1-degree polynomials can be computed in $O(n \log n)$ time. We define Z''_r to be the set of these $z''_r(\cdot, \cdot)$ 1-degree polynomials.

Sort Elements in the Sets Z_r, Z'_r, Z''_r: Obviously, each of Z_r, Z'_r, and Z''_r is of size $O(m_j \log m_j)$. A comparison between two 1-degree polynomials with unknown

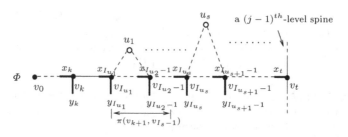

Fig. 3. $z''(u_s, k), 1 \le s \le h$

$l^*_{p(1)}$ can be resolved by solving one feasibility test. Under Valiant's comparison model [8], all the $z_r(\cdot, \cdot)$ (resp. $z'_r(\cdot, \cdot)$, $z''_r(\cdot, \cdot)$) can be sorted in $O(n \log^2 m_j)$ time by applying Megiddo's parametric-searching technique [7].

Actually, this sorting step can be speeded up by applying the result of Cole [3]. In this way, the sorting can be done in time $O(n \log m_j + m_j \log^2 m_j)$.

Finally, we have the following lemma.

Lemma 8. *The computation and sorting of elements of Z_r, Z'_r, Z''_r, that is, the preprocessing step for computing edge-partitions at all j^{th}-level spines, can be done in time $O(n \log n + m_j \log^2 m_j)$, $1 \le j \le \tau$.*

Algorithm. Our algorithm to compute the edge-partitions at j^{th}-level spines in an optimal solution consists of two steps.

The first step is to locate vertex-cuts on all j^{th}-level spines in an optimal solution. The second step is to compute the remainder of each jth-level spine.

First Step: Computing Vertex-cuts We explore a property (see Lemma 9) between two consecutive vertex-cuts in an optimal solution. Based upon this property, given a vertex-cut on a jth-level spine, we are able to locate the next vertex-cut efficiently if it exists.

Lemma 9. *Refer to Fig. 4. Assume that there is a vertex-cut on vertex v_{i-1} and edge e_i in an optimal solution. If the next vertex-cut is on vertex v_k and edge e_{k+1} ($i \le k \le t-1$), then*

$$\lceil \frac{\sum_{s=i}^{l} (x_s + y_s)}{l^*_{p(1)}} \rceil = \lceil \frac{\sum_{s=i}^{l-1} (x_s + y_s) + x_h}{l^*_{p(1)}} \rceil, \text{ where } i \le l < k,$$

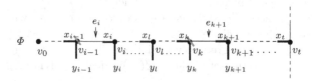

Fig. 4. Lemma 9 shows a property about the next vertex-cut on vertex v_k and edge e_{k+1} after vertex-cut on vertex v_{i-1} and edge e_i

and

$$\lceil \frac{\sum_{s=i}^{k} (x_s + y_s)}{l^*_{p(1)}} \rceil > \lceil \frac{\sum_{s=i}^{k-1} (x_s + y_s) + x_k}{l^*_{p(1)}} \rceil.$$

Lemma 9 says that we can find all vertex-cuts on a jth-level spine Φ one by one, starting from v_0 (we can assume that there is a vertex-cut on v_0 and e_1 since v_0 is a leaf vertex in T). However, this approach is inefficient and its running time is $O(m_j n)$ (in worst case, we need to solve $O(m_j)$ feasibility tests).

In order to improve the running time of computing vertex-cuts on all j^{th}-level spines, we propose a parallel approach that is described as follows. We first describe an algorithm to compute the next vertex-cut for a vertex v_i in a jth-level spine $\Phi : \{v_0, \cdots, v_t\}$ ($0 \le i < t - 1$) with the assumption that there is a vertex-cut on v_i and e_{i+1}, and then present our parallel approach to compute next vertex-cuts for all vertices in j^{th}-level spines.

Computing the next vertex-cut for a vertex v_i: refer to Fig. 3. There are $O(\log m_j)$ subtrees of T_Φ containing all the vertices in path $\pi(v_{i+1}, v_{t-1})$, i.e., $T_{u_1}, \cdots, T_{u_h}$ ($I_{u_1} = i+1 < I_{u_2} < \cdots < I_{u_h} < t$). For each subtree T_{u_s}, $1 \le s \le h$, we locate the first vertex $v_k \in V(T_{u_s})$ s.t. the following condition is satisfied.

$$C1: \lceil \frac{\sum_{a=i+1}^{k} (x_a + y_a)}{l^*_{p(1)}} \rceil > \lceil \frac{\sum_{a=i+1}^{k-1} (x_a + y_a) + x_k}{l^*_{p(1)}} \rceil.$$

Note that we might not be able to find such a vertex in some subtrees. If such vertex does not exist for any subtree T_{u_s}, $1 \le s \le h$, then next vertex-cut does not exist after v_i on spine Φ. Otherwise, let $T_{u_{s1}}$ be the first subtree in which such type of vertex, say v_{i1}, does exist. It is not difficult to see that the next vertex-cut after v_i will be on vertex v_{i1} and edge e_{i1+1}.

We next show an approach to locate the first vertex $v_k \in V(T_{u_s})$ ($1 \le s \le h$) s.t. C1 is satisfied.

We depict important information of vertices in $V(T_{u_s}), 1 \le s \le h$, using a two-dimensional diagram (see Fig. 5). For each $v_k \in V(T_{u_s})$, the horizontal coordinate of v_k in the two-dimensional diagram corresponds to $z_r(u_s, k)$ and the vertical coordinate of v_k corresponds to $z'_r(u_s, k)$. Note that these $z_r(u_s, \cdot)$ and $z'_r(u_s, \cdot)$ values are 1-degree polynomials and are already sorted in the preprocessing step.

We denote by $\mathcal{I}$ the region in the two-dimensional diagram that contains all points whose horizontal coordinates are in $[0, l^*_{p(1)} - z''_r(u_s, i)]$ and whose vertical coordinates are in $(l^*_{p(1)} - z''_r(u_s, i), \infty)$. We denote by $\mathcal{II}$ the region in the two-dimensional diagram that contains all points whose horizontal coordinates are in $(l^*_{p(1)} - z''_r(u_s, i), \infty)$ and whose vertical coordinates are in $(2l^*_{p(1)} - z''_r(u_s, i), \infty)$.

Lemma 10. *Suppose that there is no vertex-cut in $\pi(v_{i+1}, v_{I_s-1})$. Then, there is a vertex-cut in $\pi(v_{I_s}, v_{I_{s+1}-1})$ if and only if $\mathcal{I} \cup \mathcal{II}$ contains at least one point. Also, if points exist in $\mathcal{I} \cup \mathcal{II}$, let $i1$ be the smallest index among them, then the first vertex-cut in $\pi(v_{I_s}, v_{I_{s+1}-1})$ is on vertex v_{i1} and edge e_{i1+1}.*

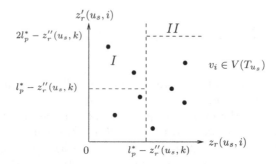

Fig. 5. Check if there is a vertex-cut in T_{u_s} and locate it if exists

Using priority search tree structure [6] to maintain the two-dimensional diagram for subtree T_{u_s}, we can check if there exists a vertex-cut and locate it if exists after $O(|V(T_{u_s})|) \subseteq O(\log m_j)$ comparisons between 1-degree polynomials with unknown $l^*_{p(1)}$ (one feasibility test needs to be resolved for each comparison). We note that the vertical and horizontal coordinates of points are 1-degree polynomials and are already sorted in the preprocessing step. Hence, we are able to create the priority search tree in $O(|V(T_{u_s})| \log |V(T_{u_s})|)$ time [6].

Computing next vertex-cuts for vertices in j^{th}-level spines: Since it is inefficient to locate vertex-cuts one by one, in order to speed up the computation, we can compute them in a parallel way. For each vertex v in a jth-level spine, we need to locate candidates for its next vertex-cut among $O(\log m_j)$ subtrees. As shown above, a candidate in a subtree (for v) can be computed in $O(\log m_j)$ steps where each step is a comparison between two 1-degree polynomials with unknown $l^*_{p(1)}$.

Therefore, the computation of all possible vertex-cuts can be done in $O(\log m_j)$ parallel steps by using $O(m_j \log m_j)$ processors (there are $O(\log m_j)$ processors associated with each vertex), where each step is a comparison between two 1-degree polynomials with unknown $l^*_{p(1)}$. By applying the idea of Cole [3], the computation can be done in time $O(n \log m_j + m_j \log^2 m_j)$.

Second Step: Computing Remainders of Spines After locating vertex-cuts on all j^{th}-level spines in an optimal solution, we are able to compute the remainder of each jth-level spine efficiently. For a jth-level spine $\Phi : \{v_0, \cdots, v_t\}$, assume that the last vertex-cut is on vertex v_k and edge e_{k+1}, $0 \leq k \leq t-1$ ($k = 0$ means that there is no vertex-cut on Φ). Then the remainder of spine Φ is the remainder of $\sum_{i=k+1}^{t} x_i + \sum_{i=k+1}^{t-1} y_i$. It is trivial that the 1-degree polynomial $\sum_{i=k+1}^{t} x_i + \sum_{i=k+1}^{t-1} y_i$ is less than $n \times l^*_{p(1)}$. We need to compute the remainders of λ 1-degree polynomials (λ is the number of jth-level spines). Similar to the computation of $z_r(\cdot, \cdot)$ (in the preprocessing step), all these remainders can be computed in $O(n \log n)$ time.

From the above discussion, the total effort (including the effort for the preprocessing step) to compute the edge-partitions in j^{th}-level spines in an optimal solution is $O(n \log n + m_j \log^2 m_j)$. It completes the proof of Theorem 1.

3   The Min-Max Continuous Edge-Partition Problem

Unfortunately, we cannot use the same approach to solve the min-max CEP problem as the one for the max-min CEP problem. However, from the linear-time feasibility test [4] for the min-max problem, we can see that all current cluster vertices can be handled in a parallel way. Due to page restrictions our algorithm to compute the edge-partitions at all current cluster vertices in an optimal solution, by solving logarithmic feasibility tests, is omitted here.

We have the following result for the min-max CEP problem.

Theorem 2. *The continuous min-max tree edge-partitioning problem can be solved in $O(nh_T \log n)$ (or $O(n \sum_{v \in V(T)} \delta_T(v)))$ time, where h_T is the height of the underlying tree network and $\delta_T(v)$ is the degree of vertex v.*

4   An Algorithm for the Ratio Search Problem

In this section we propose an algorithm for the ratio search problem (see Definition 1). Let a^*, b^* be numbers s.t. $b^* \in \Delta$, a^* is an integer number in $[g^*+1, g^*+q]$ (g^* is the value associated with b^*), and $z^* = b^*/a^*$. Without loss of any generality, we assume that $\mathcal{F}(0) \geq t$.

Notation Δ': Let Δ' be the subset of Δ s.t. $\mathcal{F}(b_i/(g_i+q)) \geq t$ for any $b_i \in \Delta'$.

Clearly, $b^* \in \Delta'$ since $\mathcal{F}(\cdot)$ is non-increasing function. For each $b_i \in \Delta'$, we denote by $a(b_i)$ the smallest integer number in $[g_i+1, g_i+q]$ with $\mathcal{F}(b/a(b_i)) \geq t$.

Notations Δ'_1, Δ'_2: Let $\Delta'_1 = \{b_i|a(b_i) > g_i+1, b_i \in \Delta'\}$ and let $\Delta'_2 = \Delta' \setminus \Delta'_1$.

Obviously, $\mathcal{F}(b_i/(g_i+1)) \geq t$ for any $b_i \in \Delta'_2$, but, $\mathcal{F}(b_i/(g_i+1)) < t$ for any $b_i \in \Delta'_1$. We can identify the elements in Δ'_1 and Δ'_2 by sorting the elements in Δ' and evaluating values of $\mathcal{F}(\cdot)$ for $O(\log|\Delta'|)$ elements in Δ'.

Notation z^*_2: Let $z^*_2 = \max_{b_i \in \Delta'_2} b_i/(g_i+1)$. It is trivial to see that $z^* \geq z^*_2$ since $\mathcal{F}(z^*_2) \geq t$. In the following, we assume that $z^* > z^*_2$.

The Case When $z^* > z^*_2$ In this case, b^* must be in Δ'_1. We define $k_b(z) = \lfloor b/z \rfloor$ for each $b \in \Delta'_1$ where z is a parameter in $(0, \infty)$. Let $f(z) = \sum_{b \in \Delta'_1} k_b(z)$. We observe that $f(z)$ is a step function with jumps, including the jump point at z^*.

Notations a', z': Let a' be the smallest integer number s.t. $\mathcal{F}((\sum_{b \in \Delta'_1} b)/a') \geq t$. Let $z' = (\sum_{b \in \Delta'_1} b)/a'$.

Lemma 11. $\sum_{b_i \in \Delta'_1} g_i + |\Delta'_1| < a' \leq \sum_{b_i \in \Delta'_1} g_i + q \times |\Delta'_1|$.

Since there are at most $(q-1) \times |\Delta'_1|$ candidate integer values for a' (Lemma 11), we are able to compute a' in time $O(k + t_{\mathcal{F}} \times (\log q + \log k))$ (note that $|\Delta'_1| \leq k$).

For any $b \in \Delta'_1$, we can see that $b/a(b) < (\sum_{b_i \in \Delta'_1} b_i)/(a' - 1)$ since $\mathcal{F}((\sum_{b_i \in \Delta'_1} b_i)/(a' - 1)) < t$ and $\mathcal{F}(b/a(b)) \geq t$. Similarly, for any $b \in \Delta'_1$, $z' < b/(a(b) - 1)$ since $\mathcal{F}(z') \geq t$ and $\mathcal{F}(b/(a(b) - 1)) < t$ (note that $a(b) > 1$).

For any $b \in \Delta'_1$, $b/a(b) < (\sum_{b_i \in \Delta'_1} b_i)/(a' - 1) \Rightarrow a(b) > b(a' - 1)/\sum_{b_i \in \Delta'_1} b_i$ $\Rightarrow a(b) > (b/z') - 1$ (since $z' = (\sum_{b \in \Delta'_1} b)/a')$; and $z' < b/(a(b) - 1) \Rightarrow a(b) < (b/z') + 1$. Therefore, $(b/z') - 1 < a(b) < (b/z') + 1$ for any $b \in \Delta'_1$. We note that $a(b)$ is an integer number and then there are at most two integer numbers between $(b/z') - 1$ and $(b/z') + 1$ (not including $(b/z') - 1, (b/z') + 1$), denoted by a^1_b and a^2_b. It is not difficult to see that z^* is in the set $\{b/a^1_b, b/a^2_b | b \in \Delta'_1\}$ if $z^* > z^*_2$.

The finding can be done by using the prune-and-search technique. First, we compute the median x of the numbers in $\{b/a^1_b, b/a^2_b | b \in \Delta'_1\}$ by the linear-time select algorithm in [2]. If $\mathcal{F}(x) \geq t$ we prune away all numbers smaller than x in this set; otherwise, we prune away all numbers larger than x. The process is repeated on the remaining numbers. Note that the size of this set is no more than $2|\Delta'_1| \leq 2k$. Therefore, the above finding of the largest real number z^*_3 in $\{b/a^1_b, b/a^2_b | b \in \Delta'_1\}$ such that $\mathcal{F}(z^*_3) \geq t$ requires $O(k + t_{\mathcal{F}} \times \log k)$ time.

Based upon the above discussion, an algorithm for the ratio search problem is presented as follows (Algorithm 1).

Algorithm 1. Ratio-Search($q, \mathcal{F}, t, \Delta$)

Input: an integer $q > 0$, a non-increasing function $\mathcal{F}$, a real number t, a set Δ of k non-negative real numbers, and each real number b_i in Δ is associated with a nonnegative integer $g_i, i = 1, \cdots, k$.

Output: the largest real number z in $\{b_i/a_i | a_i \in [g_i + 1, g_i + q], b_i \in \Delta\}$ such that $\mathcal{F}(z) \geq t$.

begin

1: $b'' \leftarrow$ the largest number in $\{b_i/(g_i + q) | b_i \in \Delta\}$ s.t. $\mathcal{F}(b'') \geq t$.

2: $\Delta' \leftarrow$ the subset of Δ that contains all the numbers $b_i \in \Delta$ with $b_i/(g_i + q) \leq b''$.

3: $z^*_2 \leftarrow$ the largest real number in $\{b_i/(g_i + 1) | b_i \in \Delta'\} \cup \{0\}$ s.t. $\mathcal{F}(z^*_2) \geq t$.

4: $\Delta'_1 \leftarrow$ the subset of Δ' that contains all the numbers $b_i \in \Delta'$ with $b_i/(g_i + 1) > z^*_2$.

5: $a' \leftarrow$ the smallest integer number s.t. $\mathcal{F}((\sum_{b \in \Delta'_1} b)/a') \geq t$.

6: $z' \leftarrow (\sum_{b \in \Delta'_1} b)/a'$.

7: $a^1_b, a^2_b \leftarrow$ the two integers between $(b/z') - 1$ and $(b/z') + 1$, for each $b \in \Delta'_1$.

8: $z^*_3 \leftarrow$ the largest real number in $\{b/a^1_b, b/a^2_b | b \in \Delta'_1\}$ such that $\mathcal{F}(z^*_3) \geq t$.

9: return $\max\{z^*_2, z^*_3\}$.

end

The time complexity of Algorithm 1 is analyzed as follows. Both Line 1 and Line 3 can be done in time $O(k + t_{\mathcal{F}} \times \log k)$ by the prune-and-search technique described above for completing Line 8. It is easy to see that the steps in Line 2, Line 4, Line 6, and Line 7 can be done in $O(k)$ time. It is known that Line 5 can be completed in time $O(k + t_{\mathcal{F}} \times (\log q + \log k))$. Therefore, the ratio search problem can be solved in time $O(k + t_{\mathcal{F}} \times (\log q + \log k))$.

Theorem 3. *The ratio search problem can be solved in time* $O(k + t_{\mathcal{F}} \times (\log q + \log k))$.

5 Summary

In this paper we study continuous tree p-edge-partition problems on a tree network, and propose efficient algorithms for the max-min and min-max problems.

Similar to the approaches developed in [4,5], our approaches are also based on the general parametric approach of Megiddo [7]. The main difference is that in their approaches one feasibility test is needed at each vertex. However, in our approach for the max-min problem, we build a spine tree decomposition structure [1] over the underlying tree and locate edge-partitions at all the spines at the same level by solving logarithmic feasibility tests. In our approach for the min-max problem, we locate the edge-partitions at all current cluster vertices by solving logarithmic feasibility tests. In this way, we are able to solve the max-min problem in sub-quadratic time and to solve the min-max problem in time $O(nh_T \log n)$.

In [4], Halman and Tamir mentioned that their algorithms for the continuous tree p-edge-partition problems can be extended to yield polynomial algorithms of the same complexity for the problems in cactus networks, that is, $O(n^2 \log(\min\{p, n\}))$. We conjecture that our algorithms for tree networks can also be extended to cactus networks.

References

1. Benkoczi, R.: Cardinality constrainted facility location problems in trees, Ph.D. Thesis, School of computing secience, SFU, Canada (2004)
2. Blum, M., Floyd, R.W., Rivest, R.L., Tarjan, R.E.: Time bounds for selection. J. Comput. Sys. Sci. 7(4), 448–461 (1973)
3. Cole, R.: Slowing down sorting networks to obtain faster sorting algorithms. J. ACM 34, 200–208 (1987)
4. Halman, N., Tamir, A.: Continuous bottleneck tree partitioning problems. Disc. App. Math. 140, 185–206 (2004)
5. Lin, J.-J., Chan, C.-Y., Wang, B.-F.: Improved algorithms for the continuous tree edge-partition problems. Disc. App. Math. (submitted, 2007)
6. McCreight, E.M.: Priority search trees. SIAM J. Comput. 14(2), 257–276 (1985)
7. Megiddo, N.: Applying parallel computation algorithms in the design of serial algorithms. J. ACM 30(4), 852–865 (1983)
8. Valiant, L.G.: Parallelism in comparison problems. SIAM J. Comput. 4, 348–355 (1975)

A Meeting Scheduling Problem
Respecting Time and Space

Florian Berger[1], Rolf Klein[1], Doron Nussbaum[2],
Jörg-Rüdiger Sack[2], and Jiehua Yi[2]

[1] Institute of Computer Science
University of Bonn
53117 Bonn, Germany
{berger,rolf.klein}@cs.uni-bonn.de
[2] School of Computer Science
Carleton University
Ottawa, Canada K1S 5B6
{nussbaum,sack,jyi}@scs.carleton.ca

Abstract. We consider the problem of determining suitable meeting times and locations for a group of participants wishing to schedule a new meeting subject to already scheduled meetings possibly held at a number of different locations. Each participant must be able to reach the new meeting location, attend for the entire duration, and reach the next meeting location on time. In particular, we give a solution to the problem instance where each participant has two scheduled meetings separated by a free time interval. In [2], we presented an $O(n \log n)$ algorithm for n participants obtained by purely geometrical arguments. Our new approach uses the concept of LP-type problems and leads to a randomized algorithm with expected running time $O(n)$.

1 Introduction

We consider the following problem: A set of people, called participants, would like to schedule a meeting. The participants are located at different sites and would like to find a common point for the meeting to be held. These sites are rather far from each other, so that the travel times to the meeting point are important. We assume that the travel distances between locations can be measured using the Euclidean distance in the plane.

Participants have individual schedules which specify: the earliest possible time they can leave their current location and the latest possible time they must arrive at their next location. We wish to solve the problem of finding a meeting point so that the time for all participants to meet is maximized.

Our general objective for this research is to derive efficient algorithms for solving general meeting scheduling problems, to find approximate solutions, where appropriate, to implement our solutions, and to integrate them into applications that allow users who are connected over a network to schedule meetings. Currently meeting scheduling systems take into consideration only time and not location/geometry, see e.g. [1,6].

R. Fleischer and J. Xu (Eds.): AAIM 2008, LNCS 5034, pp. 50–59, 2008.
© Springer-Verlag Berlin Heidelberg 2008

2   Problem Definition

We assume a situation where all participants can travel at the same, constant speed v in the plane. Since the precise value of v is not critical, we may as well assume that $v = 1$ holds. Then, the time it takes to travel from location l to location l' is given by the Euclidean distance $\|l - l'\|$.

Let $M := \{M_1, \ldots, M_n\}$ denote the set of participants. The ith participant M_i has a previous meeting at location $l_i^{pre} \in \mathbb{R}^2$ that lasts until time $t_i^{pre} \in \mathbb{R}$. He is due for a subsequent meeting that starts at time $t_i^{sub} \in \mathbb{R}$ at location $l_i^{sub} \in \mathbb{R}^2$.

We want to schedule a meeting of longest possible duration in between the previous and subsequent meetings of all participants (or report that no such meeting is possible).

To solve our problem, we have to find a location $x \in \mathbb{R}^2$ such that the time interval between the arrival of the last participants from their previous meetings at time

$$S(x) := \max_{1 \leq i \leq n} \left(t_i^{pre} + \|l_i^{pre} - x\| \right)$$

and the departure of the first participants to their subsequent meetings at time

$$E(x) := \min_{1 \leq i \leq n} \left(t_i^{sub} - \|l_i^{sub} - x\| \right)$$

is maximized.

We define $f(x) := E(x) - S(x)$. Of course, no common meeting is possible at locations x with $f(x) < 0$, but we can easily check this condition once we have computed a location maximizing f.

In the following, we can assume that there is a solution x with $f(x) > 0$, because the optimum meeting points are not changed if we increase f by a constant. For example, if we define $c := f(l_1^{pre}) - 1$ and replace all t_i^{sub} values by $t_i^{sub} - c$, this leads to a transformed instance where a meeting of positive duration is possible, because all participants can meet for one time unit at location l_1^{pre}. If we can find an optimum solution for the transformed instance, this also leads to an optimum solution for the original instance.

As an example for the meeting scheduling problem, we consider the following instance, where every participant has to return to the point of his previous meeting: $\forall i : 1 \leq i \leq n$

- $l_i^{pre} = l_i^{sub}$
- $t_i^{pre} = 0$
- $t_i^{sub} = C := 1 + 2\max_{2 \leq j \leq n} \|l_1^{pre} - l_j^{pre}\|$

The way we have chosen C ensures that $f(l_1^{pre}) = 1$, so there are feasible locations for the new meeting. The best one can do for this instance is to find a location where the participants can arrive earliest, because of the symmetry this will also be a location where they can stay longest.

Every participant can reach the location x at time $S(x)$. This means that all l_i^{pre} are within a circle with midpoint x and radius $S(x)$. The center of the *Smallest Enclosing Disk* of the l_i^{pre} corresponds to the optimum solution.

There is a randomized algorithm for computing the smallest enclosing disk of n points in the plane with expected running time $O(n)$, see Welzl [8]. LP-type problems, introduced by Sharir and Welzl [7], build an abstract framework for Welzl's algorithm for the smallest enclosing disk. We will use this concept for developing the algorithm in Section 4.

3 Geometric Interpretation of the Problem

The possible locations participant M_i can reach, once his previous meeting has ended, are given by a system of circles centered at l_i^{pre} that start expanding at time t_i^{pre}. If we use the time as a third coordinate, the expanding circles for participant M_i form a vertical cone PRE_i with apex angle $\frac{\pi}{2}$ whose bottommost point is its apex $((l_i^{pre})_X, (l_i^{pre})_Y, t_i^{pre})$. Let SUB_i denote the corresponding cone for the subsequent meeting, so SUB_i is a vertical cone whose uppermost point is its apex $((l_i^{sub})_X, (l_i^{sub})_Y, t_i^{sub})$.

We define $PRE := \cap_{i=1}^n PRE_i$, $SUB := \cap_{i=1}^n SUB_i$ and $F := PRE \cap SUB$.

We call F the set of *feasible points* because $(l_X, l_Y, t) \in F$ means that all participants can meet at location l at time t. Observe that F is convex as the intersection of convex sets. A meeting at location l is possible from time t_1 to t_2, if and only if the vertical line segment with endpoints (l_X, l_Y, t_1) and (l_X, l_Y, t_2) is fully contained in F. Consequently, we have to *find a vertical line segment of maximum length contained in F*.

The projection of the cones in PRE on the XY-plane forms a furthest site Voronoi diagram of circles. The same holds for the cones in SUB. Within each cell

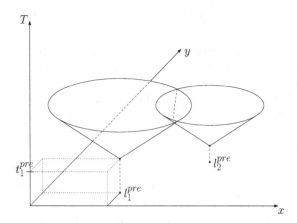

Fig. 1. PRE is the intersection of upward cones with apex angle $\frac{\pi}{2}$

of the overlay of these two Voronoi diagrams, a point corresponding to a vertical line segment of maximum length can be found in time $O(1)$. Thus, inspecting every cell of the overlay leads to an algorithm solving the problem in time $O(n^2)$.

In [2], we presented an $O(n \log n)$ algorithm to find a vertical line segment of maximum length. To this end, we use two planes sweeping upwards, one of them through the set PRE and the other one through the set SUB. The sweep consists of different phases. For each phase, there are at most 4 cones that determine the size of the longest vertical segment within this phase. This means that the the longest vertical line segment can be computed in time $O(1)$ for each phase. Because there are only $O(n)$ many phases, the sweeping can be done in time $O(n)$. Since the algorithm starts with computing furthest site Voronoi diagrams of circles, its running time is increasesd to $O(n \log n)$.

4   The LP-Approach

In this section, we provide a solution to the meeting scheduling problem using the concept of LP-type problems. We will show that the problem can be stated as an LP-type problem and its dimension equals 4. This enables us to compute a meeting of longest duration for n participants in expected time $O(n)$.

4.1   General Framework of LP-Type Problems

LP-type problems were introduced by Sharir and Welzl [7]. Here we will follow the slightly modified description by Matoušek [3]. We describe the problem as a maximization problem instead of a minimization problem.

A *maximization problem* is a pair (H, w), where H is a finite set, and $w : 2^H \to W$ is a function with values in a linearly ordered set $(W, \leq)$. The elements of H are called the *constraints*, and for a subset $G \subset H$, $w(G)$ is called the *value* of G. Intuitively, the value $w(G)$ for a subset G of constraints stands for the largest value attainable by a certain objective function while satisfying all the constraints of G. The goal is to find $w(H)$. The set W is assumed to possess a largest element denoted by ∞ (intuitively, it stands for 'optimum undefined') and usually also a smallest element $-\infty$ (with intuitive meaning 'no feasible solution exists').

The maximization problem (H, w) is called an *LP-type problem* if the following two axioms are satisfied:

Axiom 1. (Monotonicity) For any F, G with $F \subset G \subset H$, $w(F) \geq w(G)$.

Axiom 2. (Locality) For any $F \subset G \subset H$ with $w(F) = w(G) < \infty$ and any $h \in H$, $w(G \cup \{h\}) < w(G)$ implies that also $w(F \cup \{h\}) < w(F)$.

We recall some terminology from the theory of LP-type problems. A *basis* B is a set of constraints with $w(B') > w(B)$ for all proper subsets B' of B. A *basis for* a subset G of H is a basis $B \subset G$ with $w(B) = w(G)$. The maximum cardinality of any *basis* is called the *dimension* of (H, w) and denoted by $dim(H, w)$. We say that a constraint $h \in H$ violates a set G if $w(G \cup \{h\}) < w(G)$.

There are several algorithms for solving an LP-type problem of constant bounded dimension in time $O(|H|)$. The algorithms differ in the assumptions on the primitive operations available for the considered problem. The primitive operations of Sharir and Welzl are:

Violation Test. Given a basis B and a constraint $h \in H$, decide whether h violates B.

Basis Change. Given a basis B and a constraint h, return (some) basis for $B \cup \{h\}$.

Initial Basis. In the beginning, we have some basis B_0 with $w(B_0) < \infty$.

The expected number of primitive operations performed by the algorithm by Sharir and Welzl is $O(|H|)$ for any LP-type problem (H, w) with constant bounded dimension.

4.2 Preparations for Applying the General Framework

Is the meeting scheduling problem, as we stated it until now, an LP-type problem, if we consider the set of participants M as the set of constraints and the function w maps a subset $M' \subset M$ to the optimum meeting duration for all participants in M'? The answer to this question is negative, because Axiom 2 does not hold in general. Figure 2 shows a situation where Axiom 2 does not hold.

We will establish that the problem becomes an LP-type problem if we lexicographically maximize (f, S) instead of just maximizing f.

As Figure 2 indicates, we have to analyze those situations where f obtains its maximum in more than one point.

Lemma 1. $|S(x_1) - S(x_2)| = \|x_1 - x_2\|$ holds, if both locations x_1, x_2 maximize f.

Proof. Since the set of feasible points F is convex, the line segment B_S from $(x_1, S(x_1))$ to $(x_2, S(x_2))$ is completely contained in F. Analogously, the line segment B_E from $(x_1, E(x_1))$ to $(x_2, E(x_2))$ is in F.

$$l_2^{sub} = (1,2) \quad l_3^{sub} = (5,2)$$
$$t_2^{sub} = 3 \quad t_3^{sub} = 7$$

$$l_1^{pre} = (0,1) \qquad\qquad\qquad l_1^{sub} = (6,1)$$
$$t_1^{pre} = 0 \qquad\qquad\qquad t_1^{sub} = 7$$

$$l_2^{pre} = (1,0) \quad l_3^{pre} = (5,0)$$
$$t_2^{pre} = 0 \quad t_3^{pre} = 4$$

Fig. 2. Axiom 2 does not hold for the participants M_1, M_2, M_3, if we use $x \mapsto f(x)$ as objective function. For the sets $F := \{M_1\}$, $G := \{M_1, M_2\}$ and $h := M_3$, we obtain $w(F) = w(G) = 1$ and $w(G \cup \{h\}) < w(G)$, but $1 = w(F \cup \{h\}) = w(F)$.

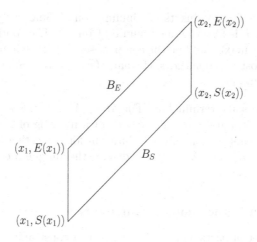

Fig. 3. Every vertical segment of maximum length contained in this parallelogram corresponds to a meeting of longest duration

If $(l, z) \in B_S$ were an inner point of F, we could schedule a meeting at l of longer duration than at x_1. But this would be a contradiction to x_1 maximizing f, so $B_S \subset \partial F$ holds. There must be some cone PRE_j whose boundary contains B_S. Otherwise, the line segment B_S could intersect every cone's boundary in at most 2 points. Finally, $B_S \subset \partial PRE_j$ implies that B_S is a line segment of slope 1. $\qquad\square$

Lemma 2. *The locations maximizing f form a line segment.*

Proof. We have to show that every 3 locations maximizing f are collinear. Let x_1, x_2, x_3 denote locations maximizing f, such that $S(x_1) \leq S(x_2) \leq S(x_3)$. By Lemma 1, we conclude $\|x_3 - x_1\| = S(x_3) - S(x_1) = S(x_3) - S(x_2) + S(x_2) - S(x_1) = \|x_3 - x_2\| + \|x_2 - x_1\|$. So, the triangle inequality holds with equality, which means that x_1, x_2, x_3 are collinear. $\qquad\square$

Lemma 3. *The maximum meeting duration is already determined by at most 2 participants, if f obtains its maximum in more than one point.*

Proof. Let x_1 and x_2 denote distinct points where f obtains its maximum and let B_S, B_E and PRE_j be defined as in the proof of Lemma 1. An analogous argument to the one showing that $B_S \subset \partial PRE_j$ leads to the fact that there is a downward cone SUB_k such that $B_E \subset \partial SUB_k$. The optimum meeting duration T is obtained for a subset of the line segment $l_j^{pre} l_k^{sub}$. But even a 2-participant-meeting just of M_j and M_k has maximum duration T. If $j = k$, which is also possible, the maximum meeting duration is already determined by only one participant. $\qquad\square$

Lemma 4. *There is a unique location lexicographically maximizing $(f(x), S(x))$.*

Proof. We define $C := \{x \in \mathbb{R}^2 : \|l_1^{pre} - x\| \leq t_1^{sub} - t_1^{pre} - f(l_1^{pre})\}$. This implies $f(x) < f(l_1^{pre})$ for all $x \in \mathbb{R}^2 \backslash C$. Due to the compactness of C and the continuity of $f|_C$, the function $f|_C$ obtains its maximum T on C. Since $T \geq f(l_1^{pre}) > f(x)$ for all $x \in \mathbb{R}^2 \setminus C$, T is also the maximum of f on $\mathbb{R}^2$. The continuous function $S|_{f^{-1}(T)}$ obtains its maximum t on the compact set $f^{-1}(T)$. Thanks to Lemma 1, there can be at most one point x, such that $f(x) = T$ and $S(x) = t$. Hence the optimum point is unique.   □

Now, we introduce some terminology. For $M' \subset M$ we refer to $OPT_{M'}$ as the lexicographically optimum pair $(f(x), S(x))$ for a meeting of all participants in M'. Let $f|_{M'}(x)$ denote the maximum duration for a meeting at location x of all participants in M'. Let $X(OPT_{M'})$ denote the unique location where the optimum meeting is held.

4.3 Application of the General Framework

We consider the set of participants M as the set of constraints. The function w maps a subset $M' \subset M$ to the pair $w(M') := OPT_{M'}$. Those pairs are compared lexicographically.

Theorem 1. *To find the optimum meeting for a set of participants is an LP-type problem.*

Proof. Clearly, a meeting, which is possible for a set of participants G, is also possible for every subset of G. Thus Axiom 1 holds.

To show that Axiom 2 holds, we consider sets $F \subset G \subset M$ and a participant $h \in M$ such that $OPT_F = OPT_G$ and $OPT_{G \cup \{h\}} < OPT_G$. According to Lemma 4, every point different from $X(OPT_F)$ leads to a solution worse than $OPT_F = OPT_G$ already for the participants in F. Hence, every point different from $X(OPT_F)$ leads also for the participants in G to a solution worse than $OPT_F = OPT_G$. We conclude $X(OPT_F) = X(OPT_G)$. The inequality $OPT_{G \cup \{h\}} < OPT_G$ means that participant h cannot stay at the location $X(OPT_G)$ for the whole time interval $[S|_G(X(OPT_G)), E|_G(X(OPT_G))]$. But, this also means that he cannot stay the whole time interval at $X(OPT_F)$ and we conclude $OPT_{F \cup \{h\}} < OPT_F$.   □

Now, we have to look at the dimension of the problem. Figure 4 shows that the dimension is at least 4.

The proof that the dimension of the problem is at most 4 uses Helly's theorem. A similar argument using Helly's theorem for a related question concerning the weighted Euclidean 1-center problem is described by Megiddo [4].

Theorem 2 (Helly's Theorem [5]). *Let $C_1, \ldots, C_n$ be convex sets in $\mathbb{R}^d$, $n \geq d + 1$. Suppose that the intersection of every $d + 1$ of these sets is nonempty. Then the intersection of all the C_i is nonempty.*

Theorem 3. *The dimension of the meeting scheduling problem is at most 4.*

Proof. Let us assume that $B = \{M_{i_1}, \ldots, M_{i_b}\}$ is a basis and $b = |B| > 4$ holds. For every 4-element subset W of $\{PRE_{i_1} \cap SUB_{i_1}, \ldots, PRE_{i_b} \cap SUB_{i_b}\}$, let T_W

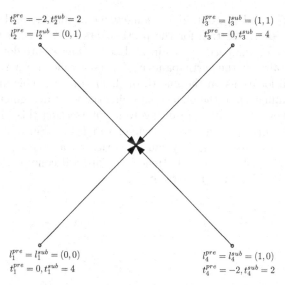

$$t_2^{pre} = -2, t_2^{sub} = 2$$
$$l_2^{pre} = l_2^{sub} = (0, 1)$$

$$l_3^{pre} = l_3^{sub} = (1, 1)$$
$$t_3^{pre} = 0, t_3^{sub} = 4$$

$$l_1^{pre} = l_1^{sub} = (0, 0)$$
$$t_1^{pre} = 0, t_1^{sub} = 4$$

$$l_4^{pre} = l_4^{sub} = (1, 0)$$
$$t_4^{pre} = -2, t_4^{sub} = 2$$

Fig. 4. Set of 4 participants $\{M_1, M_2, M_3, M_4\}$ and their way to the optimum meeting location. The start of their optimum meeting is determined by participants M_1 and M_3, while the end is determined by M_2 and M_4. Removing any participant enables a meeting with longer duration. If one, for example, removes participant M_1, the optimum meeting location becomes $(1, 1)$ and the duration of the optimum meeting increases by $\sqrt{2} - 1$.

denote the maximum length of a vertical segment in the intersection of the sets in W. Note that, by the assumption that there is a solution with $f(x) > 0$, we know that the intersection of the sets in W is nonempty for every W. We define T' to be the minimum T_W obtained for all possibilities of W. Clearly, T' is an upper bound on the length T of a maximum vertical line segment contained in the set of feasible points for the participants in B. In fact, we can even show that $T = T'$ holds. To this end, let SUB'_{i_k} denote the vertical cone with apex angle $\frac{\pi}{2}$ whose uppermost point is its apex $((l_{i_k}^{sub})_X, (l_{i_k}^{sub})_Y, t_{i_k}^{sub} - T')$ for $1 \leq k \leq b$. We observe that all sets in $\{PRE_{i_1} \cap SUB'_{i_1}, \ldots, PRE_{i_b} \cap SUB'_{i_b}\}$ are convex and, by definition of T', the intersection of every 4 of them is nonempty. By Helly's theorem, this implies that the intersection of all sets in $\{PRE_{i_1} \cap SUB'_{i_1}, \ldots, PRE_{i_b} \cap SUB'_{i_b}\}$ is nonempty. But a point in this intersection corresponds to a vertical line segment of length at least T' in the intersection of the original cones $\{PRE_{i_1}, \ldots, PRE_{i_b}, SUB_{i_1}, \ldots, SUB_{i_b}\}$. This means, that $T \geq T'$ and together with $T \leq T'$ we conclude that $T = T'$ holds.

Let V denote a 4-element subset of B such that $f|_V$ and $f|_B$ obtain the same value as maximum. If $f|_V$ obtains its maximum only in one single point x^*, this point must be the optimum meeting point for V as well as for B. But in this case $S|_V(x^*) = S|_B(x^*)$ holds, which means $OPT_V = OPT_B$ and we get contradiction to B being a basis.

So, we must consider the case that $f|_V$ obtains its maximum in more than one point. By Lemma 3, there is a subset $Z \subset V$ of at most 2 participants such that $f|_Z, f|_V$ and $f|_B$ obtain the same value as maximum. Let x_Z denote the optimum meeting location for the participants in Z and x_B the optimum meeting location for the participants in B. If $x_Z = x_B$, we are done.

By Lemma 2, already the participants in Z can schedule meetings of maximum duration only in locations on the line through x_B and x_Z. This means, we can restrict our examination to the plane containing the vertical segments at x_B and x_Z. Figure 5 shows the curves resulting from intersecting this plane with the downward cones corresponding to participants in $B \setminus Z$. Since all participants in B can reach x_B at $S|_Z(x_B)$, they can also reach x_Z at $S|_Z(x_B) + \|x_Z - x_B\|$, which equals $S_Z(x_Z)$ by Lemma 1. Hence, the upward cones corresponding to participants of $B \setminus Z$ are not important.

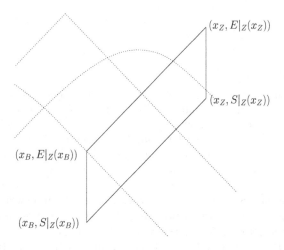

$(x_Z, E|_Z(x_Z))$

$(x_Z, S|_Z(x_Z))$

$(x_B, E|_Z(x_B))$

$(x_B, S|_Z(x_B))$

Fig. 5. The longest meeting duration that is possible for the participants in Z is the same as for the participants in B. The dotted curves correspond to downward cones in $B \setminus Z$.

The optimality of x_B ensures that there is a downward cone PRE_{i_j} which contains the point $(X_B, E|_Z(X_B))$ on its boundary and prevents improving the solution by moving from x_B into the direction of x_Z. We conclude, that the optimum meeting for the at most 3 participants in $Z \cup \{M_{i_j}\}$ is as good as for all participants in B, which is a contradiction to B being a basis.   □

Let us recall the primitive operations used for the algorithm by Sharir and Welzl. The optimum solution can be found in time $O(1)$ for at most 5 participants. To this end, we could for example use the algorithm we presented in [2]. We conclude that the primitive operations *basis change* and *violation test* can be done in time $O(1)$. Finally, any participant can be used as initial basis.

Now we can state our main result:

Theorem 4. *We can compute a meeting of longest duration (or determine that no meeting is possible) for n participants in expected time $O(n)$.*

5   Summary and Current Work

The paper addresses the problem which is finding a suitable place and time for a group of people to meet. By taking into consideration both time and space limitations of the participants, this paper distinguishes itself from the current practice of scheduling meetings only based on time limitations. We have seen that the problem we consider turned out to be an LP-type problem of small dimension.

One interesting problem we are currently working on is the following: Given a set of n participants and a parameter k, find a meeting point which maximizes the duration for a meeting of at least $n - k$ participants. An optimum meeting for all except k participants can be found by inspecting the optimum meeting point for every subset of 4 participants, but we would like to find a fast algorithm. However, a subquadratic solution in the full range of k seems unlikely, because Matoušek [3] mentions that already the problem to find the smallest circle enclosing at least q points of a set of n points in the plane, is 3SUM-hard.

The techniques described in this paper appear to be generalizable to the setting of different speeds among participants. We are currently examining this problem instance.

Acknowledgments. We would like to thank David Kirkpatrick and Emo Welzl for very helpful discussions.

References

1. BenHassine, A., Ho, T.: An agent-based approach to solve dynamic meeting scheduling problems with preferences. Engineering Applications of Artificial Intelligence 20(6), 857–873 (2007)
2. Klein, R., Yi, J., Nussbaum, D., Sack, J.-R.: How to Fit in another Meeting. In: 2nd International Conference on Collaborative Computing (CollaborateCom 2006), Atlanta, November 2006, IEEE, Los Alamitos (2006), http://ieeexplore.ieee.org
3. Matoušek, J.: On Geometric Optimization with Few Violated Constraints. In: Symposium on Computational Geometry, pp. 312–321 (1994)
4. Megiddo, N.: The weighted Euclidean 1-center problem. Math. Oper. Res. 8, 498–504 (1983)
5. Radon, J.: Mengen konvexer Körper, die einen gemeinsamen Punkt enthalten. Math. Ann. 83, 113–115 (1921)
6. Shakshuki, E., Koo, H.-H., Benoit, D., Silver, D.: A distributed multi-agent meeting scheduler. Journal of Computer and System Sciences 74(2), 279–296 (2008)
7. Sharir, M., Welzl, E.: A combinatorial bound for linear programming and related problems. In: Finkel, A., Jantzen, M. (eds.) STACS 1992. LNCS, vol. 577, pp. 569–579. Springer, Heidelberg (1992)
8. Welzl, E.: Smallest enclosing disks (balls and ellipsoids). In: Maurer, H.A. (ed.) New Results and New Trends in Computer Science. LNCS, vol. 555, pp. 359–370. Springer, Heidelberg (1991)

Fixed-Parameter Algorithms for Kemeny Scores

Nadja Betzler[1,*], Michael R. Fellows[2,**], Jiong Guo[1,***],
Rolf Niedermeier[1], and Frances A. Rosamond[2,**]

[1] Institut für Informatik, Friedrich-Schiller-Universität Jena
Ernst-Abbe-Platz 2, D-07743 Jena, Germany
{betzler,guo,niedermr}@minet.uni-jena.de
[2] PC Research Unit, Office of DVC (Research), University of Newcastle,
Callaghan, NSW 2308, Australia
{michael.fellows,frances.rosamond}@newcastle.edu.au

Abstract. The KEMENY SCORE problem is central to many applications
in the context of rank aggregation. Given a set of permutations (votes)
over a set of candidates, one searches for a "consensus permutation"
that is "closest" to the given set of permutations. Computing an optimal consensus permutation is NP-hard. We provide first, encouraging
fixed-parameter tractability results for computing optimal scores (that
is, the overall distance of an optimal consensus permutation). Our fixed-
parameter algorithms employ the parameters "score of the consensus",
"maximum distance between two input permutations", and "number of
candidates". We extend our results to votes with ties and incomplete
votes, thus, in both cases having no longer permutations as input.

1 Introduction

To aggregate inconsistent information does not only appear in classical voting
scenarios but also in the context of meta search engines and many other applications [8,6,1,5]. In some sense, herein one always deals with *consensus problems*
where one wants to find a solution to various "input demands" such that these
demands are met as well as possible. Naturally, contradicting demands cannot
be fulfilled at the same time. Hence, the consensus solution has to provide a
balance between opposing requirements. The concept of *Kemeny consensus* is
among the most classical and important research topics in this context. In this
paper, we study new algorithmic approaches based on parameterized complexity analysis [7,10,13] for computing Kemeny scores and, thus, Kemeny consensus
solutions. To describe our results, we start with introducing Kemeny elections.

Kemeny's voting scheme goes back to the year 1959. It can be described
as follows. An *election* (V, C) consists of a set V of n votes and a set C of

* Supported by the DFG, research project DARE, GU 1023/1.
** Supported by the Australian Research Council. Work done while staying in Jena as
a recipient (MF) of the Humboldt Research Award of the Alexander von Humboldt
foundation, Bonn, Germany.
*** Supported by the DFG, Emmy Noether research group PIAF, NI 369/4.

R. Fleischer and J. Xu (Eds.): AAIM 2008, LNCS 5034, pp. 60–71, 2008.

m candidates. A vote is a *preference list* of the candidates, that is, for each voter the candidates are ordered according to preference. For instance, in case of three candidates a, b, c, the order $c > b > a$ would mean that candidate c is the best-liked one and candidate a is the least-liked one for this voter.[1] A "Kemeny consensus" is a preference list that is "closest" to the preference lists of the voters: For each pair of votes p, q, the so-called *Kendall-Tau distance* (*KT-distance* for short) between p and q, also known as the number of inversions between two permutations, is defined as $\text{dist}(p, q) = \sum_{\{c,d\} \subseteq C} d_{p,q}(c, d)$, where the sum is taken over all unordered pairs $\{c, d\}$ of candidates, and $d_{p,q}(c, d)$ is 0 if p and q rank c and d in the same order, and 1 otherwise. Using divide and conquer, the KT-distance can be computed in $O(m \cdot \log m)$ time. The *score* of a preference list l with respect to an election (V, C) is defined as $\sum_{v \in V} \text{dist}(l, v)$. A preference list l with the minimum score is called *Kemeny consensus* of (V, C) and its score $\sum_{v \in V} \text{dist}(l, v)$ is the *Kemeny score* of (V, C). The problem considered in this work is as follows:

KEMENY SCORE
Input: An election (V, C) and a positive integer k.
Question: Is the Kemeny score of (V, C) at most k?

Clearly, in applications we are mostly interested in computing a Kemeny consensus of a given election. All our algorithms that decide the KEMENY SCORE problem actually provide a corresponding Kemeny consensus.

Known results. We summarize the state of the art concerning the computational complexity of KEMENY SCORE. Bartholdi et al. [2] showed that KEMENY SCORE is NP-complete, and it remains so even when restricted to instances with only four votes [8,9]. Given the computational hardness of KEMENY SCORE on the one side and its practical relevance on the other side, polynomial-time approximation algorithms have been studied. Thus, the Kemeny score can be approximated to a factor of 8/5 by a deterministic algorithm [16] and to a factor of 11/7 by a randomized algorithm [1]. Recently, a (theoretical) PTAS result has been obtained [12]. Conitzer, Davenport, and Kalagnanam [6,5] performed computational studies for the efficient exact computation of a Kemeny consensus, using heuristic approaches such as greedy and branch and bound. Finally, note that Hemaspaandra et al. [11] provided further, exact classifications of the computational complexity of Kemeny elections.

Our results. As pointed out by Conitzer et al. [5], for obvious reasons approximate solutions for election problems such as KEMENY SCORE may be of limited interest. Hence, *exact* solutions are of particular relevance in this context. Given the NP-completeness of the problem, however, it seems inevitable to live with exponential-time algorithms for solving KEMENY SCORE. Fortunately, parameterized complexity analysis as pioneered by Downey and Fellows [7,10,13] seems a fruitful approach here. This will be shown by positive results based on three natural parameterizations. Before that, note that studying the parameter "number of votes" is pointless because, as mentioned before, the problem is already

---

[1] Some definitions also allow ties between candidates—we deal with this later.

NP-complete for only four votes. First of all, using the Kemeny score k itself as the parameter, we derive an algorithm solving KEMENY SCORE in $O(1.53^k + m^2 n)$ time, where $n := |V|$ and $m := |C|$. This algorithm is based on a problem kernelization and a depth-bounded search tree. Further, we introduce a structural parameterization by studying the parameter "maximum KT-distance d between any two input votes". Note that in application scenarios such as meta search engines small d-values may be plausible. We show that KEMENY SCORE can be solved in $O((3d + 1)! \cdot d \cdot \log d \cdot m \cdot n)$ time by a dynamic programming approach. Eventually, note that by trying all possible permutations of the m candidates, we can trivially attain an efficient algorithm if m is very small. The corresponding combinatorial explosion $m!$ in the parameter m is fairly large, though. Using dynamic programming, we can improve this to an algorithm running in $O(2^m \cdot m^2 \cdot n)$ time. Finally, we extend our findings to the cases where ties within votes are allowed and to incomplete votes where not all candidates are ranked. Due to the lack of space, several proofs are deferred to the full version.

2 Preliminaries

We refer to the introductory section for some basic definitions concerning (Kemeny) elections. Almost all further concepts are introduced where needed. Hence, here we restrict ourselves to concepts of "general" importance. Let the *position* of a candidate a in a vote v be the number of candidates that are better than a in v. That is, the leftmost (and best) candidate in v has position 0 and the rightmost has position $m - 1$. Then, $\mathrm{pos}_v(a)$ denotes the position of candidate a in v. Moreover, we say that two candidates a and b form a *dirty pair* if in V there is one vote with $a > b$ and another vote with $b > a$.

We briefly introduce the relevant notions of parameterized complexity theory [7,10,13]. Parameterized algorithmics aims at a multivariate complexity analysis of problems. This is done by studying relevant problem parameters and their influence on the computational complexity of problems. The hope lies in accepting the seemingly inevitable combinatorial explosion for NP-hard problems, but confining it to the parameter. Hence, the decisive question is whether a given parameterized problem is *fixed-parameter tractable (FPT)* with respect to the parameter, say k. In other words, here we ask for the existence of a solving algorithm with running time $f(k) \cdot \mathrm{poly}(n, m)$ for some computational function f. A core tool in the development of fixed-parameter algorithms is polynomial-time preprocessing by *data reduction rules*, often yielding a *kernelization*. Herein, the goal is, given any problem instance x with parameter k, to transform it in polynomial time into a new instance x' with parameter k' such that the size of x' is bounded from above by some function only depending on k, $k' \leq k$, and (x, k) is a yes-instance if and only if (x', k') is a yes-instance. We call a data reduction rule *sound* if the new instance after an application of this rule is a yes-instance iff the original instance is a yes-instance. We also employ search trees for our fixed-parameter algorithms. Search tree algorithms work in a recursive manner. The number of recursion calls is the number of nodes in the according tree. This number is governed by linear recurrences with constant coefficients. It is well known how to solve these [13]. If the algorithm solves a problem instance of size s and calls itself recursively for problem instances of sizes $s - d_1, \ldots, s - d_i$, then $(d_1, \ldots, d_i)$ is called the *branching vector* of this recursion. It corresponds to the recurrence $T_s = T_{s-d_1} + \cdots + T_{s-d_i}$ for the asymptotic size T_s of the overall search tree.

3 Parameterization by the Kemeny Score

We present a kernelization and a search tree algorithm for KEMENY SCORE. The following lemma, whose correctness follows directly from the Extended Condorcet criterion [15], is used for deriving the problem kernel and the search tree.

Lemma 1. *Let a and b be two candidates in C. If $a > b$ in all votes $v \in V$, then every Kemeny consensus has $a > b$.*

3.1 Problem Kernel

When applied to an input instance of KEMENY SCORE, the following polynomial-time executable data reduction rules yield an "equivalent" election with at most $2k$ candidates and at most $2k$ votes with k being the Kemeny score. Note that, if we use a preference list over a *subset* of the candidates to describe a vote, then we mean that the remaining candidates are positioned arbitrarily in this vote. We apply the following data reduction rule to shrink the number of candidates in a given election (V, C).

Rule 1. Delete all candidates that are in no dirty pair.

Lemma 2. *Rule 1 is sound and can be carried out in $O(m^2n)$ time.*

Lemma 3. *After having exhaustively applied Rule 1, in a yes-instance there are at most $2k$ candidates.*

Next, we apply a data reduction rule to get rid of too many identical votes.

Rule 2. If there are more than k votes in V identical to a preference list l, then return "yes" if the score of l is at most k; otherwise, return "no".

Lemma 4. *Rule 2 is sound and works in $O(mn)$ time.*

Lemma 5. *After having exhaustively applied Rule 1 and Rule 2, in a yes-instance $((V, C), k)$ of KEMENY SCORE there are at most $2k$ votes.*

In summary, we can state the following:

Theorem 1. KEMENY SCORE *admits a problem kernel with at most $2k$ votes over at most $2k$ candidates. It can be computed in $O(m^2 \cdot n)$ time.*

3.2 Search Tree Algorithm

It is trivial to achieve an algorithm with search tree size $O(2^k)$ by simply branching on dirty pairs. For the description of an improved search tree algorithm, we need the following definition: Three candidates a, b, c form a *dirty triple* if they occur in at least two dirty pairs. The search tree algorithm first enumerates all dirty pairs of the given election (V, C) and then branches according to the dirty triples. At a search tree node, in each case of the branching, an order of the candidates involved in the dirty triples processed at this node is fixed and maintained in a set. This order represents the relative positioning of these candidates in the Kemeny consensus sought for. Then, the parameter is decreased according to this order. Since every order of two candidates in a dirty pair decreases the parameter at least by one, the height of the search tree is upper-bounded by the parameter.

Next, we describe the details of the branching. At each node of the search tree, we store two types of information:

- The dirty pairs that have not been processed by ancestor nodes are stored in a set D.
- The information about the orders of candidates that have already been determined when reaching the node is stored in a set L. That is, for every pair of candidates whose order is already fixed we store this order in L.

For any pair of candidates a and b, the order $a > b$ is *implied* by L if there is a subset of ordered pairs $\{(c_1, c_2), (c_2, c_3), \ldots, (c_{i-1}, c_i)\}$ in L which can be concatenated such that we have $a > c_1 > \cdots > c_i > b$. To add the order of a "new" pair of candidates, for example $a > b$, to L, we must check if this is *consistent* with L, that is, L does not already imply $b > a$.

At the root of the search tree, D contains all dirty pairs occurring in (V, C). For each non-dirty pair, its relative order in an optimal Kemeny ranking can be determined using Lemma 1. These orders are stored in L. At a search tree node, we distinguish three cases:

Case 1. If there is a dirty triple $\{a, b, c\}$ forming three dirty pairs contained in D, namely, $\{a, b\}, \{b, c\}, \{a, c\} \in D$, then remove these three pairs from D. Branch into all six possible orders of a, b, and c. In each subcase, if the corresponding order is not consistent with L, discard this subcase, otherwise, add the corresponding order to L and decrease the parameter according to this subcase. The worst-case branching vector of this case is $(3, 4, 4, 5, 5, 6)$, giving a branching number 1.52. To see this, note that we only consider instances with at least three votes, since KEMENY SCORE is polynomial-time solvable for only two votes. Thus, for every dirty pair $\{c, c'\}$, if there is only one vote with $c > c'$ (or $c' > c$), then there are at least two votes with $c' > c$ (or $c > c'$). A simple calculation then gives the branching vector.

Case 2. If Case 1 does not apply and there is a dirty triple $\{a, b, c\}$, then a, b, c form exactly two dirty pairs contained in D, say $\{a, b\} \in D$ and $\{b, c\} \in D$. Remove $\{a, b\}$ and $\{b, c\}$ from D. As $\{a, c\}$ is not a dirty pair, its order is determined by L. Hence, we have to distinguish the following two subcases.

If $a > c$ is in L, then branch into three further subcases, namely,

- $b > a > c$,
- $a > b > c$, and
- $a > c > b$.

For each of these subcases, we add the pairwise orders induced by them into L if they are consistent for all three pairs and discard the subcase, otherwise. The worst-case branching vector here is $(3, 3, 2)$, giving a branching number 1.53.

If $c > a$ is in L, then we also get the branching vector $(3, 3, 2)$ by branching into the following three further subcases:

- $b > c > a$,
- $c > b > a$, and
- $c > a > b$.

Case 3. If there is no dirty triple but at least one dirty pair (a, b) in D, then check whether there exists some relative order between a and b implied by L: If L implies no order between a and b, then add an order between a and b to L that occurs in at least half of the given votes; otherwise, we add the implied order to L. Finally, decrease the parameter k according to the number of votes having a and b oppositely ordered compared to the one added to L.

The search tree algorithm outputs "yes" if it arrives at a node with $D = \emptyset$ and a parameter value $k \geq 0$; otherwise, it outputs "no". Observe that a Kemeny consensus is then the *full* order implied by L at a node with $D = \emptyset$.

Combining this search tree algorithm with the kernelization given in Section 3.1, we arrive at the main theorem of this section.

Theorem 2. KEMENY SCORE *can be solved in* $O(1.53^k + m^2 n)$ *time, where* k *denotes the Kemeny score of the given election.*

Observe that the search tree algorithm also works for instances in which the votes are weighted by positive integers. If the votes have arbitrary positive weights, we can use the dynamic programming algorithm that is described in the following section.

4 Parameterization by the Maximum KT-Distance

The Kemeny score of some instances might be quite large. So, we consider a further parameterization, now with the maximum KT-distance d between any two given votes as the parameter. Formally, $d := \max_{p, q \in V} \{\text{dist}(p, q)\}$. We describe a dynamic programming fixed-parameter algorithm for this parameterization. To this end, we need the following definitions:

A *block* of size s with start position p denotes the set of candidates a satisfying $p \leq \text{pos}_v(a) \leq p + s - 1$ for at least one $v \in V$. By block(p) we denote a block of size $d + 1$ with start position p. With a *subvote* $v_{C'}$ of a vote v restricted to a subset $C' \subseteq C$ we mean the order of the candidates in C' such that the preferences of v with respect to the candidates in C' are preserved. A *subinstance* of (V, C) restricted to $C' \subseteq C$ consists of all $v_{C'}$ with $v \in V$.

Now, we state two lemmas that are crucial for the algorithm.

Lemma 6. *In an input instance with maximum KT-instance d, the positions of a candidate in two votes differ by at most d.*

Proof. Consider a candidate a and two votes v and w. Assume that $\text{pos}_v(a) = p$ and $\text{pos}_w(a) \geq p + d + 1$. Then, in w there are at least $p + d + 1$ candidates that are better than a, and in v there are p candidates that are better than a. Therefore, there are at least $d + 1$ candidates that are ranked higher than a in w and lower than a in v. This means that the KT-distance between v and w is at least $d + 1$, a contradiction. □

Lemma 7. *In an input instance with maximum KT-instance d, every block of size $d + 1$ contains at most $3d + 1$ candidates.*

Proof. Assume that there are $3d + 2$ candidates $c_1, c_2, \ldots, c_{3d+2}$ in a block of size $d + 1$ with start position p. Without loss of generality, assume that $v_1 := c_1 > \cdots > c_d > c_{d+1} > \cdots > c_{3d+2}$. Let c_i be the candidate in position p in v_1. If $i \leq d + 1$, then candidate c_{3d+2} is in v_1 in position $p + 2d + 1$ or in a higher position. However, since c_{3d+2} is in the block with start position p, it must occur in position $p + d$ or lower in some other vote, contradicting Lemma 6. If $i > d + 1$, then a similar argument applies to c_1, again contradicting Lemma 6. □

Basic Idea. For designing the algorithm we exploit that blocks of size $d + 1$ can be used to decompose an election when computing the Kemeny score. More precisely, consider a block(i) of size $d+1$, then any candidate a with $a \notin$ block(i) must fulfill either $\text{pos}_v(a) < i$ for all $v \in V$ or $\text{pos}_v(a) > i + d + 1$ for all $v \in V$. Further, we can show that there is a Kemeny consensus in which a must have a position in a range that is a function of d. This means that, if we iterate from left to right over the votes and store the Kemeny score for all "partial orders" of the candidates of a block, then we can forget all candidates that appear only left of this block.

Roughly speaking, our algorithm works as follows: It iterates from left to right over the votes and, in each iterative step, considers a block of size $d + 1$. That is, in the first iterative step, the initialization, it considers all possible orders of the candidates of block(0) and uses a table to store the scores of all such orders with respect to the subinstance of (V, C) restricted to block(0). Then, in the following iterations, it considers block(i) with $i \geq 1$. The computation of the score of the orders of block(i) is based on the table entries for block$(i - 1)$. The Kemeny score of (V, C) is the minimum of the entries for block$(m - d - 1)$. For the formal description of the algorithm we need some further definitions.

Definitions. For a subset of candidates C', let $\pi(C')$ denote an order of C'. For two candidate subsets C' and C'', the orders $\pi'(C')$ and $\pi''(C'')$ are *compatible* if the candidates of $C' \cap C''$ have the same order according to π' and π''. Further,

Initialization:

For all permutations $\pi(\text{block}(0))$

$$T(\pi(\text{block}(0)), 0) := S(\pi(\text{block}(0)))$$

Update:

For $i = 1, \ldots, m - d - 1$

$$T(\pi(\text{block}(i)), i) := \min_{\pi'(\text{block}(i-1)) \in \text{Comp}(i,\pi)} \{T(\pi'(\text{block}(i-1)), i-1)\}$$
$$+ S(\pi(\text{block}(i))) - S(\pi_{\text{block}(i-1)}(\text{block}(i)))$$

Output:

$$\min_{\pi(\text{block}(m-d-1))} \{T(\pi(\text{block}(m-d-1)), m-d-1)\}.$$

Fig. 1. Dynamic programming algorithm for KEMENY SCORE

if $C' \subseteq C''$, let $\pi_{C'}(C'')$ denote the suborder of $\pi(C'')$ restricted to C'. For an order $\pi(\text{block}(i))$ the set $\text{Comp}(i, \pi)$ contains all orders of candidates of $\text{block}(i-1)$ that are compatible with $\pi(\text{block}(i))$. For any fixed order $\pi(C')$ of $C' \subseteq C$, its *score* $S(\pi(C'))$ is the sum of its KT-distances between $\pi(C')$ and all subvotes of votes in V restricted to C'.

Algorithmic Details. Next, we define the table T used in the dynamic programming algorithm. The columns of T one-to-one correspond to the candidate subsets $\text{block}(i) \subseteq C$ with $0 \leq i \leq m - d - 1$. Each row corresponds to a possible order of the candidates in $\text{block}(i)$. By Lemma 7, the number of candidates of each block is upper-bounded by $3d + 1$. Thus, T has at most $(3d + 1)!$ rows. An entry in column $\text{block}(i)$ and row π, where π is an order of the candidates in $\text{block}(i)$, stores the minimum score of the orders π' of $C' := \bigcup_{j \leq i} \text{block}(j)$ under the condition that π' is compatible with π.

The algorithm is given in Figure 1. In the following, we first state the correctness of the dynamic programming and then analyze its running time.

Lemma 8. *The dynamic programming algorithm (see Figure 1) correctly decides* KEMENY SCORE.

Theorem 3. KEMENY SCORE *can be solved in* $O((3d+1)! \cdot d \cdot \log d \cdot m \cdot n)$ *time with d being the maximum KT-distance between two input votes.*

Proof. As argued above, the table size is bounded by $O((3d + 1)! \cdot m)$. For the initialization, we compute the score of all orders of at most $3d + 1$ candidates and, for each order, the score is computable in $O(d \log d \cdot n)$ time as argued in the introductory section. Therefore, the initialization can be done in $O((3d + 1)! \cdot d \log d \cdot n)$ time.

At $\text{block}(i)$ with $i \geq 1$, we compute

$$\min_{\pi'(\text{block}(i-1)) \in \text{Comp}(i,\pi)} \{T(\pi'(\text{block}(i-1)), i-1)\}$$

for all possible orders $\pi(\text{block}(i))$. Using a pointer data structure, this can be done in $(3d+1)! \cdot d \log d$ time by going through the $(i-1)$th column once. Thus, the entries of $T(\pi(\text{block}(i)), i)$ can be computed in $O((3d+1)! \cdot d \log d \cdot n)$ time, giving the running time of the algorithm.   □

Note that it is easy to modify the dynamic programming to return a Kemeny consensus within the same asymptotic running time. Furthermore, the algorithm can be easily adapted to deal with weighted votes and/or weighted candidates.

5   Parameterization by the Number of Candidates

Theorem 4. KEMENY SCORE *can be solved in* $O(2^m \cdot m^2 \cdot n)$ *time.*[2]

Proof. (Sketch) The algorithm goes like this: For each subset $C' \subseteq C$ compute the Kemeny score of the given election system restricted to C'. The recurrence for a given subset C' is to consider every subset $C'' \subseteq C'$ where C'' is obtained by deleting a single candidate c from C'. Let π'' be a Kemeny consensus for the election system restricted to C''. Compute the score of the permutation π' of C' obtained from π'' by putting c in first position. Take the minimum score over all π' obtained from subsets of C'. The correctness of this algorithm follows from the following claim.

Claim: A permutation π' of minimum score obtained in this way is a Kemeny consensus of the election system restricted to C'.

Proof: Let σ' be a Kemeny consensus of the election system restricted to C', and suppose that candidate c is on the first position of σ'. Consider $C'' = C' \setminus \{c\}$, and let σ'' be the length-$|C''|$ tail of σ'. The score of σ' is $t + u$, where u is the score of σ'' for the votes that are the restriction of V to C'', and where t is the "cost" of inserting c into the first position: Herein, the cost is the sum over the votes (restricted to C') of the distance of c from first position in each vote. Now suppose that π'' is a Kemeny consensus of the election system restricted to C''. Compared with the score of σ', augmenting π'' to π' by putting c in the first position increases the score of π' by exactly t. The score of π'' is at most u (since it is a Kemeny consensus for the election system restricted to C''), and so the score of π' is at most $t + u$, so it is a Kemeny consensus for the election system restricted to C'. This completes the proof of the claim.

For each of the subsets of C, the algorithm computes a Kemeny score. Herein, a minimum is taken from at most m values and the computation of each of these values needs $O(m \cdot n)$ time. Therefore, the total running time is $O(2^m \cdot m^2 \cdot n)$.   □

6   Ties and Incomplete Votes

The algorithm from Section 5 also applies to the following generalizations; we omit the respective technical details.

---

[2] We give a direct proof here. Basically the same result also follows from a reduction to FEEDBACK ARC SET ON TOURNAMENTS and a corresponding exact algorithm [14].

6.1 Kemeny Score with Ties

Introducing ties, we use the definition of Hemaspaandra et al. [11]: Each vote consists of a *ranking* of candidates that can contain ties between candidates and the Kemeny consensus can contain ties as well. In the definition of the KT-distance, we then replace $d_{p,q}(c,d)$ by

$$t_{p,q}(c,d) := \begin{cases} 0 \text{ if } p \text{ and } q \text{ agree on } c \text{ and } d, \\ 1 \text{ if } p \ (q) \text{ has a preference among } c \text{ and } d \text{ and } q \ (p) \text{ does not }, \\ 2 \text{ if } p \text{ and } q \text{ strictly disagree on } c \text{ and } d. \end{cases}$$

Additionally, we extend the KT-distance to candidate weights, that is, given a weight function $w : C \to R^+$, the KT-distance $\text{dist}(p,q)$ between two votes p, q is defined as

$$\text{dist}(p,q) = \sum_{\{c,d\} \subseteq C} w(c) \cdot w(d) \cdot t_{p,q}(c,d).$$

By increasing the Kemeny score by two if two votes strictly disagree on two candidates, the overall score becomes larger. It is not surprising that also in case of ties we can get a linear problem kernel and a search tree with respect to the parameter Kemeny score. To this end, we extend the notion of dirty pairs such that a pair of candidates $\{a, b\}$ is also dirty if we have $a = b$ in one vote and $a > b$ or $a < b$ in another vote. Then, we can obtain a linear problem kernel in complete analogy to the case without ties. Concerning the search tree, by branching for a dirty pair $\{a, b\}$ into the three cases $a > b$, $a = b$, and $a < b$, we already achieve the branching vector $(1, 2, 4)$ and, hence, the branching number 1.76. For example, having three votes in which candidates a and b are ranked equal in two of the votes and we have $a < b$ in the other vote, by branching into the case "$a = b$" we have one inversion that counts one, branching into "$a < b$" we have two inversions that count one, and branching into "$a > b$" we have two inversions counting one and one inversion counting two.

With some more effort, we can also show that KEMENY SCORE WITH TIES is fixed-parameter tractable with respect to the parameter maximum KT-distance. To use the dynamic programming of Section 4, we need some bounds on the positions a candidate can take and the size of a block in analogy to Lemma 6 and Lemma 7. In order to achieve this, we need a preprocessing step, which is based on the following lemma:

Lemma 9. *Let d denote the maximum KT-distance of an input instance. If there is a vote in which at least $d + 2$ candidates $c_1, \ldots, c_{d+2}$ are tied, then in a yes-instance the candidates $c_1, \ldots, c_{d+2}$ are tied in all votes.*

Now, we can state the following data reduction rule:

Rule 3. *If there are $d + i$ candidates for any integer $i \geq 2$ that are tied in all votes, replace them by one candidate whose weight is the sum of the weights of the replaced candidates.*

The correctness of Rule 3 follows from Lemma 9. Further, note that the dynamic programming procedure from Section 4 can be adapted to deal with candidate weights. Now, similar to Lemmas 6 and 7 we can prove the following.

Lemma 10. *In a candidate-weighted instance with maximum KT-instance d obtained after exhaustively applying Rule 3, the positions of a candidate in two votes differ by at most $2d + 1$.*

Lemma 11. *In a reduced instance with maximum KT-instance d, every block of size 2d contains at most $6d + 2$ candidates.*

Using Lemmas 10 and 11, we can show that with the dynamic programming algorithm given in Section 4, the following running time can be obtained.

Theorem 5. KEMENY SCORE WITH TIES *can be solved in $O((6d+2)! \cdot d \cdot \log d \cdot n \cdot m)$ time with d being the maximum KT-distance between two input votes.*

6.2 Incomplete Votes

We now consider KEMENY SCORE WITH INCOMPLETE VOTES, first introduced by Dwork et al. [8]. Here, the given votes are not required to be permutations of the entire candidate set, but only of candidate subsets, while the Kemeny consensus sought for should be a permutation of all candidates. In the definition of the KT-distance, we then replace $d_{p,q}(c, d)$ by

$$d'_{p,q}(c, d) := \begin{cases} 0 & \text{if } \{c,d\} \not\subseteq C_p \text{ or } \{c,d\} \not\subseteq C_q \text{ or } p \text{ and } q \text{ agree on } c \text{ and } d, \\ 1 & \text{otherwise,} \end{cases}$$

where C_v contains the candidates occurring in vote v.

A simple reduction from FEEDBACK ARC SET shows that, in contrast to KEMENY SCORE with complete votes, the parameterization by the maximum KT-distance does not lead to fixed-parameter tractability for KEMENY SCORE WITH INCOMPLETE VOTES:

Theorem 6. KEMENY SCORE WITH INCOMPLETE VOTES *is NP-hard even if the maximum KT-distance between two input votes is zero.*

By way of contrast, we now show that, parameterized by the Kemeny score k, also KEMENY SCORE WITH INCOMPLETE VOTES is fixed-parameter tractable.

Theorem 7. KEMENY SCORE WITH INCOMPLETE VOTES *is solvable in $(1.48k)^k \cdot p(n, m)$ time with k being the Kemeny score and p being a polynomial function of n and m.*

Proof. Let $((V, C), k)$ be the given KEMENY SCORE instance. The algorithm consists of three steps. First, transform the given election system into one where each vote contains only two candidates. For example, a vote $a > b > c$ is replaced by threes votes: $a > b$, $b > c$, and $a > c$. Second, find all dirty pairs and, for each of them, remove the corresponding votes and branch into two cases, in each case an order between the two candidates of this pair is fixed and the parameter k is decreased accordingly. Third, after having processed all dirty pairs, we construct an edge-weighted directed graph for each of the election systems generated by the second step. The vertices of this graph one-to-one correspond to the candidates. An arc (c, c') from vertex c to vertex c' is added if there is a vote of the form $c > c'$ and a weight equal to the number of the $c > c'$-votes is assigned to this arc. For each pair of candidates c and c' where the second step has already fixed an order $c > c'$, add an arc (c, c') and assign a weight equal to $k + 1$ to this arc. Observe that solving KEMENY SCORE on the instances generated by the first two steps is now equivalent to solving the WEIGHTED DIRECTED FEEDBACK ARC SET problems on the constructed edge-weighted directed graphs. DIRECTED FEEDBACK ARC SET is the special case of WEIGHTED FEEDBACK ARC SET with unit edge weight. The fixed-parameter algorithm by Chen et al. [4] for DIRECTED FEEDBACK ARC SET can also handle WEIGHTED DIRECTED FEEDBACK ARC SET with integer edge weights [3]. Therefore, applying this algorithm in the third step gives a total running time of $(1.48k)^k \cdot p(n, m)$ for a polynomial function p. □

References

1. Ailon, N., Charikar, M., Newman, A.: Aggregating inconsistent information: Ranking and clustering. In: Proc. 37th STOC, pp. 684–693. ACM, New York (2005)
2. Bartholdi III, J., Tovey, C.A., Trick, M.A.: Voting schemes for which it can be difficult to tell who won the election. Social Choice and Welfare 6, 157–165 (1989)
3. Chen, J.: Personal communication (December 2007)
4. Chen, J., Liu, Y., Lu, S., O'Sullivan, B., Razgon, I.: A fixed-parameter algorithm for the directed feedback vertex set problem. In: Proc. 40th STOC, ACM, New York (2008)
5. Conitzer, V., Davenport, A., Kalagnanam, J.: Improved bounds for computing Kemeny rankings. In: Proc. 21st AAAI, pp. 620–626 (2006)
6. Davenport, A., Kalagnanam, J.: A computational study of the Kemeny rule for preference aggregation. In: Proc. 19th AAAI, pp. 697–702 (2004)
7. Downey, R.G., Fellows, M.R.: Parameterized Complexity. Springer, Heidelberg (1999)
8. Dwork, C., Kumar, R., Naor, M., Sivakumar, D.: Rank aggregation methods for the Web. In: Proc. WWW, pp. 613–622 (2001)
9. Dwork, C., Kumar, R., Naor, M., Sivakumar, D.: Rank aggregation revisited (manuscript, 2001)
10. Flum, J., Grohe, M.: Parameterized Complexity Theory. Springer, Heidelberg (2006)
11. Hemaspaandra, E., Spakowski, H., Vogel, J.: The complexity of Kemeny elections. Theoretical Computer Science 349, 382–391 (2005)
12. Kenyon-Mathieu, C., Schudy, W.: How to rank with few errors. In: Proc. 39th STOC, pp. 95–103. ACM, New York (2007)
13. Niedermeier, R.: Invitation to Fixed-Parameter Algorithms. Oxford University Press, Oxford (2006)
14. Raman, V., Saurabh, S.: Improved fixed parameter tractable algorithms for two "edge" problems: MAXCUT and MAXDAG. Information Processing Letters 104(2), 65–72 (2007)
15. Truchon, M.: An extension of the Condorcet criterion and Kemeny orders. Technical report, cahier 98-15 du Centre de Recherche en Économie et Finance Appliquées (1998)
16. van Zuylen, A., Williamson, D.P.: Deterministic algorithms for rank aggregation and other ranking and clustering problems. In: Kaklamanis, C., Skutella, M. (eds.) WAOA 2007. LNCS, vol. 4927, pp. 260–273. Springer, Heidelberg (2007)

The Distributed Wireless Gathering Problem

Vincenzo Bonifaci[1,3,*], Peter Korteweg[2,**],
Alberto Marchetti-Spaccamela[3,*], and Leen Stougie[2,4,**]

[1] Università degli Studi dell'Aquila, Italy
bonifaci@dis.uniroma1.it
[2] Eindhoven University of Technology, The Netherlands
p.korteweg@tue.nl, l.stougie@tue.nl
[3] Sapienza Università di Roma, Italy
alberto@dis.uniroma1.it
[4] CWI, Amsterdam, The Netherlands
stougie@cwi.nl

Abstract. We address the problem of data gathering in a wireless network using multihop communication; our main goal is the analysis of simple algorithms suitable for implementation in realistic scenarios. We study the performance of distributed algorithms, which do not use any form of local coordination, and we focus on the objective of minimizing average flow times of data packets. We prove a lower bound of $\Omega(\log m)$ on the competitive ratio of any distributed algorithm minimizing the maximum flow time, where m is the number of packets. Next, we consider a distributed algorithm which sends packets over shortest paths, and we use resource augmentation to analyze its performance when the objective is to minimize the average flow time. If interferences are modeled as in Bar-Yehuda et al. (J. of Computer and Systems Science, 1992) we prove that the algorithm is $(1 + \epsilon)$-competitive, when the algorithm sends packets a factor $O(\log(\delta/\epsilon) \log \Delta)$ faster than the optimal offline solution; here δ is the diameter of the network and Δ the maximum degree. We finally extend this result to a more complex interference model.

1 Introduction

Wireless networks are used in many areas of practical interest, such as mobile phone communication, ad-hoc networks, and radio broadcasting. Moreover, recent advances in miniaturization of computing devices equipped with short range radios have given rise to strong interest in sensor networks for their relevance in many practical scenarios (environment control, accident monitoring etc.) [1,19].

In many applications of wireless networks data gathering is a critical operation for extracting useful information from the operating environment: information

---

* Research supported by EU ICT-FET 215270 FRONTS and MIUR-FIRB Italy-Israel project RBIN047MH9.
** Research supported by EU FET-project under contract no. FP6-021235-2 AR-RIVAL, EU COST-action 293 GRAAL, and the Dutch BSIK-BRICKS project.

collected from multiple nodes in the network should be transmitted to a sink that may process the data, or act as a gateway to other networks. In the case of wireless sensor networks sensor nodes have limited computation capabilities, thus implying that data gathering is an even more crucial operation. For these reasons, data gathering in sensor networks has received significant attention in the last few years; we restrict ourselves to cite only two contributions [1,9]. The problem finds also applications in Wi-Fi networks when many users need to access a gateway using multi-hop wireless relay-routing [5].

An instance of the *Wireless Gathering Problem* (WGP) is given by a static wireless network which consists of several stations (nodes) and one base station (the sink), modeled as vertices of a graph; over time data packets arrive at stations that have to be gathered at the base station. In the sequel we assume that time is discrete and that stations have a common clock, hence time can be divided into rounds. Following Bar-Yehuda et al. [3] we adopt the half-duplex model, in which nodes can either send or receive during a single round. Typically, not all nodes in the network can communicate with each other, hence packets have to be sent through several nodes before they can be gathered at the sink; this is called *multi-hop* routing. The crucial issue to be considered is *interference*: the communication between two pairs of nodes causes interference if either of the receiver nodes also receives the communication signal intended for the other node. In case of interference, the receiver node does not receive the packet.

Realistic models of communication among nodes depend on many parameters that influence the performance of transmissions (see for example [1,22]); this raises many combinatorial optimization problems that received significant attention in the last few years. We notice that most solution methods proposed so far are concentrating on polynomial time algorithms that achieve provably good performance in the worst case. Unfortunately these methods are not suitable for practical implementations; in fact they are either centralized and/or require solving complex combinatorial optimization problems [5,15,16]. Since sensors have limited computational power and are unable to perform sophisticated coordination activities, sophisticated algorithms that require solving complex combinatorial optimization problems are impractical for implementations and have mainly theoretical interest. Communications algorithms that have been implemented and tested in real scenarios are not only distributed but can be implemented with very limited overhead (see e.g. BMAC and DMAC [17,21]).

We notice that a formal analysis of the performance of simple algorithms suitable for practical implementation is still missing. In this paper we continue the work initiated in [3,4,7] on the problem of analyzing *simple distributed* algorithms that have *good approximation guarantees* in *realistic scenarios*. We emphasize that the algorithms analyzed are very simple and that the challenge is not in their design but in their analysis. Namely, we consider fully distributed algorithms, i.e. each node makes decisions of when to transmit independently of other nodes. In order to decrease the possibility of interference we assume clock synchronization at the nodes. Our model is more restricted than decentralized algorithms, which may allow information exchange between neighbouring nodes before transmission to decrease the possibility of interference even further.

Our work extends the work of Bar Yehuda et al. [4], where the problem of data gathering in a wireless network is considered as a subproblem in wireless routing. In [4] the authors consider the special case where all release times are zero. The nodes are partitioned into layers, and all nodes of a layer are labeled based on their distance to the sink. The labeling reduces interference, without violating the distributed nature of the algorithm. In a typical wireless network, information on distances is often available or it can be obtained using distributed algorithms [20]. Bar Yehuda et al. [4] prove that a simple randomized algorithm requires in expectation $32.27(m+\delta)\log\Delta$ rounds to gather m packets, where δ is the diameter of the network, and Δ the maximum degree of a node. Hence, their algorithm is in expectation an asymptotic $O(\log\Delta)$-approximation of optimal maximum completion time.

In Bar-Yehuda et al. [4] all packets are released at time 0 and the authors analyze the expected makespan of the schedule. We assume that packets are released over time and we analyze expected flow times, i.e. the time elapsed between the release of a packet and its arrival at the sink; it is well known that flow is a measure more suitable than completion time to describe the Quality of Service (QoS) of a system. Our main contributions are the following. First, in Section 3, we provide a lower bound on the approximability of distributed (possibly exponential-time) algorithms for WGP, minimizing the maximum flow time. We prove a lower bound of $\Omega(\log m)$ on the competitive ratio of distributed algorithms, where m is the number of packets. In Section 4 we analyse the performance of the simple on-line algorithm proposed by Bar-Yehuda et al. [4] when the goal is to minimize average flow time. We prove that the algorithm is $(1+\epsilon)$-competitive when the algorithm sends packets a factor $O(\log(\delta/\epsilon)\log\Delta)$ faster than the optimal offline solution. As ϵ approaches zero, the solution of the more powerful algorithm approaches the optimum. In Section 5 we extend the result to include the more complex interference model proposed by Bermond et al. [5].

We notice once more that the main challenge is not in the design of the algorithm but in the analysis: it is easy to find worst case instances where the algorithm has a very poor performance. The challenge is in finding the right tool for assessing its quality: we show that resource augmentation allows to bound the competitive ratio. Resource augmentation was introduced in the context of machine scheduling in [12], where the idea is to study the performance of on-line algorithms which are given processors faster than those of the adversary. Intuitively, this has been done to compensate an on-line scheduler for its lack of future information. Such an approach has led to a number of interesting results showing that moderately faster processors are sufficient to attain satisfactory performance guarantee for different scheduling problems, e.g. [10,12].

Related Work. The wireless gathering problem in a centralized setting was studied in [5], and [6,7]. Bermond et al. [5] demonstrated that the problem of minimizing maximum completion time is **NP**-hard, even when all release times are zero. In [6,7] we studied centralized and distributed algorithms for the wireless gathering problem with arbitrary release times. For the case of minimizing maximum completion times, we presented a 4-competitive algorithm, and demonstrated that no shortest paths following algorithm can be better than 4-approximate [6]. For the case of minimizing maximum flow time, we presented an algorithm which, under resource augmentation by sending packets a factor 5 faster than an offline algorithm, produces a solution with value at most the optimal off-line solution value. For this objective, we also demonstrated that no algorithm can be better than $\Omega(m^{1/3})$-approximate, without resource augmentation, unless $\mathbf{NP} = \mathbf{P}$ [7]. We remark that distributed algorithms considered in [7] require coordination among neighbouring nodes to avoid interferences during transmission; this can be accomplished through a suitable signaling protocol that is not realistic in current sensor networks.

Kumar et al. [15,16] considered decentralized algorithms for wireless packet routing. They provide near-optimal polynomial time routing and scheduling algorithms for solving throughput maximization problems in wireless ad hoc networks. However, their algorithms are not distributed in the sense that nodes use information of neighbouring nodes in order to decrease interference and in some case hinge on the solution of a linear or a convex optimization problem. For this reason the above results are mainly of theoretical interest.

As we remark there has been much work in recent years on various optimization issues in communication algorithms for wireless and sensor networks: see [13,14,18,23] and references therein. However, to the best of our knowledge a formal analysis of the worst case performance of simple distributed algorithms that are suitable for implementation in real scenarios of sensor networks is still lacking.

A different approach in the analysis of routing protocols that has been proposed in the literature is known as *adversarial queuing theory* as introduced in [2,8]. Typically, in queuing theory the input is given by a distribution function, e.g., a Poisson distribution. In the adversarial queuing theory model the input is given by an adversary, who is restricted by a bound on the average arrival rate of packets. An adversarial queuing theory model seems more appropriate than queuing theory when minimizing packet flow times in case of bursty traffic that consists of alternating periods of high and low packet arrival rates. The goal is to prove stability: a system is said to be *stable* if the average number of packets in the system is bounded [2,8,11]. Given a routing problem and an algorithm A, let λ^* and λ^A be the highest arrival rate that can be sustained, respectively, by the omniscient optimum finding adversary and by algorithm A while maintaining stability. The performance of an algorithm is measured by the ratio λ^*/λ^A. Note that the reduction in the arrival rate of the algorithm is a dual compensation to our resource augmentation to compensate the limitations imposed to the algorithm. Given a stable system, it is of interest to consider the expected flow time of packets. In case of packet routing, in [2] and [8] it is demonstrated that, even in a stable system, the maximum flow time of packets may be unbounded.

2   Preliminaries

We formulate WGP as a graph optimization problem. Given is an undirected graph $G = (V, E)$ with $|V| = n$, sink $s \in V$, and a set of data packets $M = \{1, 2, \ldots, m\}$ which arrive over time. We assume that each edge has unit length. For each pair of nodes $u, v \in V$ we define the *distance* between u and v, denoted by $d(u, v)$, as the length of a *shortest path* from u to v in G. A *layer* of nodes is a set of all nodes at distance d from sink s, for some d. Each packet $j \in M$ has an *origin* $v_j \in V$ and a release time $r_j \in \mathbb{Z}_+$ at which it enters the network;

We assume that time is discrete; we call a time unit a *round*. Packet j can be sent for the first time in round r_j. The rounds are numbered $0, 1, 2, \ldots$. During each round a node may either be *sending* a packet, be *receiving* a packet or be *inactive*. If $d(u, v) = 1$ then u can send some packet j to v during a round. If node u sends a packet j to v in some round, then the pair (u, v) is called a *call* of packet j during that round. Two calls (u, v) and (u', v') *interfere* if $d(u', v) = 1$ or $d(u, v') = 1$; otherwise the calls are *compatible*. The solution of WGP is a schedule of compatible calls such that all packets are sent to the sink.

The above definition of interference is known in the literature as primary interference; we consider in section 5 the extension to incorporate secondary interference, i.e. interference between non-adjacent but proximate links in the network.

Given a schedule, let v_j^t be the node where packet j resides at time t. $C_j := \min\{t : v_j^t = s\}$ is called the *completion time* of packet j. We define $F_j := C_j - r_j$ as the *flow time* of packet j. We assume that packets cannot be aggregated. As extra notation, let $\delta_j := d(v_j, s)$ be the minimum number of calls required for packet j to reach s, and let $\delta := \max_{j \in M} \delta_j$.

In this paper we analyze distributed algorithms under the following assumptions. A network with a shortest paths routing tree is given, and each node knows the next node on the path to the tree root (the sink), as well as $d(v_j, s)$, its distance to the root. Further, we assume that each node is equipped with a clock and the clocks are synchronized, i.e. indicate the same time. This enables nodes to synchronize packet communications. Also, each node knows an upper bound on the maximum node degree, called Δ.

3   A Lower Bound for Distributed Algorithms

It is known from previous work that WGP is **NP**-hard both when minimizing the maximum completion time and the maximum flow time [6]. The latter problem is actually **NP**-hard to approximate within a factor of $O(m^{1/3})$ [7]. Here we give an *unconditional* lower bound on the approximability of the maximum flow time in a distributed setting where routing is done through an in-tree. We consider a scenario in which the conflicts between transmissions from one layer of the tree to the next are resolved randomly: whenever several transmissions from a layer occur in the same round, only a uniformly chosen one succeeds. We call this the *random selection* model. This assumption seems natural for distributed algorithms, as they have no simple means of coordinating the transmitting nodes (or more precisely, coordinating the transmitting nodes is as hard as the original communication task).

Proposition 1. *In the random selection model, the approximation ratio of any algorithm is at least $\Omega(\log m)$ when minimizing the maximum flow time.*

Proof. Consider a constant $k \geq 1$. An adversary releases $m = 2h^2$ packets, for h some integer power of 2. In round $i \cdot k \cdot h$ it releases 2 packets, and in each round in interval $[ikh + 1, ikh + h - 2]$ it releases one packet, for $i = 1, \ldots, 2h$. The only use of k in this setting is provision of a control over the average arrival rate of packets.

We use a star network with three rays, with the sink as the center of the star. Pairs of packets released in the same round are released on two different rays of the star, say ray 1 and ray 2; all the other packets are released on ray 3.

Both the algorithm and the adversary can send only one of the packets released at time ikh in that round, and have to send the other packet in a later round. We call the packet which is released but not sent in round ikh the *target packet*. We prove the proposition by demonstrating that the expected flow time of one of the $2h$ target packets exceeds $\log h + 1$ for any algorithm, whereas the maximum flow time of the adversary is at most 2.

The adversary sends one of the packets, released at time ikh, in that same round, and the target packet in the next round. The adversary can send all other packets one round after their release dates, hence the maximum flow time of the adversary is at most 2.

Consider any algorithm in the random selection model. Clearly the algorithm can forward at most one packet to the sink each round. Because of the random selection rule, for $f \geq 1$, we have

$$\mathbf{Pr}[F_j = f + 1 \,|\, j \text{ is a target packet}] = 1/2^f.$$

So, if p is the probability that the flow time of a target packet is at most $\log h + 1$, we have $p = \sum_{f=1}^{\log h} (1/2)^f = 1 - 1/h$. Consider the set of $2h$ target packets. For $\ell = 1, \ldots, 2h$, let $X_\ell = 1$ be the event that target packet ℓ has flow time exceeding $\log h + 1$. Otherwise $X_\ell = 0$. Then $\mathbb{E}[X_\ell] = 1/h$ for all $\ell = 1, \ldots, 2h$ and therefore the expected number of target packets with flow time exceeding $\log h + 1$ is $\mathbb{E}[\sum_{\ell=1}^{2h} X_\ell] = \sum_{\ell=1}^{2h} \mathbb{E}[X_\ell] = 2$. Hence, in expectation at least one target packet has flow time at least $\log h + 1$. $\square$

It is useful to compare Proposition 1 with the above cited result, that there is no polynomial time algorithm which can approximate WGP within a factor $O(m^{1/3})$ when minimizing the maximum flow time, unless $\mathbf{NP} = \mathbf{P}$. The latter condition is not required in Proposition 1. In [7] we showed that allowing a constant increase in speed one can obtain a solution with maximum flow time which is less than that of the optimal solution for the original instance. The proof of Proposition 1 indicates that such a result is not likely to be obtainable for distributed algorithms, as they have basically no control on which packet to advance. The reason is that the lower bound is due to an adversarial selection of which packet to advance; and the probability of obtaining such a selection depends on the number of packets, and not on the speed of the algorithm.

In the next section we will analyze a distributed algorithm. It cannot determine which packet is advanced from each layer to the next. Proposition 1 and the observations above should thus explain why we focus on the analysis of the expected total flow time (as opposed to maximum flow time) and why we need to use resource augmentation.

Finally we remark the intuitive fact that, if conflicts are solved adversarially as opposed to randomly, Proposition 1 can be strengthened further. In the *worst-case selection* model, whenever there is a set of pairwise incompatible calls, only an adversarially chosen one will succeed.

Proposition 2. *In the worst-case selection model, the approximation ratio of any deterministic algorithm is at least $\Omega(m)$ when minimizing maximum flow time or average flow time.*

4   A Distributed Algorithm and Its Analysis

4.1   The Algorithm

We consider a distributed algorithm for WGP first introduced by Bar-Yehuda et al. in the context of minimizing the maximum completion time for the gathering problem without release dates [4]. We focus on flow times.

To reduce interference between nodes, the algorithm uses node *labels*. A node at distance d from the sink is assigned label $d \bmod 3$. Each node can be either *active* during a round or *inactive*; only active nodes will transmit a packet. A node will not be active if its packet buffer is empty.

Before we describe the algorithm, we introduce a basic but crucial procedure which enables communication from a set of active nodes. The procedure, first introduced and studied by Bar-Yehuda, Goldreich and Itai [3], is called DE-CAY and requires $2 \log \Delta$ rounds; the time needed for a single execution of the procedure is called a *phase*.

Algorithm 1. DECAY(u, v) [3]

---

    for $j = 1, 2, \ldots, 2 \log \Delta$ **do**
        u sends to v the oldest packet from its buffer;
        u deactivates itself for the rest of the phase with probability $1/2$.
    end for

We can now state a distributed algorithm for WGP (Algorithm 2). The protocol requires the existence of a basic communication structure. Therefore, as a first step of the algorithm a breadth-first search (BFS) tree with the sink as the root is constructed. This can be done in expected time $O((n + \delta \log n) \log \Delta)$ [4]. Since it is done only once, when starting up the system, we will not count this time in defining the cost of the schedule.

Algorithm 2. DISTRIBUTEDGREEDY (DG) [4]

Construct a breadth-first search tree with root s
for each next phase $k = 1, 2, \ldots$ **do**
    Activate each node with label k mod 3 having a nonempty packet buffer;
    Execute DECAY$(u, \text{parent}(u))$ in parallel for each active node u.
end for

Although the algorithm does not model acknowledgement of packets explicitly, it is easy to include them, e.g. by doubling the number of rounds, having communication in odd rounds and acknowledgements in even rounds, as in [4]. Using this, we can assume that successful receipt of a packet (by the parent of the sending node in the BFS tree) is acknowledged immediately. Only at that time it gets removed from the sender's buffer.

By the transmission protocol in DG, where in phase k only nodes of layer k mod 3 transmit, if two nodes transmit, then either they are at the same layer or they are at least distance 3 apart. Hence, in DG two nodes can only interfere if both sender nodes are in the same layer.

A superphase consists of three consecutive phases. Another important ingredient in the analysis of DG is the following.

Theorem 1 ([4]). *Let i be a layer of the tree containing some packet at the beginning of a superphase. There is probability at least $\mu := e^{-1}(1 - e^{-1})$ that during this superphase DG sends a packet from a node u in layer i successfully to the parent node of u in the BFS tree.*

This theorem shows that, during a superphase, each nonempty layer forwards a packet with probability μ to the following layer. Notice however that there is no guarantee on which particular packet is advanced.

4.2 The Analysis

For our analysis we define three solution models. For a given instance $\mathcal{I}$ of WGP, we construct relations between the completion times of packets in these three models. This approach is similar to the approach of Bar-Yehuda et al. [4]. However, one additional difficulty is that, because packets have release dates, the extra speed given to the algorithm does not directly translate into shorter flow times.

- In Model 1 each layer sends with probability at least μ at least one packet every superphase;
- In Model 2 each layer sends with probability $p := 1 - (1 - \mu)^a$ one packet in one of the three last rounds of every a superphases, $a \in \mathbb{N}$, and with probability $(1 - \mu)^a$ no packet;
- In Model 3 each layer sends with probability p one packet in each round, and with probability $1 - p$ no packet.

Note, that the probabilities stated above refer to the case in which the layer contains a packet at the start of the period considered (a single superphase in Model 1, a superphases in Model 2, and a single round in Model 3). This restriction is similar to the restriction on the DECAY procedure. Thus, it follows from Theorem 1 that a DG-solution is a solution which fits into Model 1. We define the completion time and the flow time of a packet j in a Model k solution, respectively, as $C_j^{(k)}$ and $F_j^{(k)}$, $k = 1, 2, 3$.

Motivated by the negative results of Theorem 1 we focus on deriving a bound on the expected average flow time of DG. We use resource augmentation to analyze the performance of DG. A σ-speed algorithm sends data packets at a speed that is σ times faster than an offline algorithm.

Through a sequence of steps we relate the expected flow times of Model 1 to those of Model 2. Subsequently, we demonstrate that the expected flow time of σ-speed DG is bounded by the expected flow time of a Model 3-solution. Finally, we relate the expected flow times of a Model 3-solution to the expected flow times of an optimal offline solution.

Lemma 1. *The expected sum of completion times of a Model 1-schedule is at most the expected sum of completion times of some Model 2-schedule, for every* WGP-*instance.*

Lemma 2. *The expected sum of flow times of σ-speed DG is at most the expected sum of flow times of a Model 3 solution, for $\sigma \geq 6a \log \Delta$.*

A deterministic tandem queue with unit processing times is a network which consists of a sink M_0, and a set of machines $M_i, i = 1, \ldots, \delta$, which are positioned in sequence. I.e. a job which has been processed on machine M_i is sent to machine $M_{i-1}, i = 1, \ldots, \delta$. The processing time of a job is 1 on each machine. Jobs arrive on some machine. We relate the expected flow time of a Model 3-schedule to the flow time of a tandem queue, in which the layers are the machines.

Lemma 3. *The expected sum of flow times of a Model 3-schedule is at most $1/p^\delta$ times the sum of flow times of a deterministic tandem queue with unit processing times.*

Proof. Consider a Model 3 schedule S for instance $\mathcal{I}$. First, assume that when in a round some layer fails to communicate, then all layers fail in that round. In this case, the probability of having no communication in a round is $1 - p^\delta$. The schedule S' that remains after removing rounds without communication is equivalent to a schedule of a deterministic tandem queue with unit processing times, where each packet j of the Model 3 instance is transformed in a job arriving at machine $M_{d(v_j,s)}$. Consider any set of k rounds of a schedule of a deterministic tandem queue with unit processing times. The expected number of rounds required to schedule these k rounds in schedule S' is $k p^\delta \cdot \sum_{i=0}^{\infty} (1 - p^\delta)^i \cdot (i + 1) = k/p^\delta$. Let F_j' be the flow time of packet j in S'. Then we have $\mathbb{E}[F_j^{(3)}] = F_j'/p^\delta$, for each packet j. Flow times are not increased if only some layers fail, but others forward a packet. □

Lemma 4. *The sum of flow times of a deterministic tandem queue with unit processing times is at most the sum of flow times of an optimal offline* WGP *schedule.*

Proof. Let S be a deterministic tandem queue schedule, and let S^* be a schedule where in each round each layer, except layer 1, can forward any number of packets, and layer 1 can forward at most one packet.

Consider schedule S^*. Let M_t^* be the set of packets which have not arrived at the sink in round t. Because the schedule can forward any number of packets over an edge, we have that if no packet is sent to the sink in S^* in round t, then no packet in M_t^* can arrive at the sink before or at round t, i.e. $r_j + \delta_j > t$ for each $j \in M_t^*$. We prove the lemma by demonstrating that if some packet is sent to the sink in S^* in round t, then also some packet is sent to the sink in S in round t. This suffices to prove the lemma, because an offline optimal WGP schedule can not send packets faster than in schedule S^*.

Suppose to the contrary that there is a first round t in which some packet is sent to the sink in S^*, but no packet is sent to the sink in S. Let $t', t' < t$, be the last round before t' in which no packet is sent to the sink in S^*. Then, there is a set of $t - t'$ packets in S^* which arrive at the sink in rounds $(t', t]$. Hence, it follows from this and the observation above that there are $t - t'$ packets j such that $t' < r_j + \delta_j \le t$. Now consider schedule S; in this schedule $t - t' - 1$ packets are sent to the sink in rounds $(t', t - 1]$, hence there is a packet j with $t' < r_j + \delta_j \le t$ which has not arrived at the sink in round t. But then, j must have been in layer $1 + i$ or higher in rounds $t - i$, $i = 1, \ldots, \delta_j$. I.e. j must have been in layer $1 + \delta_j$ or higher in round $r_j \le t - \delta_j$ which gives a contradiction. □

Theorem 2. *Let $0 < \epsilon \le 1$ and $\sigma = 6\mu^{-1} \cdot \log \Delta \cdot \ln(\delta/\epsilon)$. Then σ-speed* DG *is in expectation* $(1 + 3\epsilon)$-*competitive when minimizing the average flow time.*

Proof. It follows from Lemmas 3 and 4 that the expected sum of flow times of σ-speed DG is at most $1/p^\delta$ times the sum of flow times of an optimal offline solution, for $\sigma = 6a \log \Delta$. As σ-speed DG is an online algorithm, DG is σ-speed $p^{-\delta}$-competitive when minimizing average flow times. The probability $p = 1 - (1 - \mu)^a$ depends on the choice of the speedup a. We set $a := \mu^{-1} \ln(\delta/\epsilon)$, which gives $p = 1 - (1 - \mu)^{\mu^{-1} \ln(\delta/\epsilon)} \ge 1 - e^{-\ln(\delta/\epsilon)} = 1 - \epsilon/\delta$, so that $p^{-\delta} \le (1 - \epsilon/\delta)^{-\delta} \le e^\epsilon \le e^{2\ln(1+\epsilon)} = 1 + 2\epsilon + \epsilon^2 \le 1 + 3\epsilon$. □

It follows from the theorem that the competitive ratio of DG can be made arbitrarily close to 1 with an appropriate increase in speed.

5   An Extension

The model that we considered so far for WGP assumes that a node can only cause interference at nodes that are adjacent to it in the network. In practice, the interference caused by the radio signal can go beyond the transmission radius [22]. This can be modeled by an integer $d_I \ge 1$ that specifies the distance (expressed in number of hops) up to which the signal can cause interference: two calls (u, v) and (u', v') are now compatible if $d(u, v') \leq d_I$ and $d(u', v) \leq d_I$ (we still require that u is adjacent to v, and u' to v').

Algorithm DG can be extended to this setting by assigning to a node in layer d the label $d \bmod d_I + 2$ and then using superphases consisting of $d_I + 2$ phases each. In this way one can avoid interference between nodes from different layers of the tree. It is easy to see that Theorem 1 can be extended to this setting.

Using this fact, we can extend our analysis in the previous section to prove the following result.

Theorem 3. *Let $\epsilon > 0$ and $\sigma = \Theta(d_I^2 \cdot \log \Delta \cdot \ln(\delta/\epsilon))$. Then σ-speed DG is in expectation $(1 + \epsilon)$-competitive when minimizing the average flow time.*

We notice that one d_I factor is due to the longer superphases, and another one is due to DECAY having to cope with larger neighborhoods (of size Δ^{d_I}).

6   Conclusion and Open Problems

We considered the wireless gathering problem with the objective of minimizing the average flow time of data packets when nodes are restricted in their computational and communication capabilities. We showed that a simple on-line algorithm has favorable behavior when the objective function is minimizing the average flow: although the problem is extremely hard to approximate in general, augmenting the transmission rate allows to remain within a small factor of the cost of an optimal solution for the problem without augmentation. We also showed for max flow a lower bound on the competitive ratio of any (possibly non polynomial time) algorithm, which depends on the number of packets.

The proposed algorithm is simple; however it assumes the existence of a common clock that is used to reduce interferences. It would be interesting to extend the results by removing such an assumption. It is interesting for future research to study other objective functions and to allow other routing than through a tree, in which case it will be challenge to design and analyze congestion avoiding algorithms with better ratios than those developed for trees.

References

1. Akyildiz, I.F., Su, W., Sankarasubramaniam, Y., Cayirci, E.: Wireless sensor networks: a survey. Computer Networks 38(4), 393–422 (2002)
2. Andrews, M., Awerbuch, B., Fernández, A., Leighton, F.T., Liu, Z., Kleinberg, J.M.: Universal-stability results and performance bounds for greedy contention resolution protocols. Journal of the ACM 48(1), 39–69 (2001)
3. Bar-Yehuda, R., Goldreich, O., Itai, A.: On the time-complexity of broadcast in multihop radio networks: an exponential gap between determinism and randomization. Journal of Computer and System Sciences 45(1), 104–126 (1992)
4. Bar-Yehuda, R., Israeli, A., Itai, A.: Multiple communication in multihop radio networks. SIAM Journal on Computing 22(4), 875–887 (1993)

5. Bermond, J., Galtier, J., Klasing, R., Morales, N., Pérennes, S.: Hardness and approximation of gathering in static radio networks. Parallel Processing Letters 16(2), 165–183 (2006)
6. Bonifaci, V., Korteweg, P., Marchetti-Spaccamela, A., Stougie, L.: An approximation algorithm for the wireless gathering problem. In: Proc. of the 10th Scandinavian Workshop on Algorithms and Theory, pp. 328–338 (2006)
7. Bonifaci, V., Korteweg, P., Marchetti-Spaccamela, A., Stougie, L.: Minimzing flow time in the wireless gathering problem. In: Proc. of the 25th Int. Symp. on Theoretical Aspects of Computer Science (2008)
8. Borodin, A., Kleinberg, J.M., Raghavan, P., Sudan, M., Williamson, D.P.: Adversarial queuing theory. Journal of the ACM 48(1), 13–38 (2001)
9. Florens, F.C., McEliece, R.M.: Lower bounds on data collection time in sensory networks. IEEE J. on Selected Areas in Communications 22, 1110 (2004)
10. Chekuri, C., Goel, A., Khanna, S., Kumar, A.: Multi-processor scheduling to minimize flow time with epsilon resource augmentation. In: Proc. of the ACM Symp. on the Theory of Computing, pp. 363–372 (2004)
11. Håstad, J., Leighton, F.T., Rogoff, B.: Analysis of backoff protocols for multiple access channels. SIAM Journal on Computing 25(4) (1996)
12. Kalyanasundaram, B., Pruhs, K.: Speed is as powerful as clairvoyance. Journal of the ACM 47(4), 617–643 (2000)
13. Kar, K., Kodialam, M.S., Lakshman, T.V., Tassiulas, L.: Routing for network capacity maximization in energy-constrained ad-hoc networks. In: Proc. IEEE INFOCOM (2003)
14. Kodialam, M.S., Nandagopal, T.: Characterizing achievable rates in multi-hop wireless networks: the joint routing and scheduling problem. In: Proc. MOBICOM, pp. 42–54 (2003)
15. Kumar, V.S.A., Marathe, M.V., Parthasarathy, S., Srinivasan, A.: End-to-end packet-scheduling in wireless ad-hoc networks. In: Proc. of the ACM-SIAM Symp. on Discrete Algorithms, pp. 1021–1030 (2004)
16. Kumar, V.S.A., Marathe, M.V., Parthasarathy, S., Srinivasan, A.: Provable algorithms for joint optimization of transport, routing and MAC layers in wireless ad hoc networks. In: Proc. Dial-M/POMC (2007)
17. Lu, G., Krishnamachari, B., Raghavendra, C.S.: An adaptive energy-efficient and low-latency mac for data gathering in wireless sensor networks. In: Proc. IEEE IPDPS (2004)
18. Moscibroda, T., Wattenhofer, R.: The complexity of connectivity in wireless networks. In: Proc. IEEE INFOCOM (2006)
19. Pahlavan, K., Levesque, A.H.: Wireless information networks. Wiley, Chichester (1995)
20. Perkins, C.E.: Ad hoc networking. Addison-Wesley, Reading (2001)
21. Polastre, J., Hill, J.L., Culler, D.E.: Versatile low power media access for wireless sensor networks. In: Proc. SenSys., pp. 95–107 (2004)
22. Schmid, S., Wattenhofer, R.: Algorithmic models for sensor networks. In: Proc. IEEE IPDPS (2006)
23. Xiao, L., Johansson, M., Boyd, S.P.: Simultaneous routing and resource allocation via dual decomposition. IEEE Transactions on Communications 52(7), 1136–1144 (2004)

Approximating Maximum Edge 2-Coloring in Simple Graphs Via Local Improvement

Zhi-Zhong Chen and Ruka Tanahashi

Department of Mathematical Sciences, Tokyo Denki University,
Hatoyama, Saitama 350-0394, Japan

Abstract. We present a polynomial-time approximation algorithm for legally coloring as many edges of a given simple graph as possible using two colors. It achieves an approximation ratio of $\frac{24}{29} = 0.827586\ldots$. This improves on the previous best ratio of $\frac{468}{575} = 0.813913\ldots$.

1 Introduction

Given a graph G and a natural number t, the *maximum edge t-coloring problem* (called MAX EDGE t-COLORING for short) is to find a maximum set F of edges in G such that F can be partitioned into at most t matchings of G. Motivated by call admittance issues in satellite based telecommunication networks, Feige et al. [2] introduced the problem and proved its APX-hardness. They also observed that MAX EDGE t-COLORING is obviously a special case of the well-known maximum coverage problem (see [4]). Since the maximum coverage problem can be approximated by a greedy algorithm within a ratio of $1 - (1 - \frac{1}{t})^t$ [4], so can MAX EDGE t-COLORING. In particular, the greedy algorithm achieves an approximation ratio of $\frac{3}{4}$ for MAX EDGE 2-COLORING which is the special case of MAX EDGE t-COLORING where the input number t is fixed to 2. Feige et al. [2] has improved the trivial ratio $\frac{3}{4} = 0.75$ to $\frac{10}{13} \approx 0.769$ by an LP approach.

The APX-hardness proof for MAX EDGE t-COLORING given by Feige et al. [2] indeed shows that the problem remains APX-hard even if we restrict the input graph to a simple graph and fix the input integer t to 2. We call this restriction (special case) of the problem MAX SIMPLE EDGE 2-COLORING. Feige et al. [2] also pointed out that for MAX SIMPLE EDGE 2-COLORING, an approximation ratio of $\frac{4}{5}$ can be achieved by the following *simple algorithm*: Given a simple graph G, first compute a maximum subgraph H of G such that the degree of each vertex in H is at most 2 and there is no 3-cycle in H, and then remove one *arbitrary* edge from each odd cycle of H.

In [1], the authors have improved the ratio to $\frac{468}{575}$. Essentially, the algorithm in [1] differs from the simple algorithm only in the handling of 5-cycles where instead of removing one arbitrary edge from each 5-cycle of H, we remove a *random* edge from each 5-cycle of H. The intuition behind the algorithm is as follows: If we delete a random edge from each 5-cycle of H, then for each edge $\{u, v\}$ in the optimal solution such that u and v belong to different 5-cycles, both u and v become of degree 1 in H (after handling the 5-cycles) with a probability of $\frac{4}{25}$ and so can be added into H without losing the edge 2-colorability of H.

R. Fleischer and J. Xu (Eds.): AAIM 2008, LNCS 5034, pp. 84–96, 2008.

In this paper, we further improve the ratio to $\frac{24}{29}$. The basic idea behind our algorithm is as follows: Instead of removing a random edge from each 5-cycle of H and removing an arbitrary edge from each other odd cycle of H, we remove one edge from each odd cycle of H with more care in the hope that after the removal, a lot of edges $\{u, v\}$ (in the optimal solution) with u and v belonging to different odd cycles of H can be added to H. More specifically, we define a number of operations that modify each odd cycle of H together with its neighborhood carefully without decreasing the number of edges in H by two or more; our algorithm just performs these operations on H until none of them is applicable. The nonapplicability of these operations guarantees that H is edge 2-colorable and its number of edges is close to optimal; the analysis is quite challenging.

Kosowski et al. [7] also considered MAX SIMPLE EDGE 2-COLORING. They presented an approximation algorithm that achieves a ratio of $\frac{28\Delta-12}{35\Delta-21}$, where Δ is the maximum degree of a vertex in the input simple graph. This ratio can be arbitrarily close to the trivial ratio $\frac{4}{5}$ because Δ can be very large. In particular, this ratio is smaller than $\frac{24}{29}$ when $\Delta \geq 6$.

Kosowski et al. [7] showed that approximation algorithms for MAX SIMPLE EDGE 2-COLORING can be used to obtain approximation algorithms for certain packing problems and fault-tolerant guarding problems. Combining their reductions and our improved approximation algorithm for MAX SIMPLE EDGE 2-COLORING, we can obtain improved approximation algorithms for their packing problems and fault-tolerant guarding problems immediately.

2 Basic Definitions

A graph always means a simple undirected graph. A graph G has a vertex set $V(G)$ and an edge set $E(G)$. For each $v \in V(G)$, $N_G(v)$ denotes the set of all vertices adjacent to v in G and $d_G(v) = |N_G(v)|$ is the *degree* of v in G. If $d_G(v) = 0$, then v is an *isolated vertex* of G. For each $U \subseteq V(G)$, $G[U]$ denotes the subgraph of G induced by U.

A *path* in G is a connected subgraph of G in which exactly two vertices are of degree 1 and the others are of degree 2. Each path has two endpoints and zero or more inner vertices. An edge $\{u, v\}$ of a path P is an *inner edge* of P if both u and v are inner vertices of P. The *length* of a cycle or path C is the number of edges in C. A cycle of odd (respectively, even) length is an *odd* (respectively, *even*) cycle. A *k-cycle* is a cycle of length k. Similarly, a k^+-*cycle* is a cycle of length at least k. A *path component* (respectively, *cycle component*) of G is a connected component of G that is a path (respectively, cycle). A *path-cycle cover* of G is a subgraph H of G such that $V(H) = V(G)$ and $d_H(v) \leq 2$ for every $v \in V(H)$. A *cycle cover* of G is a path-cycle cover of G in which each connected component is a cycle. A path-cycle cover C of G is *triangle-free* if C does not contain a 3-cycle.

G is *edge-2-colorable* if each connected component of G is an isolated vertex, a path, or an even cycle. Note that MAX SIMPLE EDGE 2-COLORING is the problem of finding a maximum edge-2-colorable subgraph in a given graph.

3 The Algorithm

Throughout this section, fix a graph G and a maximum edge-2-colorable subgraph Opt of G. For convenience, for each path-cycle cover K of G, we define two numbers as follows:

- $n_0(K)$ is the number of isolated vertices in K.
- $p(K)$ is the number of path components in K.

Like the simple algorithm described in Section 1, our algorithm starts by performing the following step:

---

1. Compute a maximum triangle-free path-cycle cover H of G.

Since $|E(H)| \geq |E(Opt)|$, it suffices to modify H into an edge-2-colorable subgraph of G without significantly decreasing the number of edges in H. The simple algorithm achieves an approximation ratio of $\frac{4}{5}$ because it simply removes an arbitrary edge from each odd cycle in H. In order to improve this ratio, we have to treat 5-cycles (and other short odd cycles) in H more carefully. In more details, when removing edges from odd cycles in H, we also want to add some edges of $E(G) - E(H)$ to H. For this purpose, we will define a number of operations on H that always decrease the number of cycles in H but may decrease the number of edges in H or not. To tighten the analysis of the approximation ratio achieved by our algorithm, we set up a charging scheme that charges the net loss of edges from H (due to the operations) to some edges still remaining in H. Whenever we do this, we will always maintain the following invariants:

I1. Every edge of H is charged a real number smaller than or equal to $\frac{1}{9}$.
I2. The total charge on the edges of H equals the total number of operations performed on H that decrease the number of edges in H.
I3. No cycle component of H contains a charged edge.
I4. If a path component P of H contains a charged edge, then the length of P is at least 6.

Initially, every edge of H is charged nothing. However, as we modify H by performing operations (to be defined below), some edges of H will be charged.

We first define those operations on H that decrease the number of odd cycles in H but do not decrease the number of edges in H. In order to do this, the following three concepts are necessary:

A quadruple (x, y, P, u, v) is a *5-opener* for an odd cycle C of H if the following hold:

- $d_H(x) \leq 1$ and $y \in V(C)$.
- P is a path component of H, both u and v are inner vertices of P, and x is not a vertex of P.
- Both $\{u, x\}$ and $\{v, y\}$ are contained in $E(G) - E(H)$.

A sextuple (x, y, Q, P, u, v) is a *6-opener* for an odd cycle C of H if the following hold:

- $x \in V(C)$ and $y \in V(C)$. Moreover, if $x = y$, then Q is a cycle cover of $G[V(C) - \{x\}]$ in which each connected component is an even cycle; otherwise, Q is a path-cycle cover of $G[V(C)]$ in which one connected component is a path from x to y and each other connected component is an even cycle.
- P is a path component of H and both u and v are inner vertices of P.
- Both $\{u, x\}$ and $\{v, y\}$ are contained in $E(G) - E(H)$.

An operation (to be performed) on H is *robust* if the following holds:

- If G has no edge $\{u, v\}$ before the operation such that u is an isolated vertex in H and either v is an isolated vertex in H or v appears in a cycle component of H, then neither does it after the operation.

Based on the above concepts, we are now ready to define six robust operations on H that decrease the number of odd cycles in H but do not decrease the number of edges in H.

Type 1: Suppose that $\{u, v\}$ is an edge in $E(G) - E(H)$ such that $d_H(u) \leq 1$ and v is a vertex of some cycle C of H. Then, a *Type-1 operation* on H using $\{u, v\}$ modifies H by deleting one (arbitrary) edge of C incident to v and adding edge $\{u, v\}$. Obviously, this operation is robust and does not change $|E(H)|$. (*Comment:* If $d_H(u) = 0$ before a Type-1 operation, then $n_0(H)$ decreases by 1 and $p(H)$ increases by 1 after the operation. Similarly, if $d_H(u) = 1$ before a Type-1 operation, then neither $n_0(H)$ nor $p(H)$ changes after the operation.)

Type 2: Suppose that some odd cycle C of H has a 5-opener (x, y, P, u, v) with $\{u, v\} \in E(H)$ (see Figure 1). Then, a *Type-2 operation* on H using (x, y, P, u, v) modifies H by deleting edge $\{u, v\}$, deleting one (arbitrary) edge of C incident to y, and adding edges $\{u, x\}$ and $\{v, y\}$. Obviously, this operation is robust and does not change the number of edges in H. However, edge $\{u, v\}$ may have been charged before this operation. If that is the case, we move its charge to $\{u, x\}$. Moreover, if the path component Q of H containing edge $\{u, x\}$ after this operation is of length at most 5, then we move the charges on the edges of Q to edge $\{v, y\}$ and the edges of C still remaining in H. (*Comment:* A Type-2 operation on H maintains Invariants I1 through I4. Moreover, if $d_H(x) = 0$ before a Type-2 operation, then $n_0(H)$ decreases by 1 and $p(H)$ increases by 1 after the operation. Similarly, if $d_H(x) = 1$ before a Type-2 operation, then neither $n_0(H)$ nor $p(H)$ changes after the operation.)

Type 3: Suppose that some odd cycle C of H has a 5-opener (x, y, P, u, v) such that $E(G) - E(H)$ contains the edge $\{w, s\}$, where w is the neighbor of v in the subpath of P between u and v and s is the endpoint of P with $dist_P(s, u) < dist_P(s, v)$ (see Figure 2 in the appendix). Then, a *Type-3 operation* on H using (x, y, P, u, v) modifies H by deleting edge $\{v, w\}$, deleting one (arbitrary) edge e_u of P incident to u, deleting one (arbitrary) edge of C incident to y, and

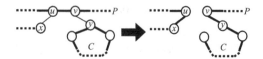

Fig. 1. A Type-2 operation, where bold edges are in H

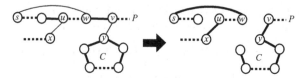

Fig. 2. A Type-3 operation, where bold edges are in H

adding edges $\{u, x\}$, $\{v, y\}$, and $\{s, w\}$. Obviously, this operation is robust and does not change the number of edges in H. Note that $\{v, w\}$ or e_u may have been charged before this operation. If that is the case, we move their charges to edges $\{u, x\}$ and $\{s, w\}$, respectively. Moreover, if the path component Q of H containing edge $\{u, x\}$ after this operation is of length at most 5, then we move the charges on the edges of Q to edge $\{v, y\}$ and the edges of C still remaining in H. (*Comment:* A Type-3 operation on H maintains Invariants I1 through I4. Moreover, if $d_H(x) = 0$ before a Type-3 operation, then $n_0(H)$ decreases by 1 and $p(H)$ increases by 1 after the operation. Similarly, if $d_H(x) = 1$ before a Type-3 operation, then neither $n_0(H)$ nor $p(H)$ changes after the operation.)

Type 4: Suppose that there is a quadruple (x, P, u, v) satisfying the following conditions (see Figure 3 in the appendix):

- x is a vertex of a cycle component C of H.
- P is a path component of H and $\{u, v\}$ is an inner edge of P.
- $E(G) - E(H)$ contains both $\{u, x\}$ and $\{s, v\}$, where s is the endpoint of P with $dist_P(s, u) < dist_P(s, v)$.

Then, a *Type-4 operation* on H using (x, P, u, v) modifies H by deleting edge $\{u, v\}$, deleting one (arbitrary) edge of C incident to x, and adding edges $\{u, x\}$ and $\{s, v\}$. Obviously, this operation is robust and does not change the number of edges in H. However, $\{u, v\}$ may have been charged before this operation. If that is the case, we move its charge to $\{u, x\}$. (*Comment:* A Type-4 operation on H maintains Invariants I1 through I4, and changes neither $n_0(H)$ nor $p(H)$.)

Type 5: Suppose that there is a quintuple (x, P, u, v, w) satisfying the following conditions (see Figure 4 in the appendix):

- x is a vertex of a cycle component C of H.
- P is a path component of H, u is an inner vertex of P, $\{v, w\}$ is an inner edge of P, and $dist_P(u, v) < dist_P(u, w)$.

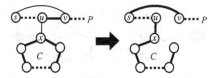

Fig. 3. A Type-4 operation, where bold edges are in H

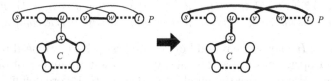

Fig. 4. A Type-5 operation, where bold edges are in H

- $E(G) - E(H)$ contains $\{u, x\}$, $\{s, w\}$, and $\{t, v\}$, where s is the endpoint of P with $dist_P(s, u) < dist_P(s, v)$ and t is the other endpoint of P.

Then, a *Type-5 operation* on H using (x, P, u, v, w) modifies H by deleting edge $\{v, w\}$, deleting one (arbitrary) edge e_u of P incident to u, deleting one (arbitrary) edge of C incident to x, and adding $\{s, w\}$, $\{t, v\}$, and $\{u, x\}$. Obviously, this operation is robust and does not change the number of edges in H. However, $\{v, w\}$ and e_u may have been charged before this operation. If that is the case, we move their charges to $\{u, x\}$ and $\{s, w\}$, respectively. (*Comment:* A Type-5 operation on H maintains Invariants I1 through I4, and changes neither $n_0(H)$ nor $p(H)$.)

Type 6: Suppose that some odd cycle C of H with length at most 9 has a 6-opener (x, y, Q, P, u, v) such that $\{u, v\} \in E(H)$ (see Figure 5 in the appendix). Then, a *Type-6 operation* on H using (x, y, Q, P, u, v) modifies H by deleting edge $\{u, v\}$, deleting all edges of C, adding edges $\{u, x\}$ and $\{v, y\}$, and adding all edges of Q. Obviously, this operation does not change the number of edges in H, and is robust because (1) it does not create a new isolated vertex in H and (2) if it creates one or more new cycles in H then $V(C') \subseteq V(C)$ for each new cycle C'. However, $\{u, v\}$ may have been charged before this operation. If that is the case, we move its charge to $\{u, x\}$. (*Comment:* A Type-6 operation on H maintains Invariants I1 through I4, and changes neither $n_0(H)$ nor $p(H)$.)

Using the above operations, our algorithm then proceeds to modifying H by performing the following step:

---
2. Repeat performing a Type-i operation on H with $1 \leq i \leq 6$, until none is applicable.

Obviously, H remains a triangle-free path-cycle cover of G. Moreover, the following fact holds:

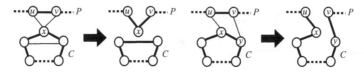

Fig. 5. A Type-6 operation, where bold edges are in H

Lemma 1. *After Step 2, G has no edge $\{u, v\}$ such that u is an isolated vertex in H and either v is an isolated vertex in H or v appears in a cycle component of H.*

Unfortunately, H may still have odd cycles after Step 2. So, we need to perform new types of operations on H that always decrease the number of odd cycles in H but may also decrease the number of edges in H. Before defining the new operations on H, we define two concepts as follows. Two cycles C_1 and C_2 of H are *pairable* if at least one of them is odd and their total length is at least 10. A quintuple (x, y, P, u, v) is an *opener* for two pairable cycles C_1 and C_2 of H if the following hold:

- x is a vertex of C_1 and y is a vertex of C_2.
- P is either a path component of H or a 4-cycle of H other than C_1 and C_2.
- u and v are distinct vertices of P with $d_H(u) = d_H(v) = 2$.
- Both $\{u, x\}$ and $\{v, y\}$ are in $E(G) - E(H)$.

Now, we are ready to define the new types of robust operations on H as follows:

Type 7: Suppose that C is an odd cycle of H with length at least 11. Then, a *Type-7 operation* on H using C modifies H by deleting one (arbitrary) edge from C. Clearly, the net loss in the number of edges in H is 1. We charge this loss evenly to the edges of C still remaining in H. In more details, if C was a k-cycle before the operation, then a charge of $\frac{1}{k-1}$ is charged to each edge of C still remaining in H after the operation. Since $k \geq 11$, the charge assigned to one edge here is at most $\frac{1}{10}$. Obviously, this operation is robust. (*Comment:* A Type-7 operation on H maintains Invariants I1 through I4, does not change $n_0(H)$, and increases $p(H)$ by 1.)

Type 8: Suppose that C_1 and C_2 are two pairable cycles of H such that there is an edge $\{u, v\} \in E(G)$ with $u \in V(C_1)$ and $v \in V(C_2)$. Then, a *Type-8 operation* on H using $\{u, v\}$ modifies H by deleting one (arbitrary) edge of C_1 incident to u, deleting one (arbitrary) edge of C_2 incident to v, and adding edge $\{u, v\}$. Note that this operation decreases the number of edges in H by 1. So, the net loss in the number of edges in H is 1. We charge this loss evenly to edge $\{u, v\}$ and the edges of C_1 and C_2 still remaining in H. In more details, if C_1 was a k-cycle and C_2 was an ℓ-cycle in H before the operation, then a charge of $\frac{1}{k+\ell-1}$ is assigned to $\{u, v\}$ and each edge of C_1 and C_2 still remaining in H after the operation. Since $k \geq 5$ and $\ell \geq 5$, the charge assigned to one edge here is at most $\frac{1}{9}$. Obviously, this operation is robust. (*Comment:* A Type-8 operation on H maintains Invariants I1 through I4, does not change $n_0(H)$, and increases $p(H)$ by 1.)

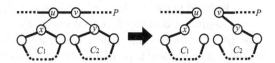

Fig. 6. A Type-9 operation, where bold edges are in H

Type 9: Suppose that two odd cycles C_1 and C_2 of H have an opener (x, y, P, u, v) with $\{u, v\} \in E(H)$ (see Figure 6 in the appendix). Then, a *Type-9 operation* on H using (x, y, P, u, v) modifies H by deleting edge $\{u, v\}$, deleting one (arbitrary) edge of C_1 incident to x, deleting one (arbitrary) edge of C_2 incident to y, and adding edges $\{u, x\}$ and $\{v, y\}$. Note that edge $\{u, v\}$ may have been charged before this operation. If that is the case, we move its charge to edge $\{u, x\}$. Moreover, the operation decreases the number of edges in H by 1. So, the net loss in the number of edges in H is 1. We charge this loss evenly to edge $\{v, y\}$ and the edges of C_1 and C_2 still remaining in H. Obviously, the charge assigned to one edge here is at most $\frac{1}{9}$. It is also clear that this operation is robust. (*Comment:* A Type-9 operation on H maintains Invariants I1 through I4, does not change $n_0(H)$, and increases $p(H)$ by 1.)

Type 10: Suppose that two odd cycles C_1 and C_2 of H have an opener (x, y, P, u, v) such that $E(G) - E(H)$ contains the edge $\{w, s\}$, where w is the neighbor of v in the subpath of P between u and v and s is the endpoint of P with $dist_P(s, u) < dist_P(s, v)$ (see Figure 7 in the appendix). Then, a *Type-10 operation* on H using (x, y, P, u, v) modifies H by deleting edge $\{v, w\}$, deleting one (arbitrary) edge e_u of P incident to u, deleting one (arbitrary) edge of C_1 incident to x, deleting one (arbitrary) edge of C_2 incident to y, and adding edges $\{u, x\}$, $\{v, y\}$, and $\{s, w\}$. Note that $\{v, w\}$ or e_u may have been charged before this operation. If that is the case, we move their charges to edges $\{u, x\}$ and $\{v, y\}$, respectively. Moreover, the operation decreases the number of edges in H by 1. So, the net loss in the number of edges in H is 1. We charge this loss evenly to edge $\{s, w\}$ and the edges of C_1 and C_2 still remaining in H. Obviously, the charge assigned to one edge here is at most $\frac{1}{9}$. It is also clear that this operation is robust. (*Comment:* A Type-10 operation on H maintains Invariants I1 through I4, does not change $n_0(H)$, and increases $p(H)$ by 1.)

After Step 2, no matter how many times we perform Type-i operations on H with $1 \leq i \leq 10$, G cannot have an edge $\{u, v\}$ such that u is an isolated vertex

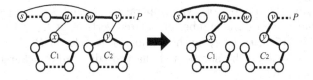

Fig. 7. A Type-10 operation, where bold edges are in H

in H and either v is an isolated vertex in H or v appears in a cycle component of H. This follows from Lemma 1 and the fact that every Type-i operation on H with $1 \leq i \leq 10$ is robust. However, after performing a Type-i operation on H with $7 \leq i \leq 10$, the following new type of robust operations on H may be applicable:

Type 11: Suppose that $\{u, v\}$ is an edge in $E(G) - E(H)$ such that $d_H(u) = 1$, $d_H(v) \leq 1$, and no connected component of H contains both u and v. Then, a *Type-11 operation* on H using $\{u, v\}$ modifies H by adding edge $\{u, v\}$. Obviously, this operation is robust and increases the number of edges in H by 1. (*Comment:* If $d_H(v) = 0$ before a Type-11 operation, then $p(H)$ does not change and $n_0(H)$ decreases by 1 after the operation. Similarly, if $d_H(v) = 1$ before a Type-11 operation, then $n_0(H)$ does not change and $p(H)$ decreases by 1 after the operation.)

Using the above operations, our algorithm then proceeds to modifying H by performing the following steps:

---

3. Repeat using a Type-i operation to modify H with $1 \leq i \leq 11$, until none is applicable.

4. For each odd cycle C of H, remove one (arbitrary) edge from C. (*Comment:* Each odd cycle modified in this step is a 5-, 7-, or 9-cycle.)

5. Output H.

---

4   Performance Analysis

For $1 \leq i \leq 4$, let H_i be the triangle-free path-cycle cover H of G immediately after Step i of our algorithm. In order to analyze the approximation ratio achieved by our algorithm, we need to define several notations as follows:

- Let n, m, n_{is}, and n_{pc} be the numbers of vertices, edges, isolated vertices, and path components in H_2, respectively. (*Comment:* $m \geq |E(Opt)|$.)
- Let m_- be the number of Type-i operations with $7 \leq i \leq 10$ performed in Step 3.
- Let $m_{+,-1}$ be the number of Type-11 operations performed Step 3 that decrease the number of isolated vertices in H by 1.
- Let $m_{+,0}$ be the number of Type-11 operations performed Step 3 that do not change the number of isolated vertices in H.
- Let $n_{0,-1}$ be the number of Type-i operations with $1 \leq i \leq 3$ performed in Step 3 that decrease the number of isolated vertices in H by 1.
- For each $i \in \{5, 7, 9\}$, let c_i be the number of i-cycles in H_3.
- Let m_c and m_{uc} be the numbers of charged edges and uncharged edges in H_3, respectively.)

Lemma 2. *The following statements hold:*

1. $m_- \leq \frac{1}{10}(m + m_{+,0} + m_{+,-1} - m_{uc})$.
2. $|E(H_4)| = m - m_- + m_{+,0} + m_{+,-1} - c_5 - c_7 - c_9$.
3. $|E(H_4)| \geq \frac{9}{10}(m + m_{+,0} + m_{+,-1}) - (\frac{1}{2}c_5 + \frac{3}{10}c_7 + \frac{1}{10}c_9)$.

Proof. By the algorithm, $|E(H_3)| = m - m_- + m_{+,0} + m_{+,-1}$. On the other hand, $|E(H_3)| = m_c + m_{uc}$ by definition. So, $m_c = m - m_- + m_{+,0} + m_{+,-1} - m_{uc}$. We also have $m_- \leq \frac{1}{9}m_c$ by Invariant I2. Thus, $m_- \leq \frac{1}{10}(m + m_{+,0} + m_{+,-1} - m_{uc})$.

By Step 3, $|E(H_4)| = |E(H_3)| - c_5 - c_7 - c_9$. So, by the first equality in the last paragraph, $|E(H_4)| = m - m_- + m_{+,0} + m_{+,-1} - c_5 - c_7 - c_9$.

By Statements 1 and 2, $|E(H_4)| \geq \frac{9}{10}(m + m_{+,0} + m_{+,-1}) + \frac{1}{10}m_{uc} - c_5 - c_7 - c_9$. We also have $m_{uc} \geq 5c_5 + 7c_7 + 9c_9$, because each edge in a cycle component of H_3 is uncharged according to Invariant I3. Combining these two inequalities, we have Statement 3.   □

Lemma 3. *The following statements hold:*

1. $n - n_0(H_3) - 2p(H_3) = m - n_{pc} - 2m_- + 2m_{+,0} + m_{+,-1} - n_{0,-1}$.
2. $p(H_3) = n_{pc} + m_- - m_{+,0} + n_{0,-1}$.

Proof. Immediately before Step 3, $n - n_0(H) - 2p(H) = m - n_{pc}$ because $p(H) = n_{pc}$ and the number of vertices on a path is 1 plus the number of edges on the path. Now, to prove the lemma, it suffices to see how the values of $n - n_0(H) - 2p(H)$ and $p(H)$ change when performing an operation in Step 3. The comment on the definition of each type of operations helps.   □

In order to analyze the algorithm, we need more definitions:

- For $i \in \{0, 1\}$, let T_i be the set of all vertices v in H_3 with $d_{H_3}(v) = i$.
- Let T_2 be the set of all vertices v in H_3 such that v appears in an odd cycle of H_3.
- Let $T = T_0 \cup T_1 \cup T_2$.
- For $i \in \{0, 1, 2\}$, let $\overline{T_i}$ be the set of vertices $u \in V(G) - T$ such that the number of edges $\{u, v\} \in E(Opt)$ with $v \in T$ is exactly i. (*Comment:* $V(G) - T = \overline{T_0} \cup \overline{T_1} \cup \overline{T_2}$.)
- Let E_{opt}^T be the set of all edges $\{u, v\}$ in Opt such that both u and v are vertices of T.
- Let $\mathcal{C}_{-2}$ be the set of all odd cycles in H_3 such that Opt contains at most $|V(C)| - 2$ edges $\{u, v\}$ with $\{u, v\} \subseteq V(C)$.

Lemma 4. $|E_{opt}^T| \leq p(H_3) + 4c_5 + 6c_7 + 8c_9 - |\mathcal{C}_{-2}| \leq n_{pc} + m_- - m_{+,0} + n_{0,-1} + 4c_5 + 6c_7 + 8c_9 - |\mathcal{C}_{-2}|$.

Proof. First, we claim that each vertex $u \in T_0$ is an isolated vertex in $G[T]$. To see this, consider an arbitrary $u \in T_0$. Because of Lemma 1 and the fact that all Type-i operations with $1 \leq i \leq 11$ are robust, there is no vertex $v \in T_0 \cup T_2$ with $\{u, v\} \in E(G)$. Moreover, since no Type-11 operation can be applied to H_3, there is no vertex $v \in T_1$ with $\{u, v\} \in E(G)$. So, the claim holds.

Next, we claim that there is no edge $\{u, v\} \in E(G)$ with $u \in T_1$ and $v \in T_2$. This follows from the fact that no Type-1 operation can be applied to H_3.

By the above two claims, each edge in E_{opt}^T is either in $G[T_1]$ or in $G[T_2]$. Since no Type-11 operation can be applied to H_3, there is no edge $\{u, v\} \in E(G)$ with $\{u, v\} \subseteq T_1$ such that u and v belong to different connected components of H_3. So, there are at most $p(H_3)$ edges in $G[T_1]$. Consequently, to show the first inequality in the lemma, it remains to show that E_{opt}^T contains at most $4c_5 + 6c_7 + 8c_9 - |\mathcal{C}_{-2}|$ edges $\{u, v\}$ with $\{u, v\} \subseteq T_2$.

Suppose that $\{u, v\}$ is an edge in E_{opt}^T with $\{u, v\} \subseteq T_2$. Since no Type-8 operation can be applied to H_3, x and y belong to the same cycle component of $H_3[T_2]$. On the other hand, since each cycle C in $H_3[T_2]$ is an odd cycle, E_{opt}^T can contain at most $|E(C)| - 1$ edges $\{u, v\}$ with $\{u, v\} \subseteq V(C)$. In particular, for each cycle $C \in \mathcal{C}_{-2}$, E_{opt}^T can contain at most $|E(C)| - 2$ edges $\{u, v\}$ with $\{u, v\} \subseteq V(C)$. Hence, E_{opt}^T contains at most $4c_5 + 6c_7 + 8c_9 - |\mathcal{C}_{-2}|$ edges $\{u, v\}$ with $\{u, v\} \subseteq T_2$. This completes the proof of the first inequality in the lemma. The second follows from the first and Statement 2 in Lemma 3. $\square$

Lemma 5. $|E_{opt}^T| \geq |E(Opt)| - 2m + 2n_{pc} + 4m_- - 4m_{+,0} - 2m_{+,-1} + 2n_{0,-1} + 10c_5 + 14c_7 + 18c_9 + |\overline{T_0}| + \frac{1}{2}|\overline{T_1}|$.

Proof. The idea behind the proof is to obtain an upper bound on $|E(Opt) - E_{opt}^T|$. A trivial upper bound is $2(n - |T|)$, because each edge in $E(Opt) - E_{opt}^T$ must be incident to a vertex in $V(G) - T$ and each vertex in $V(G) - T$ can be adjacent to at most two edges in Opt. This bound is not good enough because each edge $\{u, v\} \in E(Opt)$ with $\{u, v\} \subseteq V(G) - T$ is counted twice.

To get a better bound, we set up a savings account for each vertex in $V(G) - T$. Initially, we deposit two credits to the account of each vertex. The total credits amount to $2(n - |T|)$ (namely, the trivial upper bound). Next, for each edge $\{u, v\} \in E(Opt)$ with $\{u, v\} \subseteq V(G) - T$, we pay a half credit from the account of u and another half credit from the account of v. After this, for each edge $\{u, v\} \in E(Opt)$ with $u \in V(G) - T$ and $v \in T$, we pay one credit from the account of u. Obviously, we paid a total of $|E(Opt) - E_{opt}^T|$ credits. We want to estimate the number of credits that are still left in the accounts of the vertices in $V(G) - T$. First, for each vertex $u \in \overline{T_0}$, we paid at most one credit, because u is incident to at most two edges in $E(Opt) - E_{opt}^T$ and each of them has both of its endpoints in $V(G) - T$. So, the total number of credits still left in the accounts of the vertices in $\overline{T_0}$ is at least $|\overline{T_0}|$. Second, for each vertex $u \in \overline{T_1}$, we paid at most one and a half credits, because u is incident to at most two edges in $E(Opt) - E_{opt}^T$ and one of them has an endpoint in T. Thus, the total number of credits still left in the accounts of the vertices in $\overline{T_1}$ is at least $\frac{1}{2}|\overline{T_1}|$. In summary, we have shown that $|E(Opt) - E_{opt}^T| \leq 2(n - |T|) - |\overline{T_0}| - \frac{1}{2}|\overline{T_1}|$.

Obviously, $|T| = n_0(H_3) + 2p(H_3) + 5c_5 + 7c_7 + 9c_9$. So, by Statement 1 in Lemma 3, $|T| = n - (m - n_{pc} - 2m_- + 2m_{+,0} + m_{+,-1} - n_{0,-1}) + 5c_5 + 7c_7 + 9c_9$. In other words, $n - |T| = m - n_{pc} - 2m_- + 2m_{+,0} + m_{+,-1} - n_{0,-1} - 5c_5 - 7c_7 - 9c_9$. Hence, by the last inequality in the last paragraph, $|E(Opt) - E_{opt}^T| \leq$

$2m - 2n_{pc} - 4m_- + 4m_{+,0} + 2m_{+,-1} - 2n_{0,-1} - 10c_5 - 14c_7 - 18c_9 - |\overline{T_0}| - \frac{1}{2}|\overline{T_1}|$.
So, the lemma holds.   □

Lemma 6. $|E(H_4)| \geq \frac{37}{45}|E(Opt)|$.

Proof. Combining Lemmas 4 and 5, we have $|E(Opt)| \leq 2m - n_{pc} - 3m_- + 3m_{+,0} + 2m_{+,-1} - n_{0,-1} - 6c_5 - 8c_7 - 10c_9$. So, by Statement 2 in Lemma 2, $3|E(H_4)| - |E(Opt)| \geq m + m_{+,-1} + n_{pc} + n_{0,-1} + 3c_5 + 5c_7 + 7c_9$. Thus, $3c_5 + 5c_7 + 7c_9 \leq 3|E(H_4)| - |E(Opt)| - m$. Hence, $\frac{1}{2}c_5 + \frac{3}{10}c_7 + \frac{1}{10}c_9 \leq \frac{1}{6}(3c_5 + 5c_7 + 7c_9) \leq \frac{1}{2}|E(H_4)| - \frac{1}{6}|E(Opt)| - \frac{1}{6}m$. Therefore, by Statement 3 in Lemma 2, we have $|E(H_4)| \geq \frac{9}{10}m - (\frac{1}{2}|E(H_4)| - \frac{1}{6}|E(Opt)| - \frac{1}{6}m)$. Rearranging this inequality and using the fact that $m \geq |E(Opt)|$, we finally obtain $|E(H_4)| \geq \frac{37}{45}|E(Opt)|$.   □

Note that $\frac{37}{45} = 0.8222\ldots$ which is better than the previously best ratio. To show that our algorithm indeed achieves an even better ratio, our idea is to show that $|\mathcal{C}_{-2}| + |\overline{T_0}| + \frac{1}{2}|\overline{T_1}|$ is large. This holds basically because no Type-i operation with $i \in \{1, \ldots, 11\}$ can be applied to H_3. The proof is very complicated and will appear in the full version of this paper. In summary, we have our main result:

Theorem 1. *There is an $O(n^2m^2)$-time approximation algorithm for MAX SIMPLE EDGE 2-COLORING achieving a ratio of $\frac{24}{29}$, where n (respectively, m) is the number of vertices (respectively, edges) in the input graph.*

5 An Application

Let G be a graph. An *edge cover* of G is a set F of edges of G such that each vertex of G is incident to at least one edge of F. For a natural number k, a *$[1,\Delta]$-factor k-packing* of G is a collection of k disjoint edge covers of G. The *size* of a $[1,\Delta]$-factor k-packing $\{F_1, \ldots, F_k\}$ of G is $|F_1| + \cdots + |F_k|$. The problem of deciding whether a given graph has a $[1,\Delta]$-factor k-packing was considered in [5,6]. In [7], Kosowski *et al.* defined the *minimum $[1,\Delta]$-factor k-packing problem* (MIN-k-FP) as follows: Given a graph G, find a $[1,\Delta]$-factor k-packing of G of minimum size or decide that G has no $[1,\Delta]$-factor k-packing at all.

According to [7], MIN-2-FP is of special interest because it can be used to solve a fault tolerant variant of the guards problem in grids (which is one of the art gallery problems [8,9]). MIN-2-FP is NP-hard [7].

Lemma 7. [7] *If MAX SIMPLE EDGE 2-COLORING admits an approximation algorithm A achieving a ratio of α, then MIN-2-FP admits an approximation algorithm B achieving a ratio of $2 - \alpha$. Moreover, if the time complexity of A is $T(n)$, then the time complexity of B is $O(T(n))$.*

Theorem 2. *There is an $O(n^2m^2)$-time approximation algorithm for MIN-2-FP achieving a ratio of $\frac{34}{29}$, where n (respectively, m) is the number of vertices (respectively, edges) in the input graph.*

Theorem 2 follows from Theorem 1 and Lemma 7 immediately. Previously, the best ratio achieved by a polynomial-time approximation algorithm for MIN-2-FP was $\frac{682}{575}$ [1], although MIN-2-FP admits a polynomial-time approximation algorithm achieving a ratio of $\frac{42\Delta-30}{35\Delta-21}$, where Δ is the maximum degree of a vertex in the input graph [7].

References

1. Chen, Z.-Z., Tanahashi, R., Wang, L.: An Improved Approximation Algorithm for Maximum Edge 2-Coloring in Simple Graphs. Journal of Discrete Algorithms (to appear)
2. Feige, U., Ofek, E., Wieder, U.: Approximating Maximum Edge Coloring in Multi-graphs. In: Jansen, K., Leonardi, S., Vazirani, V.V. (eds.) APPROX 2002. LNCS, vol. 2462, pp. 108–121. Springer, Heidelberg (2002)
3. Hartvigsen, D.: Extensions of Matching Theory. Ph.D. Thesis, Carnegie-Mellon University (1984)
4. Hochbaum, D.: Approximation Algorithms for NP-Hard Problems. PWS Publishing Company, Boston (1997)
5. Jacobs, D.P., Jamison, R.E.: Complexity of Recognizing Equal Unions in Families of Sets. Journal of Algorithms 37, 495–504 (2000)
6. Kawarabayashi, K., Matsuda, H., Oda, Y., Ota, K.: Path Factors in Cubic Graphs. Journal of Graph Theory 39, 188–193 (2002)
7. Kosowski, A., Malafiejski, M., Zylinski, P.: Packing Edge Covers in Graphs of Small Degree (manuscript, 2006)
8. O'Rourke, J.: Art Gallery Theorems and Algorithms. Oxford University Press, Oxford (1987)
9. Urrutia, J.: Art Gallery and Illumination Problems. In: Handbook on Computational Geometry, Elsevier Science, Amsterdam (2000)

An Improved Randomized Approximation Algorithm for Maximum Triangle Packing

Zhi-Zhong Chen[1], Ruka Tanahashi[1], and Lusheng Wang[2]

[1] Department of Mathematical Sciences, Tokyo Denki University,
Hatoyama, Saitama 350-0394, Japan
[2] Department of Computer Science, City University of Hong Kong,
Tat Chee Avenue, Kowloon, Hong Kong

Abstract. This paper deals with the maximum triangle packing problem. For this problem, Hassin and Rubinstein gave a randomized polynomial-time approximation algorithm and claimed that it achieves an expected ratio of $\frac{89}{169}(1 - \epsilon)$ for any constant $\epsilon > 0$. However, their analysis was flawed. We present a new randomized polynomial-time approximation algorithm for the problem which achieves an expected ratio very close to their claimed expected ratio.

1 Introduction

In the *maximum triangle packing problem* (MTP for short), we are given an edge-weighted complete graph $G = (V, E)$ such that the edge weights are nonnegative and $|V|$ is a multiple of 3. The objective is to find a partition of V into $\frac{1}{3}|V|$ disjoint subsets each of size exactly 3 such that the total weight of edges whose endpoints belong to the same subset is maximized. MTP is a classic NP-hard problem; indeed, it is contained in Garey and Johnson's famous book on the theory of NP-completeness [2]. MTP is not only NP-hard but also MAX SNP-hard [5], implying that it does not admit a polynomial-time approximation scheme unless $P = NP$. A stronger hardness result has been obtained by Chlebík and Chlebíková [1]: No polynomial-time approximation algorithm can approximate MTP within a ratio of 0.9929 unless $P = NP$.

On the positive side, Hassin and Rubinstein [3] have presented a randomized polynomial-time approximation algorithm for MTP. In [3], they claimed that their algorithm achieves an expected ratio of $\frac{89}{169}(1 - \epsilon)$ for any constant $\epsilon > 0$. However, the first author of this paper pointed out a flaw in their analysis to them and they [4] have corrected the expected ratio to $\frac{43}{83}(1 - \epsilon)$.

In this paper, we obtain a new randomized polynomial-time approximation algorithm for MTP by substantially modifying the algorithm due to Hassin and Rubinstein. Like their algorithm, our algorithm starts by computing a maximum cycle cover $\mathcal{C}$ in the input graph G, then processes $\mathcal{C}$ to obtain three triangle packings of G, and finally outputs the maximum weighted packing among the three packings. Unlike their algorithm, our algorithm processes triangles in $\mathcal{C}$ in a different way than the other cycles in $\mathcal{C}$, and tries to connect the cycles in $\mathcal{C}$ by

R. Fleischer and J. Xu (Eds.): AAIM 2008, LNCS 5034, pp. 97–108, 2008.
© Springer-Verlag Berlin Heidelberg 2008

using some edges in a maximum-weight b-matching (rather than a maximum-weight matching) between the cycles. Although our algorithm may look similar to the one in [3], our algorithm needs a deeper analysis of various probabilities. By carefully analyzing the new algorithm, we show that it achieves an expected ratio of $\frac{88.85}{169}(1-\epsilon)$ for any constant $\epsilon > 0$. Although the new ratio (namely, $\frac{88.85}{169}(1-\epsilon)$) may seem to be only slightly better than the old ratio (namely, $\frac{43}{83}(1-\epsilon)$), it is of interest for the following two reasons: First, the new ratio is very close to $\frac{89}{169}(1-\epsilon)$ which is the expected ratio wrongly claimed in [3]; hence our new algorithm and its analysis can be viewed as an almost complete correction of the flaw committed in [3]. Second, the improvement (achieved by our new algorithm) from $\frac{43}{83}(1-\epsilon)$ to $\frac{88.85}{169}(1-\epsilon)$ is almost half the improvement (achieved by the algorithm in [3]) from the trivial ratio $\frac{1}{2}$ to $\frac{43}{83}(1-\epsilon)$.

Our randomized algorithm is too sophisticated to derandomize. However, we can modify it to obtain another randomized algorithm which achieves a slightly smaller expected ratio but can be derandomized using the pessimistic estimator method [6]; the resulting deterministic polynomial-time approximation algorithm for MTP still achieves a better ratio (namely, $\frac{43.1}{83}(1-\epsilon)$) than the *expected* ratio achieved by Hassin and Rubinstein's *randomized* algorithm [4]. We omit the details here.

2 Basic Definitions

Throughout the remainder of this paper, a graph means an undirected graph without parallel edges or self-loops whose edges each have a nonnegative weight. A graph G has a vertex set $V(G)$ and an edge set $E(G)$. We denote the weight of a subgraph H of G by $w(H)$. For a function b mapping each vertex v of G to a nonnegative integer, a b-*matching* of G is a subset F of $E(G)$ such that each vertex v of G is incident to at most $b(v)$ edges in F. A *path component* of G is a connected component of G that is a path.

For a random event A, $\Pr[A]$ denotes the probability that A occurs. For a random event A and one or more random events $B_1, \ldots, B_h$, $\Pr[A \mid B_1, \ldots, B_h]$ denotes the probability that A occurs given the occurrences of $B_1, \ldots, B_h$. For a random variable X, $\mathcal{E}[X]$ denotes the expected value of X. For a random variable X and one or more random events $B_1, \ldots, B_h$, $\mathcal{E}[X \mid B_1, \ldots, B_h]$ denotes the expected value of X given the occurrences of $B_1, \ldots, B_h$.

3 Sketch of Hassin and Rubinstein's Algorithm

Throughout this section, fix an instance G of MTP and an arbitrary constant $\epsilon > 0$. Moreover, fix a maximum-weight triangle packing $\mathcal{O}pt$ of G.

To compute a triangle packing of large weight, Hassin and Rubinstein's algorithm [3] (H&R-algorithm for short) starts by computing a maximum-weight cycle cover $\mathcal{C}$ of G. It then breaks each cycle $C \in \mathcal{C}$ with $|C| > \frac{1}{\epsilon}$ into cycles of length at most $\frac{1}{\epsilon}$. This is done by removing a set F of edges in C with

$w(F) \le \epsilon \cdot w(C)$ and then adding one edge between the endpoints of each resulting path. In this way, the length of each cycle in C becomes short, namely, is at most $\frac{1}{\epsilon}$. H&R-algorithm then uses C to compute three triangle packings P_1, P_2, and P_3 of G, and further outputs the packing whose weight is maximum among the three.

Computing P_1 and P_2 from C in H&R-algorithm is easy:

Lemma 1. [3] *Let* $\alpha \cdot w(C)$ *be the total weight of edges in triangles in* C. *Then,* $w(P_1) \ge \frac{1+\alpha}{2} \cdot w(C) \ge \frac{1+\alpha}{2}(1 - \epsilon) \cdot w(Opt)$.

Lemma 2. [3] *Let* $\beta \cdot w(Opt)$ *be the total weight of those edges* $\{u, v\}$ *such that some triangle in* Opt *contains both* u *and* v *and some cycle in* C *contains both* u *and* v. *Then,* $w(P_2) \ge \beta \cdot w(Opt)$.

Unlike P_1 and P_2, the computation of P_3 from C in H&R-algorithm is done by a complicated randomized subroutine which can be sketched as follows: Because of P_2, we only need to consider how to find a triangle packing of large weight when Opt contains a heavy set of edges between cycles in C. So, assume that Opt contains a heavy set of edges between cycles in C. Then, G must contain a heavy matching M consisting of edges whose endpoints belong to different cycles in C. Thus, we want P_3 to contain not only as many edges of C but also as many edges of M as possible. Towards this goal, one way is to mark each edge of C with a probability $0 < q < 1$ such that adjacent edges in C and edges in different cycles in C are marked independently at random and each cycle in C has at least one edge marked. Let R be the set of marked edges. We say that an edge e of M *survives* the marking process if both endpoints of e become of degree at most 1 in graph $C - R$. Note that each edge of M survives the marking process with probability $(2q - q^2)^2$. Let C' be the graph obtained from $C - R$ by adding the edges of M. Then, the expected weight of C' is $(1 - q)w(C) + (2q - q^2)^2 w(M)$. Moreover, each connected component of C' is either a path, or a cycle containing at least two edges of M. So, after removing one edge of M from each cycle in C', we obtain a collection of paths whose expected total weight is at least $(1 - q)w(C) + \frac{1}{2}(2q - q^2)^2 w(M)$. These paths are then patched together into a Hamiltonian cycle of G which is then cut into paths of length 2 by removing one third of its edges. The resulting paths of length 2 lead to a triangle packing whose expected weight is at least $\frac{2}{3}(1-q)w(C) + \frac{1}{3}(2q-q^2)^2 w(M)$, which is large if $w(M)$ is large.

4   New Computation of P_3

This section details our new computation of P_3, which is basically a significant refinement of the computation of P_3 in H&R-algorithm. We inherit the notations in the last section.

There are two main ideas in our new computation of P_3. First, the weight of the matching M computed in H&R-algorithm (outlined above) may be as small as $\frac{1}{3}w(Opt)$. In our new computation, we instead compute a maximum-weight

b-matching M_1 consisting of edges whose endpoints belong to different cycles in $\mathcal{C}$, where $b(v) = 2$ for every vertex v. Note that $w(M_1)$ is close to $w(\mathcal{O}pt)$ when $\mathcal{O}pt$ contains a heavy set of edges between cycles in $\mathcal{C}$. If we are lucky enough that M_1 contains no cycle of odd length, then we can partition M_1 into two matchings among which the heavier one has weight close to $\frac{1}{2}w(\mathcal{O}pt)$ and we can use it as M. Suppose that C is a cycle of odd length in M_1. The crucial point is that with a significantly high probability, at least one edge of C will not survive the marking process. Thus, with a significantly high probability, we can partition the survived edges of M_1 into two matchings among which the heavier one can be used as M.

Second, it is unnecessary to require that each triangle C in $\mathcal{C}$ have at least one edge marked because if C has no edge marked then C can be included in P_3 as it is. That is, we may distinguish the triangles in $\mathcal{C}$ from the other cycles and mark the edges in triangles in $\mathcal{C}$ with a smaller probability (so that the edges in triangles have a larger probability to remain in P_3).

Next, we detail our new computation of P_3 from $\mathcal{C}$. The first step is as follows:

1. Compute a maximum-weight b-matching M_1 in a graph G_1, where
   - $V(G_1) = V(G)$,
   - $E(G_1)$ consists of those $\{u, v\} \in E(G)$ such that u and v belong to different cycles in $\mathcal{C}$, and
   - $b(v) = 2$ for each $v \in V(G_1)$. (*Comment:* Note that $b(v)$ is an upper bound on the degree of v in G_1 and hence is not necessarily the degree of v in G_1.)

Note that $w(M_1)$ is close to $w(\mathcal{O}pt)$ when $\mathcal{O}pt$ contains a heavy set of edges between cycles in $\mathcal{C}$. So, we want to add the edges of M_1 to $\mathcal{C}$. However, adding the edges of M_1 to $\mathcal{C}$ yields a graph which may have a lot of vertices of degree 3 or 4 and is hence far from a triangle packing of G. To remedy this situation, we want to compute a set R of edges in $\mathcal{C}$ and a subset M of M_1 such that adding the edges of M to $\mathcal{C} - R$ yields a graph in which each connected component is a cycle or path.

The next two steps of our algorithm are for computing the set R. Before describing the details, we need to define several notations. Let $C_1, \ldots, C_r$ be the cycles in $\mathcal{C}$. Moreover, throughout the remainder of this section, let p be the smallest real number satisfying the inequality $\frac{27}{20}p^2 - \frac{9}{10}p^3 \geq \frac{27}{320}$; the reason why we select p in this way will become clear in Lemma 11. Note that $0.276 < p < 0.277$; hence $(1 - p)^2 > \frac{1}{2}$ and $\frac{27}{80}p - \frac{9}{80}p^2 \geq \frac{27}{320}$. Now, we are ready to describe Steps 2 and 3 of our algorithm.

2. In parallel, for each cycle C_i in $\mathcal{C}$, process C_i by performing the following steps:
   (a) Initialize R_i to be the empty set.
   (b) If $|C_i| = 3$, then for each edge e of C_i, add e to R_i with probability p.
   (c) If $|C_i| \geq 4$, then perform the following steps:
       i. Choose one edge e_1 from C_i uniformly at random.

    ii. Starting at e_1 and going clockwise around C_i, label the other edges of C_i as $e_2, \ldots, e_c$, where c is the number of edges in C_i.

    iii. Add the edges e_j with $j \equiv 1 \pmod 4$ and $j \leq c-3$ to R_i. (*Comment:* R_i is a matching of C_i and $|R_i| = \lfloor \frac{|C_i|}{4} \rfloor$.)

    iv. If $c \equiv 1 \pmod 4$, then add e_{c-1} to R_i with probability $\frac{1}{4}$. (*Comment:* R_i remains to be a matching of C_i. Moreover, $\mathcal{E}[|R_i|] = \frac{|C_i|-1}{4} + 1 \cdot \frac{1}{4} = \frac{|C_i|}{4}$.)

    v. If $c \equiv 2 \pmod 4$, then add e_{c-1} to R_i with probability $\frac{1}{2}$. (*Comment:* R_i remains to be a matching of C_i. Moreover, $\mathcal{E}[|R_i|] = \frac{|C_i|-2}{4} + 1 \cdot \frac{1}{2} = \frac{|C_i|}{4}$.)

    vi. If $c \equiv 3 \pmod 4$, then add e_{c-2} to R_i with probability $\frac{3}{4}$. (*Comment:* R_i remains to be a matching of C_i. Moreover, $\mathcal{E}[|R_i|] = \frac{|C_i|-3}{4} + 1 \cdot \frac{3}{4} = \frac{|C_i|}{4}$.)

3. Let $R = R_1 \cup \cdots \cup R_r$.

The next lemma is obvious from Step 2b:

Lemma 3. *For every triangle C_i in $\mathcal{C}$ and for every vertex v of C_i, the following hold:*

1. *v is incident to no edge of R with probability $(1-p)^2$.*
2. *v is incident to exactly one edge of R with probability $2p(1-p)$.*
3. *v is incident to exactly two edges of R with probability p^2.*

Note that our algorithm processes those cycles C_i of $\mathcal{C}$ with $|C_i| \geq 4$ as in the H&R-algorithm. So, we have the following lemma:

Lemma 4. *For every cycle C_i of $\mathcal{C}$ with $|C_i| \geq 4$, the following hold:*

1. *For every edge e of C_i, $\Pr[e \in R] = \frac{1}{4}$.*
2. *For every vertex v of C_i, v is incident to at least one edge of R with probability $\frac{1}{2}$.*

Proof. We include the proof for completeness. Consider an arbitrary cycle C_i of $\mathcal{C}$ with $|C_i| \geq 4$. By the comments on Steps 2(c)iv through 2(c)vi, we have $\mathcal{E}[|R_i|] = \frac{|C_i|}{4}$. Moreover, each edge of C_i is added to R_i with the same probability. Thus, $\Pr[e \in R_i] = \frac{1}{4}$ for every edge e of C_i, and hence each vertex of C_i is incident to at least one edge of R with probability $\frac{1}{2}$. $\qquad\square$

We next turn to the computation of the subset M of M_1. Steps 4 through 9 of our algorithm are for this purpose.

4. Let M_2 be the set of all edges $\{u, v\} \in M_1$ such that both u and v are of degree 0 or 1 in graph $\mathcal{C} - R$. Let G_2 be the graph $(V(G), M_2)$.
5. For each odd cycle C of G_2, select one edge uniformly at random and delete it from G_2.
6. Partition the edge set of G_2 into two matchings N_1 and N_2.

7. For each edge e of G_2 which alone forms a connected component of G_2, add e to the matching N_i ($i \in \{1, 2\}$) which does not contain e.
8. Select M from N_1 and N_2 uniformly at random. (*Comment:* $M \subseteq M_1$ is a matching of G.)
9. Let $\mathcal{C}'$ be the graph obtained from graph $\mathcal{C} - R$ by adding the edges in M.

The next lemma is clear from the above construction of $\mathcal{C}'$:

Lemma 5. *The following hold:*

1. *Each connected component of $\mathcal{C}'$ is a cycle or path.*
2. *Every triangle in $\mathcal{C}'$ is also a triangle in $\mathcal{C}$.*
3. *Every cycle C in $\mathcal{C}'$ with $|C| \geq 4$ contains at least two edges in M.*

The next lemma will be used to show that for each edge $e \in M_1$, e is included in the output triangle packing by our algorithm with high probability.

Lemma 6. *For each $e \in M_1$, $\Pr[e \in M \mid e \in M_2] \geq \frac{9}{20}$.*

Proof. Assume that the event $e = \{u, v\} \in M_2$ occurs. Let K be the connected component of G_2 that contains e. If K is not an odd cycle, then clearly e is in M with probability at least $\frac{1}{2} \geq \frac{9}{20}$. So, assume that K is an odd cycle. We distinguish two cases as follows:

Case 1: K is an odd cycle of length at least 5. In this case, K must contain a vertex $z \notin \{u, v\}$ such that the cycle in $\mathcal{C}$ containing z contains neither u nor v. Let B_z be the event that the degree of z in the graph $\mathcal{C} - R$ is 2. By Statement 1 of Lemma 3 and Statement 2 of Lemma 4, B_z occurs with probability $(1-p)^2$ or $\frac{1}{2}$ independently of the event $e \in M_2$. Obviously, when B_z occurs, e is contained in M with probability at least $\frac{1}{2}$. On the other hand, when B_z does not occur, e is contained in M with probability at least $\frac{|K|-1}{|K|} \cdot \frac{1}{2} \geq \frac{2}{5}$. So, the probability that e is in M is at least $\frac{1}{2} \cdot \frac{1}{2} + \frac{1}{2} \cdot \frac{2}{5} = \frac{9}{20}$.

Case 2: K is a triangle. In this case, let z be the vertex in K other than u and v. Note that u, v, and z belong to different cycles in $\mathcal{C}$. Let B_z be the event that the degree of z in the graph $\mathcal{C} - R$ is 2. By Statement 1 of Lemma 3 and Statement 2 of Lemma 4, B_z occurs with probability $(1-p)^2$ or $\frac{1}{2}$ independently of the event $e \in M_2$. Obviously, when B_z occurs, e is definitely contained in M because of Step 7. On the other hand, when B_z does not occur, e is contained in M with probability at least $\frac{1}{3}$. So, the probability that e is in M is at least $\frac{1}{2} \cdot 1 + \frac{1}{2} \cdot \frac{1}{3} = \frac{2}{3} > \frac{9}{20}$. $\square$

By Lemma 5, $\mathcal{C}'$ is a collection of disjoint paths and cycles. Our plan is to transfrom $\mathcal{C}'$ into a triangle packing of G as follows:

- First, break each nontriangle cycle in $\mathcal{C}'$ by removing one edge.
- Next, patch the path components of $\mathcal{C}'$ together into a single path Y.
- Finally, cut Y into paths of length 2 by removing one third of its edges. (*Comment:* Each path of length 2 can be trivially transformed into a triangle by adding the edge between its endpoints.)

The nontriangle cycles in C' should be broken carefully. Steps 10 through 11 of our algorithm are for this purpose.

10. Classify the cycles C of C' into three types: *superb*, *good*, or *ordinary*. Here, C is *superb* if $|C| = 3$; C is *good* if $|C| = 6$, $|E(C) \cap M| = 2$, and there are triangles C_i and C_j in C such that $|E(C_i) \cap E(C)| = 2$ and $|E(C_j) \cap E(C)| = 2$; C is *ordinary* if it is neither good nor superb.)
11. For each ordinary cycle C in C', choose one edge in $E(C) \cap M$ uniformly at random and delete it from C'.
12. For each good cycle C in C', change C back to two triangles in C as follows: Delete the two edges of $M \cap E(C)$ from C to obtain two paths Q_1 and Q_2 of length 2, add the edge between the endpoints of Q_1, and add the edge between the endpoints of Q_2. (*Comment:* Because of the maximality of C, this step does not decrease $w(C')$. Moreover, after this step, each cycle of C' is a triangle.)

We next show that each edge of M_1 remains in C' after Step 11 with high probability.

Lemma 7. *For each edge $e \in M$ such that at least one endpoint of e appears in a non-triangle in C, e survives the deletion in Step 11 with probability at least $\frac{3}{4}$.*

Proof. Without loss of generality, we may assume that C_1 is not a triangle and contains one endpoint v of e. Because the cycles in C are processed independently in Step 2, we may assume that C_1 is processed after the other cycles in C have been processed.

Consider the time point t at which our algorithm has just finished processing all the cycles in C other than C_1. Let S^t be the set of all matchings N in the graph $(V(G), M_1)$ such that each connected component of the graph obtained from $C - (E(C_1) \cup R_2 \cup \cdots \cup R_r)$ by adding the edges of N is a path or cycle. For each matching $N \in S^t$, let $p^t(N)$ be the probability that the matching M constructed in Step 8 equals N. Note that $p^t(N)$ only depends on the random choices made by our algorithm when processing C_1 and later in Steps 5 and 8. Further let B_e^t be the event that e is contained in C' immediately after Step 11.

Let S_e^t be the set of all matchings $N \in S^t$ with $e \in N$ and $p^t(N) > 0$. We claim that for each matching $N \in S_e^t$, $\Pr[B_e^t \mid M = N] \geq \frac{3}{4}$. If this claim indeed holds, then

$$\Pr[B_e^t \mid e \in M] = \sum_{N \in S_e^t} \Pr[B_e^t \mid M = N] \cdot \frac{p^t(N)}{\sum_{N' \in S_e^t} p^t(N')} \geq \frac{3}{4}$$

which implies the lemma immediately. So, it remains to prove the claim.

To prove the claim, consider an arbitrary matching $N \in S_e^t$. Assume that the event $M = N$ occurs. Let K be the graph obtained from $C - (E(C_1) \cup R_2 \cup \cdots \cup R_r)$ by adding the edges of N. Let Q be the connected component of K in which v appears. Note that Q is a path and v is an endpoint of Q. Let u be the endpoint of Q other than v. If u does not appear in C_1, then B_e^t occurs with probability 1.

So, assume that u appears in C_1. Then, because $|C_1| \geq 4$ and our new algorithm processes each non-triangle C_i in $\mathcal{C}$ in the same way as H&R-algorithm does, Lemma 2 in [3] guarantees that B_e^t occurs with probability at least $\frac{3}{4}$. This completes the proof of the claim and hence that of the lemma. $\square$

Lemma 8. *For each $e \in M_1$ such that neither endpoint of e appears in a triangle in $\mathcal{C}$, e is contained in $\mathcal{C}'$ immediately after Step 11 with probability at least $\frac{27}{320}$.*

Proof. Consider an edge $e = \{u, v\}$ in M_1 such that neither u nor v appears in a triangle in $\mathcal{C}$. Let B_u (respectively, B_v) be the event that u (respectively, v) is incident to one edge in R. Note that when both B_u and B_v occur, e is contained in M_2. So, by Lemma 6, $\Pr[e \in M \mid E_u \text{ and } E_v] \geq \frac{9}{20}$. Moreover, by the comment on Step 3, $\Pr[E_u \text{ and } E_v] = \frac{1}{4}$. Hence, $\Pr[e \in M] \geq \frac{9}{80}$. Consequently, by Lemma 7, the probability that e is contained in $\mathcal{C}'$ immediately after Step 11 is at least $\frac{9}{80} \cdot \frac{3}{4} = \frac{27}{320}$. $\square$

Lemma 9. *For each $e \in M_1$ such that exactly one endpoint of e appear in a triangle in $\mathcal{C}$, e is contained in $\mathcal{C}'$ immediately after Step 11 with probability at least $\frac{27}{320}$.*

Proof. Consider an edge $e = \{u, v\}$ in M_1 such that u appears in a triangle in $\mathcal{C}$ but v does not. Let B_1 be the event that u is incident to exactly one edge in R and so does v. Similarly, let B_2 be the event that u is incident to exactly two edges in R and v is incident to exactly one edge in R. Note that when B_1 or B_2 occurs, e is contained in M_2. So, by Lemma 6, $\Pr[e \in M \mid E_1] \geq \frac{9}{20}$ and $\Pr[e \in M \mid E_2] \geq \frac{9}{20}$. Moreover, by Statement 2 of Lemma 3 and Statement 2 of Lemma 4, $\Pr[E_1] = 2p(1-p) \cdot \frac{1}{2} = p(1-p)$ and $\Pr[E_2] = p^2 \cdot \frac{1}{2} = \frac{1}{2}p^2$. Hence, $\Pr[e \in M \text{ and } E_1] \geq \frac{9}{20}p(1-p)$ and $\Pr[e \in M \text{ and } E_2] \geq \frac{9}{40}p^2$.

Obviously, when B_1 occurs and $e \in M$, e survives the deletion in Step 11 with probability at least $\frac{3}{4}$ by Lemma 7. The crucial point is that when B_2 occurs and $e \in M$, e survives the deletion in Step 11 with probability 1. Therefore, the probability that e is contained in $\mathcal{C}'$ immediately after Step 11 is at least $\frac{9}{20}p(1-p) \cdot \frac{3}{4} + \frac{9}{40}p^2 \cdot 1 \geq \frac{27}{320}$. $\square$

Lemma 10. *Suppose that $e = \{u_1, v_1\}$ is an edge in M such that both u_1 and v_1 appear in triangles in $\mathcal{C}$ and both u_1 and v_1 are incident to exactly one edge in R. Then, the probability that e is contained in $\mathcal{C}'$ immediately after Step 11 is at least $\frac{3}{4}$.*

Proof. Without loss of generality, we may assume that C_1 (respectively, C_2) is the cycle in $\mathcal{C}$ to which u_1 (respectively, v_1) belongs. Because the cycles in $\mathcal{C}$ are processed independently in Step 2, we may assume that C_1 and C_2 are processed after the other cycles in $\mathcal{C}$ have been processed.

Consider the time point t at which our algorithm has just finished processing all the cycles in $\mathcal{C}$ other than C_1 and C_2. Let $\mathcal{S}^t$ be the set of all matchings N in the graph $(V(G), M_1)$ such that each connected component of the graph obtained from $\mathcal{C} - (E(C_1) \cup E(C_2) \cup R_3 \cup \cdots \cup R_r)$ by adding the edges of N is a path or cycle. For each matching $N \in \mathcal{S}^t$, let $p^t(N)$ be the probability that the matching M constructed in Step 8 equals N. Note that $p^t(N)$ only depends on the random choices made by our algorithm when processing C_1 and C_2 and later in Steps 5 and 8. Further let $B_{1,1}^t$ be the event that both u_1 and v_1 are incident to exactly one edge in R, and let B_e^t be the event that e is contained in $\mathcal{C}'$ immediately after Step 11.

Let $\mathcal{S}_e^t$ be the set of all matchings $N \in \mathcal{S}^t$ with $e \in N$ and $p^t(N) > 0$. We claim that for each matching $N \in \mathcal{S}_e^t$, $\Pr[B_e^t \mid M = N$ and $B_{1,1}^t] \geq \frac{3}{4}$. If this claim indeed holds, then

$$\Pr[B_e^t \mid e \in M \text{ and } B_{1,1}^t] = \sum_{N \in \mathcal{S}_e^t} \Pr[B_e^t \mid M = N \text{ and } B_{1,1}^t] \cdot \frac{p^t(N)}{\sum_{N' \in \mathcal{S}_e^t} p^t(N')} \geq \frac{3}{4}$$

which implies the lemma immediately. So, it remains to prove the claim.

To prove the claim, consider an arbitrary matching $N \in \mathcal{S}_e^t$. Assume that the events $M = N$ and $B_{1,1}^t$ occur. Then, B_e^t occurs with probability at least $\frac{1}{2}$ because each cycle C in the graph $\mathcal{C}'$ constructed in Step 9 with $E(C) \cap M_1 \neq \emptyset$ contains at least two edges of N. Our goal is to show that B_e^t occurs with probability at least $\frac{3}{4}$. First, we need several definitions:

- Let u_2 and u_3 be the vertices of C_1 other than u_1.
- Let v_2 and v_3 be the vertices of C_2 other than v_1.
- Let H^t be the graph obtained from $\mathcal{C} - (E(C_1) \cup E(C_2) \cup R_3 \cup \cdots \cup R_r)$ by adding the edges of N. (*Comment:* The degree of each vertex of $V(C_1) \cup V(C_2)$ in H^t is 0 or 1.)

Since N is a matching, there are at most two paths Q in H^t such that one endpoint of Q is in $\{u_2, u_3\}$ and the other is in $\{v_2, v_3\}$. We distinguish three cases as follows:

Case 1: There is no path Q in H^t such that one endpoint of Q is in $\{u_2, u_3\}$ and the other is in $\{v_2, v_3\}$. In this case, it is easy to see that B_e^t occurs with probability 1.

Case 2: There are two paths Q in H^t such that one endpoint of Q is in $\{u_2, u_3\}$ and the other is in $\{v_2, v_3\}$. Without loss of generality, we may assume that H^t contains a path between u_2 and v_2 and contains another path between u_3 and v_3. Note that each vertex in $V(C_1) \cap V(C_2)$ is incident to an edge in N. So, exactly one of the following four events occurs:

- $R \cap (E(C_1) \cup E(C_2)) = \{\{u_1, u_2\}, \{v_1, v_3\}, \{u_2, u_3\}, \{v_2, v_3\}\}$.
- $R \cap (E(C_1) \cup E(C_2)) = \{\{u_1, u_3\}, \{v_1, v_2\}, \{u_2, u_3\}, \{v_2, v_3\}\}$.
- $R \cap (E(C_1) \cup E(C_2)) = \{\{u_1, u_2\}, \{v_1, v_2\}, \{u_2, u_3\}, \{v_2, v_3\}\}$.
- $R \cap (E(C_1) \cup E(C_2)) = \{\{u_1, u_3\}, \{v_1, v_3\}, \{u_2, u_3\}, \{v_2, v_3\}\}$.

Obviously, when either of the first two events occurs, e does not appear in a cycle in the graph $\mathcal{C}'$ constructed in Step 9 and hence B_e^t occurs with probability 1. Moreover, if we concentrate on the random choices made when processing C_1 and C_2, the four events occur with the same probability (namely, $(1-p)^2 p^4$). Thus, B_e^t occurs with probability at least $\frac{2(1-p)^2 p^4}{4(1-p)^2 p^4} \cdot 1 + \frac{2(1-p)^2 p^4}{4(1-p)^2 p^4} \cdot \frac{1}{2} = \frac{3}{4}$.

Case 3: There is exactly one path Q in H^t such that one endpoint of Q is in $\{u_2, u_3\}$ and the other is in $\{v_2, v_3\}$. Without loss of generality, we may assume that H^t contains a path between u_2 and v_2. If N contains an edge incident to u_3 and another edge incident to v_3, then the same discussion in Case 2 applies. So, it suffices to consider the following two subcases:

Case 3.1: Exactly one of u_3 and v_3 is incident to an edge in N. Without loss of generality, we may assume that u_3 is incident to an edge in N but v_3 is not. Then, each vertex in $V(C_1) \cap V(C_2)$ other than v_3 is incident to an edge in N. So, exactly one of the following six events occurs:

- $R \cap (E(C_1) \cup E(C_2)) = \{\{u_1, u_2\}, \{v_1, v_2\}, \{u_2, u_3\}\}$.
- $R \cap (E(C_1) \cup E(C_2)) = \{\{u_1, u_2\}, \{v_1, v_2\}, \{u_2, u_3\}, \{v_2, v_3\}\}$.
- $R \cap (E(C_1) \cup E(C_2)) = \{\{u_1, u_2\}, \{v_1, v_3\}, \{u_2, u_3\}, \{v_2, v_3\}\}$.
- $R \cap (E(C_1) \cup E(C_2)) = \{\{u_1, u_3\}, \{v_1, v_2\}, \{u_2, u_3\}, \{v_2, v_3\}\}$.
- $R \cap (E(C_1) \cup E(C_2)) = \{\{u_1, u_3\}, \{v_1, v_2\}, \{u_2, u_3\}\}$.
- $R \cap (E(C_1) \cup E(C_2)) = \{\{u_1, u_3\}, \{v_1, v_3\}, \{u_2, u_3\}, \{v_2, v_3\}\}$.

Obviously, when either of the first four events occurs, e does not appear in a cycle in the graph $\mathcal{C}'$ constructed in Step 9 and hence B_e^t occurs with probability 1. Moreover, if we concentrate on the random choices made when processing C_1 and C_2, then the first and the fifth events occur with the same probability (namely, $(1-p)^3 p^3$), while the other events occur with the same probability (namely, $(1-p)^2 p^4$). Thus, B_e^t occurs with probability at least $\frac{(1-p)^3 p^3 + 3(1-p)^2 p^4}{2(1-p)^3 p^3 + 4(1-p)^2 p^4} \cdot 1 +$ $\frac{(1-p)^3 p^3 + (1-p)^2 p^4}{2(1-p)^3 p^3 + 4(1-p)^2 p^4} \cdot \frac{1}{2} \geq \frac{3}{4}$.

Case 3.2: Neither u_3 nor v_3 is incident to an edge in N. In this case, each vertex in $V(C_1) \cap V(C_2)$ other than u_3 and v_3 is incident to an edge in N. So, exactly one of the following nine events occurs:

- $R \cap (E(C_1) \cup E(C_2)) = \{\{u_1, u_2\}, \{v_1, v_2\}, \{v_2, v_3\}\}$.
- $R \cap (E(C_1) \cup E(C_2)) = \{\{u_1, u_2\}, \{v_1, v_2\}, \{u_2, u_3\}\}$.
- $R \cap (E(C_1) \cup E(C_2)) = \{\{u_1, u_2\}, \{v_1, v_2\}, \{u_2, u_3\}, \{v_2, v_3\}\}$.
- $R \cap (E(C_1) \cup E(C_2)) = \{\{u_1, u_2\}, \{v_1, v_3\}, \{u_2, u_3\}, \{v_2, v_3\}\}$.
- $R \cap (E(C_1) \cup E(C_2)) = \{\{u_1, u_3\}, \{v_1, v_2\}, \{u_2, u_3\}, \{v_2, v_3\}\}$.
- $R \cap (E(C_1) \cup E(C_2)) = \{\{u_1, u_2\}, \{v_1, v_2\}\}$.
- $R \cap (E(C_1) \cup E(C_2)) = \{\{u_1, u_2\}, \{v_1, v_3\}, \{v_2, v_3\}\}$.
- $R \cap (E(C_1) \cup E(C_2)) = \{\{u_1, u_3\}, \{v_1, v_2\}, \{u_2, u_3\}\}$.
- $R \cap (E(C_1) \cup E(C_2)) = \{\{u_1, u_3\}, \{v_1, v_3\}, \{u_2, u_3\}, \{v_2, v_3\}\}$.

Obviously, when one of the first five events occurs, e does not appear in a cycle in the graph $\mathcal{C}'$ constructed in Step 9 and hence B_e^t occurs with probability 1. As for the last four events, we can say the following:

- Suppose that $|E(Q)| = 1$. Then, when the sixth event occurs, e appears in a good cycle in the graph $\mathcal{C}'$ constructed in Step 9 and hence B_e^t occurs with probability 1. Of course, when the ith event with $i \in \{7, 8, 9\}$ occurs, e appears in an ordinary cycle K in the graph $\mathcal{C}'$ constructed in Step 9 with $|E(K) \cap N| \geq 2$ and hence B_e^t occurs with probability $\frac{1}{2}$.

- Suppose that $|E(Q)| \geq 2$. Then, Q contains at least two edges of N. So, when the ith event with $i \in \{6, \ldots, 9\}$ occurs, e appears in an ordinary cycle K with $|E(K) \cap N| \geq 3$ in the graph $\mathcal{C}'$ constructed in Step 9 and hence B_e^t occurs with probability at least $\frac{2}{3}$.

Furthermore, if we concentrate on the random choices made when processing C_1 and C_2, the first, the second, the seventh, and the eighth events occur with the same probability (namely, $p^3(1-p)^3$), the third, the fourth, the fifth, and the ninth events occur with the same probability (namely, $p^4(1-p)^2$), and the sixth event occurs with probability $p^2(1-p)^4$. Thus, if $|E(Q)| = 1$, then B_e^t occurs with probability at least $\frac{2p^3(1-p)^3 + 3p^4(1-p)^2 + p^2(1-p)^4}{4p^3(1-p)^3 + 4p^4(1-p)^2 + p^2(1-p)^4} \cdot 1 + \frac{2p^3(1-p)^3 + p^4(1-p)^2}{4p^3(1-p)^3 + 4p^4(1-p)^2 + p^2(1-p)^4}$. $\frac{1}{2} \geq \frac{3}{4}$. On the other hand, if $|E(Q)| \geq 2$, then B_e^t occurs with probability at least $\frac{2p^3(1-p)^3 + 3p^4(1-p)^2}{4p^3(1-p)^3 + 4p^4(1-p)^2 + p^2(1-p)^4} \cdot 1 + \frac{p^2(1-p)^4 + 2p^3(1-p)^3 + p^4(1-p)^2}{4p^3(1-p)^3 + 4p^4(1-p)^2 + p^2(1-p)^4} \cdot \frac{2}{3} \geq \frac{3}{4}$. This completes the proof of the claim and hence that of the lemma. $\square$

Lemma 11. *For each $e \in M_1$ such that both endpoints of e appear in triangles in $\mathcal{C}$, e is contained in $\mathcal{C}'$ immediately after Step 11 with probability at least $\frac{27}{320}$.*

Proof. Consider an edge $e = \{u, v\}$ in M_1 such that both u and v appear in a triangle in $\mathcal{C}$. Let B_0 be the event that both u and v are incident to exactly one edge in R, and let B_1 be the event that both u and v are incident to at least one edge in R and at least one of them is incident to exactly two edges in R.

Note that when B_0 or B_1 occurs, e is contained in M_2. So, by Lemma 6, $\Pr[e \in M \mid E_0] \geq \frac{9}{20}$ and $\Pr[e \in M \mid E_1] \geq \frac{9}{20}$. Moreover, by Statements 2 and 3 of Lemma 3 $\Pr[E_0] = 2p(1-p) \cdot 2p(1-p) = 4p^2(1-p)^2$ and $\Pr[E_1] = p^2 \cdot 2p(1-p) + 2p(1-p) \cdot p^2 + p^4 = 4p^3 - 3p^4$. Hence, $\Pr[e \in M$ and $E_0] \geq \frac{9}{5}p^2(1-p)^2$ and $\Pr[e \in M$ and $E_1] \geq \frac{9}{20}(4p^3 - 3p^4)$.

By Lemma 10, when B_0 occurs and $e \in M$, e survives the deletion in Step 11 with probability at least $\frac{3}{4}$. The crucial point is that when B_1 occurs and $e \in M$, e survives the deletion in Step 11 with probability 1. Therefore, the probability that e is contained in $\mathcal{C}'$ immediately after Step 11 is at least $\frac{9}{5}p^2(1-p)^2 \cdot \frac{3}{4} + \frac{9}{20}(4p^3 - 3p^4) \cdot 1 = \frac{27}{20}p^2 - \frac{9}{10}p^3 \geq \frac{27}{320}$. $\square$

By the comment on Step 12, each connected component of $\mathcal{C}'$ after Step 12 is a triangle or path. So, it is now almost trivial to transform $\mathcal{C}'$ into a triangle packing. Step 12 of our algorithm is for this purpose.

12. If $\mathcal{C}'$ has at least one path component, then perform the following steps:
    (a) Connect the path components of $\mathcal{C}'$ into a single cycle Y by adding some edges of G.
    (b) Break Y into paths each of length 2 by removing a set F of edges from Y with $w(F) \leq \frac{1}{3} \cdot w(Y)$.
    (c) For each path Q of length 2 obtained in Step 12b, add the edge between the endpoints of Q.
    (*Comment:* After this step, $\mathcal{C}'$ is a triangle packing of G.)
13. Let $P_3 = \mathcal{C}'$.

5   Analysis of the Approximation Ratio

By the comment on Step 3, the expected total weight of edges of $\mathcal{C}$ remaining in $\mathcal{C}'$ immediately after Step 11 is at least $\left((1-p)\alpha + \frac{3}{4}(1-\alpha)\right) w(\mathcal{C}) = \left(\frac{3}{4} - (p - \frac{1}{4})\alpha\right) w(\mathcal{C}) \geq \left(\frac{3}{4} - (p - \frac{1}{4})\alpha\right)(1-\epsilon)w(\mathcal{O}pt)$. Moreover, by Lemmas 8, 9, and 11, the expected total weight of edges of M_1 remaining in $\mathcal{C}'$ immediately after Step 11 is at least $\frac{27}{320}w(M_1)$. Furthermore, by the construction of M_1, $w(M_1)$ is larger than or equal to the total weight of those edges $\{u,v\}$ such that some triangle in $\mathcal{O}pt$ contains both u and v but no cycle in $\mathcal{C}$ contains both u and v. So, $w(M_1) \geq (1-\beta)w(\mathcal{O}pt)$. Now, since $w(P_3)$ is at least two-thirds of the total weight of edges in $\mathcal{C}'$ immediately after Step 11, we have

$$\mathcal{E}[w(P_3)] \geq \frac{2}{3}\left(\frac{3}{4} - (p - \frac{1}{4})\alpha\right)(1-\epsilon)w(\mathcal{O}pt) + \frac{2}{3}\cdot\frac{27}{320}(1-\beta)w(\mathcal{O}pt) \quad (1)$$

$$= \left(\frac{89}{160} - \frac{1}{2}\epsilon - \frac{2}{3}(p - \frac{1}{4})(1-\epsilon)\alpha - \frac{9}{160}\beta\right)w(\mathcal{O}pt). \quad (2)$$

So, by Lemmas 1 and 2, we have

$$\frac{4}{3}(p - \frac{1}{4})w(P_1) + \frac{9}{160}w(P_2) + w(P_3) \geq \frac{187 + 320p - (320p + 160)\epsilon}{480}\cdot w(\mathcal{O}pt).$$

Therefore, the weight of the best packing among P_1, P_2, and P_3 is at least

$$\frac{187 + 320p - (320p + 160)\epsilon}{640p + 347}\cdot w(\mathcal{O}pt) \geq \frac{187 + 320p}{347 + 640p}\cdot(1-\epsilon)w(\mathcal{O}pt).$$

In summary, we have proven the following theorem:

Theorem 1. *For any constant $\epsilon > 0$, there is a polynomial-time randomized approximation algorithm for MTP that achieves an expected ratio of $\frac{187+320p}{347+640p}$ · $(1-\epsilon) > \frac{88.85}{169}\cdot(1-\epsilon)$.*

References

1. Chlebík, M., Chlebíková, J.: Approximating Hardness for Small Occurrence Instances of NP-Hard Problems. In: Petreschi, R., Persiano, G., Silvestri, R. (eds.) CIAC 2003. LNCS, vol. 2653, pp. 152–164. Springer, Heidelberg (2003)
2. Garey, M.R., Johnson, D.S.: Computers and Intractability: A Guide to the Theory of NP-Completeness. Freeman, New York (1979)
3. Hassin, R., Rubinstein, S.: An Approximation Algorithm for Maximum Triangle Packing. Discrete Applied Mathematics 154, 971–979 (2006)
4. Hassin, R., Rubinstein, S.: Erratum to gAn approximation algorithm for maximum triangle packingh: [Discrete Applied Mathematics 154 (2006) 971-979]. Discrete Applied Mathematics 154, 2620–2620 (2006)
5. Kann, V.: Maximum Bounded 3-Dimensional Matching Is MAX SNP-Complete. Information Processing Letters 37, 27–35 (1991)
6. Raghavan, P.: Probabilistic Construction of Deterministic Algorithms: Approximating Packing Integer Programs. Journal of Computer and System Sciences 38, 683–707 (1994)

Line Facility Location in Weighted Regions*

Yam Ki Cheung and Ovidiu Daescu

Department of Computer Science,
University of Texas at Dallas,
Richardson , TX 75080, USA
{samykcheung,daescu}@utdallas.edu

Abstract. In this paper, we present approximation algorithms for the
line facility location problem in weighted regions: Given l fixed points in
a 2-dimensional weighted subdivision of the plane, with n vertices, find a
line L such that the sum of the weighted distances from the fixed points
to L is minimized. The weighted region setup is a more realistic model
for many facility location problems that arise in practical applications.
Our algorithms exploit an interesting property of the problem, that could
possibly be used for solving other problems in weighted regions.

1 Introduction

Let $R = \{R_1, R_2, \ldots, R_{n'}\}$ be a subdivision of the plane with n vertices, with
each region $R_i \in R$ having associated an integer weight $w_i \in \{1, 2, 3, \ldots\}$. Without loss of generality, we assume all regions of R are triangulated. The distance
between two points s and t within the same region $R_i \in R$ is defined as the
product of the weight w_i of the region R_i and the Euclidean length of the line
segment $\overline{st}$ [8].

In this paper we study the following problem that we call the *line facility
location in weighted regions*: Given a set of l fixed points $P = \{s_1, s_2, \ldots, s_l\} \in
R$, find a line L such that the sum $S(L)$ of the weighted distances from the
points in P to L is minimized (see Fig. 1 for an example). More formally, $S(L)$
is defined as follows. Let L_i be the orthogonal link from a point $s_i \in P$ to
L. The cost (weighted length, weighted distance) of the link L_i is defined as
$S(L, s_i) = \sum_{L_i \cap R_j \neq \emptyset} w_j * d_j(L_i)$, where $d_j(L_i)$ is the length of L_i within region
R_j. Then $S(L) = \sum_{i=1}^{l} S(L, s_i)$ is the overall cost of all orthogonal links from
P to L.

The facility location problem is a well-studied problem in the operations research and computer science literature. The problem has many formulations,
resulting in various definitions for the objective function. For example, in one
formulation, there is a predefined cost for opening a facility and also for connecting a customer to a facility, and the goal is to minimize the total cost. For
the proposed problem, one may imagine that the line L is an oil pipeline and
the points in P are oil wells. The oil field is divided into weighted regions based

---

* This research was partially supported by NSF grant CCF-0635013.

R. Fleischer and J. Xu (Eds.): AAIM 2008, LNCS 5034, pp. 109–119, 2008.

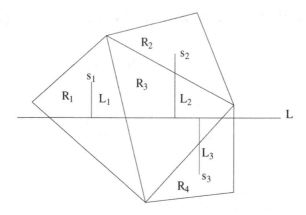

Fig. 1. A line L and the orthogonal links from $P = s_1, s_2, s_3$ to L

on its characteristics (cost of digging, ownership rights, etc.) From each well, a spur pipeline is to be connected directly to L, in straight line. Given the x- and y-coordinates of the wells, the goal is to find the optimal location for the main pipeline that minimizes the total weighted length of the spurs.

The unweighted case of the problem was studied and solved by Megiddo and Tamir more than two decades ago [7]. In this case, the goal is to find a line that minimizes the sum of the Euclidean distances from the points in P to the line. Megiddo and Tamir showed that, for the unweighted case, the optimal line could be found in $O(n^2 \log n)$ time by proving that it must pass through at least two of the points in P. However, to the best of our knowledge, there is no previous work for the weighted version discussed in this paper.

The proposed problem is also closely related to the optimal weighted link problem [2], where the goal is to minimize the weighted length of a line segment $\overline{st}$ connecting two given regions, R_s and R_t, of R. In [2], it has been proven that the problem to minimize the cost $S(\overline{st}) = \sum_{\overline{st} \cap R_j \neq \emptyset} w_i * d_i(\overline{st})$ of $\overline{st}$ can be reduced to a number of ($O(n^2)$ in worst case) global optimization problems, each of which asks to minimize a 2-variable function over a convex domain. In [4,5], it has been shown that to minimize the objective function $S(\overline{st})$, $\overline{st}$ must pass through a vertex of R. By considering each candidate vertex, they reduced the problem to solving $O(n)$ optimization problems at each vertex. The objective function can also be expressed in the form $\sqrt{1 + m^2}(d_0 + \sum_{i=1}^{k} \frac{a_i}{m + b_i})$, where d_0, a_i and b_i are constants, $k = O(n)$, m is the slope of $\overline{st}$, and D_m is a slope interval for the line passing through the vertex. Hence, each subproblem contains one variable only and can be solved by computing the roots of a polynomial of degree $O(n)$ over the given domain D_m. Unfortunately, this approach is only practical when the number of fractional terms in $S(\overline{st})$ is small. In [5], a faster approach is suggested, which aims to compute an approximate solution by bisecting the slope interval and pruning out subproblems that cannot lead to an optimal solution. Some of the results in [2] were later on generalized in [3] to more difficult problems, where the line is replaced by a parallel or cone like beam.

In this paper we first study a special case, in which the slope of L is fixed, and show that an optimal solution can be found at one of $O(ln)$ well defined locations called event points. Hence, the optimal line can be computed in $O(ln \log ln)$ time by a simple sweeping line algorithm. For the general case, we show that the objective function $S(L)$ can be expressed in the form of a summation of a number of fractional terms, that resembles those in [2,3]. Similarly to the work in [2], the problem can be reduced to a number of ($O(l^2 n^2)$ in worst case) 1-variable optimization problems, which can be solved by either computing the exact roots of polynomials or by approximation algorithms. Using point-line duality transforms [9], the feasible domain for each subproblem is an arc or a line segment on the dual plane.

2 A Special Case

We start with a special case of the proposed problem, where the slope of L is fixed. Without loss of generality, we assume that L is horizontal. Let y_{max} and y_{min} be the maximum and minimum y-coordinates of the points in P, respectively.

Lemma 1. *The y-coordinate y_L of the optimal line L ($L : y = y_L$) satisfies $y_L \in [y_{min}, y_{max}]$. In other words, if all points in P are strictly to one side of L, $S(L)$ cannot be the sought minimum.*

Proof. If $y_L \notin [y_{min}, y_{max}]$, we can always find a new line L', such that $S(L') < S(L)$. First we translate L towards P until any point in P is reached. Let the new line be L'. Since all points are strictly to one side of L, all orthogonal links decrease in length, that is, $S(L', s_i) \leq S(L, s_i)$, for $i = 1, 2, \ldots, l$, i.e. $S(L') < S(L)$. □

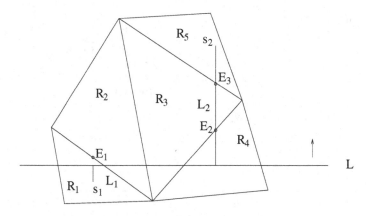

Fig. 2. A simple scenario with two points, s_1 and s_2

Lemma 2. *Let Q be the set of intersections between the vertical lines passing through the points in P and the edges of R. Then there is at least one optimal solution L for $S(L)$ such that $L \cap (P \cup Q) \neq \emptyset$, that is, L passes trough at least one point in P or Q.*

Proof. By Lemma. 1, $y_L \in [y_{min}, y_{max}]$. Consider the case that we sweep the line L upwards from y_{min} to y_{max}. We call points in $P \cup Q$ event points, since these are the points where we need to update the sweeping process. Let function

$$f(x) = S(L : y = y_{min} + x)$$

be the function of total cost of the orthogonal links from P to L, where x is the distance swept by L. Now sweep L a small distance δ upwards such that L does not pass trough any event point.
We have $f(x + \delta) = f(x) + \delta * \sum_{i=1...l, L_i \cap L \cap R_j \neq \emptyset} (w_j * N(L_i \cap L))$, where

$$N(L_i \cap L) = \begin{cases} 1 & \text{if } s_i \text{ is below } L \\ -1 & \text{otherwise} \end{cases}$$

Hence the first derivative $f'(x) = \sum_{i=1...l, L_i \cap L \cap R_j \neq \emptyset}(w_j * N(L_i \cap L))$ is a constant between two consecutive event points. This implies that $f(x)$ is piecewise linear and its slope only changes when L passes though an event point, i.e. local extreme values can only be found at event points.   □

For example, in Fig. 2, P contains two points, s_1 and s_2. Let $w_1 > w_4 > w_2 > w_3 > w_5$. If we sweep L by a small distance δ, the change of $f(x)$ is $(w_1 - w_4) * \delta$. Then $f(x)$ is linearly increasing with rate $w_1 - w_4$ until L reaches the next event point $E1$. Similarly, between event points E_1 and E_2, $f(x)$ is linearly decreasing with rate $w_2 - w_4$, and between E_2 and E_3, $f(x)$ is linearly increasing again with rate $w_2 - w_3$. Between E_3 and p_2, $f(x)$ is increasing but with a smaller rate $w_2 - w_5$. Fig. 3 shows a plot of $f(x)$ for the scenario described above.

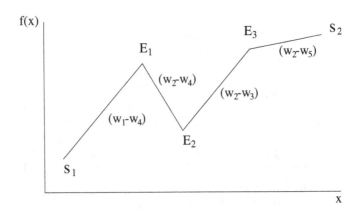

Fig. 3. A plot of $f(x)$ for the scenario shown in Fig. 2

Lemma 3. *The line L that minimizes $S(L)$ can be computed in $O(ln \log(ln))$ time.*

Proof. The optimal solution can be computed by a line sweeping algorithm: (1) Find out all event points and sort them by the y-coordinate. (2) Compute the value of $f(x)$ at one event point by brute force. (3) By the piecewise linearity of $f(x)$, the value of $f(x)$ at successive event points can be calculated in constant time. There are $O(ln)$ event points. The overall cost is $O(ln \log(ln))$, since sorting is the dominant step.   □

3   The General Case

In this section, we consider the general problem: Given a triangulated weighted planar subdivision R with n vertices and a set P of l fixed points in R, find a line L (of arbitrary slope) such that $S(L)$ is minimized.

By Lemma 2, if we fix the slope of L and sweep L across the subdivision, $S(L)$ is minimized when L passes through some fixed point in P or it intersects some orthogonal link from a point in P to L, at some edge of R.

Case 1: L with slope m passes through some fixed point $s_r \in P$. Reset the coordinate system such that s_r is the origin. Then we have $L : y = mx$. Similar to [3], we define $Seq(L_i, L)$ as the sequence of edges or lines intersected by the orthogonal link L_i, i.e. the set of edges intersected by L_i sorted in ascending order by the distance between their intersection points with L_i and the point s_i. We say that two sequences are equal, e.g. $Seq(L_i, L) = Seq(L_j, L)$, if and only if the two sequences consists of the same set of intersected edges, in the same order. For a point $s_i \in P$, let $Seq(L_i, L) = \{e_1, e_2, e_3, \ldots, e_k\}$, where $e_k = L$. Let (x_j, y_j) be the coordinates of the intersection between L_i and e_j, $j = 1, 2, \ldots, k$. Then $S(L_i) = \sqrt{1 + m_o^2} \sum_{j=1}^{k-1} w_j |x_{j+1} - x_j|$, where w_j is the weight of the region bounded by edges e_j and e_{j+1}, and $m_o = -1/m$ is the slope of the orthogonal links (see Fig. 4). Let (a_i, b_i) be the coordinates of s_i. Since the intersection point

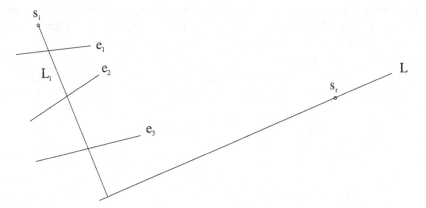

Fig. 4. An orthogonal link L_i that intersects with four edges

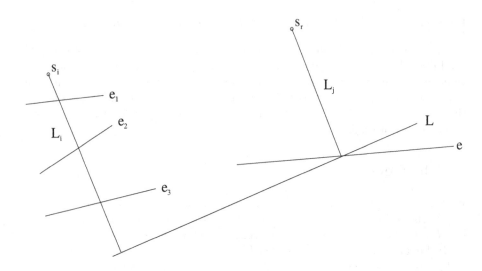

Fig. 5. An example for case 2

(x_j, y_j) lies on both L_i and e_j, we have $L_i : y = m_o x + p_i$, where $p_i = (b_i - m_o a_i)$, $y_j = m_o x_j + p_j$, $e_j : y = m_j x + p_j$, $y_j = m_j x_j + p_j$, $x_j = \frac{p_j - p_i}{m_o - m_j}$, and thus

$$S(L_i) = \sqrt{1 + m_o^2} \sum_{j=1}^{k-1} w_i \left| \frac{p_{j+1} - p_i}{m_o - m_{j+1}} - \frac{p_j - p_i}{m_o - m_j} \right| =$$
$$\sqrt{1 + m_o^2} (\sum_{j=1}^{k-2} w_i \left| \frac{p_{j+1} - p_i}{m_o - m_{j+1}} - \frac{p_j - p_i}{m_o - m_j} \right| + w_{k-1} \left| \frac{-p_i}{m_o - m} - \frac{p_{k-1} - p_i}{m_o - m_{k-1}} \right|) \quad (*)$$

Substituting $p_i = (b_i - m_o a_i)$ and $m = -1/m_o$ into (*), we have
$S(L_i) = \sqrt{1 + m_o^2} (\sum_{j=1}^{k-1} \frac{c_j}{m_o - m_j} + c_k \frac{b_i m_o + a_i}{m_o^2 + 1} + C)$, where c_j and C are constants.
Hence, $S(L) = \sum_{i=1}^{l} S(L_i) = \sqrt{1 + m_o^2} (\sum_{j=1}^{M} \frac{c_j}{m_o - m_j} + \sum_{i=1}^{l} d_i \frac{b_i m_o + a_i}{m_o^2 + 1} + C)$,
where $M = O(n)$, and c_j, d_i, and C are constants.

Case 2: If L with slope m does not pass through a fixed point in P, then L must intersect some orthogonal link L_r, originating at $s_r \in P$, on some edge $e : y = m_e x + p_e$ of R. Again, reset the coordinate system such that s_r is the origin. Let L be $y = mx + p$ (see Fig. 5).

Similar to case 1, we have
$S(L_i) = \sqrt{1 + m_o^2} (\sum_{j=1}^{k-2} w_i \left| \frac{p_{j+1} - p_i}{m_o - m_{j+1}} - \frac{p_j - p_i}{m_o - m_j} \right| + w_{k-1} \left| \frac{p - p_i}{m_o - m} - \frac{p_{k-1} - p_i}{m_o - m_{k-1}} \right|) \quad (**).$
$L_r : y = m_o x$ intersects $e : y = m_e x + p_e$ at $v = (\frac{p_e}{m_o - m_e}, \frac{m_o p_e}{m_o - m_e})$. So we have:

$$L : y - \frac{m_o p_e}{m_o - m_e} = -\frac{1}{m_o} (x - \frac{p_e}{m_o - m_e})$$
$$y = -\frac{1}{m_o} x + (\frac{p_e}{m_o(m_o - m_e)} + \frac{m_o p_e}{m_o - m_e}), \text{ and}$$
$$L_i : y - b_i = m_o(x - a_i)$$
$$y = m_o x + (b_i - m_o a_i)$$

Substituting $m = -\frac{1}{m_o}$, $p = \frac{p_e}{m_o(m_o-m_e)} + \frac{m_o p_e}{m_o-m_e}$, and $p_i = b_i - m_o a_i$ into (**), we have:

$$S(L_i) = \sqrt{1+m_o^2}(\sum_{j=1}^{k-2} w_i |\frac{p_{j+1}-p_i}{m_o-m_{j+1}} - \frac{p_j-p_i}{m_o-m_j}| + w_{k-1}|\frac{p-p_i}{m_o-m} - \frac{p_{k-1}-p_i}{m_o-m_{k-1}}|)$$

$$= \sqrt{1+m_o^2}(\sum_{j=1}^{k-2} w_i |\frac{p_{j+1}-(b_i-m_o a_i)}{m_o-m_{j+1}} - \frac{p_j-(b_i-m_o a_i)}{m_o-m_j}| +$$

$$w_{k-1}|\frac{\frac{p_e}{m_o(m_o-m_e)}+\frac{m_o p_e}{m_o-m_e}-(b_i-m_o a_i)}{m_o+\frac{1}{m_o}} - \frac{p_{k-1}-(b_i-m_o a_i)}{m_o-m_{k-1}}|)$$

$$= \sqrt{1+m_o^2}(\sum_{j=1}^{k-1} \frac{c_j}{m_o-m_j} + c_k(\frac{p_e}{m_o-m_e} - \frac{b_i m_o + a_i}{m_o^2+1}) + C), \text{ and thus}$$

$$S(L) = \sum S(L_i) = \sqrt{1+m_o^2}(\sum_{j=1}^{M} \frac{c_j}{m_o-m_j} + \sum_{i=1}^{l} d_i \frac{b_i m_o + a_i}{m_o^2+1} + C) \quad (***),$$

where $M = O(n)$ is upper bounded by the number of edges of the subdivision, and c_j, d_i, and C are all constants.

Thus, both cases have the same mathematical expression for $S(L)$. Obviously, when we change the slope of L, an orthogonal link may pass through a different set of edges, and the constants in (***) will change accordingly.

Next, we consider the problem in the dual plane. We use the duality transform which maps a line $L : y = mx + p$ on the plane onto a point $L^* : (m,p)$ on the dual plane and maps a point (a,b) onto the line $y = -ax + b$ on the dual plane. We will show how to partition the dual plane into cells so that each cell corresponds to a set of lines L such that (1) L penetrates a fixed sequence of edges, and (2) all orthogonal links L_i from each point s_i to L have the same sequence of intersected edges. That is, if L^* and L'^* belong to the same cell, $Seq(L_i, L) = Seq(L_i, L')$ for all i, where L^* and L'^* are the duals of L and L'.

First, we transform all vertices of R onto the dual plane. The arrangement of the n dual lines gives us $O(n^2)$ convex cells on the dual plane. Let A_v denote this arrangement. It is known that for every point L^* of a given cell, the corresponding line L in the primal plane intersects the same set of edges (see [2,3] for details).

To guarantee the second property, consider a special case in which a line L intersects some orthogonal link L_i exactly at some edge e. Let point L^* be the

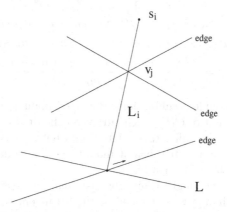

Fig. 6. Moving the intersection point between L and orthogonal link L_i along an edge

mapping of L on the dual plane. Let point L_a^* be a point on the dual plane slightly vertically above L^*. The line L_a corresponding to L_a^* on the primal plane has the same slope as L and is slightly above L. That is, with L_a replacing L, L_i does not intersect with e. Similarly, let L_b^* be a point slightly vertically below L^*. On the primal plane, when L_b replaces L, L_i intersects with e. Hence, if we transform all lines intersecting with some orthogonal link at some edge onto the dual plane, the dual plane will be divided into cells such that if L^* and L'^* belong to the same cell, the last edge in $Seq(L_i, L)$ is the same as the last edge in $Seq(L_i, L')$ for all i, i.e. in a given cell, L_i intersects with L in the same region, for all i (see Fig. 6).

Let line $L : y = mx + p$ intersect some orthogonal link L_i from some fixed point $s_i : (a_i, b_i)$ at some edge $e : y = m_e x + p_e$. We have

$$L_i : (y - b_i) = -\frac{1}{m}(x - a_i)$$

$$y = -\frac{1}{m}x + \frac{a_i}{m} + b_i$$

$$L : y = mx + p$$

Find the intersection point of L and L_i:

$$-\frac{1}{m}x + \frac{a_i}{m} + b_i = mx + p$$

$$x = \frac{b_i m - pm + a_i}{m^2 + 1}$$

L and L_i intersect at $v : (\frac{b_i m - pm + a_i}{m^2 + 1}, \frac{b_i m^2 + a_i m + p}{m^2 + 1})$. Since $v \in L_e$, we have

$$\frac{b_i m^2 + a_i m + p}{m^2 + 1} = m_e \frac{b_i m - pm + a_i}{m^2 + 1} + p_e$$

$$b_i m^2 + a_i m + p = m_e(b_i m - pm + a_i) + p_e(m^2 + 1)$$

$$(b_i - p_e)m^2 + (a_i + m_e p - m_e b_i)m + (p - m_e a_i - p_e) = 0$$

$$p = -\frac{(b_i - p_e)m^2 + (a_i - m_e b_i)m - m_e a_i - p_e}{m_e m + 1}$$

Considering all possible combinations of a fixed point s_i and an edge $e \in R$, we have an arrangement of $O(ln)$ such curves on the dual plane. Let A_c denote this arrangement. To find all intersection points between two curves, we need to solve a cubic equation, which has at most three real roots. Hence, two such curves can intersect at most 3 times.

However, this is not enough. While the last edge intersected by each orthogonal link does not change within a cell of the arrangement A_c defined above, given a cell of A_c it is still possible for some orthogonal link L_i to sweep through some vertex v_j, which would change $Seq(L_i, L)$ within the cell (see Fig. 6). This

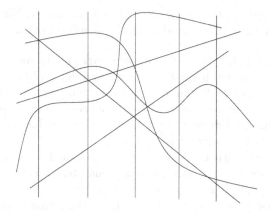

Fig. 7. Partition of the dual space

can be prevented by forbidding the line to rotate through a set of slopes. For example, in Fig. 6, L_i would sweep trough v_j. Then, we should not allow L to rotate through slope $-1/m_{ij}$, where m_{ij} is the slope of the line passing though s_i and v_j. There are ln such slopes. On the dual plane, this is equivalent to adding ln vertical lines. Let A_r denote this arrangement.

The overall partition to enforce both criteria is the overlapping of A_v, A_c and A_r (see Fig. 7), which is an arrangement of $O(ln)$ curves and $O(ln)$ lines. Notice that all curves in this partition have constant number of pairwise intersections. Using the algorithm in [1], the arrangement can be computed in $O(ln \log(ln)+k)$ time and $O(ln+k)$ space, where k is the number cells, which is $O(l^2n^2)$ in worst case.

Lemma 4. *The global minimum of $S(L)$ can be found on the boundary of the cells of A_v and A_c.*

Proof. Consider a sweeping process with a fixed slope m. We have already shown that the optimal line L, which minimize $S(L)$, must pass through an event point. On the dual plane, this corresponds to checking all intersections between the corresponding vertical line $x^* = m$ and the boundary of the cells of A_v and A_c (the lines and curves of A_v and A_c, respectively). If we set m as a variable, a global minimum should appear on the boundary of the cells of A_v and A_c. The proof follows.   □

We already showed that on the boundaries of A_v and A_c, $S(L)$ can be expressed as a function of only the slope of L. By the preceding lemma, the problem is reduced to a one variable optimization problem. From the definition of the partition, on the boundaries of A_v and A_c, the objective function $S(L)$ changes when L^* crosses some intersection point of the lines and curves defining A_v and A_c, i.e. when the sequence of intersected edges of some orthogonal link changes. Then the problem of finding the optimal L can be divided into $O(k)$ one variable subproblems, where k is the complexity of the arrangement. The corresponding objective functions can be obtained using the algorithm below:

1. On the partition $A_v \cup A_c \cup A_r$, for every curve in A_c and every line in A_v, start from the left end point $L^* = (m, p)$, and find $S(L)$ as a function of m by brute force, in $O(ln)$ time. Denote the current curve or line by h.
2. Walk on h by increasing m, (all curves in A_c and all lines in A_v are x-monotone) until it intersects with another curve or line at $L'^* = (m', p')$. Then find the minimum of S(L) over [m,m').
    - If h intersects with a curve in A_c, walking through such intersection point corresponds to same orthogonal link entering or leaving a region of R. We can update $S(L)$ in constant time.
    - If h intersects with a dual line in A_v, walking trough such intersection point corresponds to L passing trough some fixed point s_i and we can update $S(L)$ in constant time as well.
    - If l intersects with a vertical line in A_r, some orthogonal link L_i may sweep through some vertex v_j, and it takes $O(deg(v_j))$ to update $S(L)$.
3. Start from L'^*, continue the walk by repeating step 2.

Theorem 1. *It takes $O(l^2 n^2)$ time to generate all objective functions.*

We have shown that finding the optimal L can be reduced to a number of 1-dimensional optimization subproblems. The domain of each subproblem is a line segment or an arc of the arrangement constructed on the dual plane. Since each subproblem can have a large number of fractional terms, it seems unpractical to solve it by finding roots of polynomial.

Instead, following a similar idea proposed in [5], we design a prune-and-search approximation algorithm combined with the subdivision approach. The general outline is:

(1) Find upper and lower bounds for each subproblem of interest.
(2) Maintain a priority queue of subproblems based on their lower bounds.
(3) Update S^{min}, which is the minimum of upper bounds of all subproblems in the queue.
(4) For each subproblem P_i in the queue, if the lower bound of P_i is greater than S^{min}, prune P_i. (5) Get the first candidate P_i from the priority queue and find a coarse approximation by placing Steiner points and sampling.
(6) A solution L is accepted for candidate P_i, if $S(L) \leq (1 + \epsilon)S_i^{min}$, where ϵ is a parameter defining the quality of the approximation and S_i^{min} is the lower bound of p_i.
(7) Find the upper bound and lower bound of all new subproblems and add them back to the priority queue. Go to step (3).

Next, we give more details of the prune-and-search algorithm. For a given subproblem P_i, by definition, for each orthogonal link L_j, the sequence of intersected edges $Seq(L_j, L)$ does not change, i.e. L_j intersects with the same regions of R over the feasible domain. Let this set of regions be $\{R_{j_1}, R_{j_2}, \ldots, R_{j_k}\}$. Hence, we can set the lower bound, L_j^{min}, of the weighted sum of L_j, as $\sum_{q=1}^{k} w_{j_q} *$

$min(L_j \cap R_{j_q})$, over the given feasible domain (the value of $min(L_j \cap R_{j_q})$ can be computed in $O(1)$ time). The lower bound, S_i^{min}, of P_i is $\sum_{j=1}^{k} L_j^{min}$. The upper bound of P_i could be any sample value of the objective function in the given domain. If the domain is bounded, we can sample at both endpoints and choose the smaller value as the lower bound.

Lemma 5. *The prune process is safe.*

Proof. Consider a subproblem P_i. If the lower bound $S_i^{min} \geq S^{min}$, then the subproblem that produces S^{min} can give us a solution at least as good as the best possible solution given by P_i. Hence, there is no need to consider P_i for optimization.

4   Conclusion

In this paper, we discussed a facility location problem in weighted regions. We proved that if the slope of the line L is fixed, we can compute the optimal solution in $O(ln \log ln)$ time, where n is the number vertices of the weighted subdivision and l is the number of fixed input points. For the general case, in which the slope of L is not fixed, we showed that the problem can be reduced to $O(l^2 n^2)$ 1-dimensional global optimization problems, and proposed a prune-and-search scheme to effectively deal with those problems.

References

1. Amato, N.M., Goodrich, M.T., Ramos, E.A.: Computing the arrangement of curve segments: Divide-and-conquer algorithms via sampling. In: Proc. 11th Annual CAM-SIAM Symposium on Discrete Algorithms, 2000, pp. 705–706 (2000)
2. Chen, D.Z., Daescu, O., Hu, X., Wu, X., Xu, J.: Determining an optimal penetration among weighted regions in two and three dimensions. J. combinat. Optim. 5(1), 59–79 (2001)
3. Chen, D.Z., Hu, X., Xu, J.: Computing Beams in Two and Three Dimensions. Journal of Combinatorial Optimization 7, 111–136 (2003)
4. Daescu, O.: Improved optimal weighted links algorithms. In: Proceedings of ICCS, 2nd International Workshop on Computational Geometry and Applications, pp. 227–233 (2002)
5. Daescu, O., Palmer, J.: Minimum Separation in Weighted Subdivisions. The International J. Computat. Geom. & Appl. (Submitted, 2003) (Manuscript)
6. Michel, X.: Cooperative facility location games. In: Proceedings of the eleventh annual ACM-SIAM symposium on Discrete algorithms SODA 2000(2000)
7. Megiddo, N., Tamir, A.: Finding Least-Distance Lines. SIAM J. ALG. DISC. MATH. 4(2) (1983)
8. Mitchell, J.S.B., Papdimitriou, C.H.: The weighted region problem: Finding shortest paths through a weighted planer subdivision. Journal of the ACM 38(1), 18–73 (1991)
9. Preparata, F.P., Shamos, M.I.: Computational Geometry: An Introduction. Springer, New York (1985)

Algorithms for Temperature-Aware Task Scheduling in Microprocessor Systems

Marek Chrobak[1], Christoph Dürr[2], Mathilde Hurand[2],
and Julien Robert[3]

[1] Department of Computer Science, University of California,
Riverside, CA 92521
Supported by NSF grants OISE-0340752 and CCF-0729071
[2] CNRS, LIX UMR 7161, Ecole Polytechnique, France
Supported by ANR Alpage
[3] Laboratoire de l'Informatique du Parallélisme,
ENS Lyon, France

Abstract. We study scheduling problems motivated by recently developed techniques for microprocessor thermal management at the operating systems level. The general scenario can be described as follows. The microprocessor temperature is controlled by the hardware thermal management system that continuously senses the chip temperature and automatically reduces the processor's speed as soon as the thermal threshold is exceeded. Some tasks are more CPU-intensive than other and thus generate more heat during execution. The cooling system operates nonstop, reducing (at an exponential rate) the deviation of the processor's temperature from the ambient temperature. As a result, the processor's temperature, and thus the performance as well, depends on the order of the task execution. Given a variety of possible underlying architectures, models for cooling and for hardware thermal management, as well as types of tasks, this gives rise to a plethora of interesting and never studied scheduling problems.

We focus on scheduling real-time jobs in a simplified model for cooling and thermal management. A collection of unit-length jobs is given, each job specified by its release time, deadline and heat contribution. If, at some time step, the temperature of the system is t and the processor executes a job with heat contribution h, then the temperature at the next step is $(t+h)/2$. The temperature cannot exceed the given thermal threshold τ. The objective is to maximize the throughput, that is, the number of tasks that meet their deadlines. We prove that in the offline case computing the optimum schedule is NP-hard, even if all jobs are released at the same time. In the online case, we show a 2-competitive deterministic algorithm and a matching lower bound.

1 Introduction

Background. The problem of managing the temperature of processor systems is not new; in fact, the system builders had to deal with this challenge since the

R. Fleischer and J. Xu (Eds.): AAIM 2008, LNCS 5034, pp. 120–130, 2008.

inception of computers. Since early 1990s, the introduction of battery-operated laptop computers and sensor systems highlighted the related issue of controlling the energy consumption.

Most of the initial work on these problems was hardware and systems oriented, and only during the last decade substantial progress has been achieved on developing models and algorithmic techniques for microprocessor temperature and energy management. This work proceeded in several directions. One direction is based on the fact that the energy consumption is a fast growing function of the processor speed (or frequency). Thus we can save energy by simply slowing down the processor. Here, algorithmic research focussed on *speed scaling* – namely dynamically adjusting the processor speed over time to optimize the energy consumption while ensuring that the system meets the desired performance requirements. Another technique (applicable to the whole system, not just the microprocessor) involves *power-down strategies*, where the system is powered-down or even completely turned off when some of its components are idle. Since changing the power level of a system introduces some overhead, scheduling the work to minimize the overall energy usage in this model becomes a challenging optimization problem.

Models have also been developed for the processor's thermal behavior. Here, the main objective is to ensure that the system's temperature does not exceed the so-called *thermal threshold*, above which the processor cannot operate correctly, or may even be damaged. In this direction, techniques and algorithms have been proposed for using speed-scaling to optimize the system's performance while maintaining the temperature below the threshold.

We refer the reader to the survey by Irani and Pruhs [4], and references therein, for more in-depth information on the models and algorithms for thermal and energy management.

Temperature-aware Scheduling. The above models address energy and thermal management at the micro-architecture level. In contrast, the problem we study in this paper addresses the issue of thermal management at the operating systems level. Most of the previous work in this direction focussed on multi-core systems, where one can move tasks between the processors to minimize the maximum temperature [8,1,2,5,6,7,3,9]. However, as it has been recently discovered, even in single-core systems one can exploit variations in heat contributions among different tasks to reduce the processor's temperature through appropriate task scheduling [1,3,5,6,10]. In this scenario, the microprocessor temperature is controlled by the hardware dynamic thermal management (DTM) systems that continuously senses the chip temperature and automatically reduces the processor's speed as soon as the thermal threshold (maximum safe operating temperature) is exceeded. Typically, the frequency is reduced by half, although it can be further reduced to one fourth or even one eighth, if needed. Once the chip cools down to below the threshold, the frequency is increased again. In addition, the cooling system operates non-stop, reducing the deviation of the processor's temperature from the ambient temperature at an exponential rate. The general accepted model stipulates that there is a constant time T, such that after T time units, the cooling system divided by 2 the difference between the processor's temperature and the ambient temperature. In this work we assume that this constant is exactly the length of a unit length job. The general case will be considered in the full version of this paper.

Different tasks use different microprocessor units in different ways; in particular, some tasks are more CPU-intensive than other. As a result, the processor's thermal behavior – and thus the performance as well – depends on the order of the task execution. In particular, Yang *et al.* [10] point out that the accepted model for the microprocessor thermal behavior implies that, given two jobs, scheduling the "hotter" job before the "cooler" one, results in a lower final temperature. They take advantage of this phenomenon to improve the performance of the OS scheduler.

With multitudes of possible underlying architectures (for example, single- vs. multi-core systems), models for cooling and hardware thermal management, as well as types of jobs (real-time, batch, etc.), this gives rise to a plethora of interesting and never yet studied scheduling problems.

Our Model. We focus on scheduling real-time jobs in a somewhat simplified model for cooling and thermal management. The time is divided into unit time slots and each job has unit length. These jobs represent unit slices of the processes present in the OS scheduler's queue. We assume that the heat contributions of these jobs are known. This is counterintuitive, but reasonably realistic, for, as discussed in [10], these values can be well approximated using appropriate prediction methods.

In our thermal model we assume, without loss of generality, that the ambient temperature is 0 and that the heat contributions are expressed in the units of temperature (that is, by how much they would increase the chip temperature in the absence of cooling). Suppose that at a certain time the processor temperature is t and we are about to execute a job with heat contribution h in the next time slot. In reality [10], during the execution of this job, its heat contribution is spread over the whole slot and so is the effect of cooling; thus, the final temperature can be expressed using an integral function. In this paper, we use a simplified model where we first take into account the job's heat contribution, and then apply the cooling, where the cooling simply reduces the temperature by half.

Finally, we assume that only one processor frequency is available. Consequently, if there is no job whose execution does not cause a thermal violation, the processor must stay idle through the next time slot.

Our Results. Summarizing, our scheduling problem can be now formalized as follows. A collection of unit-length jobs is given, each job j with a release time r_j, deadline d_j and heat contribution h_j. If, at some time step, the temperature of the system is t and the processor executes a job j, then the temperature at the next step is $(t + h_j)/2$. The temperature cannot exceed the given thermal threshold τ. The objective is to compute a schedule which maximizes the number of tasks that meet their deadlines.

We prove that in the offline case computing the optimum schedule is NP-hard, even if all jobs are released at the same time. In the online case, we show a 2-competitive deterministic algorithm and a matching lower bound.

2 Terminology and Notation

The input consists of n unit-length jobs that we number $1, 2, ..., n$. Each job j is specified by a triple (r_j, d_j, h_j), where r_j is its release time, d_j is the deadline and h_j is its heat contribution. The time is divided into unit-length slots and each job can be executed in any time slot in the interval $[r_j, d_j]$. The system temperature is initially 0 and it changes according to the following rules: if the temperature of the system at a time u is t and the processor executes j then the temperature at time $u + 1$ is $(t + h_j)/2$. The temperature cannot exceed the given thermal threshold τ that we assume to be 1. Thus if $(t + h_j)/2 > 1$ then j cannot be executed at time u. Idle slots are treated as executing a job with heat contribution 0, that is, after an idle slot the temperature decreases by half.

Given an instance, as above, the objective is to compute a schedule with maximum *throughput*, where throughput is defined as the number of completed jobs. Extending the standard notation for scheduling problems, we denote the offline version of this problem by $1|r_i, h_i| \sum U_i$.

In the online version, denoted $1|\text{online-}r_i, h_i| \sum U_i$, jobs are available to the algorithm at their release time. Scheduling decisions of the algorithm cannot depend on the jobs that have not yet been released.

3 The NP-Completeness Proof

The special case of the problem $1|r_i, h_i| \sum U_i$, when all jobs have the same release time and same deadline is denoted as $1|h_i|C_{\max}$: The objective value $C_{\max}$ stands for minimizing the maximum completion time of the jobs when there are no deadlines to respect, and the decision version of this optimization problem is exactly deciding if there is a feasible schedule where all jobs complete before a given common deadline.

We can show that this problem is NP-complete. Why is it interesting? In fact, one common approach in designing on-line algorithms for scheduling is to compute the optimal schedule for the pending jobs, and use this schedule to make the decision as to what execute next. The set of pending jobs forms an instance where all jobs have the same release time. This does not necessarily mean that the above method cannot work (assuming that we do not care about the running time of the online algorithm), but it makes this approach much less appealing, since reasoning about the properties of the pending jobs is likely to be very difficult.

Theorem 1. *The offline problem* $1|h_i|C_{\max}$ *is NP-hard.*

Proof. The reduction is from NUMERICAL 3 DIMENSIONAL MATCHING. In the later problem we are given 3 sequences A, B, C of n non-negative integers each and an integer m. A 3-dimensional matching is a set of n triplets $(a, b, c) \in A \times B \times C$ such that each number is matched exactly once. The problem consists in deciding if there is a 3-dimensional matching, such that all matches (a, b, c) satisfy $a + b + c = m$.

Without loss of generality, we can assume (A1) that every $x \in A \cup B \cup C$ satisfies $x \leq m$ and (A2) that $\sum_{x \in A \cup B \cup C} x = mn$.

In this reduction we need two constants. Let be $\alpha = 1/13$ and $\epsilon = (1/8 - 3\alpha)/m > 0$. We also define the function $f : x \mapsto \alpha + \epsilon x$.

We construct an instance to the heating problem. In total there will be $4n + 1$ jobs, and all are released at time 0 and have deadline $4n + 1$. These jobs will be of two types:

1. First we have $3n$ jobs that correspond to the instance of NUMERICAL 3 DIMENSIONAL MATCHING.
   - For every $a \in A$, there is a job of heat contribution $8f(a)$.
   - For every $b \in B$, there is a job of heat contribution $4f(b)$.
   - For every $c \in C$, there is a job of heat contribution $2f(c)$.
   We call these respectively A-, B- and C-jobs.
2. Next, we have $n + 1$ "gadget" jobs. The first of these jobs has heat contribution 2, and the remaining ones 1.75. We call these respectively 2- and 1.75-jobs.

The idea of the construction is that the gadget jobs are so hot, that they need to be scheduled every 4-th time unit, separating the time into n blocks of 3 consecutive time slots each. Now every remaining job has a heat contribution that consists in two parts. A constant part and a tiny part depending on the instance of the matching problem. The constant part is so large that in every block there is a single A-, B- and C-job and they are scheduled in that order. This defines a 3-dimensional matching. Now the gadget jobs are so hot, that they force every matching (a, b, c) to satisfy $a + b + c \leq m$. Let's make this argument formal.

Suppose there is a solution to the matching problem. We show that there is a solution to the heating problem. Schedule the job 2 at time 0, and all other gadget jobs every 4th time slot. Now the remaining slots are grouped into blocks consisting of 3 consecutive time slots. Each i-th triplet (a, b, c) from the matching is associated to the i-th block, and the corresponding A-,B- and C-jobs are executed in there in the order A, B, C, see figure 1. By construction every job meets its deadline, it remains to show that at no point the temperature exceeds 1. The non-gadget jobs have all heat contribution smaller than 1, by assumption (A1), so every execution of a non-gadget job would preserve that the temperature is not more than 1. Now we show by induction that right after

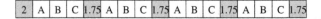

Fig. 1. The structure of the solution of the scheduling problem

the execution of a gadget job, the temperature is exactly 1. This is clearly the case after execution of the first job, since its heat contribution is 2. Now let u be the time when a 1.75-job is scheduled, and suppose that at time $u - 3$ the temperature was 1. Let (a, b, c) be the matching associated to the slots between $u - 3$ and u. Then at time u the temperature is

$$\frac{1}{8} + \frac{8f(a)}{8} + \frac{4f(b)}{4} + \frac{2f(c)}{2} = \frac{1}{8} + 3\alpha + (a + b + c)\epsilon = \frac{1}{8} + 3\alpha + m\epsilon = \frac{1}{4}.$$

This shows that at time $u + 1$ the temperature is again 1. We conclude that the schedule is feasible.

Now we show the remaining direction in the NP-hardness proof, namely that if there is a solution to the scheduling problem, then there is a solution to the matching problem. To that purpose, suppose that there is a solution to the scheduling problem. We first show that it has the form of figure 1. Since all $4n + 1$ jobs have deadline $4n + 1$, all jobs must be scheduled without any idle time between time 0 and $4n + 1$. This means that the gadget job of heat contribution 2 must be scheduled at time 0, because that is the only moment of temperature 0.

Now we claim that all 1.75-jobs have to be scheduled every 4-th time unit. For a proof by contradiction assume that at some times u and $u + 3$ two gadget jobs are scheduled. The lightest job that can be scheduled at times $u + 1, u + 2$ would be a C-job of heat contribution $2f(0) = 2\alpha$. The temperature at time $u + 1$ is at least $1.75/2 = 7/8$. Therefore the temperature at time $u + 3$ is at least

$$\frac{7}{32} + \frac{3}{2}\alpha = \frac{7}{32} + \frac{3}{26} > 1/4.$$

This contradicts that a 1.75-job is scheduled at time $u+3$, since the temperature would exceed 1 at $u + 4$. So we can conclude that the 1.75-jobs are scheduled every 4-th time unit. This partitions the remaining time slots into blocks of 3 time slots each.

We show now that every block contains exactly one A-job, one B-job and one C-job, in that order. For a proof by contradiction assume that some A-job is scheduled at the 2nd position of some block. Its heat contribution is at least $8f(0)$. The coolest jobs scheduled at position 1 and 3 have heat contribution $2f(0) = 2\alpha$ at least. Then the temperature at the end of the block is at least

$$\frac{7}{64} + \frac{\alpha}{4} + 2\alpha + \alpha = \frac{7}{32} + \frac{13}{4}\alpha = \frac{7}{32} + \frac{1}{4} > 1/4.$$

This would be too hot for the following 1.75-job. The same argument shows that A-jobs cannot be scheduled at position 3 in a block, and therefore the 1st position of a block is always occupied by an A-job. Now we show that a B-job cannot be scheduled at the 3rd position of some block. This is because otherwise the temperature at the end of the block would be again at least

$$\frac{7}{64} + \alpha + \frac{\alpha}{2} + 2\alpha = \frac{7}{32} + \frac{7}{2}\alpha > 1/4.$$

We showed that every block contains jobs that correspond to some matching $(a, b, c) \in A \times B \times C$. It remains to show that each matching satisfies $a+b+c = m$. Let (a_i, b_i, c_i) be the matching corresponding to the i-th block, for $1 \leq i \leq n$. Let be $t_i = (3\alpha - (a_i + b_i + c_i)\epsilon)/2 + 7/8$. Also let be $t_0 = 1$. Then for $1 \leq k \leq n$ the temperature at the end of time $4k + 1$ is

$$\sum_{i=1}^{k} \left(\frac{1}{16}\right)^{k-i} t_i \leq 1, \tag{1}$$

For convenience we write $p_i = 15/16 - t_i$. Assumption (A2) said that the average of $a_i + b_i + c_i$ over $1 \leq i \leq n$ is m. Therefore the average of t_i (excluding t_0) is $15/16$, and therefore we have

$$\sum_{i=1}^{n} p_i = 0. \tag{2}$$

Furthermore from (1) we have

$$\forall 1 \leq k \leq n : \sum_{i=1}^{k} 16^i p_i \leq 0. \tag{3}$$

We will show now that for every i, $p_i = 0$. This would imply that $a_i + b_i + c_i = m$, and imply that the matching problem has a solution. For a proof by contradiction suppose that (p_i) is not all zero. Let ℓ be the smallest index such that $p_\ell > 0$ and

$$p_1 + \ldots + p_\ell \geq 0. \tag{4}$$

By minimality of ℓ we have for every $2 \leq k \leq \ell - 1$

$$p_1 + \ldots + p_{k-1} \leq 0 \quad \text{and} \quad p_k + \ldots + p_\ell \geq 0$$

So we can increase the coefficients of the variables $p_k, \ldots, p_\ell$ by some same amount and preserve (4), obtain

$$16p_1 + 16^2 p_2 + \ldots 16^2 p_{\ell-1} + 16^2 p_\ell \geq 0$$

and finally

$$16p_1 + 16^2 p_2 + \ldots 16^{\ell-1} p_{\ell-1} + 16^{\ell-1} p_\ell \geq 0.$$

Now using $p_\ell > 0$, we obtain

$$16p_1 + 16^2 p_2 + \ldots 16^{\ell-1} p_{\ell-1} + 16^\ell p_\ell > 0,$$

which contradicts (4). This completes the proof.   □

4   An Online Competitive Algorithm

In this section we show that there is a 2-competitive algorithm. We will show, in fact, that a large class of deterministic algorithms is 2-competitive.

Given a schedule, we will say that a job j is *pending* at time u if j is released, not expired (that is, $r_j \leq u < d_j$) and j has not been scheduled before u. If the temperature at time u is t and u is pending, then we call j *available* if $t + h_j \leq 2$, that is, j is not too hot to be executed.

We say that job j *dominates* job k if j is both cooler and has the same or smaller deadline than k, that is $h_j \leq h_k$ and $d_j \leq d_k$. Also we say that j *dominates strictly* job k if at least one of the inequalities is strict. An online algorithm is called *reasonable* if at each step (i) it schedules a job whenever if is available (the *non-waiting property*), and (ii) it schedules a pending job that is not strictly dominated by another pending job. The class of reasonable algorithms contains, for example, the algorithm which schedules the coolest available job, and the algorithm which schedules the earliest deadline available job.

Theorem 2. *Any reasonable alg. for* $1|online\text{-}r_i, h_i| \sum U_i$ *is 2-competitive.*

Proof. We define a charging scheme that maps jobs executed by the adversary to jobs executed by the algorithm in such a way that no job in the algorithm's schedule gets more than two charges.

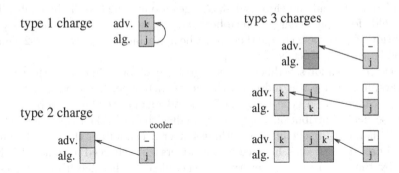

Fig. 2. the different types of charges

Fix an instance, and consider both the schedule produced by the algorithm and some arbitrary schedule, say produced by some adversary. Suppose that the adversary schedules j at time v. There will be three types of charges.

Type 1 Charges: If the algorithm schedules a job k at time v, charge j to k. Otherwise, we are idle, and we have two cases.

Type-2 Charges: Suppose that the algorithm is hotter than the adversary at time v and cooler at $v + 1$. In this case we charge j to the previous *heating* step, by which we mean a time $u < v$, where either the algorithm schedules a job which is strictly hotter than the one scheduled by the adversary, or where only the algorithm schedules something and the adversary is idle.

Type-3 Charges: Suppose now that the algorithm is hotter than the adversary at $v + 1$ or cooler at v (and therefore also at $v + 1$). Now we argue that j has been scheduled by the algorithm at some earlier time u: In the case where the algorithm was cooler at v, j would have been available and we assumed the algorithm is non-waiting. Now consider the case where the algorithm was hotter than the adversary at time $v + 1$. Its temperature is at most $1/2$ at that time, so the temperature of j can be at most 1. So again it would have been executable at time v. This observation is also true for any job cooler than j, and this will be useful later.

So let $u < v$ be the execution time of j by the algorithm. To find a job that we can charge j to, we perform *pointer chasing* of sorts: If, at time u, the adversary is idle or schedules an equal or hotter job, then we charge j to itself (its "copy" in the algorithm's schedule). Otherwise, let k be the strictly cooler job scheduled by the adversary at u. Now we claim that the algorithm schedules k at some time before v. Indeed, if it were still pending at u, since the algorithm never schedules a dominated job, its deadline is not before the one of j, in particular it is after v. By our earlier observation it would have been executable at v, so at v job k is not pending anymore. We iterate to the time the algorithm executes k. This iteration will end at some point, since we deal with cooler and cooler jobs.

Now we show that any job scheduled by the algorithm will get at most two charges. Obviously, each job in the algorithm's schedule gets at most one type-1 charge. Every chain defining a type-3 charge is uniquely defined by the first considered job, and therefore type-3 charges are assigned to distinct jobs. This also holds for type-2 charges, since between any two time steps that satisfy the condition of the type-2 charge there must be a heating step, so the type-2 charges are assigned to distinct heating steps.

Now let k be a job scheduled by the algorithm at some time v. By the previous paragraph, k can get at most one charge of each type. If the adversary is idle at v, k cannot get a type-1 charge. If the adversary schedules a job j at time v whose heat contribution is equal or smaller than k's then k cannot get the type-2 charge. If it schedules something hotter it cannot get the type-3 charge.

So in total every job scheduled by the adversary is charged to some job scheduled by the algorithm, and every job scheduled by the algorithm receives no more than 2 charges, therefore the competitive ratio is not more than 2.   □

5   A Lower Bound on the Competitive Ratio

Theorem 3. *Every deterministic online algorithm for $1|online\text{-}r_i, h_i| \sum U_i$ has competitive ratio at least 2.*

Proof. We (the adversary) release a job $j \to (r_j, d_j, h_j) = (0, 3, 1)$. If the online algorithm schedules it at time 0, we release a tight job $k \to (1, 2, 1.7)$ and schedule it followed by j. If the algorithm does not schedule j at time 0, then we do schedule it at 0 and release (and schedule) a tight job $k' \to (2, 3, 1.7)$ at time 2. In both cases we scheduled two jobs, while the algorithm scheduled only one, completing the proof.   □

6   Final Comments

Many open problems remain. Perhaps the most intriguing one is to determine the randomized competitive ratio for the problem we studied. The proof of Theorem 3 can easily be adapted to prove the lower bound of 1.5, but we have not been able to improve the upper bound of 2; this is, in fact, the main focus of our current work on this scheduling problem.

Extensions of the cooling model can be considered, where the temperature after executing j is $(t + h_j)/R$, for some $R > 1$. Even this formula, however, is only a discrete approximation for the true model (see, for example, [10]), and it would be interesting to see if the ideas behind our 2-competitive algorithm can be adapted to these more realistic cases.

In reality, thermal violations do not cause the system to idle, but only to reduce the frequency. With frequency reduced to half, a unit job will execute for two time slots. Several frequency levels may be available.

We assumed that the heat contributions are known. This is counterintuitive, but not unrealistic, since the "jobs" in our model are unit slices of longer jobs. Prediction methods are available that can quite accurately predict the heat contribution of each slice based on the heat contributions of the previous slices. Nevertheless, it may be interesting to study a model where exact heat contributions are not known.

Other types of jobs may be studied. For real-time jobs, one can consider the case when not all jobs are equally important, which can be modeled by assigning weights to jobs and maximizing the weighted throughput. For batch jobs, other objective functions can be optimized, for example the flow time.

One more realistic scenario would be to represent the whole processes as jobs, rather then their slices. This naturally leads to scheduling problems with preemption and with jobs of arbitrary processing times. When the thermal threshold is reached, the execution of a job is slowed down by a factor of 2. Here, a scheduling algorithm may decide to preempt a job when another one is released or, say, when the processor gets too hot.

Finally, in multi-core systems one can explore the migrations (say, moving jobs from hotter to cooler cores) to keep the temperature under control. This leads to even more scheduling problems that may be worth to study.

References

1. Bellosa, F., Weissel, A., Waitz, M., Kellner, S.: Event-driven energy accounting for dynamic thermal management. In: Workshop on Compilers and Operating Systems for Low Power (2003)
2. Choi, J., Cher, C.-Y., Franke, H., Hamann, H., Weger, A., Bose, P.: Thermal-aware task scheduling at the system software level. In: International Symposium on Low Power Electronics and Design, pp. 213–218 (2007)
3. Gomaa, M., Powell, M.D., Vijaykumar, T.N.: Heat-and-run: leveraging smt and cmp to manage power density through the operating system. SIGPLAN Not. 39(11), 260–270 (2004)

4. Irani, S., Pruhs, K.R.: Algorithmic problems in power management. SIGACT News 36(2), 63–76 (2005)
5. Martonosi, M., Donald, J.: Techniques for multicore thermal management: Classification and new exploration. In: Proceedings of the International Symposium on Computer Architecture, pp. 78–88 (2006)
6. Kumar, A., Shang, L., Peh, L.-S., Jha, N.K.: HybDTM: a coordinated hardware-software approach for dynamic thermal management. In: DAC 2006: Proceedings of the 43rd Annual Conference on Design Automation, pp. 548–553 (2006)
7. Kursun, E., Cher, C.-Y., Buyuktosunoglu, A., Bose, P.: Investigating the effects of task scheduling on thermal behavior. In: The 3rd Workshop on Temperature-Aware Computer Systems (2006)
8. Merkel, A., Bellosa, F.: Balancing power consumption in multiprocessor systems. SIGOPS Oper. Syst. Rev. 40(4), 403–414 (2006)
9. Moore, J., Chase, J., Ranganathan, P., Sharma, R.: Making scheduling "cool": temperature-aware workload placement in data centers. In: ATEC 2005: Proceedings of the USENIX Annual Technical Conference 2005 on USENIX Annual Technical Conference, pp. 5–5 (2005)
10. Yang, J., Zhou, X., Chrobak, M., Zhang, Y.: Dynamic thermal management through task scheduling. In: IEEE International Symposium on Performance Analysis of Systems and Software (to appear, 2008)

Engineering Comparators for Graph Clusterings*

Daniel Delling, Marco Gaertler, Robert Görke, and Dorothea Wagner

Faculty of Informatics, Universität Karlsruhe (TH)
{delling,gaertler,rgoerke,wagner}@informatik.uni-karlsruhe.de

Abstract. A promising approach to compare two graph clusterings is based on using measurements for calculating the distance between them. Existing measures either use the structure of clusterings or quality-based aspects with respect to some index evaluating both clusterings. Each approach suffers from conceptional drawbacks. We introduce a new approach combining both aspects and leading to better results for comparing graph clusterings. An experimental evaluation of existing and new measures shows that the significant drawbacks of existing techniques are not only theoretical in nature but manifest frequently on different types of graphs. The evaluation also proves that the results of our new measures are highly coherent with intuition, while avoiding the former weaknesses.

1 Introduction

Finding groups of similar elements in datasets, a technique known as clustering, is an important problem in the analysis and exploration of data. There are numerous applications such as data mining [8], network analysis [1], and biochemistry [16]. While recent research [2,3] focused on measuring the quality of a given clustering of an underlying graph, the problem of *comparing* two graph clusterings becomes more and more important.

There exists a mutual relation between the two concepts quality and distance: One could use a quality index to obtain a distance measure as shown later, while measuring the distance of a given clustering to an "optimal" clustering could yield the quality of the clustering. Current techniques for the comparison of clusterings use only qualitative aspects or transfer existing measures from the field of data mining. Both approaches have certain drawbacks: When comparing clusterings by using qualitative aspects the results are highly dependent on the used quality measure and completely different clusterings may yield the same quality value and are thus indicated as equal. Measures originating from data mining only consider the partition of nodes and ignore the structure of graphs. Due to these conceptional disadvantages, investigated below, the introduction of new measures seems inevitable, using structural and qualitative properties of the clusterings to calculate an appropriate distance. We present a new approach combining structural properties and qualitative aspects. In order to achieve this,

---

* This work was partially supported by the DFG under grant WA 654/14-3 and EU under grant DELIS (contract no. 001907).

R. Fleischer and J. Xu (Eds.): AAIM 2008, LNCS 5034, pp. 131–142, 2008.

we extend data mining measures by adding qualitative features and introduce a new promising measure having its origin in quality measurement. Due to the high complexity of comparing clusterings we focus on the case of static comparison, i.e., the graph is unchanged, but give an outlook on the dynamic case. An experimental evaluation is presented, showing that the drawbacks of data mining measures are not only theoretical in nature but manifest often.

This paper is organized as follows. Section 2 introduces preliminaries and existing measures for comparing (data-)clusterings, including their drawbacks. Two approaches for constructing new measures are presented in Section 3. An evaluation based on artificial data of all presented measures is given in Section 4, while Section 5 shows the applicability of our approach in a real-world scenario. Section 6 concludes this paper.

2   Preliminaries

We assume that $G = (V, E)$ is an undirected, unweighted and connected graph. Let $n := |V|, m := |E|$, and $\mathcal{C} := \{C_1, \ldots, C_p\}$ a partitioning of V. We call $\mathcal{C}$ a clustering and the C_i clusters of the graph. The set of all possible clusterings is $\mathcal{A}(V)$. Let $E(\mathcal{C}) := \{\{u, v\} \in E \mid u, v \in C_i\}$ be the set of *intra-cluster edges* of $\mathcal{C}$ and $\overline{E}(\mathcal{C}) := \{\{u, v\} \in E \mid u \in C_i, v \in C_j, i \neq j\}$ the set of *inter-cluster edges* of $\mathcal{C}$. The cardinalities are indicated by $m(\mathcal{C}) := |E(\mathcal{C})|$ and $\overline{m}(\mathcal{C}) := |\overline{E}(\mathcal{C})|$. We call a graph with disjoint cliques a *clustergraph* and $F_{\mathcal{C}}$, the set of edges to be added or deleted in order to transform a given graph and clustering $\mathcal{C}$ into an according clustergraph, the *cluster editing set* of $\mathcal{C}$. When comparing two clusterings we use $\mathcal{C}$ and $\mathcal{C}'$, with $k := |\mathcal{C}|, l := |\mathcal{C}'|$. With $\deg(C_i) := \sum_{v \in C_i} \deg(v)$ we indicate the sum of all degrees of nodes within a cluster. All presented measures are given in a distance version, normalized to the interval $[0, 1]$. In the following, we give a short overview of existing comparison techniques. Among them are measures based on quality and on comparing the partitions of node-sets, the latter are also called *node-structural*.

Quality-Based Distance. Quality-based measurements can be constructed by comparing the scores of the two clusterings with respect to an arbitrary quality index such as coverage, performance or modularity [1,3]. Note, that a distance measured in such a way is highly dependent on the used index. Furthermore, completely different clusterings can yield the same value. Thus, we neglect purely quality-based distances in the following and focus on measuring the distance based on the structure of the clusterings.

Counting Pairs. In [17] some techniques based on counting pairs are presented. Summarizing, every pair of nodes is categorized based on whether they are in the same (or different) cluster with respect to both clusterings. Four sets are defined: S_{11} (S_{00}) is the set of unordered pairs that are in the same (different) clusters under both clusterings, whereas S_{01} (S_{10}) contains all pairs that are in the same cluster under C (C') and in different under C' (C). In the following we present two representatives for this class: *Rand* and *adjusted Rand* measure. Rand introduced the distance function $\mathcal{R}$ given in Equation 1 in [12], it suffers from several drawbacks. For example, it is highly dependent on the number of clusters. One attempt to remedy some of these drawbacks, which is known as *adjusted Rand* $\mathcal{AR}$ and given in Equation 1, is to subtract the expected value for clusterings with a hypergeometric distribution of nodes, see [11].

$$\mathcal{R}(C, C') := 1 - \frac{2(n_{11} + n_{00})}{n(n-1)} \quad , \quad \mathcal{AR}(C, C') := 1 - \frac{n_{11} - t_3}{\frac{1}{2}(t_1 + t_2) - t_3} \quad , \quad (1)$$

where $t_1 := n_{11} + n_{10}$, $t_2 := n_{11} + n_{01}$, and $t_3 := (2t_1 t_2)/(n(n-1))$ and t_1 (t_2) is the cardinality of all pairs of nodes that are in the same cluster under C (C').

Overlaps. Another counting approach is based on the $k \times l$ confusion matrix $CM := (m_{ij})$ whose ij-entry indicates how many elements are in Cluster C_i and C'_j, formally $m_{ij} := |C_i \cap C'_j|$, for $1 \leq i \leq k$ and $1 \leq j \leq l$. Several measures are based on the confusion matrix. We restrict ourselves to the measure $\mathcal{NVD}$, introduced by van Dongen in [15], given in Equation 2. Other measures suffer from the obvious disadvantage of asymmetries, thus we exclude them. We use a normalized version to keep the measure to the interval $[0, 1]$.

$$\mathcal{NVD}(C, C') := 1 - \frac{1}{2n} \sum_{i=1}^{k} \max_{j} m_{ij} - \frac{1}{2n} \sum_{j=1}^{l} \max_{i} m_{ij} \quad (2)$$

One major drawback of $\mathcal{NVD}$ is that the distance between the two trivial clusterings, i.e., $k = 1, l = n$, only yields a value of about 0.5. In addition, this measure suffers from the drawback that only the maximum overlaps contribute, resulting counter-intuitive examples are given in [10].

Information Theory. More promising approaches are based on information theory [4]. Informally, the *entropy* $\mathcal{H}(C)$ of a clustering is the uncertainty of a randomly picked node belonging to a certain cluster. An entropy of a clustering is always positive and is bounded by $\log_2(n)$, see [13]. An extension of entropy is the *mutual information* $\mathcal{I}(C, C')$. The mutual information of two clusterings is the loss of uncertainty of one clustering if the other is given. With $P(i) := |C_i|/n$ and $P(i, j) := (|C_i \cap C'_j|)/n$, entropy and mutual information are defined as follows.

$$\mathcal{H}(C) := - \sum_{i=1}^{k} P(i) \log_2 P(i) \quad , \quad \mathcal{I}(C, C') := \sum_{i=1}^{k} \sum_{j=1}^{l} P(i, j) \log_2 \frac{P(i, j)}{P(i)P(j)} \quad (3)$$

Note that mutual information is positive and bounded by $\min\{\mathcal{H}(C), \mathcal{H}(C')\} \leq \log_2(n)$. In the following we present two representatives in this class, namely one introduced by Fred & Jain [6] and *Variation of Information*, introduced by Meila [9].

$$\mathcal{FJ}(\mathcal{C},\mathcal{C}') := \begin{cases} 1 - \frac{2\mathcal{I}(\mathcal{C},\mathcal{C}')}{\mathcal{H}(\mathcal{C})+\mathcal{H}(\mathcal{C}')} & \text{, if } \mathcal{H}(\mathcal{C}) + \mathcal{H}(\mathcal{C}') \neq 0 \\ 0 & \text{, otherwise} \end{cases} \tag{4}$$

$$\mathcal{VI}(\mathcal{C},\mathcal{C}') := \mathcal{H}(\mathcal{C}) + \mathcal{H}(\mathcal{C}') - 2\mathcal{I}(\mathcal{C},\mathcal{C}') \tag{5}$$

The first measure $\mathcal{FJ}$, given in Equation 4, is a normalized version of the mutual information and stated as a distance function. The case differentiation is used to deal with the degenerated case of two trivial clusterings, i.e.,$k = l = 1$.

The second measure $\mathcal{VI}$ is motivated by an axiomatic approach and given in Equation 5. In [10], it is shown that $\mathcal{VI}$ is the only measure fulfilling several axioms. However, these axioms seem to be inadequate in the special case of graph clustering. According to these axioms, the movement of a node v from one cluster C_i to another cluster C_j must be equivalent to first splitting v off from C_i and then merging it with C_j. Figure 1 shows an example regarding this axiom: intuitively $d(\mathcal{C},\mathcal{C}'')$ should be greater than $d(\mathcal{C},\mathcal{C}')+d(\mathcal{C}',\mathcal{C}'')$ of which both terms represent minor changes, but according to the axiom $d(\mathcal{C},\mathcal{C}'') = d(\mathcal{C},\mathcal{C}') + d(\mathcal{C}',\mathcal{C}'')$ must hold. This measure is not normalized and the two possible normalization factors, which are $1/\log_2(n)$ and $1/\log_2(\max\{k,l\})$, mapping to the intervals $[0,x], x \leq 1$ and $[0,1]$ respectively, have significant drawbacks. Nevertheless, we use the $\log_2(n)$ normalized version for comparability with the other measures.

Drawbacks of the Data Mining Approach. All node-structural measures suffer from the same drawback that they neglect the structure of the graph. The examples in Figure 2 clarify this circumstance. The figure shows four clusterings $\mathcal{C}_1,\mathcal{C}_1',\mathcal{C}_2$ and $\mathcal{C}_2'$ on two graphs G_1 and G_2. A measure d not considering the structure of the graphs fulfills $d(\mathcal{C}_1,\mathcal{C}_1') = d(\mathcal{C}_2,\mathcal{C}_2')$. Intuitively, the distance $d(\mathcal{C}_1,\mathcal{C}_1')$ has to be greater than $d(\mathcal{C}_2,\mathcal{C}_2')$ since the quality of $\mathcal{C}_1$ is almost equal to that of $\mathcal{C}_2$, but $\mathcal{C}_1'$ has far lower quality than $\mathcal{C}_2'$. This drawback can become arbitrarily grave when the edge set of the graph is allowed to change.

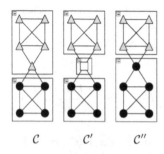

$$\mathcal{C} \qquad \mathcal{C}' \qquad \mathcal{C}''$$

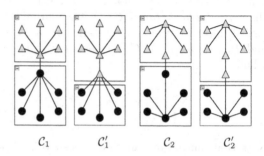

$$\mathcal{C}_1 \qquad \mathcal{C}_1' \qquad \mathcal{C}_2 \qquad \mathcal{C}_2'$$

Fig. 1. The sum of two minor changes result in a major one

Fig. 2. Two static comparisons of graph clusterings

3   Engineering Graph-Structural Comparison Measures

In order to remedy some of the disadvantages of node-structural measures, we introduce the concept of *graph-structural* measures. Since they are also based on the underlying graph structure, they can include qualitative aspects for measuring the distance of two clusterings. In the first part, Section 3.1, we extend node-structural measures, while a novel measure is introduced in the second part, Section 3.2.

3.1   Extension of Node-Structural Measures

For consistency, all extended measures should meet the following requirement: If the underlying graph is complete, then both the graph- and node-structural version should yield the same value, since then the graph structure does not provide additional information. A second objective is to adjust the three founding principles—counting pairs, overlaps and information theory—of the existing measures themselves, instead of adjusting each implementation separately.

Counting Local Pairs. Instead of categorizing every pair we only consider those pairs, that are connected by an edge. For $a, b \in \{0, 1\}$ we define $E_{ab} := S_{ab} \cap E$ and $e_{ab} := |E_{ab}|$. It is obvious that $S_{ab} = E_{ab}$ holds for complete graphs. Thus, we obtain the graph-based versions $\mathcal{R}_g$ and $\mathcal{AR}_g$ of the Rand and adjusted Rand measure given in Equation 6:

$$\mathcal{R}_g(\mathcal{C}, \mathcal{C}') := 1 - \frac{e_{11} + e_{00}}{m} \ , \quad \mathcal{AR}_g(\mathcal{C}, \mathcal{C}') := 1 - \frac{e_{11} - t_3}{\frac{1}{2}(m(\mathcal{C}) + m(\mathcal{C}')) - t_3} \ , \quad (6)$$

where $t_3 := (m(\mathcal{C})m(\mathcal{C}'))/m$. Note, that $m(\mathcal{C}) = e_{11} + e_{10}$ and $m(\mathcal{C}') = e_{11} + e_{01}$, respectively, hold.

Degree-Based Overlaps. Measures based on overlaps can be transformed into graph-structural measures by a slight modification in the definition of the confusion matrix as follows. The ij-th entry of the *degree-based confusion matrix* $CM^d := (m_{ij}^d)$ indicates the sum of the degrees of the nodes that are both in C_i and C', formally $m_{ij}^d := \deg(C_i \cap C_j')$. Note, that if G is d-regular graph, then the equality $CM = CM^d/d$ holds. In certain cases, this may lead to different normalization factors. The extension of $\mathcal{NVD}$ is given in Equation 7.

$$\mathcal{NVD}_g(\mathcal{C}, \mathcal{C}') := 1 - \frac{1}{4m} \sum_{i=1}^{k} \max_j m_{ij}^d - \frac{1}{4m} \sum_{j=1}^{l} \max_i m_{ij}^d \quad (7)$$

The equivalence of the node- and the graph-structural variant of the normalized van Dongen measure for regular graphs follows from $m = dn/2$ and $m_{ij} = m_{ij}^d/d$.

Edge Entropy. The entropy defined in Section 2 solely depends on the node-set, thus we extend it to the edge-set using the following paradigm: Instead of randomly picking a node from the graph for measuring the uncertainty, we pick the end of an edge randomly. As a consequence, a node with high degree has a greater impact on the distance. The formal definition of *edge entropy* $\mathcal{H}_E$ and *edge mutual information* $\mathcal{I}_E$ is given in Equation 8 and 9.

$$\mathcal{H}_E(\mathcal{C}) := -\sum_{i=1}^{k} P_E(i) \log_2 P_E(i) \ , \tag{8}$$

$$\mathcal{I}_E(\mathcal{C},\mathcal{C}') := \sum_{i=1}^{k}\sum_{j=1}^{l} P_E(i,j) \log_2 \frac{P_E(i,j)}{P_E(i)P_E(j)} \ , \tag{9}$$

where $P_E(i) := \deg(C_i)/2m$ and $P_E(i,j) := \deg(C_i \cap C_j')/2m$. Note that for regular graphs, the entropy and the edge entropy coincide. The extensions of $\mathcal{FJ}$ and $\mathcal{VI}$ are given in Equation 10 and 11.

$$\mathcal{FJ}_g(\mathcal{C},\mathcal{C}') := \begin{cases} 1 - \frac{2\mathcal{I}_E(\mathcal{C},\mathcal{C}')}{\mathcal{H}_E(\mathcal{C})+\mathcal{H}_E(\mathcal{C}')} & , \text{if } \mathcal{H}_E(\mathcal{C}) + \mathcal{H}_E(\mathcal{C}') \neq 0 \\ 0 & , \text{otherwise} \end{cases} \tag{10}$$

$$\mathcal{VI}_g(\mathcal{C},\mathcal{C}') := \mathcal{H}_E(\mathcal{C}) + \mathcal{H}_E(\mathcal{C}') - 2\mathcal{I}_E(\mathcal{C},\mathcal{C}') \tag{11}$$

The equivalence of the node- and the graph-structural variant for regular graphs results from the equality of entropy and edge entropy for complete graphs. Meila showed in [10] that $\mathcal{VI} \leq \log_2(n)$ also holds for weighted clusterings. Since the degree of a node can be interpreted as node-weight our $\log_2(n)$-normalization maps to the interval of $[0, 1]$.

3.2   A Novel Approach for Measuring Graph-Structural Distance

Although the extensions introduced in the previous section incorporate the underlying graph structure, they are not suitable for comparing clusterings on different graphs. As a first step to solve this task, we consider the restriction to graphs with the same node-set, but potentially different edge-sets. Motivated by the cluster editing set, we introduce the *editing set difference* defined in Equation 12.

$$\mathcal{ESD}(\mathcal{C},\mathcal{C}') = \frac{|F_\mathcal{C} \cup F_{\mathcal{C}'}| - |F_\mathcal{C} \cap F_{\mathcal{C}'}|}{|F_\mathcal{C} \cup F_{\mathcal{C}'}|} = 1 - \frac{|F_\mathcal{C} \cap F_{\mathcal{C}'}|}{|F_\mathcal{C} \cup F_{\mathcal{C}'}|} \tag{12}$$

Small cluster editing sets correspond to significant clusterings. By comparing the two clusterings with a geometric difference, we obtain an indicator for the structural difference of the two clusterings. It easy to see, that in the case of static comparison, $\mathcal{ESD}$ is a metric.

4   Experiments and Evaluation

We evaluate the introduced measures on two setups. The first focuses on structural properties of clusterings, the second concentrates on qualitative aspects:

Initial and Random Clusterings. The tests consist of two comparisons, each including clusterings with the same expected intrinsic structure of the partitions, i. e.,the expected number of clusters and the size of clusters. The first comparison uses one significant clustering and one uniformly random clustering, while the second one uses two uniformly random clusterings.

Local Minimization. The setup consists of two parts, each comparing a reference clustering with a clustering of less significance. The two parts differ in the significance of the reference clustering.

The intuition of the first test is to clarify the drawbacks of the node-structural measures, while the second setup verifies the obtained results. We use the *attractor generator* introduced in [5] which uses geometric properties based on Voronoi Diagrams to generate initial clusterings. The Voronoi cells represent clusters and the maximum Euclidean distance of two nodes being connected is determined by a perturbation parameter. All tests use $n = 1000$ nodes and are repeated until the maximal length of the 0.95-confidence intervals is not larger than 0.1.

4.1   Initial- and Random Clusterings

The generated clustering is used as a significant clustering. For the random clustering we first pick k uniformly at random between 2 and $\sqrt[3]{n}$ for the number of clusters and assign each node uniformly at random to the k clusters. Figures 3.1 and 3.2 show the measured quality by the indices coverage, performance and modularity [1,3]. The tests consists of two cases. On the one hand, the comparison of the generated and a random clustering (GvR) and on the other hand, the comparison of two random clusterings (RvR). A measure for comparing graph clusterings should differ in the two cases. For GvR, a suitable measure should indicate a decreasing distance with the loss of significance of the reference, while for RvR two interpretations are possible. On the one hand, one could claim that the distance between two random clusterings should be independent of the underlying graphs. On the other hand, the distance should decrease with the loss of significance because two random clusterings on an almost complete graph are closer to each other than on a graph with an existing significant clustering. Another interpretation seems acceptable as well: The distance of a given clustering to a random clustering should always be somehow maximal.

Figure 3 shows the results for the node- and graph-structural measures. By comparing Figure 3.3 and 3.4 it is evident that node-structural measures do not distinguish the two cases. Only Fred & Jain and adjusted Rand reflect the interpretation that the distance to a random clustering is always maximal. However, the situation changes for the graph-structural distance (Figures 3.5 and 3.6). Only Rand and $\mathcal{ESD}$ capture the difference, while the remaining measures show nearly the same behavior as their node-structural counterparts. For GvR, the distance measured by Rand is decreasing with increasing density while for RvR the distance is invariant under the density. Furthermore, the measured distance equals the node-structural measurement for RvR. $\mathcal{ESD}$ has the same behavior for GvR as Rand, whereas RvR reflects the intuition that two random clusterings become more similar with loss of significance. Under the assumption that

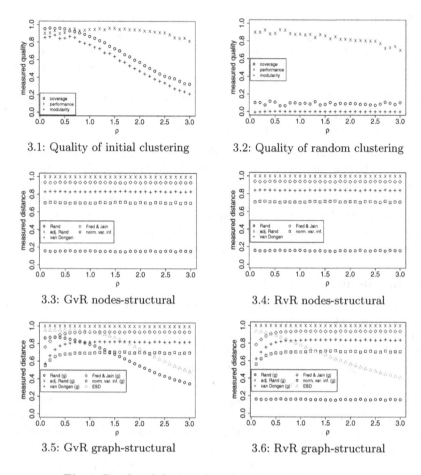

3.1: Quality of initial clustering

3.2: Quality of random clustering

3.3: GvR nodes-structural

3.4: RvR nodes-structural

3.5: GvR graph-structural

3.6: RvR graph-structural

Fig. 3. Results of the initial- and random clustering setup

a comparison to a random clustering should always be interpreted as maximal, adjusted Rand and Fred & Jain can be accepted. Nevertheless, the equivalence of the node- and the graph-structural versions of van Dongen and the normalized Variation of Information is counterintuitive. This partly originates from the fact, that attractors produce graphs that are close to regular for $\varrho > 0.5$. Furthermore, the clusters are equal in size. The strange behavior of Fred & Jain, van Dongen and the variation of Information for very small ϱ stems from the fact that for small ϱ attractors are nearly stargraphs with k centers.

4.2   Local Minimization

Since there are several possible interpretations of graph-structural distance and the structural similarity of the clusterings in Section 4.1 a second test is executed having a precise intuition for graph-structural distance. Again, as a reference

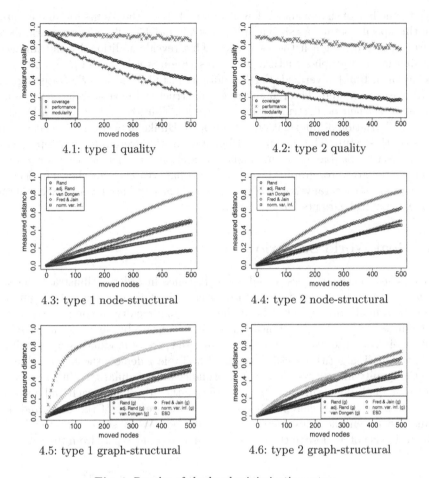

Fig. 4. Results of the local minimization setup

clustering we use the generated clustering of an attractor graph. The second clustering of less significance is obtained from the reference clustering by locally moving nodes from one cluster to another. Such a shift is executed, if it maximally decreases a given index among all possible shifts. This is done until no decrease of quality can be achieved or the number of moved nodes has reached a maximum value of $M_{\max}$.

In this setup, we use modularity as the index, the density is set to the values $\varrho = 0.5$ (type 1) and $\varrho = 2.5$ (type 2), and $M_{\max}$ increases from 0 to 500 using steps of 5. Figures 4.1 and 4.2 show the measured quality of the locally decreased clusterings on increasing number of moved nodes. Note, that for $M_{\max} = 0$ the reference and the locally decreased clustering coincide. A suitable distance measure should first of all distinguish the two cases. In addition, with increasing $M_{\max}$ the measured distance in type 2 should be smaller than in type 1, since the intuition is that in type 1 a very significant clustering is destroyed while on type 2 the loss of significance is lower. Figure 4 shows the results for all measures on this specific setup. As shown in Figures 4.3 and 4.4, all node-structural measures hardly distinguish the two cases. This reveals additional disadvantages. Evaluating the graph-structural measures (Figures 4.5 and 4.6), the intuitive behavior of Rand is verified. Furthermore, adjusted Rand and $\mathcal{ESD}$ distinguish both cases very well. The remaining graph-structural measures show the same behavior as their node-structural counterparts. Thus, the failure of van Dongen and the Variation of Information is confirmed. Unlike in Section 4.1 Fred & Jain fails on this setup. The unexpected behavior of the overlap and entropy based measures may be due to—as mentioned in Section 4.1—the fact that for $\varrho = 0.5$ and $\varrho = 2.5$ attractor graphs have a fairly regular structure. As shown in Section 2 the graph-structural versions of overlap- and entropy-based measures equal the node-structural variants for regular graphs.

5   Real-World Scenario

In this section, we discuss a real-world instance in order to illustrate the advantages of graph-structural measures over node-structural ones. As input, we use the e-mail graph (Figure 5) of the Karlsruhe faculty of computer science, introduced in [7]. As a reference clustering, we group by departments. We additionally compute two clusterings by using the greedy modularity approach [3] and the MCL algorithm, introduced in [14]. Table 1 depicts the scores achieved by the quality measures coverage, performance and modularity. With respect to all three quality measures, MCL outperforms the greedy approach and achieves a score close to the reference. Table 2 gives an overview of the measured distances between the abovementioned clusterings. We observe that the MCL-clustering is not as close to the reference than one could expect from the figures in

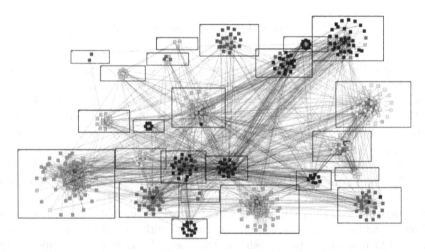

Fig. 5. Karlsruhe e-mail graph. Groups refer to the reference clustering, colors to the clustering obtained by the greedy modularity algorithm.

Table 1. Quality scores achieved by the reference clustering and those computed by the greedy approach and by MCL. The input is the Karlsruhe e-mail graph.

	reference	greedy	MCL
coverage	0.8173	0.8634	0.8182
performance	0.9387	0.8286	0.9238
modularity	0.7423	0.6725	0.7282

Table 2. Measured distances between reference and two computed clusterings. One clustering is obtained by MCL, the other one by the greedy modularity algorithm. The input is the Karlsruhe e-mail graph (cf. Figure 5).

measure type	measure	reference vs. greedy	reference vs. MCL	greedy vs. MCL
quality	modularity difference	0.0697	0.0140	0.0557
node-structural	Rand	0.1233	0.0463	0.1466
	adj. Rand	0.5765	0.3555	0.6549
	van Dongen	0.2676	0.1834	0.3465
	Fred & Jain	0.3137	0.1794	0.3876
	variation of information	0.2425	0.1658	0.2904
graph-structural	Rand	0.1963	0.1305	0.2452
	adj. Rand	0.4689	0.2820	0.5730
	van Dongen	0.2435	0.1714	0.3215
	Fred & Jain	0.2828	0.1623	0.3581
	variation of information	0.2107	0.1427	0.2549
	ESD	0.7325	0.5382	0.7796

Table 1. All graph-structural distance measures indicate a difference of more than 0.1. More interestingly, $\mathcal{ESD}$ yields a lower score than graph-structural adjusted Rand. For artificial data, the contrary is true (cf. Section 4). Most of our graph-structural measures indicate a lower distance between all clusterings than their node-structural versions. As all clusterings score similar quality values, and thus have quite a low distance with respect to quality, the graph-structural measures really incorporate qualitative aspects. Hence, they harmonize better with intuition than the purely node-structural versions. As discussed in Section 2, the node-structural Rand measure yields a very small value due to the high number of small cluster. However, this drawback appears to be remedied by the graph-structural version.

6   Conclusion

The experimental evaluation confirms the drawbacks of node-structural measures while some graph-structural measures, i. e., $\mathcal{ESD}$, adjusted Rand, and Rand perform more consistently with intuition. Furthermore, this is an indicator for the feasibility of the graph-structural distance in applications such as dynamic graph clustering. More precisely, since graph-structural measures incorporate both structural and qualitative aspects, they can be used as a foundation for clustering in dynamic scenarios. Summarizing, extensions of node-structural measures are not trivial and need not lead to intuitive results. Furthermore, our presented extensions are only suitable for comparing clusterings on the same graph. In contrast, the editing set distance only requires the same node-set. Thus, this improves the foundation for dynamic graph clusterings. Concluding, this work is a first step towards a unifying comparison framework.

References

1. Brandes, U., Erlebach, T. (eds.): Network Analysis. LNCS, vol. 3418. Springer, Heidelberg (2005)
2. Brandes, U., Gaertler, M., Wagner, D.: Experiments on Graph Clustering Algorithms. In: Di Battista, G., Zwick, U. (eds.) ESA 2003. LNCS, vol. 2832, pp. 568–579. Springer, Heidelberg (2003)
3. Clauset, A., Newman, M., Moore, C.: Finding community structure in very large networks. Physical Review E 70(066111) (2004)
4. Cover, T., Thomas, J.: Elements of Information Theory. John Wiley & Sons, Inc., Chichester (1991)
5. Delling, D., Gaertler, M., Wagner, D.: Generating Significant Graph Clusterings. In: European Conference of Complex Systems (ECCS 2006) (2006)
6. Fred, A., Jain, A.: Robust Data Clustering. In: IEEE Computer Society Conference on Computer Vision and Pattern Recognition, CVPR, pp. 128–136 (2003)
7. Gaertler, M., Görke, R., Wagner, D.: Significance-Driven Graph Clustering. In: Kao, M.-Y., Li, X.-Y. (eds.) AAIM 2007. LNCS, vol. 4508, pp. 11–26. Springer, Heidelberg (2007)
8. Jain, A., Dubes, R.: Algorithms for Clustering Data. Prentice Hall, Englewood Cliffs (1988)
9. Meila, M.: Comparing Clusterings by the Variation of Information. In: 16th Annual Conference of Computational Learning Theory (COLT), pp. 173–187 (2003)
10. Meila, M.: Comparing Clusterings - An Axiomatic View. In: 22nd International Conference on Machine Learning, Bonn, Germany, pp. 577–584 (2005)
11. Morey, R., Agresti, A.: The Measurement of Classification Agreement: An Adjustment to the RAND Statistic for Chance Agreement. Educational and Psychological Measurement 44, 33–37 (1984)
12. Rand, W.: Objective criteria for the evaluation of clustering methods. Journal of the American Statistical Association 66, 846–850 (1971)
13. Strehl, A., Ghosh, J.: Cluster ensembles – a knowledge reuse framework for combining multiple partitions. J. Mach. Learn. Res. 3, 583–617 (2003)
14. van Dongen, S.: A cluster algorithm for graphs. Tech. Report INS-R0010 (2000)
15. van Dongen, S.: Performance criteria for graph clustering and markov cluster experiments. Technical Report INS-R0012, National Research Institute for Mathematics and Computer Science in the Netherlands, Amsterdam (May 2000)
16. Vidal, M.: Interactome modeling. FEBS Lett. 579, 1834–1838 (2005)
17. Wagner, S., Wagner, D.: Comparing Clusterings – An Overview. Technical Report 2006-4, Faculty of Informatics, Universität Karlsruhe (TH) (2006)

On the Fast Searching Problem

Danny Dyer[1], Boting Yang[2], and Öznur Yaşar[1]

[1] Department of Mathematics and Statistics,
Memorial University of Newfoundland
{dyer,oznur}@math.mun.ca
[2] Department of Computer Science, University of Regina
boting@cs.uregina.ca

Abstract. Edge searching is a graph problem that corresponds to cleaning a contaminated graph using the minimum number of searchers. We define *fast searching* as a variant of this widely studied problem. Fast searching corresponds to an internal monotone search in which every edge is traversed exactly once and searchers are not allowed to jump. We present a linear time algorithm to compute the fast search number of trees. We investigate the fast search number of bipartite graphs. We also propose a general cost function for evaluating search strategies that utilizes both edge searching and fast searching.

1 Introduction

Assume that we want to secure a system of tunnels from a hidden intruder. We can model this system as a graph where junctions correspond to vertices and tunnels correspond to edges. We will launch a group of searchers into the system in order to catch the intruder. Let $G = (V, E)$ denote a graph where V is its vertex set and E is its edge set. In this paper, we only consider finite connected simple graphs. Definitions omitted here can be found in [13]. We assume that every edge of G is contaminated initially and our aim is to clean the whole graph by a sequence of steps. First we place a given set of k searchers on a subset of V (allowing multiple searchers to be placed on any vertex). Each step of an edge search strategy consists of either: removing a searcher from one vertex and placing it on another vertex (a "jump"), or sliding a searcher from a vertex along an edge to an adjacent vertex.

An *edge search strategy* is a combination of the moves defined above so that the state of all edges being simultaneously clean is achieved, in which case we say that the graph is *cleaned*. The problem becomes cleaning the graph using the fewest searchers. The least number of searchers needed to clean the graph is the *edge search number* of the graph and is denoted $s(G)$. All models of searching in this paper are variations of edge searching, so we will normally omit the term "edge".

If we are not allowed to remove a searcher from the graph, then we have an *internal search strategy*, in which each move is to slide a searcher from a vertex along an edge, to an adjacent vertex.

R. Fleischer and J. Xu (Eds.): AAIM 2008, LNCS 5034, pp. 143–154, 2008.

If a searcher slides along an edge $e = uv$ from u to v, then e is cleaned if either (i) another searcher is stationed at u, or (ii) all other edges incident to u are already clean. If a searcher is stationed at a vertex v, then we say that v is *guarded*. If a path does not contain any guarded vertex, then it is called an *unguarded path*. If there is an unguarded path that contains one endpoint of a contaminated edge and one endpoint of a cleaned edge e, then e gets *recontaminated*.

If we insist that once an edge becomes clean it must be kept clean until the end of the searching strategy, then such a strategy will be called *monotonic*. If, on the other hand, the set of clean edges induces a connected subgraph of G after each step of the strategy, then the strategy will be a *connected* one.

One definition of an efficient way to clear a graph would be to do so in the least number of steps. In particular, each edge would be traversed exactly once. We define this new version of edge searching to be *fast searching*. A *fast search strategy* for a graph $G = (V, E)$ is a sequence of $|E|$ moves that clear G. The *fast search number* of G is the least number of searchers for which a *fast search strategy* exists, and is denoted $s_f(G)$. Consequently, this must be an internal monotone search where no edge is traversed more than once.

The edge searching problem is an extensively studied graph theoretical problem. Its origins date back to the late 1960s in works of Parsons [12] and Breisch [5]. It was first faced by a group of spelunkers who were trying to find a person lost in a system of caves. They were interested in the minimum number of people they needed in the searching team.

It has been shown [4,10] that forcing a search to be monotonic does not change the search number, hence we can always assume that edge search strategies are monotonic. The relationships between the different strategies mentioned above is examined in [3]. For a recent survey on graph searching and its variants, see [8].

The problem and its variants are related to many applications such as network security [2]. It has strong connections with the cutwidth of a graph which arises in VLSI design [6]. Because of its closeness with the layout problems, it is related to graph parameters such as pathwidth [9].

This paper is organized as follows. In Section 2 we compare fast searching with other searching models. In Section 3 we show that the fast search number of trees can be found in linear time. In Section 4 we investigate the fast search number of bipartite graphs. We give a motivation for fast searching in Section 5 involving a more generalized cost function. We conclude with some related open problems in Section 6.

2   Properties

For any graph G, it is immediate from the definitions that $s(G) \leq s_f(G)$. The gap $s_f(G) - s(G)$ is zero for some graphs. For K_n, the complete graph on n vertices, it is straightforward to show that $s_f(K_n) = s(K_n) = n$ if $n \geq 4$. Furthermore, it is known that the edge search number is critical for complete graphs, that is, deleting a single edge from K_n will reduce the search number by one. We can show that this also holds for the fast search number.

Proposition 1. *For any $e \in E(K_n)$, $s_f(K_n - e) = n - 1$.*

On the other hand, $s_f(G) - s(G)$ can be arbitrarily large for some graphs. Consider $K_{1,n}$, the complete bipartite graph with bipartitions of size 1 and n. Then $s(K_{1,n}) = 2$ whereas $s_f(K_{1,n}) = \lceil \frac{n}{2} \rceil$.

Furthermore, there are graphs for which the connected fast search number may be much larger than the fast search number. Construct G' by subdividing every edge of $K_{1,n}$. Then $s(G') = 2, s_f(G') = \lceil \frac{n}{2} \rceil$ whereas the connected fast search number is $n - 1$. Notice that this graph has a much smaller order than the one in [14].

Let S denote a fast search strategy for G. Assume that λ denotes one of the searchers used in S. We say that v is the start vertex for λ, if λ is initially placed on v according to the strategy S. We say that u is the end vertex for λ, if λ stops at u (and never moves again).

Let V_o be the set of vertices in G with odd degree. Observe that for each vertex $v \in V_o$, there exists a searcher for which v is either a start or end vertex. Otherwise one of the edges incident to v would be traversed at least twice. Since it is possible that a searcher starts at a vertex in V_o and stops at another vertex in V_o, we have the following:

Lemma 1. *If V_o is the set of odd degree vertices in a graph G, then $s_f(G) \geq |V_o|/2$.*

The theory built on graph minors plays an important role in edge searching. Most variants of edge searching are minor closed. However this is not true for fast searching. Figure 1 shows an example where G has a smaller fast search number than its subgraph H. Hence fast searching is not even subgraph closed. Notice that H has more odd vertices than G and due to Lemma 1 this result is easy to achieve. On the other hand, the following example shows an instance where the subgraph has much more even vertices than the original graph, but has a much higher fast search number.

Example 1. Let K_n be a complete graph on n vertices where n is odd and $n \geq 5$. Let $G = K_n \square K_2$ be the cartesian product of K_n and K_2. Let H be the graph obtained from G by deleting $n - 1$ edges between the two maximal cliques in G. Thus, all vertices in G have odd degree n, and in H, two vertices have odd degree n and all other $2n - 2$ vertices have even degree $n - 1$. We can show that

 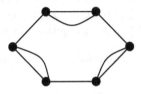

Fig. 1. The graph G on the left and its subgraph H on the right where $s_f(G) = 3$ and $s_f(H) = 5$

$n \leq s_f(G) \leq n+2$ and $s_f(H) = 2n-1$. Thus, the difference $s_f(H) - s_f(G)$ can be arbitrarily large. Even more interesting, let $G = K_n \square P_m$, where P_m is a path on $m \geq 2$ vertices, and H be the graph that is a path of m copies of K_n where any two consecutive K_n are connected by single edge. Then we can show that $s_f(G) \leq n+m$ and $s_f(H) = nm - m + 1$. Hence, the ratio $s_f(H)/s_f(G)$ can be arbitrarily large.

3   Trees

For any graph $G = (V, E)$, since $\sum_{v \in V} \deg(v) = 2|E|$, we know that the number of odd degree vertices in G must be even. For a tree T, the following algorithm TP(T) partitions T into edge disjoint paths such that the two end vertices of each path have odd degree in T.

Algorithm TP(T)

1. Initially $i = 1$.
2. Arbitrarily select a leaf u in T. Let $uv \in E(T)$ be the edge incident on u. Mark uv with color i and set $T \leftarrow T - uv$.
3. If v is an isolated vertex in T, then go to Step 4; otherwise, let $vw \in E(T)$ be an edge incident on v. Mark vw with color i, $T \leftarrow T - vw$, $v \leftarrow w$, and go to Step 3.
4. If there is no leaf in T, then stop; otherwise, set $i \leftarrow i + 1$ and go to Step 2.

Lemma 2. *The Algorithm TP(T) defines a valid coloring for T and the edges with the same color form a path such that the two end vertices of this path have odd degree.*

Proof. From Step 2, after an edge is colored, then it is deleted from the graph. Thus every edge is colored at most once. Since the algorithm keeps running as long as there is a leaf, every edge must be colored. Hence every edge is colored exactly once. Let P_i be the graph formed by all the edges with color i. From Step 3, we know that P_i is a path. From Steps 2 and 3, we know that both end vertices of P_1 have degree 1. Note that after we delete each edge of P_1 from T, the odd-even degree state of each vertex in T does not change except for the end vertices of P_1. Thus, a leaf vertex chosen by the algorithm as a start vertex for P_2 corresponds to either a leaf in the original T or a vertex with odd degree in T, and similarly, the other end vertex of P_2 found by the algorithm also corresponds to either a leaf in the original T or a vertex with odd degree in T. This holds for every $P_i, i \geq 2$. Hence both end vertices of every path have odd degree in T.   □

Using the decomposition of Algorithm TP, we can compute a fast search strategy as follows.

Algorithm FS(T)

1. Call TP(T). Let k be the number of colors used by TP(T) and P_i, $1 \leq i \leq k$, be the path formed by all edges with color i.

2. For each P_i, $1 \leq i \leq k$, place searcher λ_i on one end of P_i.

3. Arbitrarily select a searcher, say λ_i, on a leaf, and slide it along P_i until it stops at a vertex v whose degree is 1 in T, or whose degree is more than 2 in T and which contains only the searcher λ_i.

4. Update T by deleting from T all edges cleared by λ_i in Step 3 after λ_i slides along these edges of P_i. If T contains no edges, then stop; otherwise, go to Step 3.

We now prove the correctness of Algorithm FS.

Lemma 3. *In Algorithm FS(T), if T has at least one edge in Step 4, then when the procedure goes back to Step 3, there must exist a path P_i such that the searcher λ_i on a vertex $u \in V(P_i)$ can clear an edge of P_i incident on u by sliding along this edge.*

Proof. Suppose that there is one moment at which T has at least one edge and every searcher is located on an isolated vertex or a vertex of degree more than 2 which is occupied by only one searcher. For any searcher λ_i, $1 \leq i \leq k$, located on an isolated vertex, we know that all the edges of P_i has been cleared and deleted. Let F be the forest obtained from T by deleting all isolated vertices. Note that each searcher on F must be located on a vertex that is occupied by only one searcher and is incident with at least two different colored edges. Since every connected component in F contains at least one searcher, we know that each pair of leaves in F are incident on different colored edges. Thus, the number of different colors in F is greater than or equal to the number of leaves in F. Since the degree of every occupied vertex is more than 2, we know that the number of leaves in F is greater than the number of searchers on F. Hence, the number of different colors in F is greater than the number of searchers on F. This contradicts the fact that each path of edges with the same color in F must have one searcher.   □

From the above two algorithms, we can show the fast search number of a tree.

Theorem 1. *If T is a tree and V_o is the set of vertices in T with odd degree, then $s_f(T) = |V_o|/2$.*

Proof. Let k be the number of colors used by TP(T). It follows from Lemma 2 that TP(T) decomposes T into k edge disjoint paths P_i, $1 \leq i \leq k$, such that the two end vertices of each path have odd degree in T. From Lemma 3, we can use k searchers to clear T in such a way that each path P_i is cleared by one searcher sliding from one end of the path to the other end. Thus, $s_f(T) \leq k \leq |V_o|/2$. On the other hand, it follows from Lemma 1 that $s_f(T) \geq |V_o|/2$. Therefore, $s_f(T) = |V_o|/2$.   □

Theorem 2. *Both Algorithm TP(T) and Algorithm FS(T) can be implemented with linear time.*

Proof. By modifying the depth-first-search algorithm, we know that Algorithm TP(T) can be implemented with linear time. For Algorithm FS(T), Steps 1 and 2 take linear time. In Step 3, after a searcher slides along an edge, we can delete this edge immediately. Since each edge is deleted exactly once, we know that Steps 3 and 4 take linear time.   □

We can construct an infinite family of trees for which the gap between the search number and the fast search number is big.

Example 2. Let $T_1 = K_{1,3}$. Take three copies of T_1 and choose a vertex of degree one from each copy. Identify these three vertices to construct T_2. Continue in an inductive fashion. Thus T_k is constructed from three copies of T_{k-1} by identifying a vertex of degree one from each copy.

By Parson's lemma in [12], we can show that $s(T_k) = k+1$. Also, each vertex in T_k are of degree 1 or 3. Thus $V(T_k) = V_o(T_k)$. Since $|V(T_k)| = 3^k + 1$, by Theorem 1 we have $s_f(T_k) = \frac{3^k+1}{2}$.

4   Bipartite Graphs

In this section we consider complete bipartite graphs $K_{m,n}, m \leq n$ and characterize their fast search number. We start with the following initial cases where $m \leq 2$ or $m = 4$.

Lemma 4. *If* $1 \leq m \leq n$, *then*

$$s_f(K_{m,n}) = \begin{cases} \left\lceil \frac{n}{2} \right\rceil, & m = 1, \\ 2, & m = n = 2, \\ 3, & m = 2, n \geq 3, \\ 6, & m = 4, n \geq 4. \end{cases}$$

Theorem 3. *[1] If* $3 \leq m \leq n$, *then* $s(K_{m,n}) = m + 2$.

The next lemma gives an upper bound for even m where $m \geq 6$.

Lemma 5. *If* $6 \leq m \leq n$ *and* m *is even, then* $s_f(K_{m,n}) \leq m + 3$.

Proof. Let $K_{m,n}$ have bipartition $V_1 = \{v_1, v_2, \ldots, v_m\}$ and $V_2 = \{u_1, u_2, \ldots, u_n\}$. We construct a fast search strategy that uses $m + 3$ searchers.

Denote the searchers as $\lambda_1, \lambda_2, \ldots, \lambda_{m+3}$. We clean the vertices in the order $u_1, u_2, \ldots, u_{n-1}, v_1, v_2, \ldots, v_m, u_n$ when n is even, and $u_1, u_2, \ldots, u_{n-2}, v_4, v_5, \ldots, v_m, u_{n-1}, v_2, v_1, u_n, v_3$ when n is odd.

First, place λ_1 on v_1, λ_2 on v_2 and λ_3 on v_3. Then place $\lambda_4, \lambda_5, \ldots, \lambda_{m+3}$ on u_1 and then clean u_1 by these searchers so that λ_4 will be on v_1, λ_5 will be on $v_2, \ldots$, and λ_{m+3} will be on v_m. We keep $\lambda_4, \lambda_5, \ldots, \lambda_{m+3}$ on their respective vertices until just before the end of the strategy. We clean $u_2, u_3, \ldots, u_{n-1}$ by moving λ_1, λ_2 and λ_3.

Slide λ_1 along the edge $e = v_1u_2$, λ_2 along the edge $e = v_2u_2$ and λ_3 along the edge $e = v_3u_3$. Keeping λ_2 on u_2 and λ_3 on u_3, we clean all the edges between $\{v_4, v_5, \ldots, v_m\}$ and $\{u_2, u_3\}$ by λ_1. Hence λ_1 visits the following vertices in the given order: $u_2, v_4, u_3, v_5, u_2, v_6, u_3, \ldots, v_{m-1}, u_2, v_m, u_3$. After this λ_1 will end up on u_3 since the graph induced by $\{v_4, v_5, \ldots, v_m\}$ and $\{u_2, u_3\}$ contains an Eulerian path and $|\{v_4, v_5, \ldots, v_m\}|$ is odd. Next, λ_1 will slide along u_3v_1. The only dirty edge incident to u_2 is u_2v_3, hence λ_2 may slide along u_2v_3 and clean u_2. Similarly λ_3 may slide along u_3v_2 and clean u_3. Hence u_2 and u_3 are cleaned.

Next we clean u_4 and u_5 similarly. Slide λ_1 along the edge $e = v_1u_4$, λ_2 along $e = v_3u_4$ and λ_3 along the edge $e = v_2u_5$. Keeping λ_2 on u_4 and λ_3 on u_5, we clean all the edges between $\{v_4, v_5, \ldots, v_m\}$ and $\{u_4, u_5\}$ by λ_1. Then, λ_1 slides along u_5v_1, λ_2 slides along u_4v_2 and λ_3 slides along u_5v_3.

In the same way we clean u_6 and u_7, u_8 and u_9, and so on. If n is even, after cleaning u_{n-2} and u_{n-1} we let λ_i slide along $e = v_{i-3}u_n$ cleaning v_{i-3} for all $i = 4, 5, \ldots, m+3$. In this way we finish cleaning u_n. When n is odd, we follow a similar strategy. $\qquad\square$

The next theorem shows that the bound in Lemma 5 is best possible.

Theorem 4. *If* $6 \leq m \leq n$ *and* m *is even, then* $s_f(K_{m,n}) = m + 3$.

Proof. Lemma 5 states that $s_f(K_{m,n}) \leq m + 3$. From Theorem 3 we know that $m + 2 = s(K_{m,n}) \leq s_f(K_{m,n})$, we only need to show that $m + 2$ searchers do not suffice to fast search $K_{m,n}$. We will use proof by contradiction. Suppose that there exists a fast search strategy to clear $K_{m,n}$ that uses $m + 2$ searchers. Let $K_{m,n}$ have bipartition V_1 and V_2 with $|V_1| = m$ and $|V_2| = n$. Let u_1 be the first cleared vertex and t be the step at which u_1 is cleared. We now consider the case that $u_1 \in V_2$. The case when $u_1 \in V_1$ can be proved similarly.

Since at t, only u_1 is cleared and all vertices in V_1 are dirty, each vertex of V_1 must be occupied by at least one searcher. Let us denote these searchers as $\lambda_3, \lambda_4, \ldots, \lambda_{m+2}$. This accounts for m searchers. We show that no strategy can clean the graph using $m + 2$ searchers by considering each placement of the searchers λ_1 and λ_2.

CASE 1. Both of λ_1 and λ_2 are on V_1.

CASE 1.1: Suppose that λ_1 and λ_2 are on the same vertex, say v_1. Since fast searching is a monotone search and none of the searchers are located on $u_2, u_3, \cdots, u_n$, we see that all edges incident to these vertices are contaminated. Since there are no parallel edges, we can slide λ_1 and λ_2 only to different vertices in V_2, say u_2 and u_3. After this step it is not possible to move any of the searchers since they are all on vertices that have more than one contaminated edge incident to them.

CASE 1.2: Suppose that λ_1 and λ_2 are on different vertices, say v_1 and v_2 respectively. First assume that λ_1 and λ_2 slide to the same vertex, say u_2, at steps $t+1$ and $t+2$ respectively. Then λ_2 may leave u_2 and slide to any vertex other then v_1 or v_2 in V_1, say v_3 at step $t+3$. After this step λ_1 is stuck on u_2. At step $t+4$, λ_2 may slide to any vertex other then u_1, u_2 in V_2 and it is stuck at that vertex.

Otherwise, let λ_1 and λ_2 slide to different vertices, say u_2 and u_3. But since u_2 and u_3 have at least two contaminated edges incident to them, neither λ_1 nor λ_2 can move. Therefore all searchers are stuck at step $t+2$.

CASE 2. Both of λ_1 and λ_2 are on V_2.

CASE 2.1: Suppose that λ_1 and λ_2 are both on u_1. None of the searchers can move since u_1 is clean and no edge is traversed twice.

CASE 2.2: Suppose that λ_1 and λ_2 are on the same vertex $u_2 \neq u_1$. If u_2 has more than two contaminated edges incident to it, then all searchers are stuck after at most two steps since only one of the searchers, say λ_1 can leave u_2, λ_2 is stuck after this step and λ_1 is stuck at step $t + 2$. If u_2 has exactly two contaminated edges incident to it, then u_2 can be cleaned in the next two steps. At the end of step $t + 2$, λ_1 and λ_2 are located on different vertices, say v_1 and v_2 in V_1. If both of λ_1 and λ_2 move to different vertices in V_2, then they are both stuck. Hence assume that λ_1 slides along v_1u_3 and λ_2 slides along v_2u_3. Since all edges incident to u_3 were contaminated before step $t + 4$, only one searcher can leave u_3 at step $t + 5$. Hence at most six steps later all searchers are stuck. If u_2 has exactly one contaminated edge incident to it, only one searcher, say λ_1, can leave u_2 and the other searcher is stuck. Similar to the previous cases, λ_1 is also stuck after two steps.

CASE 2.3: Suppose that λ_1 and λ_2 are on different vertices, say u_2 and u_3 respectively, where u_2 and u_3 are both different from u_1. The searcher λ_1 (resp. λ_2) can leave the vertex u_2 (resp. u_3) if and only if u_2 (resp. u_3) has exactly one contaminated edge incident to it. When λ_1 and λ_2 move to the same vertex in V_1, this subcase reduces to Subcase 1.1, and all searchers are stuck after at most 4 steps. Similarly when λ_1 and λ_2 move to different vertices in V_1, it reduces to Subcase 1.2. and all searchers are stuck after at most 6 steps.

CASE 2.4: Suppose that λ_1 is on u_1 and λ_2 is on u_2, where u_2 is different from u_1. Now λ_1 can not move since all edges incident to u_1 are clean. If there is more than one contaminated edge incident to u_2, then λ_2 can not move. Otherwise, λ_2 can slide along the only contaminated edge incident to it, say $e = u_2v_1$, at step $t + 1$. At the next step, either λ_2 or λ_3, say λ_2, can slide along any contaminated edge incident to v_1, say v_1u_3. But none of the searchers can move after this step.

CASE 3. The searcher λ_1 is on $v_1 \in V_1$ and λ_2 is on a vertex in V_2.

CASE 3.1: Suppose that λ_2 is on u_1. Then λ_2 is stuck on u_1 and λ_1 can make at most two moves.

CASE 3.2: Suppose that λ_2 is on $u_2 \neq u_1$. If u_2 has only one contaminated edge incident to it, then there are two cases to consider. If that contaminated edge is v_1u_2 and it is cleaned next, then the problem reduces to Subcase 1.1 if λ_2 slides to v_1 or it reduces to Subcase 2.2 if λ_1 slides to u_2. If v_1u_2 is not cleaned in the next step, then the only searcher that can move is λ_1 and all the searchers are stuck after this step. Otherwise if the contaminated edge is v_2u_2, then λ_2 may slide along that edge and all searchers have the same positions as in Subcase 1.2. If u_2 has exactly two contaminated edges incident to it, then the searchers are stuck after at most one step in the case that none of these contaminated edges are incident to v_1 (only λ_1 can move once). Otherwise, λ_2 and λ_1 are both stuck at step $t + 3$. If there are three contaminated edges incident to u_2, then the problem reduces to Subcase 1.2. □

Now we consider the complete bipartite graphs $K_{m,n}$ where $3 \leq m \leq n$ and m is odd. Let us first give the following upper bound which can be obtained using the same strategy used in Lemma 5.

Lemma 6. *If* $3 \leq m \leq n$ *where* m *is odd and* n *is even, then* $s_f(K_{m,n}) \leq n+3$.

During the strategy given in Lemma 5, we cleaned the graph by decomposing it into $K_{2,m-3}$'s and the parity of $m-3$ allowed us to use the same searcher in the next $K_{2,m-3}$. However in this case when m is even, the parity of $m-3$ forces us to use a different searcher for all $K_{2,m-3}$ to fast search the graph if the same strategy is used.

We know that $m + 2 = s(K_{m,n}) \leq s_f(K_{m,n})$. On the other hand, when m is odd, it follows from Lemma 1 that $s_f(K_{m,n})$ is bounded below by $\frac{m+n}{2}$ when n is odd and by $\frac{n}{2}$ when n is even. This gives us another lower bound for $s_f(K_{m,n})$.

Next we give another upper bound which improves Lemma 6 for some m and n.

Lemma 7. *If* $3 \leq m \leq n$ *where* m *is odd, then*

$$s_f(K_{m,n}) \leq m + \left\lceil \frac{n}{2} \right\rceil.$$

Proof. Let $K_{m,n}$ have bipartition $V_1 = \{v_1, v_2, \ldots, v_m\}$ and $V_2 = \{u_1, u_2, \ldots, u_n\}$.

CASE 1: $n = 4k + 1$. Place m searchers on u_1 and denote them as $\lambda_2, \lambda_3, \ldots, \lambda_{m+1}$. Place a searcher, say λ_1, on v_1 and a searcher on each of u_2 and u_3, denoted as λ_{m+2} and λ_{m+3} respectively. Place a searcher on each of u_{4l+2} and u_{4l+3} for $l = 1, \ldots, k-1$. In this way we use 2 new searchers for every 4 vertices in $V_2 \backslash u_1$. In total we use $m + 1 + \frac{n-1}{2}$ searchers.

First clean u_1 so that each vertex in V_1 contains a searcher, except for v_1 which contains two searchers. Let λ_1 traverse all edges of the Eulerian graph induced by $\{v_1, v_2, \ldots, v_{m-1}\}$ and $\{u_2, u_3\}$ and hence clean it. Now λ_{m+2} may slide along $u_2 v_m$ and λ_{m+3} may slide along $u_3 v_m$ and clean u_2 and u_3.

Next let λ_{m+2} slide along $v_m u_4$ and λ_{m+3} slide along $v_m u_5$. Again, let λ_1 clean all edges of the Eulerian graph induced by $\{v_1, v_2, \ldots, v_{m-1}\}$ and $\{u_4, u_5\}$. This also cleans u_4 and u_5.

We clean the graph by repeating this procedure for all of u_{4l+2} and u_{4l+3} where $l = 1, \ldots, k-1$. First clean the Eulerian graph induced by $\{v_1, v_2, \ldots, v_{m-1}\}$ and $\{u_{4l+2}, u_{4l+3}\}$ with λ_1. Move the searcher on u_{4l+2} along $u_{4l+2} v_m$, $v_m u_{4l+4}$ and the searcher on u_{4l+3} along $u_{4l+3} v_m$, $v_m u_{4l+5}$. Then clean the Eulerian graph induced by $\{v_1, v_2, \ldots, v_{m-1}\}$ and $\{u_{4l+4}, u_{4l+5}\}$ with λ_1.

CASE 2: $n = 4k+2$. Place the searchers as in Case 1 hence $m + 1 + \frac{n-2}{2} = m + \frac{n}{2}$ searchers are placed on the graph. Clean all vertices in V_2 except for u_n with the same strategy used in Case 1. Observe that the only contaminated edges are the ones incident to u_n and there is a searcher located on each vertex in V_1. We let λ_2 slide along $v_1 u_n$ and clean v_1. Then let λ_3 slide along $v_2 u_n$ and clean v_2. Similarly we clean all vertices in V_1 and finally clean u_n.

CASE 3: $n = 4k + 3$. Once again place the searchers as in Case 1. Place another searcher on v_m. Hence we used $m + 1 + \frac{n-3}{2} + 1 = m + \frac{n+1}{2}$ searchers. Use the same strategy as in Case 1 to clean every vertex in V_2 except for u_{n-1}

and u_n. Now there is a searcher on every vertex in V_1 except for v_1 and v_m on which there are two searchers. Then we let the two searchers on v_m slide along the only contaminated edges incident to v_m which are $v_m u_{n-1}$ and $v_m u_n$, hence cleaning v_m. Finally, λ_1 cleans the Eulerian graph induced by $\{v_1, v_2, \ldots, v_{m-1}\}$ and $\{u_{n-1}, u_n\}$. In this way the graph is cleaned totally.

CASE 4: $n = 4k$. Place a searcher on every vertex in $\{v_1, v_2, \ldots, v_{m-1}, u_1, u_2, \ldots, u_{2k}\}$ and place a second searcher, say λ_1, on v_1. Hence we use $m + \frac{n}{2}$ searchers. We let λ_1 clean the Eulerian graph induced by $\{v_1, v_2, \ldots, v_{m-1}\}$ and $\{u_1, u_2, \ldots, u_{2k}\}$. Then let each searcher λ_{m+i} located on $u_i \in \{u_1, u_2, \ldots, u_{2k}\}$ slide along $u_i v_m$ and clean $\{u_1, u_2, \ldots, u_{2k}\}$. Next let each λ_{m+i} slide along $v_m u_{2k+i}$ and clean v_m. Now each vertex in $\{u_{2k+1}, u_{2k+2}, \ldots, u_n\}$ contains a searcher. Then we let λ_1 clean the Eulerian graph induced by $\{v_1, v_2, \ldots, v_{m-1}\}$ and $\{u_{2k+1}, u_{2k+2}, \ldots, u_n\}$. This cleans every vertex in $K_{m,n}$.   □

5   Cost Function

Given a graph $G = (V, E)$, let s be the number of searchers used in a search strategy to clean G. Certainly, $s(G) \leq s$. For each value of s, there is a strategy that clears G in the minimum number of steps; that is, the minimum time. We define the minimum number of steps for each s to be $t(s)$. Certainly, $|E| \leq t(s)$.

In some real-life scenarios, the cost of a searcher may be relatively low in comparison to the cost of allowing an intruder to be free for a long period of time. Thus, it may be beneficial to use more searchers than $s(G)$. Since the minimum number of steps to clear a graph is $|E|$ (each edge must be cleared), it will never be necessary to use more searchers than the minimum needed to clear the graph in $|E|$ steps. So we can bound the number of searchers above by $s_f(G)$, giving $s \leq s_f(G)$.

When we construct a cost function for searching we may consider the following parameters: α, cost per searcher; β, cost per searcher per step; and γ, cost per step. Several combinations of these parameters above can be considered. Fomin and Golovach [7] introduced a cost function in the node searching problem, which is the sum of the number of searchers in every step of the node search process. For a fixed graph G, we choose to consider the following cost function in the edge searching problem:

$$C_G(s, t) = \alpha s + \beta s t + \gamma t, \tag{1}$$

where $s(G) \leq s \leq s_f(G)$ and $t = t(s) \geq |E|$. Instead of trying to minimize s, which corresponds to (edge) searching, or to minimize t, which corresponds to fast searching, we may attempt to minimize C_G. However, in order to formulate these problems (and bound s) it is necessary to know both the search number and the fast search number.

Consider the n-star, $K_{1,n}$, where $n \geq 3$. We recall that $s(K_{1,n}) = 2$, and that $s_f(K_{1,n}) = \lceil \frac{n}{2} \rceil$. It is not difficult to see that if $2 \leq s \leq \lceil \frac{n}{2} \rceil$, then the minimum number of steps for such a strategy is $t(s) = 2n - 2s$. (The s searchers would begin at distinct leaves of the graph, and then traverse the edges to the central vertex, using s moves. They would then use 2 moves, moving out from and then back towards the central vertex, for each of $n - 2s$ pendant edges. Finally, each searcher would move out from the central vertex along one of the s remaining pendant edges.)

Substituting into (1), we obtain

$$C_{K_{1,n}}(s) = \alpha s + \beta s(2n - 2s) + \gamma(2n - 2s) = -2\beta s^2 + (\alpha + 2n\beta - 2\gamma)s + 2n\gamma. \quad (2)$$

But this is clearly a quadratic function in s, which has a maximum value at its critical point. We obtain its minimum value at the minimum or maximum value for s, that is when either $s = 2$ or $s = \lceil \frac{n}{2} \rceil$. (Both are possible depending on choice of α, β, and γ.)

More informally, the cost of cleaning the n-stars is minimized by either treating it as an edge searching problem, when searchers are expensive and time is cheap, or by treating it as a fast searching problem when searchers are cheap and time is expensive.

6   Conclusion and Open Problems

In this paper we introduced fast searching as a variant of edge searching and proposed a cost function that generalizes the searching problem. We show that the fast search number of trees can be computed in linear time.

We have proven several results on the fast search number of complete bipartite graphs. When m is even and $6 \leq m \leq n$, we have $s_f(K_{m,n}) = m + 3$. When m is odd, n is even and $3 \leq m \leq n$, we have

$$\max\left\{m + 2, \frac{n}{2}\right\} \leq s_f(K_{m,n}) \leq \min\left\{n + 3, m + \frac{n}{2}\right\} \quad (3)$$

When m and n are odd and $3 \leq m \leq n$, we have

$$\max\left\{m + 2, \frac{m + n}{2}\right\} \leq s_f(K_{m,n}) \leq m + \frac{n + 1}{2} \quad (4)$$

The gaps in Equations 3 and 4 may be large depending on m and n. It remains open to reduce these gaps to a small constant which does not depend on m and n.

It will also be interesting to investigate the fast search number of other classes of graphs, such as grids and cartesian products. Besides the fact that much is known about the search number of these graphs, knowing the fast search number will allow us to examine cost functions as well.

Given a graph G and an integer k, the problem of deciding whether $s(G)$ is at most k is NP-complete [4,10,11]. We conjecture that the problem of deciding whether $s_f(G)$ is at most k is also NP-complete.

References

1. Alspach, B., Dyer, D., Hanson, D., Yang, B.: Lower bounds on edge searching. In: Chen, B., Paterson, M., Zhang, G. (eds.) ESCAPE 2007. LNCS, vol. 4614, pp. 516–527. Springer, Heidelberg (2007)

2. Barrière, L., Flocchini, P., Fraigniaud, P., Santoro, N.: Capture of an intruder by mobile agents. In: Proceedings of the 14th annual ACM symposium on Parallel algorithms and architectures (SPAA 2002), pp. 200–209 (2002)
3. Barrière, L., Fraigniaud, P., Santoro, N., Thilikos, D.: Searching is not Jumping. In: Bodlaender, H.L. (ed.) WG 2003. LNCS, vol. 2880, pp. 34–45. Springer, Heidelberg (2003)
4. Bienstock, D., Seymour, P.: Monotonicity in Graph Searching. Journal of Algorithms 12, 239–245 (1991)
5. Breisch, R.L.: An intuitive approach to speleotopology. Southwestern Cavers 6, 72–78 (1967)
6. Chung, F.: On the cutwidth and the topological bandwidth of a tree. SIAM J. Algebraic Discrete Methods 6, 268–277 (1985)
7. Fomin, F.V., Golovach, P.A.: Interval completion and graph searching. SIAM Journal on Discrete Mathematics 13, 454–464 (2000)
8. Fomin, F.V., Thilikos, D.M.: An annotated bibliography on guaranteed graph searching. Theoretical Computer Science, Special Issue on Graph Searching (submitted on November 2007)
9. Kinnersley, N.G.: The Vertex Separation Number of a graph equals its path-width. Information Processing Letters 42, 345–350 (1992)
10. LaPaugh, A.S.: Recontamination does not help to search a graph. Journal of ACM 40, 224–245 (1993)
11. Megiddo, N., Hakimi, S.L., Garey, M.R., Johnson, D.S., Papadimitriou, C.H.: The Complexity of Searching a Graph. Journal of ACM 35, 18–44 (1988)
12. Parsons, T.: Pursuit-evasion in a graph. In: Theory and Applications of Graphs. Lecture Notes in Mathematics, pp. 426–441. Springer, Heidelberg (1976)
13. West, D.B.: Introduction to Graph Theory. Prentice Hall, Englewood Cliffs (1996)
14. Yang, B., Dyer, D., Alspach, B.: Sweeping graphs with large clique number. In: Fleischer, R., Trippen, G. (eds.) ISAAC 2004. LNCS, vol. 3341, pp. 908–920. Springer, Heidelberg (2004)

Confidently Cutting a Cake into Approximately Fair Pieces

Jeff Edmonds[1], Kirk Pruhs[2,*], and Jaisingh Solanki[1]

[1] Department of Computer Science and Engineering
York University
jeff@cse.yorku.ca
[2] Computer Science Department
University of Pittsburgh
kirk@cs.pitt.edu

Abstract. We give a randomized protocol for the classic cake cutting problem that guarantees approximate proportional fairness, and with high probability uses a linear number of cuts.

1 Introduction

The classic cake cutting problems originated in the 1940's in the Polish mathematics community and involves fairly apportioning valuable resources (the cake) when there is not an agreed upon value of the resources. The analogy of cake was used because of the well known phenomenon of people valuing frosting and cake differently. Cake cutting is widely studied within the social sciences because of the obvious importance of fairly dividing resources, and is widely studied in the mathematical sciences because of the elegance of the problems. There are several books written on cake cutting, and related fair allocation problems (See, for example, [3,12]). Because of the inherent interest of cake cutting problems to a wide audience, cake cutting is often taught in discrete mathematics courses, and often appears in the media in shows that try to popularize mathematics. For example, in the "One Hour" episode of the TV show Numb3rs, the lead FBI agent uses his understanding of cake cutting algorithms to deduce the portion of the ransom received by the head of a kidnapping conspiracy. [1]

The setting for the cake cutting problem involves a continuous resource modeled by the unit interval, n players, a value function V_p for each player p, and a referee protocol. The value function for each player specifies how much that player values each subinterval of the cake. A piece is a union of disjoint subintervals, and the value function is additive, so that the value of a piece is the sum of the values of the underlying subintervals. The value functions are initially unknown to the referee. The standard operation is a cut query, in which the

---

[*] Supported in part by NSF grants CNS-0325353, CCF-0514058 and IIS-0534531.
[1] Although, as is often the case in this show, the application of mathematics is designed to aid the plot development, and is not necessarily designed to satisfy mathematician's sensibilities.

R. Fleischer and J. Xu (Eds.): AAIM 2008, LNCS 5034, pp. 155–164, 2008.
© Springer-Verlag Berlin Heidelberg 2008

referee asks the player to identify the shortest subinterval with a fixed value and a fixed left endpoint. We assume here that the players answer queries honestly (for more discussion of this issue, see section 2). After the cut queries, the referee partitions the resulting subintervals among the players. The referee's goal is to fairly apportion the cake among the players. There are several notions of fairness in the literature, but the original, and most basic, notion is that of proportional fairness. An apportionment is *proportionally fair*, or simply *proportional* if each player believes that his piece is worth at least $1/n$ of the total value of the cake, according to that player's value function. In this paper, we consider the notion of approximate fairness. We will say that an apportionment is *c-fair* fair if each player believes that his piece is worth at least $1/(cn)$ of the total value of the cake, according to that player's value function. We will say that a referee protocol is *approximately fair* if there exists some constant $c \geq 1$ such that the protocol guarantees a c-fair apportionment. We are interested in the query complexity of a referee protocol, which is the worst-case number of queries required to achieve a fair allocation for each player.

A deterministic proportional protocol with query complexity $\Theta(n^2)$ was given in 1948 by Steinhaus in [14]. In 1984, Even and Paz [6] gave a deterministic divide and conquer proportional protocol that has query complexity $\Theta(n \log n)$. Recently, there has been several papers [7,13,4] that give lower bounds on the query complexity for proportional cake cutting. Sgall and Woeginger [13] showed that every proportional protocol (deterministic or randomized) has query complexity $\Omega(n \log n)$ if each player must receive a contiguous piece of the cake. Edmonds and Pruhs [4] show that the query complexity of every deterministic approximately-fair protocol is $\Omega(n \log n)$.

This left open the question of whether approximate-fairness was achievable by a randomized protocol with query complexity $O(n)$. In the subsequent paper [5], Edmonds and Pruhs was settled this question in the affirmative by giving a randomized approximately-fair protocol with expected query complexity $O(n)$. This protocol was based on the following theorem:

Lemma 1 (Balanced Allocation Lemma [5]). *Let $\alpha \geq 17$ be some sufficiently large constant. Assume each of the n players conceptually partitions the unit interval into αn disjoint candidate subintervals/pieces of equal value. Then assume that each player independently picks $d' = 2d = 4$ of her candidate pieces uniformly at random, with replacement. Then there is an efficient apportionment method that, with probability $\Omega(1)$, assigns each player one final piece of her d' candidate pieces, so that every point on the cake is covered by at most 2 players.*

Once overlap of $O(1)$ is achieved for every point of the cake, one can achieve approximate fairness with linearly more queries by using any proportional algorithm to apportion the portions of cake where there is contention among the final pieces. By applying the Edmonds and Pruhs Balanced Allocation Theorem until a successful apportionment is possible, one get expected query complexity of $O(n)$. This is because each application of the Balanced Allocation Theorem has complexity $O(n)$, and the number of applications until a success is a geometrically distributed random variable with probability of success $\Omega(1)$.

However, several referees of [5] complained that in order to have high confidence of success one would have to apply the Balanced Allocation Theorem $\Omega(\log n)$ times, and thus if one accepts the requirement that a randomized algorithm should succeed with high probability, then this randomized algorithm would have no better query complexity than the Evan and Paz deterministic algorithm. In the context of this paper, high probability of success means that the probability of failure, as a function of n, should approach zero as n increases.

In this paper, we answer the referees of [5] by showing the following high confidence version of the Balanced Allocation Lemma in [5].

Lemma 2 (High Confidence Balanced Allocation Lemma). *Let $\alpha \geq 17$ be some sufficiently large constant. Assume each of the n players conceptually partitions the unit interval into αn disjoint candidate subintervals of equal value. Each player independently picks $d' = k \times 2 \times 2d = 8k$ of her pieces uniformly at random, with replacement. Then there is an efficient method that, with probability $(1 - O(\frac{1}{n^k}))$, picks one of the d' subintervals for each player, so that every point on the cake is covered by at most 4 players.*

The rest of the paper is organized as follows. In section 2 we discuss some other related results, and explain how the standard balls and bins model is a special case of cake cutting. In section 3 we briefly explain how [5] obtained a (low confidence) proof of the Balanced Allocation Lemma for cake. In section 4 we give a protocol, which turns out to be a rather simple modification of the protocol in [5], that establishes the High Confidence Balanced Allocation Lemma. We discovered the simple protocol in section 4 after much effort on another approach. We briefly discuss this other approach in section 5 because we believe that it poses some interesting open questions.

2 Other Related Work

The lower bound proofs in [13,4] also allow the referee to make evaluation queries, which ask a player to state their value for a particular piece. Edmonds and Pruhs [4] also showed that every randomized approximately-fair protocol has query complexity $\Omega(n \log n)$ if answers to the queries asked by protocol are approximations to actual answers. Approximately fair protocols were introduced by Robertson and Webb [11]. There is deterministic protocol [11,10,15] that achieves approximate-fairness with $\Theta(n)$ cuts and $\Theta(n^2)$ evaluations. There are several other notions of fairness studied in the cake cutting setting, most notably envy-free fairness. There are known finite complexity protocols for envy-free divisions, but no bounded complexity protocols are known (See, for example, [3] for details). The cake cutting problem is often defined so that the players do not need to answer the queries truthfully. For deterministic protocols, lying is generally a non-issue since it is easy to catch any form of lying that would mess up the standard protocols (See [3] for details). But for randomized protocols, it seems much more difficult to catch cheaters.

In the multiple-choice balls and bins model, d of αn discrete bins are selected for each ball uniformly at random. Then we select one bin out of d bins such that maximum number of balls in any bin is minimized. There is an efficient procedure, essentially a matching algorithm for a bipartite graph, that picks one of the d bins for each player so that maximum number of balls in any bin is $O(1)$, with probability $(1-O(\frac{1}{poly(n)}))$. [2]. The balls and bins model is equivalent to the special case of the cake model in which all the players value the cake uniformly. Analysis of the balls and bins model has found wide applications in areas such as load balancing [8]. In these situations, a ball represents a job that can be assigned to various bins/machines/servers. Roughly speaking, load balancing of identical machines is to balls and bins, as load balancing on unrelated machines is to cake cutting. In the unrelated machine model, the speed that a machine runs a job depends on the job. So the jobs may not agree on the values of the various machines. Unrelated machines is one of the standard models in the load balancing literature [1].

In the balls and bins model, the maximum number of balls in any bin is $\theta(\frac{\log n}{\log \log n})$ with probability $(1 - O(\frac{1}{poly(n)}))$ [9]. Assume n balls are thrown sequentially into n bins, each ball is placed in the least full bin at the time of the placement, among d bins, $d \geq 2$, chosen independently and uniformly at random. Then after all the balls are placed, the maximum number of balls in any bin is $\theta(\frac{\log \log n}{\log d})$ with probability $\Omega(1 - \frac{1}{poly(n)})$. [2].

3   The Original Balanced Allocation Lemma for Cake

We now outline the protocol and analysis that establishes the Balanced Allocation Lemma in [5]. Note this protocol uses two graphs, the implication graph, and the same-player-vee graph, and some graph theoretic definitions, that we will define after the protocol.

Edmonds-Pruhs Protocol:
- **Step 1:** Independently, for each player $p \in [1, n]$ and each $r \in [0, 1]$, randomly choose $d = 2$ of the candidate pieces $c_{\langle p,i \rangle}$ to be in the quarterfinal bracket $A_{\langle p,r \rangle}$. Thus each player has two quarterfinals brackets, each containing two intervals.
- **Step 2:** In each quarterfinal bracket $A_{\langle p,r \rangle}$, pick as the semifinal piece $a_{\langle p,r \rangle}$, the piece that intersects the fewest other candidate pieces $c_{\langle q,j \rangle}$. Thus each player is left with two semifinal intervals.
- **Step 3:** Form the implication graph and same-player-vee graph for the semifinal pieces
- **Step 4:** If implication graph contains a pair path of length greater than or equal to 3, then admit failure.
- **Step 5:** If same-player-vee graph is not $w = 2$ colorable, then admit failure.
- **Step 6:** Let S_h be the subgraph of the implication graph containing only those players colored h, in the same-player-vee graph. This ensures that implication graph restricted to S_h contains no pair paths of length 2.

– **Step 7:** For each S_h, pick the final piece for each player involved in S_h by applying the Final Piece Selection Algorithm to S_h.

We now turn to the graph theoretic notions used in this protocol. The vertices of the *implication graph IG* are the $2n$ pieces $a_{\langle p,r \rangle}$, $1 \leq p \leq n$ and $0 \leq r \leq 1$, and if piece $a_{\langle p,r \rangle}$ intersects piece $a_{\langle q,s \rangle}$, then there is a directed edge from piece $a_{\langle p,r \rangle}$ to piece $a_{\langle q,1-s \rangle}$, and similarly from $a_{\langle q,s \rangle}$ to $a_{\langle p,1-r \rangle}$. The intuition behind the this definition is that if a player p gets $a_{\langle p,r \rangle}$ as her final piece, then player q must get piece $a_{\langle q,1-s \rangle}$ if p's and q's pieces are not to overlap. Similarly if q gets $a_{\langle q,s \rangle}$, then p must get $a_{\langle p,1-r \rangle}$. As an example, Figure 1 gives a subset of the semifinal pieces selected from the candidate pieces. The corresponding implication graph is also given in Figure 1. A *pair path* in an implication graph is a directed path between two pieces for one player.

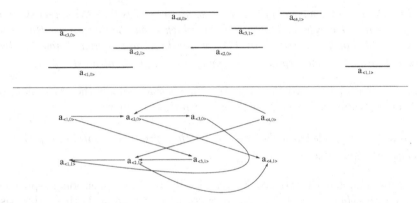

Fig. 1. Players' two selected pieces and corresponding implication graph

In Figure 1, there are two pair paths of length three from the first player's left semifinal piece to her right and two pair paths of length two from the fourth player's left semifinal piece to her right. Pair paths are problematic because they effectively imply that if the first player gets her left semifinal piece as his final piece then she must get her right piece too. Edmonds and Pruhs [5] prove that if the implication graph IG does not contain pair paths then the following algorithm selects a final piece for each player in such a way that these final pieces are disjoint.

Final Piece Selection Algorithm: We repeatedly pick an arbitrary player p that has not selected a final piece. We pick the piece $a_{\langle p,0 \rangle}$ as the final piece for p. Further, we pick as final pieces all those pieces in IG that are reachable from $a_{\langle p,0 \rangle}$ in IG.

If edges in the implication graph were independent, then we could bound the probability of a pair path as in the balls and bins case:

$$\text{Prob}[IG \text{ contain pair paths}] \approx \sum_{z=2}^{n} \binom{n}{z} \left(\frac{1}{\alpha n}\right)^z \approx \frac{1}{\alpha(\alpha - 1)}.$$

Unfortunately, edges in the implication graph are not statistically independent. For example, if all of player B and player C's pieces are contained in one candidate piece P for player A, then the existence of an edge involving A and B would mean that player A picked candidate piece P, and thus thus there must be an edge involving player A and player C. Nevertheless, [5] show that, in spite of the statistical dependencies of edges, the above calculation of the probability of a pair path does give approximately the right answer.

First, [5] observed the vital difference between pair paths of length two and pair paths of length three or more. Note that a pair path occurs when there is a *vee* among the semifinal pieces. [5] defined a *vee* to consist of a triple of pieces, one *center* piece and two *base* pieces, with the property that the center piece intersects both of the base two pieces. [5] proved the following lemma that bounds the expected number of vees in the implication graph.

Lemma 3. *If each player only chooses 2 semifinal pieces then the expected number of vees in IG can be as high as $\Theta(n^2)$, which would be disastrous. However, if two brackets of $d = 2$ pieces are chosen and these are narrowed down to two semifinal piece then the expected number of vees in IG is at most $\frac{16d^3}{\alpha^2}n$.*

Using Lemma 3 [5] proved the following lemma that bounds the probability of implication graph having pair paths of length three or more.

Lemma 4. *The probability that the implication graph IG contains a pair path of length at least three is at most $\frac{32d^5}{\alpha^2(\alpha-4d^2)}$.*

A pair path of length two occurs if and only if the implication graph contains a *same-player-vee*. A *same-player-vee* is a vee where both of the base pieces belong to the same player. That is, there is a center piece $a_{\langle p,r \rangle}$ and two bases $a_{\langle q,0 \rangle}$ and $a_{\langle q,1 \rangle}$. For example, see pieces $a_{\langle 4,0 \rangle}$, $a_{\langle 2,0 \rangle}$ and $a_{\langle 2,1 \rangle}$ in Figure 1. To get around the problem of same-player-vees, they introduced the *same-player-vee graph*. The vertices of the *same-player-vee graph* SG are the n players p, $1 \le p \le n$, and if player p and player q are involved in same-player-vee with player p in the center then there is a directed edge from p to q. [5] show how to partition the players into two groups such that there is no same-player-vee involving two players in the same partition. [5] proved Lemma 5 by bounding the probability of same-player-vee graph having a path of length two.

Lemma 5. *The probability that the same-player-vee graph is not $w = 2$ colorable is at most $\frac{16d^3}{\alpha^3} + \frac{8d^2}{\alpha^2}$.*

Finally, because the implication graph on S_h contains no pair paths of any length, the Edmonds-Pruhs protocol ensures that the final piece of at most one player from S_h covers this point. We can then conclude that for any point in the cake, the final pieces of at most $w = 2$ players cover this point.

4   The High Confidence Balanced Allocation Lemma

In this section we give a modification of the Edmonds-Pruhs protocol that will establish the High Confidence Balanced Allocation Lemma for cake. A key concept in the proof of correctness of the Edmonds-Pruhs protocol is the concept of a bad player. A player p is *bad* if a pair path of length three or more starting with p exists in the implication graph, or a path of length two or more starting with p exists in the same-player-vee graph. Edmonds and Pruhs [5] proved that for some constant c, larger than 1, the probability that a particular player is bad is at most $\frac{1}{cn}$. We modify the Edmonds-Pruhs protocol in [5] in the following ways:

Modified Edmonds-Pruhs Protocol:
  - We make two independent runs of the protocol.
  - In each run, after the formation of the implication graph and the same-player-vee graph, we remove all bad players. Because of this modification, some players (the ones that are bad in both runs) may not be assigned a candidate piece.

After these two separate runs, some players may be assigned two final pieces, one from each run. In this case, the player need only keep the final piece from the first run. But the key fact is that no point of the cake is covered by more than four candidate pieces, since each run guarantees contention at most two. As before we can then use any proportionally fair protocol to divided the portions of the cake where the final pieces overlap. A flow chart of this protocol is given in Figure 2.

We can then simply calculate the probability that one iteration of this modified protocol fails to assign a final piece to every player:

$$\text{Prob}[Modified\ Edmonds\ Pruhs\ protocol\ fails]$$
$$\leq \text{Prob}[there\ exists\ a\ player\ p\ that\ is\ bad\ in\ both\ runs]$$
$$\leq n \times \text{Prob}[player\ p\ is\ bad\ in\ both\ runs]$$
$$\leq n \times (\text{Prob}[player\ p\ is\ bad\ in\ one\ run])^2$$
$$\leq n \times O(\tfrac{1}{n})^2$$
$$= O(\tfrac{1}{n})$$

Thus the probability that the Modified-Edmonds-Pruhs protocol does not succeed after k applications is at most $O(\frac{1}{n^k})$.

5   Initial Unsuccessful Approach

To us, the most obvious way to modify the Edmonds-Pruhs protocol to obtain a high confidence result was to show that the same-player-vee graph is $O(1)$-colorable with high probability, and to show that the implication graph does not have a pair-path with high probability. We were able to accomplish the former in Lemma 6 by slightly modifying the protocol. But the latter is unfortunately false, there can be pair-paths with high probability.

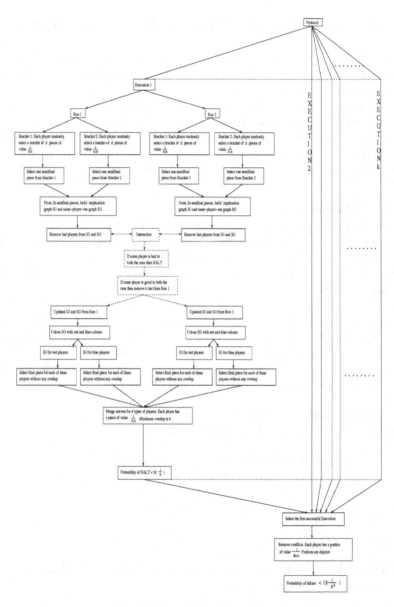

Fig. 2. Flowchart of Our Protocol

Lemma 6. *Let $\alpha \geq 10$ be some sufficiently large constant. Assume that each player conceptually partitions the unit interval into αn disjoint candidate subintervals/pieces of equal value. Each player then independently picks $d' = 3d = 6$ of her pieces uniformly at random, with replacement. There is then an efficient method that, with probability at least $(1 - O(\frac{1}{n}))$, chooses three of the d' pieces for each player, and then narrows down two pieces for each player, so that same-player-vee graph build from these chosen pieces can be colored by at most two colors.*

After some reflection, it is clear that a pair path in the implication graph, in and of itself, is not a problem. A pair path from $a_{\langle p,r \rangle}$ to $a_{\langle p,1-r \rangle}$ just implies that we should select $a_{\langle p,1-r \rangle}$ for player p. However, a directed path in both directions, from $a_{\langle p,0 \rangle}$ to $a_{\langle p,1 \rangle}$ and from $a_{\langle p,1 \rangle}$ to $a_{\langle p,0 \rangle}$ is problematic since the selection of either piece requires the selection of the other. This leads us to the following definition.

Pair Cycle: A pair cycle in the implication graph is a directed cycle containing both semifinal pieces $a_{\langle p,0 \rangle}$ and $a_{\langle p,1 \rangle}$ for some player p.

It is not to difficult to see that if the implication graph IG does not contain pair cycle then one can modify the Final Piece Selection Algorithm so that it can select a disjoint final piece for each player. Further, if the edges in the implication graph were independent, using our standard calculations, we find that the probability of a pair-cycle would be $O(\frac{1}{n})$:

$$\text{Prob}[IG \text{ contain a pair cycle}] \approx \sum_{z=2}^{n} \binom{n}{z-1} \left(\frac{1}{\alpha n}\right)^z \approx \Theta\left(\frac{1}{n}\right)$$

The reason that this calculation gives $\Theta(1)$ for a pair-path and $\Theta(\frac{1}{n})$ for a pair-cycle is that a pair-cycle has one more edge relative to the number of vertices than does a pair-path. While we know that the edges are not independent, this calculation did give approximately the right probability for a pair-path, and we the saw no reason why this calculation shouldn't also give approximately the right calculation for a pair-cycle. This led us to the following natural conjecture:

Conjecture 1. The probability that for some player p, we have pair paths of length at least three from $a_{\langle p,0 \rangle}$ to $a_{\langle p,1 \rangle}$ and from $a_{\langle p,1 \rangle}$ to $a_{\langle p,0 \rangle}$ in the implication graph IG is at most $O(\frac{1}{n})$.

Recall that pair cycle requires that at least one player's both semifinal pieces have to be present in it. In Lemma 7, we were able to prove Conjecture 1 in the case where exactly one player's both semifinal pieces are present in the pair cycle.

Lemma 7. *The probability that for some player p we have pair paths of length at least three from $a_{\langle p,0 \rangle}$ to $a_{\langle p,1 \rangle}$ and from $a_{\langle p,1 \rangle}$ to $a_{\langle p,0 \rangle}$ in the implication graph IG and except player p there is no common player involved in both the pair paths, is at most $O(\frac{1}{n})$.*

Our intuition is that the probability of having more than one player repeating in a pair cycle is not so high. But we are unable to prove this. More generally, we wonder whether the selection of the semifinal pieces from the quarterfinal pieces (by throwing out pieces that overlap the most other pieces) leave with a collection of pieces that essentially have the same graph properties as in the balls and bins model.

References

1. Azar, Y.: On-line Load Balancing. In: Online Algorithms - The State of the Art, pp. 178–195. Springer, Heidelberg (1998)
2. Azar, Y., Broder, A., Karlin, A., Upfal, E.: Balanced Allocations. SIAM Journal of Computing 29, 180–200 (2000)
3. Brams, S.J., Taylor, A.D.: Fair Division – From cake cutting to dispute resolution. Cambridge University Press, Cambridge (1996)
4. Edmonds, J., Pruhs, K.: Cake cutting really isn't a piece of cake. In: Proceedings of the 17th Annual ACM-SIAM Symposium on Discrete Algorithms (SODA 2006) (2006)
5. Edmonds, J., Pruhs, K.: Balanced Allocations of Cake. In: Proceedings of the 47th Annual IEEE Symposium on Foundations of Computer Science (FOCS 2006), pp. 623–634 (2006)
6. Even, S., Paz, A.: A note on cake cutting. Discrete Applied Mathematics 7, 285–296 (1984)
7. Magdon-Ismail, M., Busch, C., Krishnamoorthy, M.S.: Cake cutting is not a piece of cake. In: Alt, H., Habib, M. (eds.) STACS 2003. LNCS, vol. 2607, pp. 596–607. Springer, Heidelberg (2003)
8. Mitzenmacher, M., Richa, A., Sitaraman, R.: The power of two random choices: A survey of the techniques and results. In: Pardalos, P., Rajasekaran, S., Rolim, J. (eds.) Handbook of Randomized Computing, Kluwer, Dordrecht (2000)
9. Johnson, N., Kotz, S.: Urn Models and Their Application. John Wiley and Sons, Chichester (1977)
10. Krumke, S.O., Lipmann, M., de Paepe, W., Poensgen, D., Rambau, J., Stougie, L., Woeginger, G.J.: How to cut a cake almost fairly. In: Proceedings of the 13th Annual ACM-SIAM Symposium on Discrete Algorithms (SODA 2002), pp. 263–264 (2002)
11. Robertson, J.M., Webb, W.A.: Approximating fair division with a limited number of cuts. Journal of Combinatorial Theory, Series A 72, 340–344 (1995)
12. Robertson, J.M., Webb, W.A.: Cake-cutting algorithms: Be fair if you can. A.K. Peters Ltd. (1998)
13. Sgall, J., Woeginger, G.J.: A Lower Bound for Cake Cutting. In: Di Battista, G., Zwick, U. (eds.) ESA 2003. LNCS, vol. 2832, pp. 459–469. Springer, Heidelberg (2003)
14. Steinhaus, H.: The problem of fair division. Econometrica 16, 101–104 (1948)
15. Woeginger, G.J.: An approximation scheme for cake division with a linear number of cuts. In: Möhring, R.H., Raman, R. (eds.) ESA 2002. LNCS, vol. 2461, pp. 896–901. Springer, Heidelberg (2002)

Copeland Voting Fully Resists Constructive Control*

Piotr Faliszewski[1], Edith Hemaspaandra[2],
Lane A. Hemaspaandra[1], and Jörg Rothe[3]

[1] Dept. of Computer Science, University of Rochester, Rochester, NY 14627, USA
[2] Dept. of Computer Science, RIT, Rochester, NY 14623, USA
[3] Inst. für Informatik, Universität Düsseldorf, 40225 Düsseldorf, Germany

Abstract. Control and bribery are settings in which an external agent seeks to influence the outcome of an election. Faliszewski et al. [9] proved that Llull voting (which is here denoted by Copeland^1) and a variant (here denoted by Copeland^0) of Copeland voting are computationally resistant to many, yet not all, types of constructive control and that they also provide broad resistance to bribery. We study a parameterized version of Copeland voting, denoted by Copeland$^\alpha$, where the parameter α is a rational number between 0 and 1 that specifies how ties are valued in the pairwise comparisons of candidates in Copeland elections. For each rational α, $0 < \alpha < 1$, and each previously studied control scenario, we either prove that Copeland$^\alpha$ is computationally vulnerable to control in that scenario (i.e., we give a P-time algorithm that determines whether control is possible, and if so, determines exactly how to exert the control) or we prove that Copeland$^\alpha$ is computationally resistant to control in that scenario (i.e., we prove that control problem to be NP-hard). In particular, we prove that Copeland$^{0.5}$, the system commonly referred to as "Copeland voting," provides full resistance to constructive control. Among systems with a polynomial-time winner problem, this is the first natural election system proven to have full resistance to constructive control. Looking at rational α, $0 < \alpha < 1$, we give a broad set of results on bribery and on the fixed-parameter tractability of bounded-case control for Copeland$^\alpha$ (previously only Copeland^0 and Copeland^1 had been studied), and we introduce and obtain fixed-parameter tractability results even in a new, more flexible model of control (that we dub "extended control").

Keywords: Computational social choice theory; preference aggregation; multiagent systems.

* Supported in part by DFG grants RO-1202/9-3 and RO-1202/11-1, NSF grants CCR-0311021, CCF-0426761, and IIS-0713061, the Alexander von Humboldt Foundation's TransCoop program, and two Friedrich Wilhelm Bessel Research Awards. URLs: www.cs.rochester.edu/u/{pfali, lane} (Piotr Faliszewski and Lane A. Hemaspaandra), www.cs.rit.edu/~eh (Edith Hemaspaandra), and ccc.cs.uni-duesseldorf.de/~rothe (Jörg Rothe). Corresponding author: Jörg Rothe.

1   Introduction

Preference aggregation by voting procedures has been the focus of much attention within the field of multiagent systems. Agents (called voters in the context of voting) may have different, often conflicting individual preferences over the given alternatives (or candidates). Voting rules (or, synonymously, election systems) provide a useful method for them to come to a "reasonable" decision on which alternative to choose. One key issue here is that there might be attempts to influence the outcome of elections. Settings in which such influence on elections can be implemented include manipulation [3,17], electoral control [2,9,10,11,17], and bribery [7,9].

Bribery and control are settings in which an external actor seeks to influence the outcome of an election. In bribery, the briber tries to reach his or her goal via bribing some of the voters to change their preferences (i.e., via bribing them to cast their votes as the briber says). In control scenarios, the external actor—often referred to as "the chair"—attempts to influence the outcome of elections by modifying their structure, via such actions as adding, deleting, or partitioning either candidates or voters. In contrast, there is no external actor in manipulation; rather, a group of voters, the manipulators, attempt to cast their votes insincerely in order to achieve their desired goal.

Although reasonable election systems typically are susceptible to these kinds of influence (for manipulation this is, essentially, universally true, via the Gibbard–Satterthwaite and Duggan–Schwartz Theorems), computational complexity can be used to provide some protection in each such setting. We study the extent to which the Copeland election system [4] (see also [18,15]; a similar system was also studied by Zermelo [19]) resists, computationally, control and bribery attempts.

An election is specified by a finite set C of candidates and a finite collection V of voters, where each voter has preferences over the candidates. We consider both rational and irrational voters. The preferences of a rational voter are expressed by a preference list of the form $a > b > c$ (assuming $C = \{a, b, c\}$), where the underlying relation $>$ is a strict linear order that is transitive. The preferences of an irrational voter are expressed by a preference table that for any two distinct candidates specifies which of them is preferred to the other by this voter. An election system is a rule that determines the winner(s) of each given election (C, V). In this paper we consider a parameterized version of Copeland's election system [4], denoted Copeland$^\alpha$, where the parameter α is a rational number between 0 and 1 that specifies how ties are rewarded in the head-to-head majority-rule contests between any two distinct candidates. This parametrized version of Copeland's system is due to Faliszewski et al. [9].

Definition 1 ([9]). *Let α, $0 \leq \alpha \leq 1$, be a fixed rational number. In a Copeland$^\alpha$ election the voters indicate which among any two distinct candidates they prefer. For each such head-to-head contest, if some candidate is preferred by a strict majority of voters then he or she obtains one point and the other candidate obtains zero points, and if a tie occurs then both candidates obtain α*

points. Let $E = (C, V)$ be an election. For each $c \in C$, $score_E^\alpha(c)$ is the sum of c's Copeland$^\alpha$ points in E. Every candidate c with maximum $score_E^\alpha(c)$ wins.

Let Copeland$_{Irrational}^\alpha$ denote the same election system but with voters allowed to be irrational.

In the context of the above definition, the system widely referred to in the literature as "Copeland elections" is Copeland$^{0.5}$, where tied candidates receive half a point each (see, e.g., Merlin and Saari [18,15]; the definition used by Conitzer et al. [3] can be scaled to be equivalent to Copeland$^{0.5}$). Copeland^0, where tied candidates come away empty-handed, has sometimes also been referred to as "Copeland elections" (see, e.g., [16,9]). An election system proposed by the Catalan philosopher and theologian Ramon Llull in the 13th century (see, e.g., the references in [9]) is in this notation nothing other than Copeland^1, where tied candidates are awarded one point each, just like winners of head-to-head contests.

Bartholdi, Tovey, and Trick [2] were the first to study the computational aspects of control: How hard is it, computationally, for the chair to exert control? In their seminal paper they introduced a number of fundamental control scenarios involving (what is now called) *constructive* control, i.e., scenarios where the chair's goal is to make some designated candidate win. Other papers studying control include [10,17,11,9], which in addition to constructive control also consider *destructive* control, where the chair tries to preclude some designated candidate from winning. The notion of bribery in elections was introduced by Faliszewski et al. [7] and was also studied in [9]. Faliszewski et al. [9] studied the systems Copeland^0 and Copeland^1 with respect to their (computational) resistance and vulnerability to bribery and procedural control.

At first glance one might be tempted to think that the definitional perturbation due to the parameter α in Copeland$^\alpha$ elections is negligible. However, as noted in [9], "... it can make the dynamics of Llull's system quite different from those of [Copeland^0]. Proofs of results for Llull differ considerably from those for [Copeland^0]." This statement notwithstanding, we show that in most cases it is possible to obtain a unified—though sometimes rather involved—construction that works for both systems, and even for Copeland$^\alpha$ with respect to every rational α, $0 \le \alpha \le 1$. In particular, we establish resistance or vulnerability results for Copeland$^{0.5}$ (which is the system commonly referred to as "Copeland") in every previously studied control scenario.[1] In doing so, we provide an example of a control problem where the complexity of Copeland$^{0.5}$ differs from that of both Copeland^0 and Copeland^1: While the latter two systems are vulnerable to constructive control by adding (an unlimited number of) candidates, Copeland$^{0.5}$ is resistant to this control type (see Theorem 3).

Thus Copeland (i.e., Copeland$^{0.5}$) is the first natural election system with a polynomial-time winner problem that is proven to be resistant to every type of constructive control that has been proposed in the literature to date. Closing

---

[1] Also, our new results apply, e.g., to events such as the group stage of FIFA world-cup finals, which is, in essence, a series of Copeland$^\alpha$ tournaments with $\alpha = 1/3$.

a 15-year-long quest, this result resolves a question that has been open since Bartholdi, Tovey, and Trick's 1992 paper [2].

This paper is in the worst-case model. It will be interesting to see what holds in the average or typical case. (However, most of the papers that look at average-case or typical-case complexity regarding election problems are limited to specific distributions or make strong assumptions, so the state-of-the-art in that important direction is not as advanced as many believe (see the discussion of this in [12, Section 1] and [6]).)

Due to space, we omit most of our proofs. A representative proof is given in Section 2.3 and all proofs, and additional discussion and results, can be found in the in-preparation full version of the paper [8].

2   Control

2.1   Definition of Control Problems

We now define the control problems we consider, in both the constructive and the destructive version. Let $\mathcal{E}$ be an election system. In our case, $\mathcal{E}$ will be either Copeland$^\alpha$ or Copeland$^\alpha_{\text{Irrational}}$, where α, $0 \le \alpha \le 1$, is a fixed rational number. The types of control we consider here are well-known from the literature (see, e.g., [2,9,10]) and we will content ourselves with the definition of those problems whose proofs we give in the main body of the paper.

We define control via adding candidates. Note that there are two versions of this control type. The *unlimited* version (which, for the constructive case, was introduced by Bartholdi, Tovey, and Trick [2]) asks whether the election chair can add (any number of) candidates from a given pool of spoiler candidates in order to either make his or her favorite candidate win the election (in the constructive case), or prevent his or her despised candidate from winning (in the destructive case):

Name: $\mathcal{E}$-CCAC$_u$ and $\mathcal{E}$-DCAC$_u$.
Given: Disjoint candidate sets C and D, a collection V of voters represented via their preference lists (or preference tables in the irrational case) over the candidates in $C \cup D$, and a distinguished candidate $p \in C$.
Constructive Question ($\mathcal{E}$-CCAC$_u$): Does there exist a subset D' of D such that p is a winner of the $\mathcal{E}$ election with candidates $C \cup D'$ and voters V?
Destructive Question ($\mathcal{E}$-DCAC$_u$): Does there exist a subset D' of D such that p isn't a winner of the $\mathcal{E}$ election with candidates $C \cup D'$ and voters V?

The only difference in the *limited* version of constructive and destructive control via adding candidates ($\mathcal{E}$-CCAC and $\mathcal{E}$-DCAC, for short) is that the chair needs to achieve his or her goal by adding at most k candidates from the given set of spoiler candidates. This version of control by adding candidates was proposed in [9] to synchronize the definition of control by adding candidates with the definitions of control by deleting candidates, adding voters, and deleting voters.

As seen in the above definition example, we use the following naming conventions for control problems. The name of a control problem starts with the election system used (when clear from context, it may be dropped), followed by CC for "constructive control" or by DC for "destructive control," followed by the acronym of the type of control: AC for "adding (a limited number of) candidates," AC_u for "adding (an unlimited number of) candidates," DC for "deleting candidates," PC for "partition of candidates," RPC for "run-off partition of candidates," AV for "adding voters," DV for "deleting voters," and PV for "partition of voters," and all the partitioning cases (PC, RPC, and PV) are followed by the acronym of the tie-handling rule used in subelections, namely TP for "ties promote" (i.e., all winners of a given subelection are promoted to the final round of the election) and TE for "ties eliminate" (i.e., if there is more than one winner in a given subelection then none of this subelection's winners is promoted to the final round of the election).

Note that our definitions focus on *a winner*, i.e., they are in the *nonunique-winner model*. The *unique-winner* analogs of these problems can be defined by requiring the distinguished candidate p to be the unique winner (or to not be a unique winner in the destructive case).

Let $\mathcal{E}$ be an election system and let Φ be a control type. We say $\mathcal{E}$ is *immune to Φ-control* if the chair can never reach his or her goal (of making a given candidate win in the constructive case, and of blocking a given candidate from winning in the destructive case) via asserting Φ-control. $\mathcal{E}$ is said to be *susceptible to Φ-control* if $\mathcal{E}$ is not immune to Φ-control. $\mathcal{E}$ is said to be *vulnerable to Φ-control* if it is susceptible to Φ-control and there is a polynomial-time algorithm for solving the control problem associated with Φ. $\mathcal{E}$ is said to be *resistant to Φ-control* if it is susceptible to Φ-control and the control problem associated with Φ is NP-hard. The above notions were introduced by Bartholdi, Tovey, and Trick [2] (see also, e.g., [10,17,11,9]).

2.2   Overview of Results on Control

Our main result regarding control is Theorem 1 below.

Theorem 1. *For each rational α, $0 \le \alpha \le 1$, Copeland$^\alpha$ elections are resistant and vulnerable to control as shown in Table 1, both for rational and irrational voters and in both the nonunique-winner model and the unique-winner model.*

Boldface results in Table 1 are new to this paper and nonboldface results are due to Faliszewski et al. [9]. Note that the notion widely referred to in the literature simply as "Copeland elections," which we here for clarity call Copeland$^{0.5}$, possesses all ten of our basic types of constructive resistance and, in addition, even has constructive AC_u resistance. These resistances should be compared with the results known for the other notion that in the literature is occasionally referred to as "Copeland elections," namely Copeland^0, and with the results known for Llull elections, which are here denoted by Copeland^1, see [9]. While Copeland^0 and Copeland^1 possess all ten of our basic types of constructive resistance, they both are vulnerable to this eleventh type of constructive control, the incongruous

Table 1. Resistance (R) and vulnerability (V) of Copeland$^\alpha$ elections

	Copeland$^\alpha$					
	$\alpha = 0$		$0 < \alpha < 1$		$\alpha = 1$	
Control type	CC	DC	CC	DC	CC	DC
AC$_u$	V	V	**R**	V	V	V
AC	R	V	**R**	V	R	V
DC	R	V	**R**	V	R	V
RPC-{TP,TE}	R	V	**R**	V	R	V
PC-{TP,TE}	R	V	**R**	V	R	V
PV-{TP,TE}	R	R	**R**	**R**	R	R
AV	R	R	**R**	**R**	R	R
DV	R	R	**R**	**R**	R	R

but historically resonant notion of constructive control by adding an unlimited number of candidates (i.e., CCAC$_u$).

The next section discusses two particularly interesting subcases of Theorem 1, namely those regarding constructive control via adding candidates. Afterward, we turn to the issue of fixed-parameter tractability of our control problems.

2.3 Two Subcases of Theorem 1: Constructive Control Via Adding Candidates

Let us now focus on the problem of constructive control via adding candidates. In the now standard version of this problem (i.e., in the constructive control via adding a limited number of candidates problem) we obtain the following result (a special case of Theorem 1).

Theorem 2. *For each rational number α such that $0 \leq \alpha \leq 1$, Copeland$^\alpha$ is resistant to constructive control via adding candidates (CCAC).*

This result generalizes and strengthens the result of [9] in a natural way. However, if we consider the *unlimited* version of the problem (CCAC$_u$), the complexity of the problem changes dramatically depending on the parameter α. In particular, Faliszewski et al. [9] showed that for $\alpha \in \{0,1\}$ this problem is in P, whereas we show that for all rational values of α strictly between 0 and 1 it is NP-hard.

Theorem 3. *For each rational number α, $0 < \alpha < 1$, Copeland$^\alpha$ is resistant to constructive control via adding an unlimited number of candidates (CCAC$_u$).*

The proof of this theorem follows via a reduction from the vertex cover problem: Given an undirected graph $G = (V(G), E(G))$ and a nonnegative integer k, does there exist a set W such that $W \subseteq V(G)$, $\|W\| \leq k$, and for every edge $e = \{u, v\}$, $e \in E(G)$, it holds that $e \cap W \neq \emptyset$? However, before we proceed with the reduction we need the following two lemmas that simplify constructing complicated instances of Copeland$^\alpha$ elections.

Lemma 1. *Let α be a rational number such that $0 \leq \alpha \leq 1$. For each positive integer n, there is an election $\text{Pad}_n = (C, V)$ such that $\|C\| = 2n + 1$ and, for each candidate $c \in C$, it holds that $score^\alpha_{\text{Pad}_n}(c) = n$.*

Lemma 2. *Let $E = (C, V)$ be an election where $C = \{c_1, \ldots, c_n\}$, and let α be a rational number such that $0 \leq \alpha \leq 1$. For each candidate c_i, we denote the number of head-to-head ties of c_i in E by t_i. Let $k_1, \ldots, k_n$ be a sequence of n nonnegative integers such that for each k_i we have $0 \leq k_i \leq n$. There is an election $E' = (C', V')$ such that: (a) $C' = C \cup D$, where $D = \{d_1, \ldots, d_{2n^2}\}$; (b) for each i, $1 \leq i \leq n$, $score^\alpha_{E'}(c_i) = 2n^2 - k_i + \alpha t_i$; (c) for each i, $1 \leq i \leq 2n^2$, $score^\alpha_{E'}(d_i) \leq n^2 + 1$.*

Given these two results we can prove Theorem 3.

Proof of Theorem 3. We give a reduction from the vertex cover problem. Let (G, k) be an instance of the vertex cover problem, where G is an undirected graph and k is the bound on the size of the vertex cover that we seek. Let $\{e_1, \ldots, e_m\}$ be the set of G's edges and let $\{1, \ldots, n\}$ be the set of G's vertices. We assume that both n and m are nonzero and that $k \leq \min(n, m)$ as otherwise clearly G has a vertex cover of size k and our reduction could output a fixed instance of CCAC_u with a positive answer.

Let t_1 and t_2 be two positive integers such that $\alpha = \frac{t_1}{t_2}$ and such that their greatest common divisor is 1. Clearly, two such numbers exist because α is rational and greater than 0. We set ε to be $\frac{1}{t_2}$. By elementary number-theoretic arguments, there are two positive integer constants, k_1 and k_2, such that $k_1\alpha = k_2 - \varepsilon$.

Using Lemma 2 we build an election $E' = (C, V')$ such that: (a) $\|C\| = 2\ell^2 + \ell$, where $\ell \geq 2n + 2m$ and ℓ is polynomially bounded in $\max(n, m)$; (b) $\{p, r, e_1, \ldots, e_m\} \subseteq C$ (the remaining candidates are used for padding); (c) $score^\alpha_{E'}(p) = 2\ell^2 - 1$; (d) $score^\alpha_{E'}(r) = 2\ell^2 - 1 - k + k\alpha$ in the nonunique-winner case $(score^\alpha_{E'}(r) = 2\ell^2 - 1 - k + k\alpha - \varepsilon$ in the unique-winner case[2]); (e) for each $e_i \in C$, $score^\alpha_{E'}(e_i) = 2\ell^2 - 1 + \alpha$ in the nonunique-winner case $(score^\alpha_{E'}(e_i) = 2\ell^2 - 1$ in the unique-winner case); (f) the scores of all candidates in C other than $p, r, e_1, \ldots, e_m$ are at most $2\ell^2 - n - 2$.

We form election $E = (C \cup D, V)$ via combining E' with candidates $D = \{1, \ldots, n\}$ and appropriate voters such that the results of the head-to-head contests are:

1. p ties with all candidates in D;
2. for each e_j, if e_j is incident with some $i \in D$ then candidate i defeats candidate e_j, and otherwise they tie;
3. all other candidates in C defeat each of the candidates in D.

We will now show that G contains a vertex cover of size at most k if and only if there is a set $D' \subseteq D$ such that p is a winner (the unique winner) of Copeland$^\alpha$

---

[2] Note that via the second paragraph of the proof it is easy to build an election where r has a score of this form.

election $(C \cup D', V)$. It is easy to see that if D' corresponds to a vertex cover of size at most k then p is a winner (the unique winner) of Copeland$^\alpha$ election $(C \cup D', V)$. The reason is that adding each member of D' increases p's score by α, increases r's score by one, and for each e_j, adding $i \in D'$ increases e_j's score by α if and only if e_j is not incident with i. Thus, via a simple calculation of the scores of the candidates, it is easy to see that p is a winner (the unique winner) of this election.

On the other hand, assume that p can become a winner (the unique winner) of Copeland$^\alpha$ election $(C \cup D', V)$ via adding some subset D' of candidates from D. First, note that $\|D'\| \leq k$, since otherwise r would end up with more points than (at least as many points as) p and so p would not be a winner (would not be the unique winner). We claim that D' corresponds to a vertex cover of G. For the sake of contradiction, assume that there is some edge e_j incident to vertices u and v such that neither u nor v is in D'. However, if this was the case then candidate e_j would have more points than (at least as many points as) p and so p would not be a winner (would not be the unique winner). Thus, D' must form a vertex cover of size at most k.   ❑   Theorem 3

2.4   FPT Algorithm Schemes for Bounded-Case Control

The study of fixed-parameter complexity (see, e.g., [5]) has been expanding explosively since it was parented as a field by Downey, Fellows, and others in the late 1980s and the 1990s. Although the area has built a rich variety of complexity classes regarding parameterized problems, for the purpose of the current paper we need focus only on one very important class, namely, the class FPT. Briefly put, a problem parameterized by some value j is said to be *fixed-parameter tractable* (equivalently, to belong to the class FPT) if there is an algorithm for the problem whose running time is $f(j)n^{O(1)}$.

Notation. In our context, we consider two parameterizations: bounding the number of candidates and bounding the number of voters. We use the same notations used throughout this paper to describe problems, except we postpend a "-BV$_j$" to a problem name to state that the number of voters may be at most j, and we postpend a "-BC$_j$" to a problem name to state that the number of candidates may be at most j. In each case, the bound applies to the full number of such items involved in the problem. For example, in the case of control by adding voters, the j must bound the total of the number of voters in the election added together with the number of voters in the pool of voters available for adding.

Fixed–Parameter Tractability Results. In their seminal paper on NP-hard winner-determination problems, Bartholdi, Tovey, and Trick [1] suggested considering hard election problems for the cases of a bounded number of candidates or a bounded number of voters, and they obtained efficient-algorithm results for such cases. Within the study of elections, this same approach—seeking efficient fixed-parameter algorithms—has also been used, for example, within the study of bribery [7]. Faliszewski et al. [9] showed that the 16 resistance results for constructive and destructive voter control within Copeland^0 and Copeland^1 (see Table 1) are in FPT (i.e., they each are fixed-parameter tractable) if the number of candidates is bounded, and also if the number of voters is bounded. They also showed that these results hold even when the multiplicities of preference lists in a given election are represented succinctly (by a binary number).

We generalize these results in Theorems 4 and 5 below. To state those theorems concisely, we borrow a notational approach from transformational grammar, and use square brackets as an "independent choice" notation. So, for example, the claim $\begin{bmatrix} \text{It} \\ \text{She} \\ \text{He} \end{bmatrix} \begin{bmatrix} \text{runs} \\ \text{walks} \end{bmatrix}$ is a shorthand for six assertions: It runs; She runs; He runs; It walks; She walks; and He walks. A special case is the symbol "∅" which, when it appears in such a bracket, means that when unwound it should be viewed as no text at all. For example, "$\begin{bmatrix} \text{Succinct} \\ \emptyset \end{bmatrix}$ Copeland is fun" asserts both "Succinct Copeland is fun" and "Copeland is fun."

Theorem 4. *For each rational α, $0 \leq \alpha \leq 1$, and each choice from the independent choice brackets below, the specified parameterized (as j varies over $\mathbb{N}$) problem is in* FPT:

$$\begin{bmatrix} \text{succinct} \\ \emptyset \end{bmatrix} - \begin{bmatrix} \text{Copeland}^\alpha \\ \text{Copeland}^\alpha_{\text{Irrational}} \end{bmatrix} - \begin{bmatrix} C \\ D \end{bmatrix} C \begin{bmatrix} \text{AV} \\ \text{DV} \\ \text{PV-TE} \\ \text{PV-TP} \end{bmatrix} - \begin{bmatrix} \text{BV}_j \\ \text{BC}_j \end{bmatrix}.$$

Theorem 5. *For each rational α, $0 \leq \alpha \leq 1$, and each choice from the independent choice brackets below, the specified parameterized (as j varies over $\mathbb{N}$) problem is in* FPT:

$$\begin{bmatrix} \text{succinct} \\ \emptyset \end{bmatrix} - \begin{bmatrix} \text{Copeland}^\alpha \\ \text{Copeland}^\alpha_{\text{Irrational}} \end{bmatrix} - \begin{bmatrix} C \\ D \end{bmatrix} C \begin{bmatrix} \text{AC}_u \\ \text{AC} \\ \text{DC} \\ \text{PC-TE} \\ \text{PC-TP} \\ \text{RPC-TE} \\ \text{RPC-TP} \end{bmatrix} - \text{BC}_j.$$

The proofs of Theorems 4 and 5, which in particular employ Lenstra's [14] algorithm for bounded-variable-cardinality integer programming, are omitted here.

FPT and Extended Control. We now introduce and look at extended control. By that we do not mean changing the basic control notions of adding/deleting/partitioning candidates/voters. Rather, we mean generalizing past merely looking at the constructive (make a distinguished candidate a winner) and the destructive (prevent a distinguished candidate from being a winner) cases. In particular, we are interested in control where the goal can be far more flexibly specified, for example (though in the partition cases we will be even more flexible than this), we will allow as our goal region any (reasonable—there are some time-related conditions) subcollection of "Copeland output tables" (specifications of who won/lost/tied each head-to-head contest).

Since from a Copeland output table, in concert with the current α, one can read off the Copeland$^\alpha_{\text{Irrational}}$ scores of the candidates, this allows us a tremendous range of descriptive flexibility in specifying our control goals, e.g., we can specify a linear order desired for the candidates with respect to their Copeland$^\alpha_{\text{Irrational}}$ scores, we can specify a linear-order-with-ties desired for the candidates with respect to their Copeland$^\alpha_{\text{Irrational}}$ scores, we can specify the exact desired Copeland$^\alpha_{\text{Irrational}}$ scores for one or more candidates, we can specify that we want to ensure that no candidate from a certain subgroup has a Copeland$^\alpha_{\text{Irrational}}$ score that ties or beats the Copeland$^\alpha_{\text{Irrational}}$ score of any candidate from a certain other subgroup, etc.

All the FPT algorithms given in Theorems 4 and 5 regard, on their surface, the standard control problem, which tests whether a given candidate can be made a winner (constructive case) or can be precluded from being a winner (destructive case). We note that the general approaches used to prove those results in fact yield FPT schemes even for the far more flexible notions of control that we just mentioned.

Resistance Results. In contrast with the FPT results in [9] for Copeland^0 and Copeland^1, Faliszewski et al. [9] showed that, for $\alpha \in \{0,1\}$, Copeland$^\alpha_{\text{Irrational}}$ remains resistant to all types of candidate control even for two voters. We generalize these results by showing that even for each rational α, $0 \leq \alpha \leq 1$, for Copeland$^\alpha_{\text{Irrational}}$ all 19 candidate-control cases that we showed earlier in this paper (i.e., without bounds on the number of voters) to be resistant remain resistant even for the case of bounded voters (nonsuccinct). This resistance holds even when the input is not in succinct format, and so it certainly also holds when the input is in succinct format.

It remains open whether Table 1's resistant, rational-voter, candidate-control cases remain resistant for the bounded-voter case.

3 Bribery

We now turn to the issue of bribery (see [7]), where the briber seeks to reach his or her goal via bribing certain voters to make them change their preferences.

Name: $\mathcal{E}$-bribery.
Given: A set C of candidates, a collection V of voters represented via their preference lists (or preference tables in the irrational case) over C, a distinguished candidate $p \in C$, and a nonnegative integer k.
Question: Does there exist a voter collection V' over C, where V' results from V by modifying at most k voters, such that p wins the $\mathcal{E}$ election (C, V')?

For $\mathcal{E}$-destructive-bribery, the destructive bribery problem for $\mathcal{E}$, we require p to be *not* a winner. We say $\mathcal{E}$ is *vulnerable to constructive (destructive) bribery* if $\mathcal{E}$-bribery ($\mathcal{E}$-destructive-bribery) is in P. We say $\mathcal{E}$ is *resistant to constructive (destructive) bribery* if $\mathcal{E}$-bribery ($\mathcal{E}$-destructive-bribery) is NP-hard. Theorem 6 carries to the case of all rationals α, $0 \leq \alpha \leq 1$, a result for Copeland^0 and Copeland^1 of [9].

Theorem 6. *For each rational* α, $0 \le \alpha \le 1$, *Copeland*$^\alpha$ *is resistant to both constructive and destructive bribery in both the rational-voters case and the irrational-voters case.*

We also can show that for all rationals α, $0 \le \alpha \le 1$, Copeland$^\alpha_{\text{Irrational}}$ is vulnerable to destructive microbribery. Informally put, microbribery [9] means that the briber pays separately for each preference-table entry flip of irrational voters.

4   Conclusions

We studied Copeland$^\alpha$ elections with respect to their resistance and vulnerability to control and bribery. Among the election systems whose winners can be determined in polynomial time, we identified the first natural election system, Copeland (i.e., Copeland$^{0.5}$), that provides full resistance to constructive control. In addition, we proved results for bribery and for fixed-parameter tractability of bounded-case control for Copeland$^\alpha$ for each rational α, $0 < \alpha < 1$.

References

1. Bartholdi III, J., Tovey, C., Trick, M.: Voting schemes for which it can be difficult to tell who won the election. Social Choice and Welfare 6(2), 157–165 (1989)
2. Bartholdi III, J., Tovey, C., Trick, M.: How hard is it to control an election? Mathematical and Computer Modeling 16(8/9), 27–40 (1992)
3. Conitzer, V., Sandholm, T., Lang, J.: When are elections with few candidates hard to manipulate? Journal of the ACM 54(3) Article 14(2007)
4. Copeland, A.: A "reasonable" social welfare function. Mimeographed notes from a Seminar on Applications of Mathematics to the Social Sciences, University of Michigan (1951)
5. Downey, R., Fellows, M.: Parameterized Complexity. Springer, Heidelberg (1999)
6. Erdélyi, G., Hemaspaandra, L., Rothe, J., Spakowski, H.: Frequency of correctness versus average polynomial time and generalized juntas (manuscript, December 2007)
7. Faliszewski, P., Hemaspaandra, E., Hemaspaandra, L.: The complexity of bribery in elections. In: Proc. AAAI 2006, pp. 641–646. AAAI Press, Menlo Park (2006)
8. Faliszewski, P., Hemaspaandra, E., Hemaspaandra, L., Rothe, J.: Llull and Copeland voting computationally resist bribery and control. Technical report, Department of Computer Science, University of Rochester, Rochester, NY (in preparation)
9. Faliszewski, P., Hemaspaandra, E., Hemaspaandra, L., Rothe, J.: Llull and Copeland voting broadly resist bribery and control. In: Proc. AAAI 2007, pp. 724–730. AAAI Press, Menlo Park (2007)
10. Hemaspaandra, E., Hemaspaandra, L., Rothe, J.: Anyone but him: The complexity of precluding an alternative. Artificial Intelligence 171(5-6), 255–285 (2007)
11. Hemaspaandra, E., Hemaspaandra, L., Rothe, J.: Hybrid elections broaden complexity-theoretic resistance to control. In: Proc. IJCAI 2007, pp. 1308–1314. AAAI Press, Menlo Park (2007)

12. Homan, C., Hemaspaandra, L.: Guarantees for the success frequency of an algorithm for finding Dodgson-election winners. Journal of Heuristics (to appear). Full version available as [13]
13. Homan, C., Hemaspaandra, L.: Guarantees for the success frequency of an algorithm for finding Dodgson-election winners. Technical Report TR-881, Department of Computer Science, University of Rochester, Rochester, NY, September 2005. Revised (June 2007)
14. Lenstra Jr., H.: Integer programming with a fixed number of variables. Mathematics of Operations Research 8(4), 538–548 (1983)
15. Merlin, V., Saari, D.: Copeland method II: Manipulation, monotonicity, and paradoxes. Journal of Economic Theory 72(1), 148–172 (1997)
16. Procaccia, A., Rosenschein, J., Kaminka, G.: On the robustness of preference aggregation in noisy environments. In: Proc. AAMAS 2007, pp. 416–422. ACM Press, New York (2007)
17. Procaccia, A., Rosenschein, J., Zohar, A.: Multi-winner elections: Complexity of manipulation, control, and winner-determination. In: Proc. IJCAI 2007, pp. 1476–1481. AAAI Press, Menlo Park (2007)
18. Saari, D., Merlin, V.: The Copeland method I: Relationships and the dictionary. Economic Theory 8(1), 51–76 (1996)
19. Zermelo, E.: Die Berechnung der Turnier-Ergebnisse als ein Maximumproblem der Wahrscheinlichkeitsrechnung. Mathematische Zeitschrift 29(1), 436–460 (1929)

The Complexity of Power-Index Comparison

Piotr Faliszewski* and Lane A. Hemaspaandra**

Department of Computer Science
University of Rochester
Rochester, NY 14627

Abstract. We study the complexity of the following problem: Given two weighted voting games G' and G'' that each contain a player p, in which of these games is p's power index value higher? We study this problem with respect to both the Shapley-Shubik power index [16] and the Banzhaf power index [3,6]. Our main result is that for both of these power indices the problem is complete for probabilistic polynomial time (i.e., is PP-complete). We apply our results to partially resolve some recently proposed problems regarding the complexity of weighted voting games. We also show that, unlike the Banzhaf power index, the Shapley-Shubik power index is not #P-parsimonious-complete. This finding sets a hard limit on the possible strengthenings of a result of Deng and Papadimitriou [5], who showed that the Shapley-Shubik power index is #P-metric-complete.

Keywords: Weighted voting games; power indices; computational complexity.

1 Introduction

In an abstract, direct democracy, each member in a certain sense has equal potential for impact on the decisions that the society makes. However, in many practical decision-making scenarios it is reasonable to give up this noble idea and consider weighted voting instead. Here are a few motivating examples. In a country divided into districts it makes sense to give each district voting power proportional to its population (consider, e.g., the US House of Representatives or various decision making processes within the European Union). In fact, the power that various apportionment methods give to the US states in its House of Representatives has been studied in terms of how well it is proportional to the sizes of the states [10]. In a business setting, stockholders in a company might hope to have voting power proportional to the amount of stock they own. Within computer science, Dwork et al. [7] suggested building a meta search engine for the web via treating other search engines as voters in an election. It would only

---

* Work done in part while visiting Heinrich-Heine-Universität Düsseldorf, Germany. Supported in part by grant NSF-CCF-0426761.
** Supported in part by grant NSF-CCF-0426761, a TransCoop grant, and a Friedrich Wilhelm Bessel Research Award.

R. Fleischer and J. Xu (Eds.): AAIM 2008, LNCS 5034, pp. 177–187, 2008.
© Springer-Verlag Berlin Heidelberg 2008

be natural to weigh the participating search engines with their (quantified in some way) quality. Naturally, one can provide many other examples.

The focus of this paper is on the computational complexity of the following issue: Given an individual and two weighted voting scenarios (in each of them our individual might have different weight and each scenario might involve different sets of voters with different weights), in which one of them is our individual more influential? (We provide a formal definition of this problem in Section 1.1.) This problem has a very natural motivation. For example, consider a company that wishes to join some business consortium and has a choice among several consortia (e.g., consider an airline deciding which airline alliance to join). It is natural to assume that within each consortium companies make decisions via weighted voting, with companies weighted, e.g., via their size or revenue or some combination thereof. In a political context, members of the European Union sometimes try to promote new schemes of distributing vote weights among EU members. It is important for the countries involved to see which scheme is better for them. One can easily give many other applications of the issue we study.

Formally, we model the above problem via comparing the values of power index functions—in our case those of Shapley and Shubik [16] and of Banzhaf ([3], see also [6])—of a particular player within two given weighted voting games. Our main result is that this problem is PP-complete for both the Shapley-Shubik power index and the Banzhaf power index.

An extended version of this paper, with several different proofs and with additional results, appears as [8].

1.1   The Power-Index Comparison Problem

We model weighted voting via so-called weighted voting games. An n-player weighted voting game is a sequence of n nonnegative integer weights, $w_1, \ldots, w_n$, together with a quota q. We denote it as $(w_1, \ldots, w_n; q)$. We refer to the player with weight w_i as the i'th player. Weighted voting games model the following scenario: The players are given a yes/no question (e.g., should we lower the taxes? should we buy out our competitors?) and each player either agrees (answers *yes*) or disagrees (answers *no*). If the total weight of the voters who agree is at least as high as the quota then the result of the game is *yes* and otherwise it is *no*.

Let G be a voting game $(w_1, \ldots, w_n; q)$. Any subset of $\{1, \ldots, n\}$ is a coalition in G. We say that a coalition S is successful if $\sum_{i \in S} w_i \geq q$. We define $\mathrm{succ}_G(S)$ to be 1 if S is a successful coalition for G and to be 0 otherwise.

Interestingly, the relation between the effective power of a player within a voting game and his or her weight is not as simple as one might think. Consider game $G = (8, 7, 2; 9)$, i.e., a game with quota $q = 9$ and three players with weights 8, 7, and 2, respectively. It is easy to see that in this game any coalition of at least two players is successful. In effect, each of the players can influence the final result of the game to exactly the same degree, regardless of the fact that their weights differ significantly. Thus when analyzing weighted voting games it is standard to measure players' power using, e.g., the Shapley-Shubik power index [16] or the Banzhaf power index [3,6].

In essence, these power indices measure the probability that, assuming some coalition formation model, our designated player is critical for the forming coalition. By *critical* we mean here that the coalition is successful with our designated player but is not successful without him or her.

Let $G = (w_1, \ldots, w_n; q)$ be a voting game, let i be a player in this game, and let $N = \{1, \ldots, n\}$ be the set of all players of G. The value of the Banzhaf power index of i in G is defined as $\mathrm{Banzhaf}(G, i) = \frac{\mathrm{Banzhaf}^*(G,i)}{2^{n-1}}$, where $\mathrm{Banzhaf}^*(G, i)$ is the *raw* version of the index,

$$\mathrm{Banzhaf}^*(G, i) = \sum_{S \subseteq N - \{i\}} (\mathrm{succ}_G(S \cup \{i\}) - \mathrm{succ}_G(S)).$$

The Shapley-Shubik power index of player i in game G is defined as $\mathrm{SS}(G, i) = \frac{\mathrm{SS}^*(G,i)}{n!}$, where $\mathrm{SS}^*(G, i)$ is the raw version of the index,

$$\mathrm{SS}^*(G, i) = \sum_{S \subseteq N - \{i\}} \|S\|!(n - \|S\| - 1)!(\mathrm{succ}_G(S \cup \{i\}) - \mathrm{succ}_G(S)).$$

Intuitively, $\mathrm{Banzhaf}(G, i)$ gives the probability that a randomly chosen coalition of players in $N - \{i\}$ is not successful but would become successful had player i joined in. The intuition for the Shapley-Shubik index is that we count the proportion of permutations for which a given player is pivotal. Given a permutation π of $\{1, \ldots, n\}$, the $\pi(i)$'th player is pivotal if it holds that the coalition $\{\pi(1), \pi(2), \ldots, \pi(i)\}$ is successful and the coalition $\{\pi(1), \pi(2), \ldots, \pi(i-1)\}$ is not. This permutation-based intuition is motivated by the view of the successful-coalition formation as the process of players joining in in random order. Naturally, the first player that makes the coalition successful is crucial and so the idea is to measure power via counting how often our player-of-interest is pivotal.

The focus of this paper is on the computational complexity analysis of the following problem.

Definition 1. *Let f be either the Shapley-Shubik or the Banzhaf power index. By* PowerCompare$_f$ *we mean the problem where the input (G', G'', i) contains two weighted voting games, $G' = (w'_1, \ldots, w'_n, q')$ and $G'' = (w''_1, \ldots, w''_n, q'')$, and an integer i, $1 \leq i \leq n$, and where we ask whether $f(G', i) > f(G'', i)$.*

Note that in the above definition we assume that both games have the same number of players. At first this might seem to be a weakness but it is easy to see that given two games with different numbers of players we can easily pad the smaller one with weight-0 players. On the other hand, the assumption that both games have the same number of players allows us to solve the problem via comparing the raw values of the index: The scaling factor for both games is the same and thus it does not affect the result of the comparison.

1.2   Computational Complexity

We assume that the reader is familiar with the standard notions of computational complexity theory, as presented, e.g., in the textbooks of Papadimitriou [14] and

Bovet and Crescenzi [4]. Below we review some of them, focusing on the less frequently known ones.

We fix the alphabet $\Sigma = \{0,1\}$ and we assume that all the problems we consider are encoded in a natural, efficient manner over Σ. By $|\cdot|$ we mean the length function. We assume $\langle \cdot, \cdot \rangle$ to be a standard, natural pairing function such that $|\langle x, y \rangle| = 2(|x| + |y|) + 2$.

The main result of this paper, Theorem 1, says that the power-index comparison problem is PP-complete both for the Shapley-Shubik power index and for the Banzhaf power index. The class PP, probabilistic polynomial time, was defined by Simon [17] and Gill [9]. A language $L \subseteq \Sigma^*$ belongs to PP if and only if there exists a polynomial p and a polynomial-time computable relation R such that

$$x \in L \iff \|\{w \in \Sigma^{p(|x|)} \mid R(x, w) \text{ holds}\}\| > 2^{p(|x|)-1}.$$

PP captures the set of languages having a probabilistic Turing machine that on precisely the elements of the set has strictly more than 50% probability of acceptance. Let us mention that PP is a very powerful class. For example, it is well-known that NP is a subset of PP (as are even various larger classes). Via Toda's Theorem [18], we know that PH $\subseteq$ P^{PP}. That is, PP is at least as powerful as polynomial-time hierarchy, give or take the flexibility of polynomial-time Turing reductions. Many other properties of PP have been established in the literature.

Let us now recall the definition of the class #P [19]. For each NP machine N (i.e., for each nondeterministic polynomial-time machine N), by #acc$_N(x)$ we mean the number of accepting computation paths of N running with input x. A function f, $f : \Sigma^* \to \mathbb{N}$, belongs to #P if and only if there is an NP machine N such that $(\forall x \in \Sigma^*)[f(x) = \#\text{acc}_N(x)]$. #P is, in some sense, a functional counterpart of PP. For example, P$^{\#\text{P}}$ = P^{PP} [2]. More typically, #P is described as the counting analogue of NP.

As is usual, we say that a language L is hard for a complexity class C if every language in C polynomial-time many-one reduces to L. If in addition L belongs to C then we say that L is C-complete. A language A polynomial-time many-one reduces to a language B if there exists a polynomial-time computable function f such that for each string $x \in \Sigma^*$ it holds that $x \in A \iff f(x) \in B$. On the other hand, there is no one agreed-upon notion of completeness for function classes. For example, Valiant [19] in his seminal paper used Turing reductions but other people have preferred notions such as Krentel's metric reductions [12], Zankó's many-one reductions (for functions) [20], and Simon's [17] parsimonious reductions.

In the context of power index functions, Prasad and Kelly [15] (implicitly) showed that the (raw) Banzhaf power index is #P-parsimonious-complete and Deng and Papadimitriou [5] established that the (raw) Shapley-Shubik power index is #P-metric-complete (regarding the complexity analysis of power indices, we also mention the paper of Matsui and Matsui [13]). We now review the parsimonious and metric reductions that yield the parsimonious-completeness and metric-completeness notions.

Definition 2 ([17,12]). *We say that a function $f : \Sigma^* \to \mathbb{N}$ metric reduces to a function $g : \Sigma^* \to \mathbb{N}$ if there exist two polynomial-time computable functions, φ and ψ, such that $(\forall x \in \Sigma^*)[f(x) = \psi(x, g(\varphi(x)))]$. We say that f parsimoniously reduces to g if there is a polynomial-time computable function φ such that $(\forall x \in \Sigma^*)[f(x) = g(\varphi(x))]$.*

Note that if f parsimoniously reduces to g, then f metric reduces to g. Given a function class C, we say that a function f is C-parsimonious-complete if $f \in C$ and each function in C parsimonious reduces to f. C-metric-completeness is defined analogously. Typically, parsimonious-complete functions are easier to work with than functions that are only metric-complete. In particular, our proof of Theorem 4 is more involved than our proof of Theorem 2 because, as we show, the Shapley-Shubik power index is not parsimoniously complete.

2   Main Results

Our main result, Theorem 1, says that the power index comparison problem is PP-complete. This section is devoted to building infrastructure for Theorem 1's proof and giving that proof.

Theorem 1. *Let f be either the Banzhaf or the Shapley-Shubik power index. The problem PowerCompare$_f$ is PP-complete.*

We start via showing PP-membership of a problem closely related to our PowerCompare$_{\text{Banzhaf}}$ and PowerCompare$_{\text{SS}}$ problems. Let f be a #P function and let Compare$_f$ be the language $\{\langle x,y \rangle \mid x,y \in \Sigma^* \wedge f(x) > f(y)\}$. (PowerCompare$_{\text{Banzhaf}}$ and PowerCompare$_{\text{SS}}$ are essentially, up to a minor definitional issue, incarnations of Compare$_f$ for appropriate functions f.)

Lemma 1. *Let f be a #P function. The language Compare$_f$ is in PP.*

Proof. Let f be an arbitrary #P function and let N be an NP machine such that $f = \#\text{acc}_N$. Without the loss of generality, we assume that there is a polynomial q such that for each input $x \in \Sigma^*$ all computation paths of N make exactly $q(|x|)$ binary nondeterministic choices. Thus each computation path of N on input x can be represented as a string w in $\Sigma^{q(|x|)}$.

In order to show that Compare$_f$ is in PP we need to provide a polynomial-time computable relation R and a polynomial p such that for each string $z = \langle x, y \rangle$ it holds that:

$$z \in \text{Compare}_f \iff \|\{w \in \Sigma^{p(|z|)} \mid R(z,w) \text{ holds}\}\| > 2^{p(|z|)-1}.$$

We now define such R and p. Let us fix two strings, x and y, and let $z = \langle x, y \rangle$ and $n = |z|$. We define $p(n) = q(n) + 1$ and, for each string $w = w_0 w_1 \ldots w_{p(n)-1} \in \Sigma^{p(n)}$, we define $R(z,w)$ as follows:

Case 1. If $w_0 = 0$ then $R(z,w)$ is *true* exactly if the string $w_1, \ldots, w_{q(|x|)}$ spells an accepting computation path of N on x and the symbols $w_{q(|x|)+1}$ through $w_{p(n)-1}$ are all 0. $R(z,w)$ is false otherwise.

Case 2. If $w_0 = 1$ then $R(z, w)$ is *false* exactly if the string $w_1, \ldots, w_{q(|y|)}$ spells an accepting computation path of N on y and the symbols $w_{q(|x|)+1}$ through $w_{p(n)-1}$ are all 0. $R(z, w)$ is true otherwise.

Via analyzing the above two cases it is easy to see that there are exactly $f(x) + (2^{p(n)-1} - f(y)) = f(x) - f(y) + 2^{p(n)-1}$ strings $w \in \Sigma^{p(n)}$ for which $R(z, w)$ is true. This value is greater than $2^{p(n)-1}$ if and only if $f(x) > f(y)$. Thus, the relation R and the polynomial p jointly witness that Compare$_f$ belongs to PP. $\qquad \Box$

Lemma 1 gives an upper bound on the complexity of Compare$_f$ (assuming that $f \in \#P$). We now prove a matching lower bound, PP-completeness, for the case that f is #P-parsimonious-complete.

Lemma 2. *Let f be a #P-parsimonious-complete function. The language* Compare$_f$ *is PP-complete.*

Proof. Let f be a #P-parsimonious-complete function. Via Lemma 1 we know that Compare$_f$ is in PP and thus to show PP-completeness it remains to show PP-hardness. We do so via reducing an arbitrary PP language L to Compare$_f$.

Let L be an arbitrary PP language. By definition, there exists a polynomial-time relation R and a polynomial p such that for each string $x \in \Sigma^*$ it holds that $x \in L \iff \|\{y \in \Sigma^{p(|x|)} \mid R(x, y) \text{ holds}\}\| > 2^{p(|x|)-1}$. We define two functions, g_1 and g_2, such that $g_1(x) = \|\{y \in \Sigma^{p(|x|)} \mid R(x, y) \text{ holds}\}\|$ and $g_2(x) = 2^{p(|x|)-1}$. It is easy to see that both g_1 and g_2 are in #P. g_1 can be computed via a an NP machine that on input x guesses a binary string y of length $p(|x|)$ and accepts if and only if $R(x, y)$ holds. g_2 can be computed via a machine that on input x guesses a binary string of length $2^{p(|x|)-1}$ and then accepts. Naturally, $x \in L$ if and only if $g_1(x) > g_2(x)$.

Since f is #P-parsimonious-complete, both g_1 and g_2 parsimoniously reduce to f. Let φ_1 be the reduction function for g_1 and let φ_2 be the reduction function for g_2. We have that for each string x it holds that $g_1(x) = f(\varphi_1(x))$ and $g_2(x) = f(\varphi_2(x))$.

Our reduction from L to Compare$_f$ works as follows. On input x we output the string $z = \langle \varphi_1(x), \varphi_2(x) \rangle$. Clearly, this can be done in polynomial time. To show correctness it is enough to recall that $x \in L$ if and only if $g_1(x) > g_2(x)$, which is equivalent to testing whether z is in Compare$_f$. Since L was chosen as an arbitrary PP language, this proves PP-completeness. $\qquad \Box$

We are almost ready to show that PowerCompare$_{\text{Banzhaf}}$ is PP-complete. However, in order to do so, we need to justify the claim that the raw version of the Banzhaf power index is #P-parsimonious-complete. (This was shown implicitly in the work of Prasad and Kelly [15], but we feel that it is important to explicitly outline the proof.)

One of our important tools here (and later on) is the function #X3C. The input to the X3C problem is a set $B = \{b_1, \ldots, b_{3k}\}$ and a family $\mathcal{S} = \{S_1, \ldots, S_n\}$ of 3-element subsets of B. The X3C problem asks whether there exists a collection of exactly k sets in $\mathcal{S}$ whose union is B. #X3C$(B, \mathcal{S})$ is the number of solutions of the X3C instance $(B, \mathcal{S})$.

Hunt et al. [11] showed that #X3C is parsimonious complete for #P. This is very useful for us as the standard reduction from #X3C to #SubsetSum (see, e.g., [14, Theorem 9.10]; #SubsetSum is the function that accepts as input a target integer K and a set S of nonnegative integers and returns the number of subsets of S that sum up to K) is parsimonious and Prasad and Kelly's reduction from #SubsetSum to Banzhaf* (the raw version of Banzhaf's power index) is parsimonious as well. Since Banzhaf* is in #P, Banzhaf* is #P-parsimonious-complete. Thus the following theorem is, essentially, a direct consequence of Lemma 2.

Theorem 2. PowerCompare$_{\text{Banzhaf}}$ *is* PP-*complete.*

Proof. The raw version of the Banzhaf power index is #P-parsimonious-complete and so, via Lemma 2, Compare$_{\text{Banzhaf}*}$ is PP-complete. Via a slight misuse of notation, we can say that Compare$_{\text{Banzhaf}*}$ accepts as input two weighted voting games, G' and G'', and two players, p' and p'', such that p' participates in G' and p'' participates in G'' and accepts if and only if Banzhaf*(G', p') > Banzhaf*(G', p''). We give a reduction from Compare$_{\text{Banzhaf}*}$ to PowerCompare$_{\text{Banzhaf}}$.

Let G', p' and G'', p'' be our input to the Compare$_{\text{Banzhaf}*}$ problem. We can assume that G' and G'' have the same number of players. If G' and G'' do not have the same number of players then it is easy to see that the game with fewer players can be padded with players whose weight is equal to this game's quota value. Such a padding leaves the raw Banzhaf power index values of the game's original players unchanged. (The reason for this is that any coalition that includes any of the padding candidates is already winning and so none of the original players is critical to the success of the coalition, and so the coalition does not contribute to original players' power index values.)

We form two games, K' and K'', that are identical to games G' and G'', respectively, except that K' lists player p' as first and G'' lists player p'' as first. Our reduction's output is $(K', K'', 1)$.

Naturally, Banzhaf$(K', 1)$ > Banzhaf$(K'', 1)$ if and only if Banzhaf*(G', p') > Banzhaf*(G'', p''). Also, clearly, K' and K'' can be computed in polynomial time. Thus we have successfully reduced Compare$_{\text{Banzhaf}*}$ to PowerCompare$_{\text{Banzhaf}}$. This shows PP-hardness of PowerCompare$_{\text{Banzhaf}}$. PP-membership of PowerCompare$_{\text{Banzhaf}}$ is, essentially, a simple consequence of Lemma 1. This completes the proof.   □

Let us now focus on the computational complexity of the power index comparison problem for the case of Shapley-Shubik. It would be nice if the raw Shapley-Shubik power index were #P-parsimonious-complete. Had it been the case, we could establish PP-completeness of PowerCompare$_{\text{SS}}$ in essentially the same way as we have done for PowerCompare$_{\text{Banzhaf}}$. However, Prasad and Kelly [15] at the end of their paper, after—in effect—showing #P-parsimonious-completeness of the raw Banzhaf power index (their Theorem 4) write: "Such a straightforward approach does not seem possible with the Shapley-Shubik [power index]." We reinforce their intuition by now proving that the raw Shapley-Shubik power index in fact is *not* #P-parsimonious-complete.

Theorem 3. *The raw Shapley-Shubik power index, the function* SS*, *is not #P-parsimonious-complete.*

Proof. For the sake of contradiction, let us assume that SS* is #P-parsimonious-complete. Thus for each natural number k there is a weighted voting game G and a player i within G such that $SS^*(G, i) = k$. This is the case because the function $f(x) = x$ belongs to #P (we assume that the "output x" is an integer obtained via a standard bijection between Σ^* and $\mathbb{N}$) and if SS* is #P-parsimonious-complete then there has to be a parsimonious reduction from f to SS*.

Let G be an arbitrary voting game with $n \geq 4$ players and let i be a player in G. By definition, $SS^*(G, i)$ is a sum of terms of the form $k!(n - k - 1)!$, where k is some value in $\{0, \ldots, n - 1\}$. Since $n \geq 4$, each such term is even and thus $SS^*(G, i)$ is even. The raw Shapley-Shubik power index of any player in a game with at most 3 players is at most $3! = 6$ and thus there is no input on which SS* yields the value 7. This contradicts the assumption that SS* is #P-parsimonious-complete and completes the proof. $\square$

So the well-known result of Deng and Papadimitriou [5] that the raw Shapley-Shubik power index is #P-metric-complete cannot be strengthened to #P-parsimonious-completeness. Theorem 3 prevents us from directly using Lemma 2 to show that PowerCompare$_{SS}$ is PP-complete. Nonetheless, via the following lemma, we will be able to obtain our result.

Lemma 3. *Let f and g be two arbitrary #P functions. There exists a polynomial-time computable function* $cmp_{f,g}(x, y)$ *such that*

$$(\forall x, y \in \Sigma^*)[f(x) > g(y) \iff cmp_{f,g}(x, y) \in \text{PowerCompare}_{SS}].$$

Proof. Let f and g be as in the lemma and let x and y be two arbitrary strings. Since both f and g are in #P and #X3C is #P-parsimonious-complete, there exist functions ϕ_f and ϕ_g that compute parsimonious reductions from f to #X3C and from g to #X3C, respectively.

Let $(B_x, \mathcal{S}_x) = \phi_f(x)$ and $(B_y, \mathcal{S}_y) = \phi_g(y)$. We can, without the loss of generality, assume that $\|\mathcal{S}_x\| = \|\mathcal{S}_y\| = r$ and that $\|B_x\| = \|B_y\| = 3k$, where r and k are two nonnegative integers. If this is not the case, we can use the following algorithm.

If, say, $\|B_x\|$ is not a multiple of three then we have $\#X3C(B_x, \mathcal{S}_x) = 0$ and we can add one or two extra elements to B_x without introducing any new solutions but ensuring that $\|B_x\|$ is a multiple of three. Naturally, the same trick works for B_y. If both $\|B_x\|$ and $\|B_y\|$ are multiples of three and, for example, $\|B_x\| < \|B_y\|$ then we can add three new elements, b', b'', b''' to B_x and include the set $\{b', b'', b''\}$ in $\mathcal{S}_x$. This doesn't change the number of solutions of $(B_x, \mathcal{S}_x)$, but it increases the cardinality of B_x by three. We can keep repeating this operation until the cardinalities of the sets B_x and B_y match. Then, via similar trickery, without breaking the $\|B_x\| = \|B_y\|$ equality we can ensure that $\|\mathcal{S}_x\| = \|\mathcal{S}_y\|$. We omit the details of this process but the reader can easily recreate them.

Let ϕ_s be the standard, parsimonious reduction from #X3C to #SubsetSum (see, e.g., [14, Theorem 9.10]). ϕ_s has the property that given an instance $(B, \mathcal{S})$

where $\|B\| = 3m$ for some nonnegative integer m, $\phi_s(B, \mathcal{S})$ produces an instance $(s_1, \ldots, s_n; q)$ of SubsetSum such that $n = \|\mathcal{S}\|$ and every subset of $\{s_1, \ldots, s_n\}$ that sums up to q has exactly m elements. Based on such an instance of SubsetSum we can build voting game $G = (1, s_1, \ldots, s_n; q + 1)$. Deng and Papadimitriou [5, Theorem 9] observed that the raw Shapley-Shubik power index of the first player in this game is exactly

$$(n - m)!m! \cdot \#\text{SubsetSum}(s_1, \ldots, s_n; q).$$

Since ϕ_s is parsimonious, this value is equal to $(n - m)!m! \cdot \#\text{X3C}(B, \mathcal{S})$.

We now describe our function $\text{cmp}_{f,g}$. Given instance $(B_x, \mathcal{S}_x)$ we build voting game G_x as follows. We compute SubsetSum instance $\phi_s(B_x, \mathcal{S}_x) = (t_1, \ldots, t_r; q_x)$ and we let $G_x = (1, t_1, \ldots, t_r; q_x + 1)$. Analogously, based on instance $(B_y, \mathcal{S}_y)$, we build voting game G_y. (Note that since $\|\mathcal{S}_x\| = \|\mathcal{S}_y\| = r$, both games have the same number of players.) We define $\text{cmp}_{f,g}(x, y)$ to output $(G_x, G_y, 1)$. Via the discussion above it holds that:

$$\text{SS}^*(G_x, 1) = (r - k)!k! \cdot \#\text{X3C}(B_x, \mathcal{S}_x) = (r - k)!k! f(x)$$
$$\text{SS}^*(G_y, 1) = (r - k)!k! \cdot \#\text{X3C}(B_y, \mathcal{S}_y) = (r - k)!k! g(y).$$

Thus, $f(x) > f(y)$ if and only if $\text{SS}(G_x, 1) > \text{SS}(G_y, 1)$ and so it is clear that function $\text{cmp}_{f,g}$ does what the theorem claims. Naturally, $\text{cmp}_{f,g}$ can be computed in polynomial time. □

Theorem 4. PowerCompare$_{\text{SS}}$ *is* PP-*complete*.

Proof. Via Lemma 1 it is easy to see that PowerCompare$_{\text{SS}}$ is in PP. Let h be some #P-parsimonious-complete function. PP-hardness of PowerCompare$_{\text{SS}}$ follows via a reduction from PP-complete problem Compare$_h$ (see Lemma 2). As a reduction we can use, e.g., the function $\text{cmp}_{h,h}$ from Lemma 3. This completes the proof. □

3   Conclusions and Open Problems

We have shown that the problem of deciding in which of the two given voting games our designated player has higher power index value is PP-complete for both Banzhaf and Shapley-Shubik power indices. For the case of Banzhaf, we have used the fact that Banzhaf power index is #P-parsimonious-complete. For the case of Shapley-Shubik, we have shown that Shapley-Shubik power index is not #P-parsimonious-complete, but using the index's properties we were able to show PP-completeness of PowerCompare$_{\text{SS}}$. We believe that these results are interesting and practically important. Below we mention one particular application.

In the context of multiagent systems, the Shapley-Shubik power index is often used to distribute players' payoffs, i.e., each player's payoff is proportional to his or her power index value. Recently Bachrach and Elkind [1] asked about the exact complexity of the following problem: Given a weighted voting game $G = (w_1, w_2, \ldots, w_n; q)$, is it profitable for players 1 and 2 to join? That is, if $G' = (w_1 + w_2, w_3, \ldots, w_n; q)$, is it the case that $SS(G', 1) > SS(G, 1) + SS(G, 2)$. Using Lemma 3 and the fact that #P is closed under addition we can easily show that this problem reduces to PowerCompare$_{SS}$ and thus is in PP. We believe that Bachrach and Elkind's problem is, in fact, PP-complete and that the techniques presented in this paper will lead to the proof of this fact. However, at this point the exact complexity of the problem remains open.

Acknowledgments. We thank Jörg Rothe and Edith Elkind for helpful discussions on the topic of weighted voting games and Jörg Rothe for hosting a visit during which this work was done in part.

References

1. Bachrach, Y., Elkind, E.: Divide and conquer: False-name manipulations in weighted voting games. In: Proceedings of the 7th International Conference on Autonomous Agents and Multiagent Systems, May 2008 (to appear, 2008)
2. Balcázar, J., Book, R., Schöning, U.: The polynomial-time hierarchy and sparse oracles. Journal of the ACM 33(3), 603–617 (1986)
3. Banzhaf, J.: Weighted voting doesn't work: A mathematical analysis. Rutgers Law Review 19, 317–343 (1965)
4. Bovet, D., Crescenzi, P.: Introduction to the Theory of Complexity. Prentice-Hall, Englewood Cliffs (1993)
5. Deng, X., Papadimitriou, C.: On the complexity of comparative solution concepts. Mathematics of Operations Research 19(2), 257–266 (1994)
6. Dubey, P., Shapley, L.: Mathematical properties of the Banzhaf power index. Mathematics of Operations Research 4(2), 99–131 (1979)
7. Dwork, C., Kumar, R., Naor, M., Sivakumar, D.: Rank aggregation methods for the web. In: Proceedings of the 10th International World Wide Web Conference, pp. 613–622. ACM Press, New York (2001)
8. Faliszewski, P., Hemaspaandra, L.: The complexity of power-index comparison. Technical Report TR-929, Department of Computer Science, University of Rochester, Rochester, NY (January 2008)
9. Gill, J.: Computational complexity of probabilistic Turing machines. SIAM Journal on Computing 6(4), 675–695 (1977)
10. Hemaspaandra, L., Rajasethupathy, K., Sethupathy, P., Zimand, M.: Power balance and apportionment algorithms for the United States Congress. ACM Journal of Experimental Algorithmics 3(1) (1998), http://www.jea.acm.org/1998/HemaspaandraPower
11. Hunt, H., Marathe, M., Radhakrishnan, V., Stearns, R.: The complexity of planar counting problems. SIAM Journal on Computing 27(4), 1142–1167 (1998)
12. Krentel, M.: The complexity of optimization problems. Journal of Computer and System Sciences 36(3), 490–509 (1988)
13. Matsui, Y., Matsui, T.: NP-completeness for calculating power indices of weighted majority games. Theoretical Computer Science 263(1–2), 305–310 (2001)
14. Papadimitriou, C.: Computational Complexity. Addison-Wesley, Reading (1994)
15. Prasad, K., Kelly, J.: NP-completeness of some problems concerning voting games. International Journal of Game Theory 19(1), 1–9 (1990)

16. Shapley, L., Shubik, M.: A method of evaluating the distribution of power in a committee system. American Political Science Review 48, 787–792 (1954)
17. Simon, J.: On Some Central Problems in Computational Complexity. PhD thesis, Cornell University, Ithaca, N.Y, Available as Cornell Department of Computer Science Technical Report TR75-224 (January 1975)
18. Toda, S.: PP is as hard as the polynomial-time hierarchy. SIAM Journal on Computing 20(5), 865–877 (1991)
19. Valiant, L.: The complexity of computing the permanent. Theoretical Computer Science 8(2), 189–201 (1979)
20. Zankó, V.: #P-completeness via many-one reductions. International Journal of Foundations of Computer Science 2(1), 76–82 (1991)

Facility Location Problems: A Parameterized View*

Michael Fellows[1] and Henning Fernau[1,2]

[1] The University of Newcastle
University Drive, Callaghan, NSW 2308, Australia
mfellows@newcastle.edu.au
[2] Universität Trier, FB IV—Abteilung Informatik, 54286 Trier, Germany
fernau@uni-trier.de

Abstract. Facility Location can be seen as a whole family of problems which have many obvious applications in economics. They have been widely explored in the Operations Research community, from the viewpoints of approximation, heuristics, linear programming, etc. We add a new facet by initiating the study of some of these problems from a parametric point of view. Moreover, we exhibit some less obvious applications of these algorithms in the processing of semistructured documents and in computational biology.

1 Introduction

The basic task of all variants of facility location problems is the following: a company wants to open up a number of facilities to serve their customers. Both the opening of a facility at a specific location and the service of a particular customer through a facility incurs some cost. The goal is to minimize the overall cost associated to a specific way of opening up facilities and serving customers. We will formalize this task in the following more precisely.

Definitions. We study the following problem of FACILITY LOCATION and variants thereof:

Given: A bipartite graph $B = (F \uplus C, E)$, consisting of a set F of potential facility locations, a set C of customers, and an edge relation E, where $\{f, c\} \in E$ indicates that c can be served from the facility (at) f; and weight functions $\omega_F : F \to \mathbb{N}_{\geq 1}$ and $\omega_E : E \to \mathbb{N}_{\geq 1}$ (both called ω if no confusion may arise), $k \in \mathbb{N}$

Question: Is there a set $F' \subseteq F$ of facility locations and a set $E' \subseteq E$ of ways to serve customers such that $((1) \; \forall f \in F(f \in F' \iff \exists e \in E'(f \in e))$, (2) $\forall c \in C \exists e \in E'(c \in e)$, and (3) $\sum_{f \in F'} \omega_F(f) + \sum_{e \in E'} \omega_E(e) \leq k$?

The first condition links the choice of facility locations with the choice of customer services, while the second condition expresses the necessity that every customer can be served through the choice of facility locations. The third

---

* This research has been supported by the Australian Research Council through the Australian Centre of Excellence in Bioinformatics.

R. Fleischer and J. Xu (Eds.): AAIM 2008, LNCS 5034, pp. 188–199, 2008.

condition formalizes that the overall costs (weights) of opening up facilities and serving customers should be bounded by k. In the literature, the problem formulated above is mostly known as UNCAPACITATED DISCRETE FACILITY LOCATION PROBLEM, see [3] for a good recent overview. Notice that we subsumed the usually separated customer demands and service costs (per demand unit) into the weight function ω_E. We will also discuss variants (e.g., allowing real number costs) in this paper.

Alternatively, and sometimes more convenient, this problem can be formulated in terms of a "matrix problem:"

FACILITY LOCATION (MATRIX FORMULATION)

Given: A matrix $M \in \mathbb{N}_{\geq 1}^{(n+1) \times m}$, indexed as $M[0 \ldots n][1 \ldots m]$, $k \in \mathbb{N}$

Question: Is there a set $C \subseteq \{1, \ldots, m\}$ of columns and a function $s : \{1, \ldots, n\} \to C$ such that $\sum_{f \in C}(M[0, f] + \sum_{c:s(c)=f} M[c, f]) \leq k$?

In the matrix formulation, the columns play the role of the potential facility locations and the rows represent the customers to be served (except for row 0). Since "missing edges" in the bipartite graph formulation can be expressed as edges which have "infinite weight" (corresponding in turn to a weight larger than k in the decision problem formulation), we can assume that the graph is indeed a complete graph. Then, the matrix $M[1 \ldots n][1 \ldots m]$ contains the weights of the edges, while $M[0][1 \ldots m]$ stores the weights associated to potential facility locations.

In the following, we will use terminology from both formulations interchangeably, according to convenience.

Fixed Parameter Tractability. $\mathcal{NP}$-hard computational problems are ubiquitous in economics. One approach to overcoming this difficulty is to devise algorithms that can solve arbitrary instances of such a problem under the restriction that a certain entity, called the parameter, is small. This concept is usually formalized as follows: Problem instances are elements of $\Sigma^* \times \mathbb{N}$, and an instance $I = (w, k)$ is to be decided in time $\mathcal{O}(p(|w|)f(k))$, where p is a polynomial (whose degree does not depend on the parameter k) and f is an arbitrary function. Problems that can be solved within such a time restriction are called *fixed parameter tractable*, or in $\mathcal{FPT}$, for short. Equivalently, a problem is in $\mathcal{FPT}$ iff there exists a polytime computable self-reduction that maps an instance $I = (w, k)$ onto an(other) instance $I' = (w', k')$ of the same problem whose overall size is limited by a function $g(k)$, i.e., $|w'| + k' \leq g(k)$. Then, I' is also called a *problem kernel* for I. There is also a complementing hardness theory in this field, basically reflected by the so-called W-hierarchy $\mathcal{FPT} \subseteq W[1] \subseteq W[2] \subseteq \ldots$, where W[1]-hardness is the notion that corresponds best to $\mathcal{NP}$-hardness in classical complexity theory. Further details can be found in the textbook [8].

Our Contribution. Facility location problems have been quite extensively studied in the literature under many perspectives (hardness, approximation, heuristics, etc.), and this is of course also thoroughly done within the operations research and management science communities. Our aim is to initiate systematic research on these problems from the viewpoint of parameterized complexity. We will show parameterized tractability for the described problem formulation and many variants thereof, but also give some related intractability results. In doing so, we also provide an introduction of how $\mathcal{FPT}$-results can be obtained in a rather systmematic fashion. Moreover, we will exhibit connections to many application areas outside economics, which might stir up research between different communities on this topic.

Related Work and Variations. It is often useful to relate the costs of serving customers with an underlying metric space, i.e., $\omega_E(c, f)$ can be described by the distance of c and f times the demand that is incurred by c. This problem variant is also referred as the (metric) uncapacitated facility location problem. A more general setting is given in the capacitated facility location problem, where each facility can only serve up to a given maximum load; then, there are again two variants considering the possibility that a demand of one specific costumer might be satisfied from only one facility or possibly split among different facilities (which is of course more flexible). Good overviews of approximation algorithms for many variants can be found in [3,13,14].

2   FACILITY LOCATION Is in $\mathcal{FPT}$

We describe several approaches to the statement of the headline. This can be also taken as a short introduction to the whole field of parameterized algorithmics in a nutshell, along with this specific example.

One of the first things that has to be decided is what parameters (as a secondary measurement of the whole input) are to be considered. In our case, natural choices could be: (1) the number n of customers, (2) the number m of potential facility locations, (3) an upperbound k on the cost, or (4) an upperbound ℓ on the number of facilities that could be opened.

A trivial brute-force approach immediately yields:

Theorem 1. FACILITY LOCATION *can be solved in time* $\mathcal{O}^*(2^m)$.

Parameter m is not particularly interesting, as long as we are mainly interested in classification results. For the other parameters suggested, the situation is less clear and will be discussed in the remainder of this paper. We put special emphasis on k, since this is the choice of parameter that is usually considered natural for minimization problems. The $\mathcal{O}^*$-notation suppresses polynomial factors.

2.1   Finding Reduction Rules

Reduction rules are heuristics that help shrink the search space. Often, a collection of reduction rules define a so-called kernelization, i.e., a poly-time self-reduction of a problem instance (I, k) to (I', k') such that the size of the new instance (I', k') is only upperbounded by a function in k. This type of reductions is crucial for the $\mathcal{FPT}$ methodology since it characterizes the class $\mathcal{FPT}$. So, the quest for finding reduction rules is essential for parameterized algorithmics.

The following observation is easy but crucial:

Reduction Rule 1. *If a given instance (M, k) with $M \in \mathbb{N}_{\geq 1}^{(n+1) \times m}$ obeys $n > k$, then return* NO.

Lemma 1. *Rule 1 is valid.*

Proof. Each customer must be served. Since edges have positive integer weights, each customer thus incurs "serving costs" of at least one unit. Hence, no more than k customers can be served.   □

Lemma 2. *After having exhaustively applied Rule 1, the reduced instance (M, k) will have no more than k rows.*

Notice that the preceding lemma builds a close link between the parameters k and n.

A facility location f is described by the vector $v_f = M[0 \ldots n][f]$. These vectors can be compared componentwisely.

Reduction Rule 2. *If for two facility locations f and g, $v_f \leq v_g$, then delete g; the parameter stays the same.*

Lemma 3. *Rule 2 is sound.*

Proof. Obviously, a solution to the Rule-2-reduced instance is also a solution to the original instance. If we had a solution S to the originally given instance which contains a facility g and there is another facility f such that $v_f \leq v_g$, then a solution S' obtained from S by choosing facility f instead of g (and choosing to serve any customer served by f in S to be served by g in S') will also be a valid solution which comes at no greater cost. This way, we can gradually transform S into a solution which is also a valid solution to the Rule-2-reduced instance and has no larger costs than S.   □

2.2   Kernelization through Well-Quasi Orderings

Central to the complexity class $\mathcal{FPT}$ is the concept of kernelization, since it characterizes $\mathcal{FPT}$. We first provide a "quick classification" of FACILITY LOCATION in $\mathcal{FPT}$, based on well-quasi orderings.

Theorem 2. FACILITY LOCATION *is fixed-parameter tractable.*

Proof. We show that, after having exhaustively applied Rules 1 and 2, we are left with a problem of size $f(k)$. Let (M, k) be reduced with respect to Rules 1 and 2. By Lemma 2, we know that M contains no more than k rows, since (M, k) is Rule-1-reduced. Each facility is therefore characterized by a $(k + 1)$-dimensional vector. Since (M, k) is Rule-2-reduced, all these vectors are pairwise uncomparable. According to Dickson's Lemma they could be only finitely many, upperbounded by some function $g(k)$. Hence, M is a matrix with no more than $(k + 1)g(k)$ entries.   □

The function $f(k)$ derived for the kernel size in the previous theorem is huge, yet it provides the required classification.

2.3   Kernelization Refinements

To obtain a better algorithm, observe that a solution can be viewed as a partition of the set of all customers into groups such that customers within the same group get served by the same facility. In actual fact, a solution specified by the selected facilities and selected serving connections can be readily transformed into this sort of partition. Also the converse is true: given a partition of the set of customers, we can compute in polynomial time which the cheapest way to serve this group by a certain facility is, so that an optimal solution (given the mentioned partition) in the sense of specifying the selected facilities and the chosen serving connections can be obtained.

This model immediately allows to derive the following result:

Lemma 4. *On an instance* (M, k), *where* $M \in \mathbb{N}_{\geq 1}^{(n+1) \times m}$, FACILITY LOCATION *can be solved in time* $\mathcal{O}(k^k p(g(k)) + nm)$, *where* p *is some polynomial and* $g(k)$ *bounds the number of facilities.*

Proof. The kernelization rules 1 and 2 can be applied in time $\mathcal{O}(nm)$. Then, we have to check all partitions; there are actually $o(k^k)$ many of them (more precisely, this is described by Bell's number whose asymptotics is due to de Brujn (1958), see http://mathworld.wolfram.com/BellNumber.html). For each partition, we have to compute its cost incurred by the assumption that each group in the partition is served by one facility. Hence, per partition $p(g(k))$ computations have to be performed.   □

Still, this algorithm is practically useless due to the huge constants. Let us now develop a better algorithm. Let us first focus on kernelization.

Reduction Rule 3. *Consider an instance* $((B, \omega_E, \omega_F), k)$ *of* FACILITY LOCA-TION. *The following modifications will not affect the parameter.*
Forall *facilities* f **do**
  (1) If $\omega_F(f) \geq k$, *then delete* f.
  Forall *customers* c **do**
    (2) If $\omega_F(f) + \omega_E(c, f) > k + 1$, *then set* $\omega_E(c, f) := k + 1 - \omega_F(f)$.

The following is obvious.

Lemma 5. *Rule 3 is sound.*

Lemma 6. *An instance* (M, k), $M \in \mathbb{N}_{\geq 1}^{(n+1) \times m}$, *of* FACILITY LOCATION *obeys* $nm \leq (k + 1)^{k+2}$ *if it is reduced with respect to Rules 1, 2 and 3.*

Proof. Let (M, k) be such a reduced instance. Due to Rule 1, there are at most k customers. Due to Rule 3, there are at most $(k+1)^{k+1}$ different facility vectors. Du to Rule 2, vectors of different facilities must be different. Hence, there are at most $(k+1)^{k+1}$ facilities. Since each facility vector has at most $k+1$ entries, M has no more than $(k + 1)^{k+2}$ many entries.   □

The estimate of the preceding lemma is probably not very sharp, since Rule 2 was used in a very rough way; the number of facilities is rather upperbounded by the maximum number of antichains in the space of $(k + 1)$-dimensional vectors with entries from 1 to $k + 1$.

2.4   Improving on Brute Force by Dynamic Programming

The idea of "dynamic programming on subsets" improves the running time.

Algorithm 1. A dynamic programming algorithm for facility location: FLdp

---

Require: a bipartite graph $B = (F \uplus C, E)$ with weights ω_E and ω_F
Ensure: an implicit facility location strategy incurring minimum weight

> **if** $|C| > k$ **then**
>   return NO;
> $s(\emptyset) := 0$;
> **for all** $X \subseteq C$ **do**
>   compute $os(X)$;{in time $\mathcal{O}(|X| \cdot |F|)$}
> **for** $i := 1, \ldots, |C|$ **do**
>   **for all** $X \subseteq C, |X| = i$ **do**
> $s(X) := \min_{\emptyset \subsetneq Y \subseteq X}(os(Y) + s(X \setminus Y))$
>   {Every customer belongs either to Y, $X \setminus Y$ or to $C \setminus X$;}
>   {Convention: $\min_{\emptyset} \cdots = \infty$.}

Theorem 3. FACILITY LOCATION *can be solved in time* $\mathcal{O}(2^k m + 3^k)$ *on a given instance* (M, k) *with* $M \in \mathbb{N}_{\geq 1}^{(n+1) \times m}$.

Proof. We start with Rule 1. In a preprocessing phase, we compute the cost incurred by a certain set of customers when being served by a single facility, storing the results in a table "one-serve" (os) with 2^k entries. This also includes the costs for opening up that facility. Then, we can compute the minimal costs of serving a certain group of customers by some facilities by dynamic programming, combining two subsets at a time. If we have stored the results of the preprocessing in table os, each step in the dynamic programming will take only constant time, so that we arrive at the following formula for dynamic programming:

$$s(X) := \min_{\emptyset \subsetneq Y \subseteq X}(os(Y) + s(X \setminus Y)) \qquad (1)$$

This amounts in $\mathcal{O}(3^k)$ operations. Namely, there are basically three possilities for a customer: it either belongs to Y, $X \setminus Y$ or to $C \setminus X$. □

Remark 1. Notice that the bound $k > n$ immediately implies an $\mathcal{O}^*(3^n)$ estimate for Alg. 1. Alternatively, we can first kernelize (with respect to k) and then run Alg. 1. on the kernel, which would amount in an $\mathcal{O}^*((2k)^k)$ estimate due to Lemma 6.

2.5   Further Improvements

In a very recent paper, A. Björklund, T. Husfeldt, P. Kaski and M. Koivisto [1] described how to further speed up dynamic programming on subsets in just a situation as encountered with our problem. Their approach is based on fast subset convolution, also known as (fast) Möbius transform. More specifically, the recursion from Eq. (1) can be seen as a subset convolution (in the min-sum semiring). Let us shortly explain this approach (with n customers and m facilities). By computing first the ranked Möbius transform (which basically takes 2^n steps), this convolution operation can be performed in the transformed space with $2n + 1$ operations, and the inverse Möbius transform is of similar cost as the Möbius transform itself.

Corollary 1. *If all the integer weights lie between 1 and N for a given instance of* MINIMUM FACILITY LOCATION, *the problem can be solved in time* $\mathcal{O}(2^n n^3 (nm)^2 \log(N))$.

The assumption of a bounded range of integer weights is not so unrealistic in many scenarios; for example, in our parameterized setting, we can assume (by our reduction rules) that those weights lie between 1 and $k + 1$.

3   Variants of FACILITY LOCATION

Median / Means Problem. The k-MEDIAN PROBLEM and the k-MEANS PROBLEM are defined just as the FACILITY PROBLEM, except for the fact that there are no costs for opening up facilities (and mostly, it is required that exactly k facilities should open). Means and median problems only differ in the metric, i.e., the way point distances are measured (sums of squares of distances vs. sums of distances). All corresponding problems are $\mathcal{NP}$-hard, even for Euclidean spaces. Obviously, these problems have quite a geometric flavor. Again, it is not hard to see by a simple analysis of our results of the preceding section that also the k-MEDIAN PROBLEM and the k-MEANS PROBLEM are in $\mathcal{FPT}$: namely, even the common generalization of these problems, viewed as variants of FACILITY LOCATION, where the opening cost of a facility might be zero, is in $\mathcal{FPT}$.

Many approximation algorithms are known for this problem, see [5]. Notice that often (also there) it is assumed that we actually deal with a something like the uncapacitated (metric) facility location problem. This means that the costs for connecting facilities and customers are distances in a metric space. If in addition, the facility locations are not explicitly distinguished from the customer location, but should be rather selected in a first step, this models the problem of finding a minimum cost clustering in a metric space. Notice that our preceding arguments also provide $\mathcal{FPT}$-membership in this case.

Rational Weights. In all versions discussed up to now, including the one introduced in the very beginning, one could also allow rational (or even "real-number") weights bigger than or equal to one (modelling actual costs more realistically). Apart from numerical considerations, this would not change anything except for the Section dealing with the fast subset convolution; that approach would not work here.

Cheap Serves. In view of our discussion it might be also an idea to allow for (some) services of zero cost. Irrespectively of whether or not zero costs for opening up facilities are allowed, the complexity picture now changes dramatically:

Theorem 4. FACILITY LOCATION, *parameterized by k, becomes W[2]-complete when zero weight services are allowed.*

Proof. (Sketch) Consider the special case when opening any facility costs one unit, and all services are for free. Then, the problem corresponds to selecting (at most) k of the facility to serve all customers. In more graph-theoretic form, this exactly corresponds to RED-BLUE DOMINATING SET, which in turn can be seen to be equivalent to HITTING SET [8]. Hence, any HITTING SET instance can be seen as a special instance of the variant of FACILITY LOCATION that we are considering. Since HITTING SET is well-known to be W[2]-hard, this property transfers to our problem at hand.

Conversely, membership in W[2] can be seen, e.g., by showing how to solve our problem with a short multi-tape nondeterministic Turing machine, see [4].   □

A quick analysis of the proof of the previous theorem yields:

Theorem 5. FACILITY LOCATION, *parameterized by ℓ, is W[2]-complete.*

4   Applications: The MDL Principle

FACILITY LOCATION and its variants have numerous applications, even when only allowing integer costs. We shall now describe some of the less obvious ones that one way or the other apply the Minimum Description length (MDL) principle.

4.1   Automizing the Production of XML Documents

The web standard exchange format XML is (possibly first) described in: Extensible Markup Language (XML) 1.0, T. Bray, J. Paoli, and C. M. Sperberg-McQueen, 10 February 1998, available at http://www.w3.org/TR/REC-xml. In his notes on this standard (see www.xml.com/axml/notes/Any1.html), T. Bray wrote, commenting on the construction of document type descriptors (DTD), a kind of context-free grammars used to specify syntactic characteristics of XML documents:

"Suppose you're *given an existing well-formed XML document* and you want to build a DTD for it. One way to do this is as follows:
1. Make a list of all the element types that actually appear in the document, and *build a simple DTD* which declares each and every one of them as ANY. Now you've got a DTD (not a very useful one) and a valid document.
2. Pick one of the elements, and work out how it's actually used in the document. Design a rule, and *replace the ANY declaration with a ... content declaration. This, of course, is the tricky part*, particularly in a large document.
3. Repeat step 2, working through the elements one by one, until you have a useful DTD."

Hence, various systems—usually called *DTD generators*—were designed to automize the process of designing DTD's, best from known examples. The main problem is that of *generalization*: when given a number of sample documents which should fit the envisaged DTD, at what "moment" and in which way is the DTD generator supposed to switch from a mode where it only produces a trivial DTD (this could be either a very specific one that can only parse the given samples or a very general one, as attempted by Bray with his ANY declaration proposal) to a mode where it actually tries to cleverly guess the syntactical structure in the given documents. Since most DTD generators internally work from "most specific" to "general," this process of making "intelligent guesses" is known as *generalization*.

Garofalakis *et al.* proposed for this purpose the system XTRACT [11]. This DTD generator is based on the *Minimum Description Length (MDL) principle*, which can be viewed as a formalization of Occam's razor. In actual fact, this principle in connection with grammar induction (DTD generators are a special case of this field) was discussed earlier by D. Conkley and I. H. Witten in [7].

The *length of a description* consists of two parts:

1. the length of the theory (in bits) and
2. the length of the data (in bits) when encoded with the help of the theory.

The system XTRACT uses MDL to evaluate regular expression hypotheses. This yields the combinatorial problem MDL-OPTIMAL-CODING:

Given: a set $R = \{r_1, \ldots, r_n\}$ of regular expressions (over the basic alphabet Σ) and a set of strings $S = \{s_1, \ldots, s_m\} \subset \Sigma^+$, $k \in \mathbb{N}$

Question: Is it possible to find a subset R' of R such that

$$\sum_{r \in R'} |c(r)| + \sum_{s \in S} |c(s|R')| \leq k \quad ?$$

The coding function c is described in detail in [11].

As an example (taken from [11]), consider the strings

$$S = \{ab, abab, ac, ad, bc, bd, bbd\}.$$

As hypotheses "covering" some of these examples, we consider, besides the elements from S itself, the following regular expressions

$$\{(a|b)^*, b^*d, b^*e, b^*(d|e), (a|b)(c|d), (ab)^*\}.$$

Graphically, this covering relation can be depicted as a bipartite dominating set problem, see Fig. 1. To give an idea how the coding function c works, consider the three colored expressions. For example, to describe *abab* by the selected theory, i.e., by $\{(ab)^*, b^*(d|e), (a|b)(c|d)\}$, we need to describe that we take the first (out of the three expressions) and that we use *two* iterations. Likewise, *bbbbe* can be described by the second expression using *four* iterations.

Supplementing [11], it has been shown:

Theorem 6. *(see [10])* MDL-OPTIMAL-CODING *is $\mathcal{NP}$-complete.*

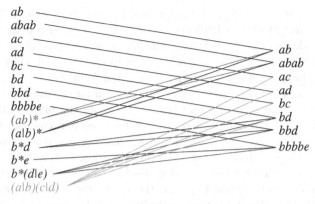

Fig. 1. How to apply the MDL principle

Garofalakis *et al.* propose using the related facility location problem, using the following translation: (*a*) place *facilities* (in our case: the selected theory) (*b*) into some *locations* (here: the regular expressions to choose from) (*c*) in order to minimize the *costs* incurred by the facilities (here: the number of bits needed to encode the theory) and by serving a given set of customers (in this case: the number of bits for encoding the given strings).

Our results imply that we can solve MDL-OPTIMAL-CODING as a parameterized problem to optimality. This is particularly important if MDL is actually used to *evaluate* hypotheses. Using here algorithms which only provide a logarithmic approximation guarantee will eventually mean that hypotheses will be rejected or preferred not according to their quality but according to the "quality" of the evaluation algorithm.

Since the number of bits (the costs) could be seen to be bounded by the input length itself, and all these numbers are integers, we can infer:

Corollary 2. MDL-OPTIMAL-CODING *can be solved in time* $\mathcal{O}^*(2^n)$ *(or also* $\mathcal{O}^*(2^k)$*), where* n *denotes the number of input strings to be coded.*

Observe that $n < m$ in this particular application, since the input strings themselves are also always seen as possible regular expressions. So, the time bound derived by the preceding corollary is always superior to the trivial approach from Thm. 1.

Notice that similar methods are also used for data compression purposes. There, the "theory part" has sometimes to be explicitly encoded (as *side information*) and sometimes it is statically known to the (de)coder, which means that the "costs" of "opening up a facility" are zero. Hence, we face (again) the MEDIAN PROBLEM discussed in Sec. 3, so that this problem is also parameterized tractable, when parameterized by an upper bound on the number of bits used for the compressed data.

4.2 Computational Biology

Koivisto *et al.* described in [12] a method of applying ideas originating in the minimum description length principle to identify so-called *haplotype blocks* and to compare the strength of block boundaries. For our exposition, it is sufficient to know that intuitively, "a haplotype block can be considered to represent a sequence of ordered markers such that, for those markers, most of the haplotypes in the population cluster into a small number of classes. Each class consists of identical or almost identical haplotypes." (Quoted from [12].)

They propose a dynamic programming algorithm for the problem of computing an optimal block structure and then to estimate the probabilities of each block boundary. This method relies on knowing a certain cost function whose computation actually is $\mathcal{NP}$-hard. However, this cost function (measuring the quality of the clusters/blocks) can be modeled with a k-MEANS PROBLEM and hence can be solved by the methods described in this paper. Since the involved weights grow at most exponential with the input length (as it appears to be generally true when dealing with weights in facility location problems that are derived from the MDL scenario), the fast subset convolution method can be used to further reduce the run times.

Again, there might be a broader connection to data compression; there, good k-means algorithms are crucial, e.g., in vector compression. The popular Lloyd algorithm (and variants thereof) may get stuck at local minima, so it might sometimes be a good idea to look for global maxima. Notice that exact algorithms might be a sensible approach here due to the slow convergence of Lloyd's algorithm in practical situations, see [9].

5   Conclusions and Further Research

We have started a systematic study of facility location problems and variants. However, although this start looks quite promising, many things are still to be done. We sketch some of these in the following. (1) Even though many problems appear to be in $\mathcal{FPT}$, the algorithms we provided can be only seen as a starting point. Are there better algorithms for these important problems? (2) What about developing exact algorithms, possibly measured in $N = n + m$. Notice that the $\mathcal{O}^*(2^n)$- and $\mathcal{O}^*(2^m)$-algorithms shown in this paper can be easily abused in a WIN-WIN scenario to derive an $\mathcal{O}^*(\sqrt{2}^N)$-algorithm. (2) It is not quite clear if the basic assumption in parameterized algorithmics, namely, that the parameter is only moderately large, is met in this set of problems. Are there different parameterizations that are more suitable, at least in some applications? (3) Just to give two examples of a different, possibly additional parameter: (3a) In [6], a modified scenario was considered where some "outliers" were permitted, i.e., customers that could not be served (at decent costs). This could deliver a natural small parameter. (3b) In many situation, a company will already have opened a number of facilities, and the question is how to optimally improve the situation for the customers (and the company's budget) by opening up a small number of new facilities.

Another aspect that has been neglected so far is geometry. Can we exploit metricity of costs to obtain better parameterized algorithms, as has been done in the case of approximation algorithms?

It is quite natural to assume that only (relatively) few facilities are within the "natural reach" of a single customer (it might be different from the viewpoint of the facilities, though). This implies a sort of degree-restriction on the side of the customer (in the underlying bipartite graph) which might yield a better runtime estimate of the dynamic programming algorithm, see [2].

References

1. Björklund, A., Husfeldt, T., Kaski, P., Koivisto, M.: Fourier meets Möbius: fast subset convolution. In: Symposium on Theory of Computing STOC, pp. 67–74. ACM Press, New York (2007)
2. Björklund, A., Husfeldt, T., Kaski, P., Koivisto, M.: Trimmed Möbius inversion and graphs of bounded degree. In: Symposium on Theoretical Aspects of Computer Science STACS, IBFI, Schloss Dagstuhl, Germany, pp. 85–96 (2008)
3. Bumb, A.: Approximation algorithms for facility location problems. PhD Thesis, Univ. Twente, The Netherlands (2002)
4. Cesati, M.: The Turing way to parameterized complexity. Journal of Computer and System Sciences 67, 654–685 (2003)
5. Charikar, M., Guha, S., Tardos, E., Shmoys, D.S.: A constant-factor approximation algorithm for the k-median problem. Journal of Computer and System Sciences 65, 129–149 (2002)
6. Charikar, M., Khuller, S., Mount, D.M., Narasimhan, G.: Algorithms for facility location problems with outliers. In: Symposium on Discrete Algorithms SODA, pp. 642–651. ACM Press, New York (2001)
7. Conklin, D., Witten, I.H.: Complexity-based induction. Machine Learning 16, 203–225 (1994)
8. Downey, R.G., Fellows, M.R.: Parameterized Complexity. Springer, Heidelberg (1999)
9. Du, Q., Emelianenko, M., Ju, L.: Convergence of the Lloyd algorithm for computing centroidal Voronoi tessellations. SIAM Journal on Numerical Analysis 44, 102–119 (2006)
10. Fernau, H.: Extracting minimum length Document Type Definitions in NP-hard. In: Paliouras, G., Sakakibara, Y. (eds.) ICGI 2004. LNCS (LNAI), vol. 3264, pp. 277–278. Springer, Heidelberg (2004)
11. Garofalakis, M., Gionis, A., Rastogi, R., Seshadri, S., Shim, K.: XTRACT: learning document type descriptors from XML document collections. Data Mining and Knowledge Discovery 7, 23–56 (2003)
12. Koivisto, M., Manila, H., Perola, M., Varilo, T., Hennah, W., Ekelund, J., Lukk, M., Peltonen, L., Ukkonen, E.: An MDL method for finding haplotype blocks and for estimating the strength of block boundaries. In: Pacific Symposium on Biocomputing 2003 (PSB 2003), pp. 502–513 (2002), http://psb.stanford.edu/psb-online/proceedings/psb03/
13. Shmoys, D.B., Tardos, É., Aardal, K.: Approximation algorithms for facility location problems (extended abstract). In: Symposium on Theory of Computing STOC, pp. 265–274. ACM Press, New York (1997)
14. Young, N.E.: k-medians, facility location, and the Chernoff-Wald bound. In: Symposium on Discrete Algorithms SODA, pp. 86–95. ACM Press, New York (2000)

Shortest Path Queries in Polygonal Domains*

Hua Guo, Anil Maheshwari, and Jörg-Rüdiger Sack

School of Computer Science, Carleton University, 1125 Colonel By Drive, Ottawa,
K1S 5B6, Canada
hua.guo.canada@gmail.com, {anil,sack}@scs.carleton.ca

Abstract. We consider shortest path queries in a polygonal domain P
having n vertices and h holes. A skeleton graph is a subgraph of a Voronoi
diagram of P. Our novel algorithm utilizes a reduced skeleton graph of P
to compute a tessellation of P. It builds a data structure of size $O(n^2)$ in
$O(n^2 \log n)$ time to support distance queries for any pair of query points
in P in $O(h \log n)$ time.

1 Introduction

We consider the *All Pairs Query* (APQ) problem in a polygonal domain P which
is a scene scattered with h polygonal obstacles (holes) with disjoint interiors. P
has totally n obstacle vertices. Paths that avoid the interior of obstacles are
called *obstacle-avoiding* paths. We focus on finding a shortest obstacle-avoiding
path (i.e., *shortest path*) and/or distance between two arbitrary query points in
P. The length of the shortest path between the query points is called *distance*
between them. If one query point is pre-selected, then the APQ problem is
simplified to the *Single Source Query* (SSQ) problem. To place our work in the
context of the literature, we state some relevant results next.

Solving the SSQ and APQ problem in the presence of multiple obstacles has
proven to be challenging. In a polygonal domain P with h holes having n obstacle
vertices, Hershberger and Suri [1] presented an optimal algorithm for computing a shortest path map (SPM) from a fixed point to all other points in P in
$O(n \log n)$ time and space using the continuous Dijkstra paradigm. The SPM
can be used to answer single-source shortest path queries to any destination in
$O(\log n)$ time. Kapoor et al. [2] proposed an algorithm which computes a SPM
for a given source in $O(n + h^2 \log n)$ time using visibility graphs.

The APQ problem in polygonal domains is considered to be more challenging than in the case of simple polygons. Only a few results are known for this
problem instance. For any pair of query points a and b in a polygonal domain P,
Chen et al. [3] have shown that using $O(n^2)$ space and $O(n^2 \log n)$ time in preprocessing, one can answer an APQ query in $O(\min(Q_a, Q_b) \log n)$ time, where
Q_a (Q_b) is the number of vertices visible from a (b, respectively). Chiang and
Mitchell [4] proposed an algorithm which uses $O(n^{11})$ space and preprocessing
time to construct a data structure supporting optimal $O(\log n)$ time queries. Alternatively, their algorithm requires $O(n^{10} \log n)$ space and preprocessing time

---

* Research supported by NSERC and SUN Microsystems.

R. Fleischer and J. Xu (Eds.): AAIM 2008, LNCS 5034, pp. 200–211, 2008.
© Springer-Verlag Berlin Heidelberg 2008

to support $O(\log^2 n)$ time queries. They also sketched strategies to make their algorithms sensitive to h but incur significant preprocessing cost.

Throughout the paper, let P be a polygonal domain having n vertices and h holes. Let (a, b) be any pair of points in P. Let $\pi(a, b)$ be a shortest path between a and b in P and $\text{dist}(a, b)$ be the length of $\pi(a, b)$ (i.e., *distance*) between them.

New Results: We present a novel algorithm for solving the APQ problem for P in $O(h \log n)$ query time. Our algorithm builds a data structure of size $O(n^2)$ in $O(n^2 \log n)$ time. In particular,

1. Using a reduced skeleton graph of P as a tool, we tessellate P into a set of triangles and simple polygons. The merit of our approach is that (i) the reduced skeleton graph is *static* and it captures the topology of P; (ii) this tessellation of P is unique.
2. Our algorithm is sensitive to the number, h, of holes.

Our results improve the previous result by Chen et al. [3] in the following way. Our algorithm is sensitive to the number of holes. In particular, it achieves an $O(\log n)$ query time for simple polygons or polygonal regions with a constant number of holes. Chen et al.'s algorithm [3] is not sensitive to the number of holes, but depends on the minimum number of vertices inside the visibility regions of the two query points. The preprocessing time and space of our algorithm matches the result in [3].

2   Preliminaries

In this section, we review some notation and theorems used.

Polygonal Domain: A *polygonal domain*, P, having n vertices and h holes, is a nonempty, bounded, multiply connected region whose boundary, ∂P, is a union of n open line segments, called *edges*, and their endpoints, called *vertices* (cf. Fig. 2). The boundary ∂P consists of $h + 1$ closed polygonal cycles. The complement of the set of holes inside P is the *free space*. A vertex of the boundary ∂P is called *convex* if the internal angle (in the free space) at the vertex is less than π and is called *reflex* otherwise.

Funnels: As appeared in, e.g., [5], a funnel represents shortest paths from a source point s to points on a polygonal edge or diagonal associated with the funnel. Let $l = p_i p_j$ be a diagonal in a simple polygon Q (Fig. 1 (a)). The union $\mathfrak{f}(s, l) = \pi(s, p_i) \cup \pi(s, p_j)$ is called the *funnel* associated with the diagonal l in Q. The shortest path $\pi(s, p_i)$ may share a subpath with the shortest path $\pi(s, p_j)$. At some vertex v, called *apex*, the two paths part and proceed separately to their respective destinations. If v is not s, the subpath from s to v is called *string* of the funnel. For any point $q \in l$, there is a unique tangent from q to the portion of the funnel between the apex and an endpoint of l. Let $\triangle p_i p_j p_k$ be a triangle in Q (Fig. 1 (b)). Let $l = p_i p_j$, $l_1 = p_i p_k$ and $l_2 = p_k p_j$. The funnel $\mathfrak{f}(s, l)$ *splits* into two funnels when it reaches the vertex p_k, i.e., $\mathfrak{f}(s, l) \cup \triangle p_i p_j p_k = \mathfrak{f}(s, l_1) \cup \mathfrak{f}(s, l_2)$.

The funnel $\mathfrak{f}(s,l)$ is *closed* if no point on l is visible from the source s. Otherwise, it is *open*.

Hourglass: Let p and q be any two points in Q. The triangles (assume they are distict) containing p and q in the underlying triangulation associated with Q determine the sequence of diagonals crossed by a shortest path $\pi(p,q)$ between p and q. These diagonals that each split Q into two parts, one containing p and the other q, are called *separating diagonals* for p and q [5]. A shortest path $\pi(p,q)$ crosses these separating diagonals exactly once in a particular order. An *hourglass* represents all shortest paths between two diagonals of Q [5]. Let $l_i = p_i p_{i+1}$, $l_j = p_j p_{j+1}$ be two diagonals in the triangulation of Q, labeled so that $p_i, p_j, p_{j+1}, p_{i+1}$ is a subsequence of the vertex sequence of Q. The union of the two shortest paths $\pi(p_i, p_j)$ and $\pi(p_{i+1}, p_{j+1})$ is the hourglass, $H(l_i, l_j)$, associated to Q for l_i and l_j (cf. Fig. 1 (c)). Let l_p and l_q be the first and last diagonals in the sequence of separating diagonals between p and q. Given the hourglass $H(l_p, l_q)$, the shortest path $\pi(p,q)$ can be computed via tangents to the hourglass that pass through p and q [5]. Consider p and q as length zero diagonals. Concatenating the hourglasses $H(p, l_p), H(l_p, l_q)$ and $H(l_q, q)$ produces a single closed hourglass which is only a string. The string is the shortest path between p and q.

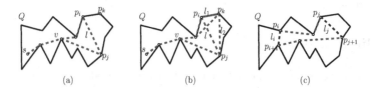

(a)   (b)   (c)

Fig. 1. An illustration of (a) a funnel $\mathfrak{f}(s,l)$; (b) a funnel $\mathfrak{f}(s,l)$ splitting into funnels $\mathfrak{f}(s,l_1)$ and $\mathfrak{f}(s,l_2)$; (c) an hourglass $H(l_i, l_j)$ between diagonals l_i, l_j

Conventionally, hourglasses (resp., funnels) as defined in a simple polygon represent *global* shortest paths between two diagonals (resp., from a source to a diagonal) [5]. Next, we define hourglasses (resp., funnels) in a polygonal domain P. Given a subpolygon Q of P and two diagonals $l_i = p_i p_{i+1}$ and $l_j = p_j p_{j+1}$ in Q, where Q is a simple polygon, an *hourglass* represents all *local* shortest paths within Q between l_i and l_j. We denote it by $H(l_i, l_j)$ in Q for l_i and l_j if no ambiguity arises. A funnel represents *local* shortest paths within Q from a source point s to points on a polygonal edge or diagonal associated with the funnel. Here, the string, i.e. the subpath from the source s to the apex v in Q, is part of the funnel. An open funnel has an empty string. A funnel with a non-empty string is a closed funnel.

Voronoi diagram: We define the Voronoi diagram of a polygonal domain P next (cf. [6]). The vertices and edges of P are the sites of the Voronoi diagram of P. Let $\mathcal{S} = \{v_1, v_2, \ldots, v_n; e_1, e_2, \ldots, e_n\}$ be these sites. The Voronoi cell corresponding to a site $\mathfrak{s}_i \in \mathcal{S}$, denoted by $\mathcal{V}(\mathfrak{s}_i)$, $1 \leq i \leq n$, is the set of all points closer to $\mathfrak{s}_i$ than to any other sites in $\mathcal{S}$. It has been shown (cf. [6]) that in this setting, a Voronoi cell is bounded by a connected subset of lines, half-lines, line segments and parabolic arcs. A *Voronoi vertex* is a point equidistant from three or more sites. A *Voronoi edge* consists of the loci of all points equidistant from two sites. Voronoi edges of P are connected line segments and parabolic arcs. Let dist(p, q) be the Euclidean distance between two points p and q. Formally, a *Voronoi diagram of a polygonal domain P*, denoted by Vor(P), is defined as the subdivision of the free space of P into Voronoi cells, one for each site in $\mathcal{S}$, with the property that each point p lies within the cell corresponding to a site, $\mathfrak{s}_i$, if and only if dist$(p, \mathfrak{s}_i) <$ dist$(p, \mathfrak{s}_j)$ for each site $\mathfrak{s}_j \in \mathcal{S}$, $\mathfrak{s}_j \neq \mathfrak{s}_i$ (cf. [6]).

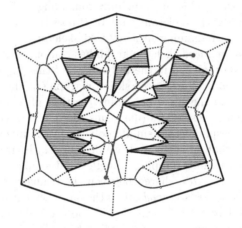

Fig. 2. An illustration of a polygonal domain P, the skeleton graph (thin, solid lines and parabolic arcs) and the Voronoi diagram of P (thin, dotted and solid lines and parabolic arcs). Edges of P are solid segments. The shortest path $\pi(a, b)$ between query points a and b passes through a sequence of reflex vertices.

Theorem 1. *[7] Given a polygonal domain P having n vertices and h holes, the Voronoi diagram of P consists of at most $n + r + 2h - 2$ Voronoi vertices and at most $2(n + r) + 3h - 3$ Voronoi edges, where r is the number of reflex vertices of P.*

3   A Tessellation of P

In this section, we present a tessellation of P that is required in our algorithm. Using a so-called *reduced skeleton graph* of P, we decompose P into a set of triangles and simple polygons.

A *skeleton graph*, $G_s = (V_s, E_s)$, of P is a subgraph of the Voronoi diagram of P, derived by removing all Voronoi edges incident to vertices of P from the diagram. Nodes and edges of a skeleton graph, called *skeleton nodes* and *skeleton edges*, respectively, are Voronoi vertices and edges. We assume that no four vertices of P are co-circular. Skeleton nodes are of degrees 1, 2 or 3. Given a skeleton graph G_s, a path joining a pair of skeleton nodes, (s_i, s_j), of G_s, is called a *skeleton path* between s_i and s_j. We denote it by $\phi(s_i, s_j)$. For any pair of points, (a, b), in P, let a_s and b_s be the points closest to points a and b, respectively, on skeleton edges of G_s. A skeleton path $\phi(a, b)$ between a pair of points (a, b) in P is defined by the skeleton path $\phi(a_s, b_s)$ between a_s and b_s in G_s. There exists a skeleton path for any pair of points in P. The *orientation* of $\phi(a, b)$ is defined as from a to b along $\phi(a, b)$ if a and b are incident to $\phi(a, b)$, or from a_s to b_s otherwise.

We construct a *reduced* skeleton graph next. In Step 1, repeatedly delete each degree 1 node of a skeleton graph G_s with its incident edge until there are no degree 1 nodes left. In Step 2, for each degree 2 node, replace it and its two incident edges with a single edge inside the free space with respect to the removed skeleton edges, unless it is the only remaining node of the graph. The resulting graph is a *reduced* skeleton graph of P possibly with cycles, which has $h + 1$ faces. All nodes remaining in a reduced skeleton graph are of degree 3. Due to planarity, the reduced skeleton graph has complexity $O(h)$. Details with respect to this will be provided in the full version of this paper.

For each skeleton node of a reduced skeleton graph G_r, there exists a triangle whose vertices are from the sites (i.e., points/vertices of holes) associated with the skeleton node and whose edges (diagonals) do not intersect the edges of P. Each edge of G_r corresponds to a simple polygon in the free space of P. Using a reduced skeleton graph G_r as a tool, we decompose the free space of P into *regions*, i.e., a set of non-overlapping triangles and simple polygons. Edges of each triangle are called *gates*, shared between the triangle and its neighboring simple polygons. Two gates are *neighbors* if they are incident to the same region. We compute a tessellation of P next. Let P' be a tessellation of P. First, for each node of G_r, construct its corresponding triangle in the free space of P. Add the triangle to P'. Second, for each edge $e = (u, v)$ of G_r, let $\triangle_u$ and $\triangle_v$ be the triangles corresponding to nodes u and v, respectively; identify a simple polygon, P_e, corresponding to e such that P_e neither overlaps with $\triangle_u$ nor with $\triangle_v$. Identify the gates shared between P_e and $\triangle_u$ and $\triangle_v$. Add P_e to P'. We summarize the above in the next lemma.

Lemma 1. *The free space of P can be tessellated into a set of non-overlapping triangles and simple polygons in $O(n)$ time. The complexity of a tessellation P' is bounded by $O(h)$.*

A skeleton graph of P not only captures the topology of P but also facilitates the tessellation of P. A different tessellation of P based on the dual graph of a triangulation of P appeared in [2,4]. Our approach constructs a unique tessellation of P as apposed to the methods described in [2,4].

4   Local Shortest Paths and Critical Sites

Let G_s be a skeleton graph of P. The skeleton graph G_s is a graph possibly containing cycles. Let $i \in \{a, b\}$, where a and b are query points. Point i is contained in a region, denoted by $R(i)$, of the tessellation P'. The region $R(i)$ is either a triangle or a simple polygon of P'. A shortest path between a and b in P may not be unique. There could be many *local* shortest paths between a and b in P (as defined below) such that each local shortest path intersects a unique sequence of regions of the tessellation.

We define local shortest paths next. A skeleton path $\phi(a, b)$ intersects a sequence of diagonals in the underlying triangulation of P. There exists a path joining a and b in P such that the path intersects these diagonals in sequence. We call such path with the minimum Euclidean length a *local shortest path* connecting a and b associated with the skeleton path $\phi(a, b)$. We denote the path by $\pi_\phi(a, b)$. A region in P with respect to a skeleton path ϕ is the union of Voronoi cells associated with the skeleton path ϕ. Walking along the skeleton path ϕ, we encounter these associated Voronoi cells in an order. There exists a unique order for all *left* Voronoi cells located to the left of ϕ and *right* cells to the right of ϕ, respectively. We can define an order for left and right cells associated with ϕ. The reflex-vertex sites associated with the cells are called *associated vertices* corresponding to the skeleton path. Given an orientation of ϕ, there exists a unique order for all associated vertices for ϕ. Vertices on a local shortest path $\pi_\phi(a, b)$ must be associated vertices for the skeleton path $\phi(a, b)$. The sequence of vertices on a local shortest path, $\pi_\phi(a, b)$, with respect to a skeleton path $\phi(a, b)$ is a subsequence of the associated vertices related to the skeleton path $\phi(a, b)$. Any vertex on a local shortest path is visible only to the vertices immediately before and after it. Two vertices on a local shortest path that are not successively adjacent can not see each other.

For simplicity, let us assume that a pair of points a and b are invisible to each other. In case that a and b are mutually visible, the straight line segment $\overline{ab}$ joining a and b is the (global) shortest path between them. We address this case in Section 5. Given an orientation of a skeleton path $\phi(a, b)$, the last visible vertex from a, called *critical site*, divides the sequence of associated vertices corresponding to $\phi(a, b)$ into two ordered sets containing the vertices that are *visible* from a and those that are *invisible* from a. The visible set from a may contain associated vertices that are invisible from a, e.g., vertices located in a pocket. Since points a and b are invisible to each other, there must exist one critical site for a with respect to each skeleton path $\phi(a, b)$. Next, we present the key lemmas of this section.

Lemma 2 (Critical Site). *The local shortest path $\pi_\phi(a, b)$ must pass through a critical site for the point a.*

Lemma 3. *Let $\mathfrak{f}_\phi(s, l)$ be a closed funnel from the source s to a polygon edge or diagonal l, lying in the free space of P with respect to a skeleton path ϕ. The first vertex (other than the source s if s is a vertex) on the string of the funnel $\mathfrak{f}_\phi(s, l)$ is the critical site on the local shortest path from s to points on l, with respect to ϕ.*

5 Our Algorithm

In this section, we present an algorithm which builds a data structure of size $O(n^2)$ in $O(n^2 \log n)$ time to support $O(h \log n)$ time shortest path queries between a pair of points, (a, b), in P. First, we describe the main idea of our algorithm. Let C_a be the set of critical sites for a and C_b for b. If a and b are mutually visible, then we compute and report the distance $\text{dist}(a, b)$ directly. Otherwise, our task is reduced to identifying the critical sites C_a and C_b such that each local shortest path leaves a via a site in C_a, and reaches b via a site in C_b (Lemma 3). For each critical site $c \in C_a$, we search the shortest path map of c and compute the distances $\text{dist}(c, a)$ and $\text{dist}(c, b)$. The sum, $\text{dist}(c, a) + \text{dist}(c, b)$, is the length of the shortest path from a to b via c. The distance $\text{dist}(a, b)$ is the minimum of these path lengths for all $c \in C_a \cup C_b$. We summarize this in the next lemma.

Lemma 4. *The distance* $\text{dist}(a, b)$ *can be computed by* $\min_{c \in C_a \cup C_b}(\text{dist}(a, c) + \text{dist}(c, b))$.

Let $C_{\min}$ be the smaller set between C_a and C_b. We can reduce our determination of $\text{dist}(a, b)$ to a smaller size, i.e. $|c_{\min}|$, as opposed to $|C_a| + |C_b|$, since it is sufficient to investigate either all prefixes of local shortest paths leaving a, or all suffixes of local shortest paths reaching b. We summarize this next.

Corollary 1. *The shortest distance* $\text{dist}(a, b)$ *can be correctly computed by* $\text{dist}(a, b) = \min_{c \in C_{\min}}(\text{dist}(a, c) + \text{dist}(c, b))$.

For the sake of simplicity, let $C_{\min}$ be C_a. Then, we compute all closed funnels from a and collect critical sites from the strings of these funnels.

Next, we present our preprocessing algorithm (Table 1) which takes P as input and outputs a data structure. The data structure, denoted by APQ_PD(P), consists of (i) a skeleton graph G_s, a reduced skeleton graph G_r and a tessellation P'; (ii) for each simple polygon $R(l_1, l_2)$ of P' with gates l_1 and l_2, a hierarchy of hourglasses, $HH(R)$ and an hourglass $H(l_1, l_2)$ (iii) a point-location data structure. The following lemma summarizes the complexity of the preprocessing algorithm.

Lemma 5. *Algorithm* APQ_PD_Preprocessing(P) *builds a data structure of size* $O(n^2)$ *in* $O(n^2 \log n)$ *time.*

Next, we present the key ideas of our query algorithm. First, we introduce notation used. Let $i \in \{a, b\}$. Let $R(i)$ be the region containing i in the tessellation P'. Let $F_o(i)$ (resp., $F_c(i)$) be a set of open (resp., closed) funnels from i to the explored gates in P'. The set $F_o(i)$ is a balanced binary search tree (BST). Funnels from a to the gates of the region $R(a)$ are called *initial funnels* from a.

Our query algorithm is a process of *funnel propagation*. Let $R(l_1, l_2)$ be a simple polygon of P' with gates l_1 and l_2. Propagating a funnel $f(a, l_1)$ to its neighboring region $R(l_1, l_2)$, i.e., to its neighboring gate l_2, is the concatenation of $f(a, l_1)$ with the hourglass $H(l_1, l_2)$ associated with $R(l_1, l_2)$. If the neighboring region of $f(a, l_1)$ is a triangle, $\triangle l_1 l_2 l_3$, then propagating $f(a, l_1)$ to $\triangle l_1 l_2 l_3$ involves

Table 1. The preprocessing algorithm APQ_PD_Preprocessing(P)

ALGORITHM: APQ_PD_Preprocessing(P)
Input: A polygonal domain P having n vertices and h holes
Output: Data structure APQ_PD(P)
Step 1. For each vertex v of P, compute and store a shortest path map SPM(v) in P using the algorithm of [1].
Step 2. Compute and store a Voronoi diagram Vor(P) of P using the algorithm of [6]. Derive a skeleton graph G_s from Vor(P) by removing Voronoi edges incident to vertices of P. Compute a reduced skeleton graph of G_s.
Step 3. Compute a tessellation P' of P (cf. Section 3).
Step 4. For each simple polygon $R(l_1, l_2)$ in P' with gates l_1 and l_2, construct and store a hierarchy of hourglasses, $HH(R)$, applying the algorithm of [5]. Also, store the hourglass $H(l_1, l_2)$.
Step 5. Using the algorithm of [8], compute a point-location data structure based on the triangulation computed in Step 4 by using the algorithm of [5].

a funnel-split. The termination condition of the propagation process is that either a sees b, then dist(a, b) can be computed directly; or, all funnels from a are closed. We find the set C_a (i.e., $C_{\min}$) from the strings of closed funnels.

Next, we present key observations with respect to open funnels from a. There exists a portion of a gate l visible from a for each open funnel $\mathfrak{f}(a, l)$. The portion is called a *visible interval* associated with the funnel $\mathfrak{f}(a, l)$. There is a one-one mapping between each open funnel and its associated visible interval. Due to visibility from a, two open funnels from a do not cross each other. All visible intervals associated with open funnels from a are angularly sorted with respect to a. Two consecutive open funnels from a, whose visible intervals are consecutive in an angularly sorted order with respect to a, account for one hole. We maintain a binary search tree (BST) for all open funnels from a with keys as the angularly sorted order of intervals associated with these funnels. For each funnel propagating to a gate, the gate stores a pointer pointing to a BST node which knows where the funnel is stored.

We build a DAG, denoted by D, that is the reduced skeleton graph except each edge becomes a directed edge, to record dependency of visiting order of gates by funnel propagations from a. Each node of D stores a pointer pointing to a gate visited by a funnel. A directed edge $(u(l_1), v(l_2))$ of D represents the funnel propagation from the gate l_1 to the gate l_2. Two nodes $v_1(l_1)$ and $v_2(l_2)$ are *siblings* with the same parent $u(l)$ if the corresponding gates l, l_1 and l_2 form a triangle of P'. A node is marked if its parent and sibling are marked. Marked nodes, called *ready* nodes, are put in a *ready* queue. While the ready queue is not empty, we remove a node, $v(l)$, from the ready queue and propagate $O(1)$ funnels associated with the gate l to a neighboring gate of l. Gates stored in the ready nodes, called *ready* gates, form a "wavefront" of the propagation process. Only ready gates are processed. The *ready* attribute associated with gates with respect to a and to b are irrelevant. A DAG $D(a)$ with respect to a is irrelevant to a DAG $D(b)$ with respect to b.

Next, we elaborate on a key idea in propagating $O(1)$ funnels for each gate. Let $\mathfrak{f}_1(a, l_j), \ldots, \mathfrak{f}_k(a, l_j)$ be a sequence of all open funnels from a to a gate l_j such that their corresponding (visible) intervals associated with the gate l_j, denoted by indices $j_1, \ldots, j_k$, are angularly sorted with respect to a. The pair of funnels corresponding to two extreme intervals, e.g., l_{j_1} and l_{j_k}, is called *end-funnel* pair, denoted by $\mathfrak{f}_-(a, l_j)$. Let l be a ready gate associated with k open funnels. If $k = O(1)$, all funnels associated with the gate l are computed. Otherwise, we proceed as follows. Call this gate l as the *base* gate. We compute, store, and propagate (1) an end-funnel pair, $\mathfrak{f}_-(a, l)$, associated with l to a neighboring gate of l; (2) a set, $S_a(l)$, of hourglasses, initially consisting of one hourglass associated with the pair $\mathfrak{f}_-(a, l)$. This hourglass starts from the gate l and follows the propagation trajectory of the pair $\mathfrak{f}_-(a, l)$ in the following way. When the end-funnels of the pair $\mathfrak{f}_-(a, l)$ split and propagate to different regions, e.g., both neighboring a triangle, we need to compute new end-funnels paired with the survived ones. Meanwhile, the hourglass corresponding to the pair $\mathfrak{f}_-(a, l)$ splits into two, each corresponds to a new pair of end-funnels associated with the same gate. The set $S_a(l)$ now has two hourglasses. This process continues. An hourglass in the set $S_a(l)$ starts and ends at base gates. The set $S_a(l)$ records the propagation trajectory from l to each ending gate of an hourglass in the set, as a set of non-crossing paths in the reduced skeleton graph. We repeat this computation for each base gate.

Next, we introduce a *pulling* operation used in our algorithm. Let l_0 be the previous base gate for the current gate l_1. Let $\mathfrak{f}_-(a, l_1)$ be the end-funnel pair associated with l_1. Assume that at a triangle $\triangle l_1 l_2 l_3$, the pair $\mathfrak{f}_-(a, l_1)$ splits and propagates to different neighboring gates l_2 and l_3. Now, each gate l_2 and l_3 has one end-funnel. We need to compute a new end-funnel paired with the survived one for each gate. We search funnels from the previous base l_0 and *pull* them to the gates l_2 and l_3 as follows. Either we find a funnel splitting at $\triangle l_1 l_2 l_3$ or two funnels. Each resulting funnel is a new end-funnel required for gates l_2 and l_3, respectively. We concatenate each funnel with the hourglass associated with the pair $\mathfrak{f}_-(a, l_1)$, then propagate the resulting one to l_2 and l_3, respectively. We call this operation *pulling*. A pulling operation takes at most $O(\log n)$ time [5]. When one or both of the pair $\mathfrak{f}_-(a, l_1)$ is closed after a propagation, we need to pull the next consecutive funnel from the previous base as a new end-funnel, and replace the closed one with the new one. This step is technical and details will be provided in the full version of this paper.

Next, we present an algorithm called FindMinCriticalSites which is a subroutine of our query algorithm. Algorithm FindMinCriticalSites (Table 5) takes as input a query pair (a, b), sets of open and closed funnels from a and from b, i.e., $F_o(a), F_c(a), F_o(b), F_c(b)$, DAGs $D(a), D(b)$ and the tessellation P'. It terminates when either a sees b, or one of the open-funnel sets $F_o(a)$ and $F_o(b)$ is empty. The output is either the set $C_{\min}$ or the distance $dist(a, b)$.

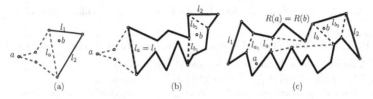

Fig. 3. An illustration of (a) a diagonal l_a associated with a funnel $\mathfrak{f}(a, l_a)$ that is incident to a triangle; or (b) a simple polygon; (c) the case that $R(a) = R(b)$

The key step (Step 2) of the algorithm loops through the BSTs, $F_o(a)$ and $F_o(b)$, until one of them is empty. Consider the point a first. In Step 2.1, for each ready gate l_a in the ready queue, two cases arise. Case 1: the gate l_a is incident to a neighboring triangle $\triangle l_a l_1 l_2$. Case 2: l_a is incident to a neighboring simple polygon R with gates l_1 and l_2, $l_a = l_1$. In Case 1, if $\triangle l_a l_1 l_2$ contains b (Fig. 3 (a)), determine if a sees b by computing a tangent from b to the open funnel associated with l_a. If yes, compute and return dist(a, b). Pull either a funnel to split at $\triangle l_a l_1 l_2$ from the previous base gate or two neighboring funnels which are end-funnels, one for each gate, l_1 or l_2. In Case 2, if R contains b (Fig. 3 (b)), identify the hourglass $H(l_a, l_{b_1})$ in R. Propagate the end-funnels of l_a and the associated hourglass to l_{b_1} by concatenating them with $H(l_a, l_{b_1})$, respectively. If there exists an open resulting funnel, determine if a sees b. If yes, compute and return dist(a, b). Otherwise, propagate the open funnel to b. The resulting funnel $\mathfrak{f}(a, b)$ is a string. In Step 2.2, propagate the end-funnels at l_a and their associated hourglass to each neighboring gate. Remove each closed funnel from the BST $F_o(a)$ and add it to the set $F_c(a)$. Update the DAG $D(a)$. Repeat Steps 2.1 and 2.2 for b.

The algorithm reaches Step 3 when a does not see b and one of the sets $F_o(a)$ and $F_o(b)$ is empty. W.l.o.g., let this set be $F_o(a)$. In Step 3, for each funnel in $F_c(a)$, collect the first vertex (other than a if a is also a vertex) on the string of the funnel (Lemma 3). Store it in $C_{\min}$. The algorithm outputs $C_{\min}$ in Step 3. We summarize the complexity of this algorithm in the next lemma.

Lemma 6. *For any query pair (a, b) in P, Algorithm* FindMinCriticalSites *either correctly finds the minimum set $C_{\min}$ of critical sites or detects that a sees b. This algorithm executes in the worst case in $O(h \log n)$ time.*

Proof. We prove the correctness of the algorithm first. Let $i \in \{a, b\}$. Upon termination, the algorithm either detects that a sees b or the set $F_o(i)$ is empty. For each funnel in $F_c(i)$, the first vertex (other than the source i) on the string of the funnel is the critical site in $C_{\min}$ (Lemma 3 and Corollary 1). Next, we analyze the complexity of the algorithm. The set of paths (in the reduced skeleton graph) forming the open funnels from i at any stage of the algorithm is planar. The total number of funnels encountered is bounded by $O(h)$. Then, the total number of funnels removed from $F_o(i)$ is bounded $O(h)$. We perform $O(h)$ concatenations, $O(1)$ for each gate, and $O(h)$ pulling operations. Thus,

Table 2. The algorithm FindMinCriticalSites($a, F_o(a), F_c(a), D(a), b, F_o(b), F_c(b), D(b), P'$)

ALGORITHM: FindMinCriticalSites($a, F_o(a), F_c(a), D(a), b, F_o(b), F_c(b), D(b), P'$)
Input: A query pair (a, b), sets of open and closed funnels $F_o(a), F_c(a), F_o(b), F_c(b)$, DAGs $D(a), D(b)$ and a tessellation P'. *Output:* A set $C_{\min}$ of critical sites if a does not see b or dist(a, b) otherwise.
Step 1. Let $C_{\min}$ be empty.
Step 2. While none of $F_o(a)$ or $F_o(b)$ is empty, do Steps 2.1 - 2.3.
2.1   For any *ready* gate l_a with respect to a, two cases arise. *Case 1:* The gate l_a is incident to a neighboring triangle, $\triangle l_a l_1 l_2$. If $\triangle l_a l_1 l_2$ contains b (Fig. 3 (a)), determine if a sees b by computing a tangent from b to the open funnel associated with l_a. If yes, compute and return dist(a, b). Pull either a funnel to split at $\triangle l_a l_1 l_2$ from the previous base gate or two neighboring funnels which are end-funnels, one for each gate l_1 and l_2.
Case 2: The gate l_a is incident to a simple polygon R with gates $l_1, l_2, l_a = l_1$. If R contains b (Fig. 3 (b)), identify the hourglass $H(l_a, l_{b_1})$ in R. Propagate the end-funnels of l_a and the associated hourglass to l_{b_1} by concatenating them with $H(l_a, l_{b_1})$, respectively. If there exists an open resulting funnel, determine if a sees b. If yes, compute and return dist(a, b). Otherwise, propagate the open funnel to b.
2.2   Propagate the end-funnels of l_a and the associated hourglass (starting from the previous base gate) to each neighboring gate. Pull a new end-funnel from the previous base gate if one of the end-funnels is closed. Remove each closed funnel from $F_o(a)$ and add it to $F_c(a)$. Update the DAG $D(a)$.
2.3   Repeat Steps 2.1 and 2.2 for b.
Step 3: W.l.o.g., let $F_o(a)$ be empty between $F_o(a)$ and $F_o(b)$. For each funnel in $F_c(a)$, collect the first vertex on its string; store it in $C_{\min}$. Return $C_{\min}$.

the number of iterations in Step 2 is bounded by $O(h)$, which dominates the execution of the algorithm. Each iteration of Step 2 takes at most $O(\log n)$ time [5]. The claim follows.   □

Finally, we present our query algorithm. The algorithm takes input as a query pair (a, b) and outputs the distance dist(a, b). Using the point-location data structure, first find triangles f_a and f_b containing a and b, respectively. Next, identify the region $R(a)$ containing a and $R(b)$ containing b in the tessellation P'. Two cases arise. Case 1: $R(a) = R(b)$; Case 2: $R(a) \neq R(b)$. We consider Case 1. If $R(a)$ is a triangle, than a sees b. Compute and return dist(a, b). Otherwise, $R(a)$ is a simple polygon. Let l_a and l_b be the separating diagonals in $R(a)$ close to a and to b, respectively (Fig. 3 (c)). Construct an hourglass $H(l_a, l_b)$ for l_a and l_b using the hierarchy of hourglasses, $HH(R(a))$, associated with $R(a)$, applying the algorithm of [5]. Concatenate $H(a, l_a), H(l_a, l_b)$ and $H(l_b, b)$. The resulting hourglass $H(a, b)$ is closed and it is the local shortest path between a

and b in $R(a)$. If there is no other vertices on $H(a, b)$ except a and b, then a sees b. Compute and return dist(a, b). Otherwise, add $H(a, b)$ to $F_c(a)$. W.l.o.g., let l_1 be the gate of $R(a)$ that is closer to a and let l_2 be the gate closer to b (Fig. 3 (c)). Construct hourglasses $H(l_{a_1}, l_1)$ and $H(l_{b_2}, l_2)$ from the hierarchy $HH(R(a))$. Compute an initial funnel $\mathsf{f}(a, l_1)$ from a to l_1 by finding tangents from a to the hourglass $H(l_{a_1}, l_1)$. Analogously, compute an initial funnel $\mathsf{f}(b, l_2)$ from b to l_2. We now consider Case 2. For $i \in \{a, b\}$, if $R(i)$ is a triangle, compute initial funnels from i to the gates of $R(i)$. Otherwise, $R(i)$ is a simple polygon with gates l_1 and l_2. Construct hourglasses $H(l_{i_1}, l_1)$ and $H(l_{i_2}, l_2)$ from the hierarchy $HH(R(i))$. Compute initial funnels $\mathsf{f}(i, l_1), \mathsf{f}(i, l_2)$ from i to l_1 and to l_2 in $R(i)$. In both cases, add each resulting funnel to $F_o(i)$ if it is open or to $F_c(i)$ otherwise. Update the DAG $D(i)$. Finally, propagate open funnels in $F_o(a)$ and $F_o(b)$ by calling Algorithm FindMinCriticalSites which returns either dist(a, b) if a seeing b is detected or the set $C_{\min}$ of critical sites otherwise. Return dist(a, b) if a seeing b is true. Otherwise, compute dist(a, b) applying Corollary 1. We summarize the complexity of the query algorithm in the next lemma, followed by our main theorem.

Lemma 7. *The distance* dist(a, b) *between any pair of query points,* (a, b), *can be computed in* $O(h \log n)$ *time.*

Theorem 2. *Given a polygonal domain P having n vertices and h holes, one can compute a data structure of size $O(n^2)$ in $O(n^2 \log n)$ time and answer shortest path queries between any pair of points in P in the worst case in $O(h \log n)$ time.*

References

1. Hershberger, J., Suri, S.: An optimal algorithm for Euclidean shortest paths in the plane. SIAM J. Computing 28(6), 2215–2256 (1999)
2. Kapoor, S., Maheshwari, S.N., Mitchell, J.S.B.: An efficient algorithm for Euclidean shortest paths among polygonal obstacles in the plane. Discete Comput. Geom. 18, 377–383 (1997)
3. Chen, D.Z., Daescu, O., Klenk, K.S.: On geometric path query problems. In: Rau-Chaplin, A., Dehne, F., Sack, J.-R., Tamassia, R. (eds.) WADS 1997. LNCS, vol. 1272, pp. 248–257. Springer, Heidelberg (1997)
4. Chiang, Y.J., Mitchell, J.S.B.: Two-point Euclidean shortest path queries in the plane. In: SODA 1999: Proceedings of the tenth annual ACM-SIAM symposium on Discrete algorithms, Philadelphia, PA, USA. Society for Industrial and Applied Mathematics, pp. 215–224 (1999)
5. Guibas, L., Hershberger, J.: Optimal shortest path queries in a simple polygon. J. Comput. Syst. Sci. (JCSS) 39(2), 126–152 (1989)
6. Srinivasan, V., Nackman, L.R.: Voronoi diagram for multiply-connected polygonal domains I: algorithm. IBM Journal of Research and Development 31(3), 361–372 (1987)
7. Yang, C.L., Wang, J.Y., Meng, X.X.: Upper bounds on the size of inner Voronoi diagrams of multiply connected polygons. J. Software 17(7), 1527–1534 (2006)
8. Kirkpatrick, D.G.: Optimal search in planar subdivisions. SIAM J. Comput. 12, 28–35 (1983)

A Fast 2-Approximation Algorithm for the Minimum Manhattan Network Problem[*]

Zeyu Guo[1], He Sun[2], and Hong Zhu[3]

[1] Department of Computer Science and Engineering
Fudan University, China
[2] Shanghai Key Laboratory of Intelligent Information Processing
Fudan University, China
[3] Shanghai Key Laboratory of Trustworthy Computing
East China Normal University, China
{gzy,sunhe,hzhu}@fudan.edu.cn

Abstract. Given a set T of n points in $\mathbb{R}^2$, a Manhattan Network G is a network with all its edges horizontal or vertical segments, such that for all $p, q \in T$, in G there exists a path (named a Manhattan path) of the length exactly the Manhattan distance between p and q. The Minimum Manhattan Network (MMN) problem is to find a Manhattan network of the minimum length, *i.e.*, the total length of the segments of the network is to be minimized. In this paper we present a 2-approximation algorithm with time complexity $O(n^2)$, which improves the 2-approximation algorithm with time complexity $\Omega(n^8)$, proposed by Chepoi, Nouioua *et al.*. To the best of our knowledge, this is the best result on this problem.

1 Introduction

A *rectilinear path* between two points $p, q \in \mathbb{R}^2$ is a path connecting p and q with all its edges horizontal or vertical segments. Furthermore, a *Manhattan path* between p and q is a rectilinear path with its length exactly $\mathrm{dist}(p, q) := |p.x - q.x| + |p.y - q.y|$, *i.e.*, the Manhattan distance between p and q.

Given a set T of n points in $\mathbb{R}^2$, a network G is said to be a *Manhattan network* on T, if for all $p, q \in T$ there exists a Manhattan path between p and q with all its segments in G. For the given network G, let the length of G, denoted by $L(G)$, be the total length of all segments of G. For the given point set T, the *Minimum Manhattan Network* (MMN) Problem is to find a Manhattan network G on T with minimum $L(G)$.

From the problem description, it is easy to show that there is a close relationship between the MMN problem and planar t-spanners. For $t \geq 1$, if there exists a planar graph G such that for all $p, q \in T$, there exists a path in G connecting p and q of length at most t times the distance between p and q, G is said to be

[*] This work is supported by Shanghai Leading Academic Discipline Project(Project Number:B412) and National Natural Science Fund (grant #60496321 and #60703091). Correspondence author: He Sun

R. Fleischer and J. Xu (Eds.): AAIM 2008, LNCS 5034, pp. 212–223, 2008.
© Springer-Verlag Berlin Heidelberg 2008

a *t-spanner* of T. The MMN Problem for T is exactly the problem to compute the 1-spanner of T under the L_1-norm.

Motivation: The spanners and the MMN problem have applications in city planning, network layout, distributed algorithms and VLSI circuit design. Lam, Alexandersson et al. [8] showed an approximation algorithm for Manhattan networks on alignment graphs that are relevant to sequence alignments in computational biology.

Related works: The MMN problem was first introduced by Gudmundsson, Levcopoulos et al. [6], and until now, it is open whether this problem belongs to the complexity class P. Gudmundsson et al. [6] proposed an $O(n^3)$-time 4-approximation algorithm, and an $O(n \log n)$-time 8-approximation algorithm. Kato, Imai et al. [7] presented an $O(n^3)$-time 2-approximation algorithm. However, the proof of their algorithm correctness is incomplete [4]. In spite of that, their paper still provided a valuable idea, that it suffices for G to be a Manhattan network if for each of $O(n)$ certain pairs there exists a Manhattan path connecting its two points. Thus it is not necessary to enumerate all the pairs in $T \times T$. Following this idea, Benkert, Wolff et al. [1,2] proposed an $O(n \log n)$-time 3-approximation algorithm. They also described a mixed-integer programming (MIP) formulation of the MMN problem. After that, Chepoi, Nouioua et al. [4] proposed a 2-approximation rounding algorithm by solving the linear programming (LP) relaxation of the MIP. In [9], S. Seibert and W. Unger proposed a 1.5-approximation algorithm. Unfortunately, their proof is incorrect. Therefore 2-approximation is, to our best knowledge, the lowest approximation ratio for this problem. However, since the algorithm of Chepoi, Nouioua et al. is based on linear programming, its time complexity, as pointed out in [1], may be as high as $\Omega(n^8)$.

Outline of our approach: In a high-level overview, our algorithm is as follows: partition the input into several blocks (ortho-convex regions) that can be solved independently of each other. For the blocks, some can be trivially solved optimally, whereas only one type of blocks is difficult to solve. For such a non-trivial block there are some horizontal and vertical strips which can be solved by horizontal and vertical nice covers plus switch segments to connect neighboring points in the same strip. Furthermore, we add some missing connections in regions called staircases. We implement this step using the dynamic programming technique to guarantee that staircases are locally optimal. Though a similar method can be found in previous work, we define a new form of staircase, called revised staircase (RS), and use the speed-up technique of the dynamic programming introduced in [5]. For the approximation analysis, the ratio of 2 follows simply from the fact that the strips and staircases are solved locally optimally, and though these regions may overlap, the overlapping segments are counted at most twice.

Our results: In this paper we present a 2-approximation algorithm with time complexity $O(n^2)$. Furthermore, for the given set T, let H and W be the height and the width of the minimum rectangle covering all points in T with its sides parallel to the axes. Let G^* be an MMN on T and G be the network constructed by our algorithm. From the algorithm description, it is proven that $L(G) \leq 2L(G^*) - H - W$.

2 Preliminaries

Basic notations: For $p = (p.x, p.y) \in \mathbb{R}^2$, let $\mathcal{Q}_k(p)$ denote the k-th closed quadrant with respect to the origin p, e.g., $\mathcal{Q}_1(p) := \{q \in \mathbb{R}^2 \mid p.x \leq q.x, p.y \leq q.y\}$.

Define $R(p, q)$ as a closed rectangle (possibly degenerate) where $p, q \in \mathbb{R}^2$ are its two opposite corners. $B_V(p, q)$ is defined as the vertical closed band bounded by p, q, whereas $B_H(p, q)$ denotes the horizontal closed band bounded by p, q.

For the given point set T, let Γ be the union of vertical and horizontal lines which pass through some point in T. In addition, we use $[c, d]$ to represent the vertical or horizontal segment with endpoints c and d, as Fig. 1 shows.

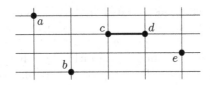

Fig. 1. $T = \{a, b, c, d, e\}$. The vertical and horizontal lines compose Γ

Pareto envelope: The Pareto envelope, originally proposed by Cheopi *et al.* [4], plays an important role in our algorithm and we give a brief introduction.

Given the set of points T, a point p is said to be *dominated* by q if $(\forall t \in T : \mathrm{dist}(q, t) \leq \mathrm{dist}(p, t)) \wedge (\exists t \in T : \mathrm{dist}(q, t) < \mathrm{dist}(p, t))$. A point is said to be an *efficient point* if it is not dominated by any point in the plane. The *Pareto envelope* of T is the set of all efficient points, denoted by $\mathcal{P}(T)$. Fig. 2 shows an example of $\mathcal{P}(T)$. It is not hard to prove that $\mathcal{P}(T) = \bigcap_{u \in T} \bigcup_{v \in T} R(u, v)$. Chalmet *et al.* [3] demonstrated that $\mathcal{P}(T)$ can be built in $O(n \log n)$ time. They also presented some other properties of $\mathcal{P}(T)$. In particular, $\mathcal{P}(T)$ is *ortho-convex*, i.e., the intersection of $\mathcal{P}(T)$ with any vertical or horizontal line is continuous, which is equivalent to the fact that for any two points $p, q \in \mathcal{P}(T)$, there exists a Manhattan path in $\mathcal{P}(T)$ between p and q.

In [4] Chepoi *et al.* also showed that the Pareto envelope is the union of some ortho-convex (possibly degenerate) rectilinear polygons (called *blocks*). Two blocks can overlap at only one point which is called a *cut vertex*. We denote by C the set of cut vertices, and let $T^+ := T \cup C$. A vertical sweeping line ℓ moving from left to right may intersect two or more blocks only when $\ell \subseteq \Gamma$, since only such kind of ℓ passes through the cut vertices where blocks overlap. This fact yields $W = \sum_{B \subseteq \mathcal{P}(T)} W_B$, and similarly $H = \sum_{B \subseteq \mathcal{P}(T)} H_B$, where H_B and W_B are the height and the width of the minimum rectangle covering

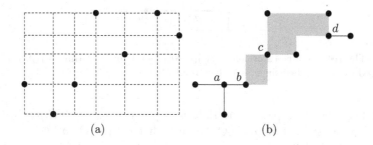

(a)   (b)

Fig. 2. An example of a Pareto envelope. The black points in (a) are the set T. The two separate grey regions in (b) are non-degenerate blocks, whereas the black lines are degenerate blocks. All these blocks form the Pareto envelope $\mathcal{P}(T)$.

the block B. For a block B let $T_B := T^+ \cap B$. We say B is *trivial* if B is a rectangle (or degenerate to a segment) such that $|T_B| = 2$. It is known that the two points in T_B must be two opposite corners of B when it is trivial. In Fig. 2, $C = \{a, b, c, d\}$ and only the block between c and d is non-trivial.

Chepoi *et al.* [4] proved that an MMN on T^+ is also an MMN on T, and to obtain an MMN on T^+, it suffices to build an MMN on T_B for each $B \subseteq \mathcal{P}(T)$. The MMN in any trivial block B can be built by simply connecting the two points in T_B using a Manhattan path. So we have reduced the MMN problem on T to MMN on non-trivial blocks.

For a non-trivial block B denote its border by ∂B and let $\Gamma_B := \Gamma \cap B$. We call a corner p in ∂B a *convex corner* if the interior angle at p equals to $\frac{\pi}{2}$, otherwise p is called a *concave corner*.

Lemma 1. *[4] For any non-trivial block B and any convex corner p in ∂B, it holds that $p \in T_B$.*

Lemma 2. *[4] For any non-trivial block B, there exists an MMN G_B on T_B such that $G_B \subseteq \Gamma_B$. Furthermore, any MMN $G_B \subseteq \Gamma_B$ on T_B contains ∂B.*

Strips and staircases: Informally, for $p, q \in T_B, p.y < q.y$, we call $R(p,q)$ a *vertical strip* if it does not contain any point in the region $B_V(p,q)$ except the vertical lines $\{(x,y)|x = p.x, y \le p.y\}$ and $\{(x,y)|x = q.x, y \ge q.y\}$. Similarly, for the points $p, q \in T_B, p.x < q.x$, we call $R(p,q)$ a *horizontal strip* $R(p,q)$ if it does not contain any point in the region $B_H(p,q)$ except the horizontal lines $\{(x,y)|x \le p.x, y = p.y\}$ and $\{(x,y)|x \ge q.x, y = q.y\}$. Especially, we say a vertical or horizontal strip $R(p,q)$ is *degenerate* if $p.x = q.x$ or $p.y = q.y$. Fig. 3 gives an example of a horizontal strip.

The other notion which plays a critical role in our algorithm is the staircase. There are four kinds of staircases specified by a parameter $k \in \{1, \cdots, 4\}$, and without loss of generality we only describe the one with $k = 1$. Suppose $R(p,q)$ is a vertical strip and $R(p',q')$ is a horizontal strip, such that $q \in \mathcal{Q}_1(p)$, $q' \in \mathcal{Q}_1(p')$, $p, q \in B_V(p',q')$, $p', q' \in B_H(p,q)$, i.e., they cross in the way as Fig. 4

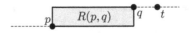

Fig. 3. The rectangle is a horizontal strip. Any point in T_B within $B_H(p,q)$ can only be placed on the dashed linee, *e.g.*, the point t.

shows. Let o be the topmost and rightmost point in $R(p,q) \cap R(p',q')$. Denote by $T_{pp'|qq'}$ the set of v in $T_B \cap Q_1(o)$ such that $(Q_3(v) \backslash Q_3(o)) \cap T_B = \{v\}$. If $T_{pp'|qq'} \neq \emptyset$, then let $S_{pp'|qq'} = \bigcup_{v \in T_{pp'|qq'}} R(o,v)$, which is said to be a *staircase* (see Fig. 4). In this figure, no point in T_B is located in the dark area and the light grey regions (*i.e.*, the staircase and the two unbounded half-bands) except those in $T_{pp'|qq'}$.

As was observed in [1,4], the interiors of two staircases are disjoint, and the interior of a staircase and a strip are disjoint. For a strip $R(p,q)$, (p,q) is called a *strip pair*. For each staircase $S_{pp'|qq'}$, let v be any point in $T_{pp'|qq'}$, and (v,p) (also (v,p')) is called a *staircase pair*.

Theorem 1. *[4] For a network G_B, if for any strip pair or staircase pair (p,q), $p, q \in T_B$, there exists a Manhattan path in G_B connecting p and q, then G_B is a Manhattan network on T_B.*

3   Algorithm Description

Following [1] we give the following definitions: a union of vertical segments C_V is said to be *a vertical cover* if for any horizontal line ℓ and any vertical strip R that ℓ intersects, it holds that $\ell \cap R \cap C_V \neq \emptyset$. A union of horizontal segments C_H is said to be *a horizontal cover* if for any vertical line ℓ and any horizontal strip R that ℓ intersects, it holds that $\ell \cap R \cap C_H \neq \emptyset$. Furthermore, *a minimum vertical cover* (MVC) is the vertical cover of the minimum length, as Fig. 5 shows, whereas *a minimum horizontal cover* (MHC) is the horizontal cover of the minimum length. In addition, an MVC/MHC is said to be *nice* if any of its segments contains at least one point in T_B. A *nice cover* is the union of a nice MVC and a nice MHC.

It can be shown that a nice cover, expressed by E_C, always exists [1,2]. Benkert *et al.* [1,2] proposed an $O(n \log n)$-time algorithm that computes a nice cover E_C for the set of points T_B, $|T_B| = n$, such that $E_C \subseteq \Gamma$. For a non-degenerate strip $R(p,q)$, we use $[p,a]$ and $[q,b]$ to express the segments of E_C covering $R(p,q)$. Especially, $a = p$ if no such a segment contains p, and $b = q$ if no such a segment contains q. Note that $[p,a] \cup [q,b] \cup [a,b]$ is a Manhattan path. We say $[a,b]$ is a *switch segment*. Let E_S denote the union of switch segments, and it has been proven that $L(E_S) \leq H_B + W_B$.

Theorem 2. *There exists a procedure* CreateNC *to construct a nice cover E_C such that $E_C \subseteq \Gamma_B$.*

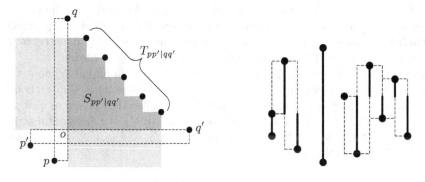

Fig. 4. A staircase consisting of a dark grey region and its borderline

Fig. 5. A nice MVC consisting of black lines

Proof. The procedure `CreateNC` consists of two steps. In the initial step, we run the algorithm proposed by Benkert *et al.* [1,2] to get a nice cover E_C.

In the second step, we modify E_C to guarantee that $E_C \subseteq \Gamma_B$. Clearly a segment e of E_C which is not entirely laid on B must be used to cover a non-degenerate strip. Without loss of generality, we use e to cover $R(p,q)$, a vertical strip with p the lower left corner and q the upper right corner. Assume that e contains q. From $q \in B$ we know ∂B crosses e. From $p \in B$, ∂B turns at some point p' down to p, as Fig. 6 shows. Let q' be the bottom endpoint of e. It turns out that the Manhattan path between p and q switches at the y-ordinate of q' which is less than that of p'.

We move the switch segment up to the y-ordinate of p' (note that switch segments are not added yet so the movement is just done in concept). Accordingly, the part of e lying outside of B is moved left onto ∂B. Since no point of T_B is put outside of B, the part that has been moved is not used to cover any other strip, which shows that E_C is still a nice cover after the movement. We repeat the modification until no such e exists. Then, we have $E_C \subseteq \Gamma_B$. □

Corollary 1. *For $|T_B| = n$, the time complexity of `CreateNC` is $O(n \log n)$.*

Proof. From [1], the time complexity of creating a nice cover is $O(n \log n)$.

Now we analyze the running time of the second step. Since each point of T_B is exactly in one vertical segment and one horizontal segment in E_C, there exist at most $2n$ segments in E_C. Due to the fact that one modification step decreases one segment that is not entirely on B and the number of segments is $O(n)$, the time complexity of the second step is $O(n)$.

In summary, the running time of `CreateNC` is $O(n \log n)$. □

Lemma 3. *After the procedure `CreateNC`, $\partial B \subseteq E_C$.*

Proof. For $p, q \in \partial B$, let ∂B_{pq} be the part of ∂B from p to q in the counter-clockwise direction (p, q excluded). It suffices to prove that $\partial B_{pq} \subseteq E_C$ whenever $p, q \in \partial B$, $T_B \cap \partial B_{pq} = \emptyset$, i.e., p, q are two neighboring points. This is obviously true if $p.x = q.x$ or $p.y = q.y$. For the other cases, we know there is no convex corner in ∂B_{pq} since if such a convex corner exists, then it is not in T_B, which contradicts Lemma 1. Thus there is a concave corner on ∂B_{pq}, denoted by t. Without loss of generality, assume that $p, q \in \mathcal{Q}_4(t)$, as Fig. 7 shows.

Among the points in T_B let p' be the leftmost one in $\mathcal{Q}_1(p)$ (choose the bottommost if more than one exist) and q' be the topmost point in $\mathcal{Q}_3(q)$ (choose the rightmost if more than one exist). Note it is possible that $p' = q$ or $q' = p$.

It is clear that $R(p, p')$ is a vertical strip and $R(q, q')$ is a horizontal strip. From $E_C \subseteq B$ we have $[p, t] \subseteq E_C$ to cover $R(p, p')$, and $[q, t] \subseteq E_C$ to cover $R(q, q')$. Therefore $\partial B_{pq} \subseteq E_C$.   □

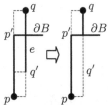
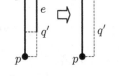
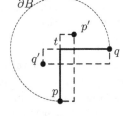

Fig. 6. Modifying the nice cover   **Fig. 7.** A concave corner

Now we add E_S to G and complete the procedure constructing the Manhattan paths of all the strip pairs. Let M_{pq} denote the Manhattan path constructed between p and q. Next we deal with the staircase pairs.

Definition 1 (revised staircase). *For a staircase $S_{pp'|qq'}$, let o' be the intersection of M_{pq} and $M_{p'q'}$ (if more than one intersections exist then choose any one). Let the revised staircase (RS) $S^\star_{pp'|qq'}$ be the part of $\bigcup_{v \in T_{pp'|qq'}} R(o', v) \backslash \partial B$ enclosed by $M_{pq}, M_{p'q'}$ ($M_{pq}, M_{p'q'}$ excluded).*

Fig. 8 shows how to compute an RS from a staircase. From the definition, notice that ∂B does not belong to the revised staircase. It is not hard to prove that no point of T_B falls into the interior of any RS. This fact implies that the interiors of two RS are disjoint, and the interior of the RS $S^\star_{pp'|qq'}$ does not overlap any strip except $R(p, q)$ and $R(p', q')$.

Given the staircase $S_{pp'|qq'}$, we will describe how to build the Manhattan paths of all the staircase pairs of $S_{pp'|qq'}$. Without loss of generality, assume that $R(p, q)$ is a vertical strip and $R(p', q')$ is a horizontal strip where $q, q' \in \mathcal{Q}_1(o)$, as Fig. 4 shows. The other cases are symmetric. Since in the following analysis we need to deal with each revised staircase, we omit the subscript and write $S := S_{pp'|qq'}, S^\star := S^\star_{pp'|qq'}$ and $T_S := T_{pp'|qq'}$.

Let $m := |T_S| + 1, v_0 := q, v_m := q'$. Express the points in T_S as $v_1, v_2, \cdots,$ v_{m-1} in the order from the topmost and leftmost one to the bottommost and rightmost one. From Lemma 2 we know there exists a Manhattan path in B for every two consecutive points v_i, v_{i+1}. These Manhattan paths, together with

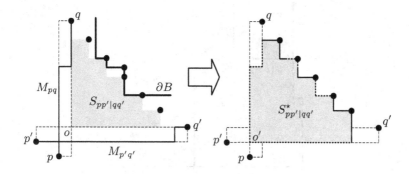

Fig. 8. From a staircase to an RS. The black line in the left picture is ∂B.

M_{pq} and $M_{p'q'}$ which are also in B, enclose a closed area covering $S^\star$. So we have $S^\star \subseteq B$, which yields that all segments added within $S^\star$ are also within B.

Theorem 3. *There exists a procedure* CompOptNet *such that for the given RS* $S^\star$ *with the point set* T_s, $|T_S| = n$, CompOptNet *takes* $O(n^2)$ *time and space to compute an optimal network* E *such that* $E \cup \partial B$ *connects each point in* T_S *to either* M_{pq} *or* $M_{p'q'}$.

Proof. We use dynamic programming technique to implement this procedure.

Suppose $[x_i, v_i]$ is the vertical segment of $S \cap \partial B$ connecting v_i, and $[y_i, v_i]$ is the horizontal one, *i.e.*, $[x_i, v_i]$ and $[y_i, v_i]$ are the segments we subtracted from $R(o', v_i)$ ($x_i = v_i$ or $y_i = v_i$ if no such a vertical or horizontal segment exists), as Fig. 9 shows. Let $S(0, m) := S^\star$ and we use $c(0, m)$ to express the length of the optimal network $E \subseteq S(0, m)$, such that $E \cup \partial B$ connects $v_1, v_2, \cdots, v_{m-1}$ to either the left boundary or bottom boundary of $S(0, m)$.

For the region $S(0, m)$, we need to choose a suitable k, $0 \le k < m$, and split $S(0, m)$ into two small areas $S(0, k)$ and $S(k + 1, m)$ according to the value of k such that the function $c(0, m)$ is minimized. For $0 \le i \le j \le m$, the notion of $S(i, j)$ is defined recursively as described in the following paragraph. Fig. 9 gives an intuitive way to understand this partition method.

For the region $S(i, j)$ and chosen k, $i \le k < j$, we connect y_k left using a horizontal segment, denoted by $w(i, k)$, and express the area above $w(i, k)$ by $S(i, k)$ if $k > i$, otherwise $S(i, k) = \emptyset$. If $k + 1 \ne j$, we connect x_{k+1} down using a vertical segment $h(k + 1, j)$ and denote the area on the right side of $h(k + 1, j)$ by $S(k + 1, j)$. It is not difficult to see that

$$c(i, j) = \min_{i \le k < j} \left\{ c(i, k) + c(k + 1, j) + w(i, k) + h(k + 1, j) \right\}$$

where $c(i, j)$ is defined recursively and $c(i, i) = w(i, i) = h(i, i) = 0$.

It is easy to show that $w(i, j)$ and $h(i, j)$ satisfy the quadrangle inequalities, *i.e.* $w(i, j) + w(i', j') \le w(i, j') + w(i', j)$, and $h(i, j) + h(i', j') \le h(i, j') + h(i', j)$,

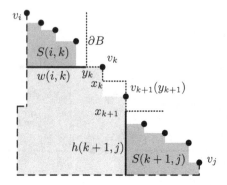

Fig. 9. The intuition behind the dynamic programming

where $i \leq i' \leq j \leq j'$. Furthermore, by Lemma 6 in the appendix we can conclude that the function c also satisfies the quadrangle inequality. Relying on the dynamic programming speed-up technique proposed by [5], all the values of c can be computed in $O(n^2)$ time.   □

In the following, we give the algorithm description:

Input: T
Algorithm:
1. Compute $\mathcal{P}(T)$.
2. For each trivial block $B \subseteq \mathcal{P}(T)$,
      connect the two points in T_B with a Manhattan path.
3. For each non-trivial block $B \subseteq \mathcal{P}(T)$ do the following:
      3.1. Run `CreateNC` to get E_C satisfying $E_C \subseteq \Gamma_B$.
      3.2. Add the union of switch segments E_S.
      3.3. For each RS $S^*_{pp'|qq'}$,
            run `CompOptNet` to compute the optimal network $E_{pp'|qq'}$.
Output: $G := E_C \cup E_S \cup_{S^*_{pp'|qq'}} E_{pp'|qq'}$.

Theorem 4. *For the given point set T of size n, the algorithm above takes $O(n^2)$ time and space to compute a Manhattan network G on T.*

Proof. As Chalmet *et al.* [3] described, we can use $O(n \log n)$ time to compute the Pareto envelope $\mathcal{P}(T)$. Step 2 takes $O(n)$ time since at most $O(n)$ blocks exist. Now we analyze the time complexity of Step 3. By Corollary 1, the time complexity of Step 3.1 is $O(n \log n)$, whereas by Theorem 3, the time complexity of Step 3.3 is

$$\sum_{S^*_{pp'|qq'}} O\left(|T_{pp'|qq'}|^2\right) = O(n^2)$$

Note that at most $O(n)$ switch segments exist, so the running time of Step 3.2 is $O(n)$.

In conclusion, the overall time complexity of the algorithm is $O(n^2)$.   □

4   Approximation Analysis

Let $G^\star \subseteq \Gamma$ be the optimal Manhattan network on T such that $\partial B \subseteq G^\star$ whenever B is non-trivial. For any block $B \subseteq \mathcal{P}(T)$ let $G_B := G \cap B$ and $G_B^\star := G^\star \cap B$, where G is the network we obtained by the approximation algorithm. So we need to give the upper bound of $L(G)$ with respect to $L(G^\star)$.

Given the non-trivial block B, we know $E_{pp'|qq'}$ is the local optimal network within a single RS $S_{pp'|qq'}^\star$. Let E_R be the union of these local optimal networks. E_C, E_R and E_S are different (perhaps not disjoint) parts of G_B. From our algorithm, we know that

$$L(G_B) \leq L(E_C) + L(E_R) + L(E_S) \tag{1}$$

We use $E_D^\star$ to express the set of segments in $G_B^\star$ shared by two different RSes. Fig. 10 shows such a segment.

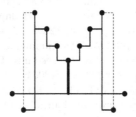

Fig. 10. A segment shared by two different RSes. These segments form the set $E_D^\star$.

Lemma 4. $L(E_R) \leq L(G_B^\star) + L(E_D^\star) - 2H_B - 2W_B$.

Proof. Let $E_{pp'|qq'}^\star$ be the part of $G_B^\star$ in a single RS $S_{pp'|qq'}^\star$.

$$
\begin{aligned}
L(E_R) &\leq \sum_{S_{pp'|qq'}^\star} L(E_{pp'|qq'}) \\
&\leq \sum_{S_{pp'|qq'}^\star} L(E_{pp'|qq'}^\star) \\
&\leq L(G_B^\star \backslash \partial B) + L(E_D^\star) \\
&= L(G_B^\star) + L(E_D^\star) - 2H_B - 2W_B
\end{aligned}
$$

The second inequality holds from the optimality of dynamic programming. In the third inequality, ∂B is subtracted because it does not intersect any RS, and $L(E_D^\star)$ appears because the segments of $E_D^\star$ are counted twice.   □

Lemma 5. $L(E_C) \leq L(G_B^\star) - L(E_D^\star)$

Proof. It is easy to prove that segments in $E_D^\star$ do not intersect any non-degenerate strip, and only intersect perpendicular degenerate strips. Therefore $G_B^\star \backslash E_D^\star$ still has the property that, for any strip pair (p, q), there exists a Manhattan path between p and q in it. It is known that any network with this property has length at least $L(E_C)$. This proves the lemma. ◻

Theorem 5. *For any block B,* $L(G_B) \leq 2L(G_B^\star) - H_B - W_B$.

Proof. If B is trivial, $L(G_B) = L(G_B^\star) = H_B + W_B = 2L(G_B^\star) - H_B - W_B$ holds. Otherwise, combining the fact $L(E_S) \leq H_B + W_B$ with (1), Lemma 4 and Lemma 5, we obtain $L(G_B) \leq 2L(G_B^\star) - H_B - W_B$. ◻

Corollary 2. $L(G) \leq 2L(G^\star) - H - W$ ◻

References

1. Benkert, M., Wolff, A., Widmann, F.: The minimum Manhattan network problem: a fast factor-3 approximation. Technical Report 2004-16, Fakultät für Informatik, Universität Karlsruhe. In: Proceedings of the 8th Japanese Conference on Discrete and Computational Geometry, pp. 16–28 (2005) (A sort version appeared)
2. Benkert, M., Shirabe, T., Wolff, A.: The minimum Manhattan network problem: approximations and exact solution. In: Proceedings of the 20th European Workshop on Computational Geometry, pp. 209–212 (2004)
3. Chalmet, G., Francis, L., Kolen, A.: Finding efficient solutions for rectilinear distance location problems efficiently. European Journal of Operations Research 6, 117–124 (1981)
4. Chepoi, V., Nouioua, K., Vaxès, Y.: A rounding algorithm for approximating minimum Manhattan networks. Theoretical Computer Science 390, 56–69 (2008) Preliminary version appeared. In: Proceedings of the 8th International Workshop on Approximation Algorithms for Combinatorial Optimization, pp. 40–51 (2005)
5. Frances Yao, F.: Efficient dynamic programming using quadrangle inequalities. In: Proceedings of the 12th Annual ACM Symposium on Theory of Computing, pp. 429–435 (1980)
6. Gudmundsson, J., Levcopoulos, C., Narasimhan, G.: Approximating a minimum Manhattan network. Nordic Journal of Computing 8, 219–232 (2001); Preliminary version appeared. In: Proceedings of the 2nd International Workshop on Approximation Algorithms for Combinatorial Optimization, pp. 28–37 (1999)
7. Kato, R., Imai, K., Asano, T.: An improved algorithm for the minimum Manhattan network problem. In: Proceedings of the 13th International Symposium on Algorithms and Computation, pp. 344–356 (2002)
8. Lam, F., Alexandersson, M., Pachter, L.: Picking alignments from (Steiner) trees. Journal of Computational Biology 10, 509–520 (2003)
9. Seibert, S., Unger, W.: A 1.5-approximation of the minimal Manhattan network problem. In: Proceedings of the 16th International Symposium on Algorithms and Computation, pp. 246–255 (2005)

A   Function c Satisfies the Quadrangle Inequality

Lemma 6. *For the dynamic programming procedure, the function*

$$c(i,j) = \min_{i \leq k < j} \left\{ c(i,k) + c(k+1,j) + w(i,k) + h(k+1,j) \right\}$$

satisfies the quadrangle inequality.

Proof. In order to prove c satisfies the quadrangle inequality, we only need to show that $c(i,j) + c(i',j') \leq c(i,j') + c(i',j)$ for any $i \leq i' \leq j \leq j'$. We prove this lemma by induction on the length of $\ell := |j' - i|$. Obviously, the inequality holds when $\ell = 0$.

Assume the inequality $c(i,j) + c(i',j') \leq c(i,j') + c(i',j)$ holds for any $i \leq i' \leq j \leq j'$ with parameter ℓ. We prove the inequality holds for the parameter $\ell + 1$.

We divide the proof into three cases.

Case 1: $i = i'$ or $j = j'$. Then, obviously $c(i,j) + c(i',j') = c(i,j') + c(i',j)$.

Case 2: $i < i' = j < j'$. Then, we need to prove $c(i,j) + c(i',j') \leq c(i,j')$. Choose $z \in [i, j'-1]$ such that $c(i,j') = c(i,z)+c(z+1,j')+w(i,z)+h(z+1,j')$. Assume $z \in [i, j-1]$ (The other cases can be proven symmetrically since $z \in [i', j'-1]$), then we have

$$\begin{aligned}
c(i,j) + c(i',j') &\leq c(i,z) + c(z+1,j) + w(i,z) + h(z+1,j) + c(i',j') \\
&\leq c(i,z) + c(z+1,j') + w(i,z) + h(z+1,j') \\
&= c(i,j')
\end{aligned}$$

The second inequality holds because (1) $h(z+1,j) \leq h(z+1,j')$ since $[z+1,j] \subseteq [z+1,j']$, and (2) $c(z+1,j) + c(i',j') \leq c(z+1,j')$ by induction hypothesis.

Case 3: $i < i' < j < j'$. Choose $y \in [i, j'-1]$ and $z \in [i', j-1]$ such that $c(i,j') = c(i,y) + c(y+1,j') + w(i,y) + h(y+1,j'), c(i',j) = c(i',z) + c(z+1,j) + w(i',z) + h(z+1,j)$. Without loss of generality, we assume $y \leq z$. Due to the fact that $z < j$, we obtain $y \in [i, j-1]$ and

$$\begin{aligned}
c(i,j) + c(i',j') &\leq c(i,y) + c(y+1,j) + w(i,y) + h(y+1,j) \\
&\quad + c(i',z) + c(z+1,j') + w(i',z) + h(z+1,j') \\
&\leq c(i,y) + c(y+1,j') + w(i,y) + h(y+1,j') \\
&\quad + c(i',z) + c(z+1,j) + w(i',z) + h(z+1,j) \\
&= c(i,j') + c(i',j)
\end{aligned}$$

The second inequality holds since (1) the function h and w satisfy the quadrangle inequality, and (2) $c(y+1,j) + c(z+1,j') \leq c(y+1,j') + c(z+1,j)$ by induction hypothesis.

Combining the three cases, we conclude that c the satisfies quadrangle inequality. $\qquad\square$

Minimum Cost Homomorphism Dichotomy for Oriented Cycles

Gregory Gutin[1], Arash Rafiey[2], and Anders Yeo[1]

[1] Department of Computer Science
Royal Holloway, University of London
Egham, Surrey TW20 0EX, UK
gutin(anders)@cs.rhul.ac.uk
[2] School of Computing Science
Simon Fraser University
Burnaby, B.C., Canada, V5A 1S6
arashr@cs.sfu.ca

Abstract. For digraphs D and H, a mapping $f : V(D) \to V(H)$ is a homomorphism of D to H if $uv \in A(D)$ implies $f(u)f(v) \in A(H)$. If, moreover, each vertex $u \in V(D)$ is associated with costs $c_i(u), i \in V(H)$, then the cost of the homomorphism f is $\sum_{u \in V(D)} c_{f(u)}(u)$. For each fixed digraph H, we have the *minimum cost homomorphism problem for* H (abbreviated MinHOM(H)). In this discrete optimization problem, we are to decide, for an input graph D with costs $c_i(u)$, $u \in V(D), i \in V(H)$, whether there exists a homomorphism of D to H and, if one exists, to find one of minimum cost. We obtain a dichotomy classification for the time complexity of MinHOM(H) when H is an oriented cycle. We conjecture a dichotomy classification for all digraphs with possible loops.

1   Introduction

For directed (undirected) graphs G and H, a mapping $f : V(G) \to V(H)$ is a *homomorphism of G to H* if uv is an arc (edge) implies that $f(u)f(v)$ is an arc (edge). Let H be a fixed directed or undirected graph. The *homomorphism problem* for H asks whether a directed or undirected input graph G admits a homomorphism to H. The *list homomorphism problem* for H asks whether a directed or undirected input graph G with lists (sets) $L_u \subseteq V(H), u \in V(G)$ admits a homomorphism f to H in which $f(u) \in L_u$ for each $u \in V(G)$.

Suppose G and H are directed (or undirected) graphs, and $c_i(u)$, $u \in V(G)$, $i \in V(H)$ are nonnegative *costs*. The *cost of a homomorphism f of G to H* is $\sum_{u \in V(G)} c_{f(u)}(u)$. If H is fixed, the *minimum cost homomorphism problem*, MinHOM(H), for H is the following discrete optimization problem. Given an input graph G, together with costs $c_i(u)$, $u \in V(G)$, $i \in V(H)$, we wish to find a minimum cost homomorphism of G to H, or state that none exists.

The minimum cost homomorphism problem was introduced in [9], where it was motivated by a real-world problem in defence logistics. We believe it offers a practical and natural model for optimization of weighted homomorphisms.

R. Fleischer and J. Xu (Eds.): AAIM 2008, LNCS 5034, pp. 224–234, 2008.
© Springer-Verlag Berlin Heidelberg 2008

The problem's special cases include the homomorphism and list homomorphism problems [15] and the general optimum cost chromatic partition problem, which has been intensively studied [12,17].

There is an extensive literature on the minimum cost homomorphism problem, e.g., see [4,5,6,7,8,9]. These and other papers study the time complexity of MinHOM(H) for various families of directed and undirected graphs. In particular, Gutin, Hell, Rafiey and Yeo [5] proved a dichotomy classification for all undirected graphs (with possible loops): If H is a reflexive proper interval graph or a proper interval bigraph, then MinHOM(H) is polynomial time solvable; otherwise, MinHOM(H) is NP-hard. It is an open problem whether there is a dichotomy classification for the complexity of MinHOM(H) when H is a digraph with possible loops. We conjecture that such a classification exists and, moreover, the following assertion holds:

Conjecture 1. Let H be a digraph with possible loops. Then MinHOM(H) is polynomial time solvable if H has either a Min-Max ordering or a k-Min-Max ordering for some $k \geq 2$. Otherwise, MinHOM(H) is NP-hard.

For the definitions of a Min-Max and k-Min-Max ordering see Section 3, where we give theorems (first proved in [9,8]) showing that if H has one of the two orderings, then MinHOM(H) is polynomial time solvable. So, it is the NP-hardness part of Conjecture 1 which is the 'open' part of the conjecture.

Very recently Gupta, Hell, Karimi and Rafiey [4] obtained a dichotomy classification for all reflexive digraphs that confirms this conjecture. They proved that if a reflexive digraph H has no Min-Max ordering, then MinHom(H) is NP-hard. Gutin, Rafiey and Yeo [7,8] proved that if a semicomplete multipartite digraph H has neither Min-Max ordering nor k-Min-Max ordering, then MinHom(H) is NP-hard.

In this paper, we show that the same result (as for semicomplete multipartite digraphs) holds for oriented cycles. This provides a further support for Conjecture 1. In fact, we prove a graph-theoretical dichotomy for the complexity of MinHom(H) when H is an oriented cycle. The fact that Conjecture 1 holds for oriented cycles follows from the proof of the graph-theoretical dichotomy. In the proof, we use a new concept of a (k, l)-Min-Max ordering introduced in Section 3. Our motivation for Conjecture 1 partially stems from the fact that we initially proved polynomial time solvability of MinHOM(H) when $V(H)$ has a (k, l)-Min-Max ordering by reducing it to the minimum cut problem. However, we later proved that (k, l)-Min-Max orderings can simply be reduced to p-Min-Max orderings for $p \geq 1$ (see Section 3).

Homomorphisms to oriented cycles have been investigated in a number of papers. Partial results for the homomorphism problem to oriented cycles were obtained in [10] and [16]. A full dichotomy was proved by Feder [2]. Feder, Hell and Rafiey [3] obtained a dichotomy for the list homomorphism problem for oriented cycles. Notice that our dichotomy is different from the ones in [2] and [3].

2   Levels of Vertices in Oriented Paths and Cycles

In this paper $[p]$ denotes the set $\{1, 2, \ldots, p\}$. Let D be a digraph. We will use $V(D)$ $(A(D))$ to denote the vertex (arc) set of D. We say that xy $(x, y \in V(D))$ is an *edge* of D if either xy or yx is an arc of D. A sequence $b_1 b_2 \ldots b_p$ of distinct vertices of D is an *oriented path* if $b_i b_{i+1}$ is an edge for every $i \in [p-1]$. If $b_1 b_2 \ldots b_p$ is an oriented path, we call $C = b_1 b_2 \ldots b_p b_1$ an *oriented cycle* if $b_p b_1$ is an edge. An edge $b_i b_{i+1}$ (here $b_p b_{p+1} = b_p b_1$) of an oriented path P or cycle C is called *forward (backward)* if $b_i b_{i+1} \in A(D)$ $(b_{i+1} b_i \in A(D))$.

Let $P = b_1 b_2 \ldots b_p$ be an oriented path. We assign *levels* to the vertices of P as follows: we set $\text{level}_P(b_1) = 0$, and $\text{level}_P(b_{t+1}) = \text{level}_P(b_t) + 1$, if $b_t b_{t+1}$ is forward and and $\text{level}_P(b_{t+1}) = \text{level}_P(b_t) - 1$, if $b_t b_{t+1}$ is backward. We say that P is *of type r* if $r = \max\{\text{level}_P(b_i) : i \in [p]\} = \text{level}_P(b_p)$ and $0 \le \text{level}_P(b_t) \le r$ for each $t \in [p]$.

An oriented cycle C is *balanced* if the number of forward edges equals the number of backward edges; if C is not balanced, it is called *unbalanced*. Note that the fact whether C is balanced or unbalanced does not depend on the choice of the vertex b_1 or the direction of C.

Let $C = b_1 b_2 \ldots b_p b_1$ be an oriented cycle. It has two *directions*: $b_1 b_2 \ldots b_p b_1$ and $b_1 b_p b_{p-1} \ldots b_1$. In what follows, we will always consider the direction in which the number of forward arcs is no smaller than the number of backward arcs. We can assign *levels* to the vertices of C as follows: $\text{level}(b_1) = k$, where k is a non-negative integer, and $\text{level}(b_{t+1}) = \text{level}(b_t) + 1$, if $b_t b_{t+1}$ is forward and and $\text{level}(b_{t+1}) = \text{level}(b_t) - 1$, if $b_t b_{t+1}$ is backward. Clearly, the value of each $\text{level}(b_i)$, $i \in [p]$, depends on both k and the choose of the initial vertex b_1. Feder [2] proved the following useful result.

Proposition 1. *The integer k and initial vertex b_1 in an oriented cycle C can be chosen such that $\text{level}(b_1) = 0$ and $\text{level}(b_i) \ge 0$ for every $i \in [p]$. If C is unbalanced, then k and b_1 can be chosen such that $\text{level}(b_1) = 0$ and $\text{level}(b_i) > 0$ for every $i \in [p] \setminus \{1\}$.*

Since the proposition was proved in [2], we will not give its complete proof. Instead, we will outline a procedure for finding appropriate k and b_1 and remark on how the procedure can be used in showing the proposition.

Let $C = b_1 b_2 \ldots b_p b_1$ be an oriented cycle. We may assume that b_1 is chosen in such a way that if C has a backward edge, then $b_p b_1$ is a backward edge. Compute m_i, the number of the forward arcs minus the number of backward arcs in the oriented path $b_1 b_2 \ldots b_i$, for each $i \in [p]$. Set $k = |\min\{m_i : i \in [p]\}|$. Assign the level to each vertex of C using the level definition and starting from assigning level k to b_1. By the definition of k, the level of each vertex b_j is non-negative and there are vertices b_i of level zero. Choose such a vertex b_i with maximum index i and reassign the levels to the vertices of C as follows. Consider $C' = b_i b_{i+1} \ldots b_p b_1 b_2 \ldots b_i$ and set $\text{level}(b_i) = 0$ and the rest of the levels according to the order of vertices given in C'.

This procedure can be turned into a proof of the proposition by observing that if C' is unbalanced, then the level of b_1 in C' will be greater than the level

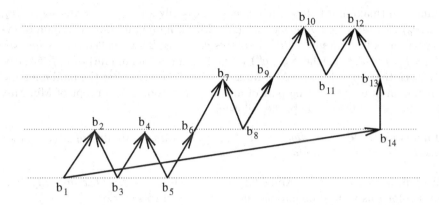

Fig. 1. A level diagram of oriented cycle $Z = b_1 b_2 \ldots b_{14} b_1$, where $b_1 b_2, b_3 b_2, b_3 b_4, b_5 b_4,$ $b_5 b_6, b_6 b_7, b_8 b_7 b_8 b_9, b_9 b_{10}, b_{11} b_{10}, b_{11} b_{12}, b_{13} b_{12}, b_{14} b_{13}, b_1 b_{14}$ are arcs

of b_1 in C. Thus, the levels of all vertices v_j, $j \in [i-1]$ will be greater than their levels in C, implying that the only level zero vertex in C' is b_i.

Thus, in the rest of the paper, we may assume that the 'first' vertex of b_1 of an oriented cycle $C = b_1 b_2 \ldots b_p b_1$ is chosen in such a way that the levels of all vertices of C satisfy Proposition 1.

We will extensively use the following notation:
$VL(C) = \{b_t : \text{level}(b_t) = 0, t \in [p]\}$, $h(C) = \max\{\text{level}(b_j) : j \in [p]\}$, and $VH(C) = \{b_t : \text{level}(b_t) = h(C), t \in [p]\}$. Note that for unbalanced cycles C we have $|VL(C)| = 1$.

The concepts of this section are illustrated on Figure 1. In particular, Z is balanced with forward edges $b_1 b_2, b_3 b_4, \ldots$ and backward edges $b_2 b_3, b_4 b_5, \ldots$. We have level$(b_1) = $ level$(b_3) = $ level$(b_5) = 0$, level$(b_2) = $ level$(b_4) = $ level$(b_6) = $ level$(b_8) = $ level$(b_{14}) = 1$, level$(b_7) = $ level$(b_9) = $ level$(b_{11}) = $ level$(b_{13}) = 2$ and level$(b_{10}) = $ level$(b_{12}) = 3$. Thus, $h(Z) = 3$, $VL(Z) = \{b_1, b_3, b_5\}$, $VH(Z) = \{b_{10}, b_{12}\}$.

3 k-Min-Max and (k, l)-Min-Max Orderings

All known polynomial cases of MinHOM(H) can be formulated in terms of certain vertex orderings. In fact, all known polynomial cases can be partitioned into two classes: digraphs H admitting a Min-Max ordering of their vertices and digraphs H having a k-Min-Max ordering of their vertices ($k \geq 2$). Both types of orderings are defined in this section, where we also introduce a new type of ordering, a (k, l)-Min-Max ordering. It may be surprising, but we prove that the new type of ordering can be reduced to the two known orderings.

Let H be a digraph and let $(v_1, v_2, \ldots, v_p)$ be an ordering of the vertices of H. Let $e = v_i v_r$ and $f = v_j v_s$ be two arcs in H. The pair $v_{\min\{i,j\}} v_{\min\{s,r\}}$ ($v_{\max\{i,j\}} v_{\max\{s,r\}}$) is called the *minimum* (*maximum*) of the pair e, f. (The minimum (maximum) of two arcs is not necessarily an arc.) An ordering $(v_1, v_2, \ldots, v_p)$ is a *Min-Max ordering of* $V(H)$ if both minimum and maximum of every two arcs in H are in $A(H)$. Two arcs $e, f \in A(H)$ are called a *crossing pair* if $\{e, f\} \neq \{g', g''\}$, where g' (g'') is the minimum (maximum) of e, f. Clearly, to check that an ordering is Min-Max, it suffices to verify that the minimum and maximum of every crossing pair of arcs are arcs, too. The concept of Min-Max ordering is of interest due to the following:

Theorem 1. *[9] If a digraph H has a Min-Max ordering of $V(H)$, then* $MinHOM(H)$ *is polynomial-time solvable.*

We will sometimes call a Min-Max ordering also a *1-Min-Max ordering*. The reason for this will become apparent in the rest of this section.

A collection $V_1, V_2, \ldots V_k$ of subsets of a set V is called a *k-partition* of V if $V = V_1 \cup V_2 \cup \cdots \cup V_k$, $V_i \cap V_j = \emptyset$ provided $i \neq j$.

Let $H = (V, A)$ be a digraph and let $k \geq 2$ be an integer. We say that H has a *k-Min-Max ordering* of $V(H)$ if there is a k-partition of V into subsets $V_1, V_2, \ldots V_k$ and there is an ordering $\boldsymbol{V}_i = (v_1^i, v_2^i, \ldots, v_{\ell(i)}^i)$ of V_i for each i such that

(i) Every arc of H is an arc from V_i to V_{i+1} for some $i \in [k]$ and

(ii) $(\boldsymbol{V}_i, \boldsymbol{V}_{i+1}) = (v_1^i, v_2^i, \ldots, v_{\ell(i)}^i v_1^{i+1} v_2^{i+1}, \ldots, v_{\ell(i+1)}^{i+1})$ is a Min-Max ordering of the subdigraph of H induced by $V_i \cup V_{i+1}$ for each $i \in [k]$. (All indices are taken modulo k.)

In such a case, $(\boldsymbol{V}_1, \boldsymbol{V}_2, \ldots, \boldsymbol{V}_k)$ is a *k-Min-Max ordering* of $V(H)$; k-Min-Max orderings are of interest due to the following:

Theorem 2. *[8] If a digraph H has a k-Min-Max ordering of $V(H)$, then* $MinHOM(H)$ *is polynomial-time solvable.*

Our study of MinHOM(H) for oriented cycles H has led us to the following new concept.

Definition 1. *Let $H = (V, A)$ be a digraph and let $k \geq 2$ and l be integers. For $l < k$ we say that H has a (k, l)-Min-Max ordering if there is a $(k + l - 2)$-partition of V into subsets $V_1, V_2, \ldots, V_k, U_2, U_3, \ldots, U_{l-1}$ (set $U_1 = V_1$, $U_l = V_k$) and there is an ordering $\boldsymbol{V}_i = (v_1^i, v_2^i, \ldots, v_{\ell^v(i)}^i)$ of V_i for each $1 \leq i \leq k$ and there is an ordering $\boldsymbol{U}_i = (u_1^i, u_2^i, \ldots, u_{\ell^u(i)}^i)$ of U_i for each $1 \leq i \leq l$ such that*

(i) *Every arc of H is an arc from V_i to V_{i+1}) for some $i \in [k-1]$, or is an arc from U_j to U_{j+1} for some $j \in [l-1]$.*

(ii) *$(\boldsymbol{V}_i, \boldsymbol{V}_{i+1})$ is a Min-Max ordering of the subdigraph $H[V_i \cup V_{i+1}]$ for all $i \in [k-1]$.*

(iii) *$(\boldsymbol{U}_i, \boldsymbol{U}_{i+1})$ is a Min-Max ordering of the subdigraph $H[U_i \cup U_{i+1}]$ for all $i \in [l-1]$.*

(iv) *$(\boldsymbol{V}_1, \boldsymbol{V}_2, \boldsymbol{U}_2)$ is a Min-Max ordering of the subdigraph $H[V_1 \cup V_2 \cup U_2]$.*

(v) (U_{l-1}, V_{k-1}, V_k) *is a Min-Max ordering of the subdigraph* $H[V_{k-1} \cup U_{l-1} \cup V_k]$.

It turns out that (k, l)-Min-Max orderings can be reduced to p-Min-Max orderings as follows from the next assertion:

Theorem 3. *If a digraph H has a (k, l)-Min-Max ordering, then MinHOM(H) is polynomial-time solvable.*

Proof. Let H have a (k, l)-Min-Max ordering as described in Definition 1. Let $d = k - l$. We will show that H has a d-Min-Max ordering, which will be sufficient because of Theorems 1 and 2. (Recall that a 1-Min-Max ordering is simply a Min-Max ordering.) Let us consider two cases.

Case 1: $d = 1$. It is not difficult to show that the ordering

$$(V_1, V_2, U_2, V_3, U_3, \ldots, V_{k-2}, U_{k-2}, V_{k-1}, V_k)$$

is a Min-Max ordering. Indeed, all crossing pairs of arcs are only in the subgraphs given in (ii)-(v) of Definition 1. According to the definition, the maximum and minimum of every crossing pair is in H.

Case 2: $d \geq 2$. Let $s_i = \max\{p : i + pd \leq l\}$ for each $i \in [d]$. Consider the following orderings for each $2 \leq i \leq d$

$$W_i = (V_i, U_i, V_{i+d}, U_{i+d}, V_{i+2d}, U_{i+2d}, \ldots, V_{i+s_i d}, U_{i+s_i d}, V_{i+(s_i+1)d})$$

and the ordering

$$W_1 = (V_1, V_{1+d}, U_{1+d}, V_{1+2d}, U_{1+2d}, \ldots, V_{1+s_1 d}, U_{1+s_1 d}, V_{1+(s_1+1)d}).$$

Observe that $W_1, W_2, \ldots, W_d$ form a partition of $V(H)$ and that every arc is from W_i to W_{i+1} for some $i \in [d]$, where $W_{d+1} = W_1$. As in Case 1, it is not difficult to see that $(W_1, W_2, \ldots, W_d)$ is a d-Min-Max ordering of H. □

Remark 1. Notice that not always a p-Min-Max ordering can be reduced to a (k, l)-Min-Max ordering. As an example, consider the directed cycle C_5.

4   Balanced Oriented Cycles

We say that a balanced oriented cycle $C = b_1 b_2 \ldots b_p b_1$ is *of the form* $(l^+ h^+)^q$ with $q \geq 1$ if $P = C - b_p b_1$ can be written as $P = P_1 R_1 P_2 R_2 \ldots P_q R_q$, where

$$V(P_i) \cap VL(C) \neq \emptyset, \ V(P_i) \cap VH(C) = \emptyset,$$
$$V(R_i) \cap VL(C) = \emptyset, \ V(R_i) \cap VH(C) \neq \emptyset$$

for each $i \in [q]$. We write $l^+ h^+$ instead of $(l^+ h^+)^1$. For example, the cycle in Figure 1 is of the form $l^+ h^+$. Balanced oriented cycles C of the form $l^+ h^+$ are considered in the following:

Theorem 4. *Let $C = b_1b_2 \ldots b_pb_1$ be a balanced oriented cycle of the form l^+h^+. Then MinHOM(C) is polynomial time solvable.*

Proof. Let $q = \min\{j : b_j \in VH(C)\}$, let $m = h(C)$ and let $V = V(C)$. Consider the following ordering $\boldsymbol{V} = (b_p, b_{p-1}, \ldots, b_{q+1}, b_1, b_2 \ldots, b_q)$ of V. We can define the following natural $(m+1)$-partition of V: $V_1, V_2, \ldots V_{m+1}$, where $V_j = \{b_s \in V(C) : \text{level}(b_s) = j - 1\}$. Note that every arc of C is an arc from V_j to V_{j+1} for some $j \in [m]$.

Let $\boldsymbol{V}_j$ be the ordering of V_j obtained from $\boldsymbol{V}$ by deleting all vertices not in V_j and let $\boldsymbol{V}_j = (s_1^j, s_2^j, \ldots, s_{b(j)}^j)$. Observe that the digraph $C[V_j \cup V_{j+1}]$ has no crossing pair of arcs for any $j \in [m]$ since, for every pair $s_\alpha^j s_\beta^{j+1}$, $s_\gamma^j s_\delta^{j+1}$ of arcs in the digraph, we have that either $\alpha \leq \gamma$ and $\beta \leq \delta$, or $\alpha \geq \gamma$ and $\beta \geq \delta$. Thus, C has an $(m+1)$-Min-Max ordering of vertices and, by Theorem 2, MinHOM(C) is polynomial-time solvable. $\qquad\square$

The following lemma was first proved in [11]; see also [2,18] and Lemma 2.36 in [15].

Lemma 1. *Let P_1 and P_2 be two oriented paths of type r. Then there is an oriented path P of type r that maps homomorphically to P_1 and P_2 such that the initial vertex of P maps to the initial vertices of P_1 and P_2 and the terminal vertex of P maps to the terminal vertices of P_1 and P_2. The length of P is polynomial in the lengths of P_1 and P_2.*

We need a modified version of Lemma 1, Lemma 2. We say that an oriented path $b_1b_2 \ldots b_p$ of type r is *of the form* $(l^+h^+)^k$ if the balanced oriented cycle

$$b_1b_2 \ldots b_pa_{r-1}a_{r-2} \ldots a_2a_1b_1$$

is of the form $(l^+h^+)^k$, where $b_1a_1a_2 \ldots a_{r-2}a_{r-1}b_p$ is a directed path.

Lemma 2. *Let P_1 and P_2 be two oriented paths of type r. Let P_1 be of the form h^+l^+ and let P_2 be of the form $(l^+h^+)^k$, $k \geq 1$. Then there is an oriented path P of type r that maps homomorphically to P_1 and P_2 such that the initial vertex of P maps to the initial vertices of P_1 and P_2 and the terminal vertex of P maps to the terminal vertices of P_1 and P_2. The length of P is polynomial in the lengths of P_1 and P_2, and P is of the form $(l^+h^+)^k$.*

Proof. We will show that our construction implies that $|V(P)| \leq |V(P_1)| \cdot |V(P_2)|$.

We first prove the lemma for the case when $k = 1$. The proof is by induction on $r \geq 0$. If $0 \leq r \leq 1$, the claim is trivial. Assume that $r \geq 2$. Let $P_1 = a_1a_2 \ldots a_p$, let $P_2 = b_1b_2 \ldots b_q$, let $s_1 = \min\{i : \text{level}_{P_1}(a_i) = r\}$ and let $s_2 = \min\{i : \text{level}_{P_2}(b_i) = r\}$. Let $\beta_1 = \min\{\text{level}_{P_1}(a_i) : s_1 \leq i\}$ and $\beta_2 = \min\{\text{level}_{P_2}(b_i) : s_2 \leq i\}$. Without loss of generality assume that $\beta_1 \leq \beta_2$ and let $t_1 = \min\{i : \text{level}_{P_1}(a_i) = \beta_1 \text{ and } i \geq s_1\}$ and let $s_2 = \max\{i : \text{level}_{P_2}(b_i) = \beta_1 \text{ and } i \leq s_2\}$. Note that $\beta_1 > 1$ as P_1 is of form h^+l^+.

By the induction hypothesis, there is an appropriate oriented path P' that can be mapped homomorphically to $a_1a_2\ldots a_{s_1-1}$ and $b_1b_2\ldots b_{s_2-1}$. There is also an oriented path P'' that can be mapped homomorphically to $a_{s_1}a_{s_1+1}\ldots a_{t_1}$ and $b_{s_2}b_{s_2-1}\ldots b_{t_2}$ (by reversing the two paths and then reversing the path we get by the induction hypothesis). Furthermore there is an oriented path P''' that can be mapped homomorphically to $a_{t_1+1}a_{t_1+2}\ldots a_p$ and $b_{t_2+1}b_{t_2+2}\ldots b_p$. Let $P = P'P''P'''$ (where the arc between the last vertex of P' to the first vertex of P'' is oriented from P' to P'' and similarly the arc between P'' and P''' is oriented in that direction). Note that P is of type r and form h^+l^+ and maps homomorphically to P_1 and P_2 such that the initial vertex of P maps to the initial vertices of P_1 and P_2 and the terminal vertex of P maps to the terminal vertices of P_1 and P_2. Furthermore $|V(P)| \le (s_1-1)(s_2-1) + (t_1 - s_1+1)(s_2-t_2+1) + (p-s_1)(q-t_2)$. As $(s_1-1) + (t_1-s_1+1) + (p-s_1) = p$ and $(s_2-1),(s_2-t_2+1),(q-t_2) \le q$ we have $|V(P)| \le pq \le |V(P_1)| \times |V(P_2)|$.

Now we proceed by induction on $k \ge 1$. The base case has already been proved. Assume that $k \ge 2$ and let $P_1 = a_1a_2\ldots a_p$ and $P_2 = b_1b_2\ldots b_q$. Let $t = \max\{i : \text{level}_{P_2}(b_i) = 0\}$ and let $s = \max\{i : i < t, \text{level}_{P_2}(b_i) = r\}$. By the induction hypothesis, there is an appropriate oriented path P' that can be mapped homomorphically to P_1 and $b_1b_2\ldots b_s$. Also, there is an appropriate oriented path P'' (P''') that can be mapped homomorphically to $a_pa_{p-1}\ldots a_1$ and $b_sb_{s+1}\ldots b_t$ (P_1 and $b_tb_{t+1}\ldots b_q$). Now obtain a new oriented path P by identifying the terminal vertex of P' with the initial vertex of P'' and the terminal vertex of P'' with the initial vertex of P'''. Observe that P satisfies the required properties. $\square$

Consider the oriented cycle $C_4^0 = 12341$ with arcs $12, 32, 14, 34$. Observe that $1, 2, 3, 4$ is a Min-Max ordering of $V(C_4^0)$ and, thus, $\text{MinHOM}(C_4^0)$ is polynomial-time solvable.

Theorem 5. *Let $C = b_1b_2\ldots b_pb_1$ be a balanced oriented cycle of the form $(l^+h^+)^k$, $k \ge 2$, and let $C \ne C_4^0$. Then $\text{MinHOM}(C)$ is NP-hard.*

Proof. Let $C \ne C_4^0$. Let $s = \min\{j : b_j \in VH(C)\}$, $q = \min\{j : j > s, b_j \in VL(C)\}$, $t = \min\{j : j > q, b_j \in VH(C)\}$ and $m = h(C)$. Let $P_1 = b_1b_2\ldots b_s$, $P_2 = b_qb_{q-1}\ldots b_s$, $P_3 = b_qb_{q+1}\ldots b_t$ and $P_4 = b_1b_pb_{p-1}\ldots b_t$. Note that each P_j is of type m. By Lemma 2 there is a path Q_1 of type m which is mapped homomorphically to P_4, P_2 and P_3. There is also a path Q_2 of type m and which is mapped homomorphically to P_1 and P_3. Since $C \ne C_4^0$ and the end-vertices vertices of Q_1 are mapped to the end-vertices vertices of P_4, the path Q_1 contains more than two vertices. Furthermore, by Lemma 2 we may assume that Q_1 is of the form $(l^+h^+)^{k-1}$ and Q_2 is of the form l^+h^+.

Let x (y) be the terminal vertex of Q_1 (Q_2). Form a new oriented path $Q = q_1q_2\ldots q_l$ by identifying x with y and let $1 \le r \le l$ be defined such that $Q_1 = q_1q_2\ldots q_r$ and $Q_2 = q_lq_{l-1}\ldots q_r$. As Q_1 contains more than two vertices we have that $r \ge 3$.

Let D be an arbitrary digraph. We will now reduce the problem of finding a maximum independent set in D (i.e. in the underlying graph of D) to

MinHOM(C). Replace every arc ab of D by a copy of Q identifying q_1 with a and q_l with b, and denote the obtained digraph by D'. For every path Q in D' which we added in the construction of D' we define the cost function c as follows, where M is a number greater than $|V(D)|$:

(i) $c_{b_1}(q_1) = c_{b_1}(q_l) = 0$ and $c_{b_q}(q_1) = c_{b_q}(q_l) = 1$;

(ii) $c_{b_s}(q_r) = c_{b_t}(q_r) = 0$ and $c_b(q_r) = M$ for all $b \in V(C) - \{b_s, b_t\}$;

(iii) $c_{b_2}(q) = M$ for all $q \in \{q_2, q_3, \ldots, q_{r-1}\}$ ($\neq \emptyset$, as $r \geq 3$);

(iv) if $s = 2$ then $c_{b_1}(q_{r-1}) = M$;

(v) all other costs of mapping vertices of D' to H are zero.

Consider a mapping g from $V(Q)$ to $V(C)$, where $g(q_1) = b_q$, $g(q_l) = b_q$, and Q_1 and Q_2 are both homomorphically mapped to P_3 (i.e., $g(q_r) = b_t$). Observe that g is a homomorphism from Q to C of cost 2. This implies, in particular, that there is a homomorphism from D' to H of cost less than M. We now consider three other homomorphisms from Q to C:

(a) $f(q_1) = b_1$ and $f(q_l) = b_q$, and Q_1 is mapped to P_4 and Q_2 is mapped to P_3 homomorphically (i.e., $f(q_r) = b_t$). The cost of f is 1.

(b) $f'(q_1) = b_q$ and $f'(q_l) = b_1$, and f' maps Q_1 to P_2 and Q_2 to P_1 homomorphically (i.e., $f'(q_r) = b_s$). The cost of f' is 1.

(c) $f''(q_1) = b_1$ and $f''(q_l) = b_1$. We will show that the cost of f'' is at least M. If this is not the case then $f''(q_r) \in \{b_s, b_t\}$ by (ii). First assume that $f''(q_r) = b_s$. By (iii) no vertex of $V(Q_1) - \{q_r\}$ is mapped to b_2 and if $s = 2$ then by (iv) q_{r-1} is not mapped to b_1. However as Q_1 is of the form $(l^+ h^+)^{k-1}$ and the path $b_1 b_p b_{p-1} \ldots b_s$ is of the form $(l^+ h^+)^k$ we get a contradiction to f'' mapping Q_1 to C and $f''(q_1) = b_1$ and $f''(q_r) = b_s$. So now assume that $f''(q_r) = b_t$. However Q_2 is of the form $(l^+ h^+)$ and there is no path in C from b_1 to b_t of the form $(l^+ h^+)$, a contradiction. Therefore the cost of f'' is at least M.

By the above a minimum cost homomorphism $h : D' \to C$ maps all vertices from a maximum independent set in D to b_1 and all other vertices from D to b_q. As finding a maximum independent set in a digraph is NP-hard we see that MinHOM(C) is NP-hard.   □

5   Dichotomy and Unbalanced Oriented Cycles

We are ready to prove the following main result:

Theorem 6. *Let C be an oriented cycle. If C is unbalanced or C is balanced of the form $l^+ h^+$ or $C = C_4^0$, then MinHOM(C) is polynomial-time solvable. Otherwise, MinHOM(C) is NP-hard.*

By Theorems 4 and 5, to show Theorem 6 it suffices to prove the following:

Theorem 7. *Let $C = b_1 b_2 \ldots b_p b_1$ be an unbalanced oriented cycle. Then MinHOM(C) is polynomial-time solvable.*

Proof. It is well-known that the minimum cost homomorphism problem to a directed cycle is polynomial-time solvable (see, e.g., [6]). Thus, we may assume that C is not a directed cycle. By Proposition 1, we may assume that level$(b_1) = 0$ and level$(b_i) > 0$ for all $i \in [p] \setminus \{1\}$. Let $q = \max\{j : b_j \in VH(C)\}$.

Consider the oriented path $P = b_1 b_2 \ldots b_q$. Let $V_{i+1} = \{b_j \in V(P) : \text{level}(b_j) = i\}$ for all $i \in [k] \cup \{0\}$, where $k = \text{level}(b_q)$. Now consider the oriented path $Q = b_1 b_p b_{p-1} \ldots b_q$. Assign levels to the vertices of Q stating from level$_Q(b_1) = 0$ and continuing as described in Section 1. Observe that all vertices of Q get nonnegative levels. Let $U_{i+1} = \{b_j \in V(Q) : \text{level}_Q(b_j) = i\}$, $i \in [l] \cup \{0\}$, where $l = \text{level}_Q(b_q)$. Clearly, $V_1 = U_1 = \{b_1\}$; set $U_{l+1} = V_{k+1}$.

Consider the ordering $\boldsymbol{U} = (b_1, b_p, b_{p-1}, \ldots, b_q)$ of the vertices of Q. For $i \in [l+1]$, the ordering $\boldsymbol{U}_i$ is obtained from $\boldsymbol{U}$ by deleting all vertices not in U_i. Consider the ordering $\boldsymbol{V} = (b_1, b_2, \ldots, b_q)$ of the vertices of P. For $i \in [k+1]$, the ordering $\boldsymbol{V}_i$ is obtained from $\boldsymbol{V}$ by deleting all vertices not in V_i. Observe that the ordering $(\boldsymbol{V}_1, \boldsymbol{V}_2, \boldsymbol{U}_2)$ of the vertices of $C[V_1 \cup V_2 \cup U_2]$ has no crossing arcs. Similarly, the ordering $(\boldsymbol{U}_l, \boldsymbol{V}_k, \boldsymbol{V}_{k+1})$ of the vertices of $C[U_l \cup V_k \cup V_{k+1}]$ has no crossing arcs, and orderings $(\boldsymbol{V}_i, \boldsymbol{V}_{i+1})$ and $(\boldsymbol{U}_j, \boldsymbol{U}_{j+1})$ ($i \in [k], j \in [l]$) have no crossing arcs. Thus, C has a $(k+1, l+1)$-ordering of vertices. Now we are done by Theorem 3. □

References

1. Bang-Jensen, J., Gutin, G.: Digraphs: Theory, Algorithms and Applications. Springer, London (2000)
2. Feder, T.: Homomorphisms to oriented cycles and k-partite satisfiability. SIAM J. Discrete Math. 14, 471–480 (2001)
3. Feder, T., Hell, P., Rafiey, A.: List homomorphism to balanced digraphs (submitted)
4. Gupta, A., Hell, P., Karimi, M., Rafiey, A.: Minimum cost homomorphisms to reflexive digraphs. In: LATIN 2008 (to appear, 2008)
5. Gutin, G., Hell, P., Rafiey, A., Yeo, A.: Minimum Cost Homomorphisms to Proper Interval Graphs and Bigraphs. Europ. J. Combin. (to appear)
6. Gutin, G., Rafiey, A., Yeo, A.: Minimum Cost and List Homomorphisms to Semicomplete Digraphs. Discrete Appl. Math. 154, 890–897 (2006)
7. Gutin, G., Rafiey, A., Yeo, A.: Minimum Cost Homomorphisms to Semicomplete Multipartite Digraphs. Discrete Applied Math. (to apear)
8. Gutin, G., Rafiey, A., Yeo, A.: Minimum Cost Homomorphisms to Semicomplete Bipartite Digraphs (submitted)
9. Gutin, G., Rafiey, A., Yeo, A., Tso, M.: Level of repair analysis and minimum cost homomorphisms of graphs. Discrete Appl. Math. 154, 881–889 (2006); Preliminary version appeared. In: Megiddo, N., Xu, Y., Zhu, B. (eds.): AAIM 2005. LNCS, vol. 3521, pp. 427–439. Springer, Heidelberg (2005)
10. Gutjahr, W.: Graph Colourings, PhD thesis, Free University, Berlin (1991)
11. Häggkvist, R., Hell, P., Miller, D.J., Neumann-Lara, V.: On multiplicative graphs and the product conjecture. Combinatorica 8, 63–74 (1988)

12. Halldorsson, M.M., Kortsarz, G., Shachnai, H.: Minimizing average completion of dedicated tasks and interval graphs. In: Goemans, M.X., Jansen, K., Rolim, J.D.P., Trevisan, L. (eds.) RANDOM 2001 and APPROX 2001. LNCS, vol. 2129, pp. 114–126. Springer, Heidelberg (2001)
13. Feder, T., Hell, P., Huang, J.: List homomorphisms and circular arc graphs. Combinatorica 19, 487–505 (1999)
14. Hell, P., Nešetřil, J.: On the complexity of H-colouring. J. Combin. Theory B 48, 92–110 (1990)
15. Hell, P., Nešetřil, J.: Graphs and Homomorphisms. Oxford University Press, Oxford (2004)
16. Hell, P., Zhu, X.: The existence of homomorphisms to oriented cycles. SIAM J. Discrete Math. 8, 208–222 (1995)
17. Jansen, K.: Approximation results for the optimum cost chromatic partition problem. J. Algorithms 34, 54–89 (2000)
18. Zhu, X.: A simple proof of the multipilicatively of directed cycles of prime power length. Discrete Appl. Math. 36, 333–345 (1992)

Minimum Leaf Out-Branching Problems

Gregory Gutin[1], Igor Razgon[2], and Eun Jung Kim[1]

[1] Department of Computer Science
Royal Holloway, University of London
Egham, Surrey TW20 0EX, UK
gutin(eunjung)@cs.rhul.ac.uk
[2] Department of Computer Science
University College Cork, Ireland
i.razgon@cs.ucc.ie

Abstract. Given a digraph D, the Minimum Leaf Out-Branching problem (MinLOB) is the problem of finding in D an out-branching with the minimum possible number of leaves, i.e., vertices of out-degree 0. We prove that MinLOB is polynomial-time solvable for acyclic digraphs. In general, MinLOB is NP-hard and we consider three parameterizations of MinLOB. We prove that two of them are NP-complete for every value of the parameter, but the third one is fixed-parameter tractable (FPT). The FPT parametrization is as follows: given a digraph D of order n and a positive integral parameter k, check whether D contains an out-branching with at most $n - k$ leaves (and find such an out-branching if it exists). We find a problem kernel of order $O(k \cdot 2^k)$ and construct an algorithm of running time $O(2^{O(k \log k)} + n^3)$, which is an 'additive' FPT algorithm.

1 Introduction

We say that a subgraph T of a digraph D is an *out-tree* if T is an oriented tree with only one vertex s of in-degree zero (called *the root*). The vertices of T of out-degree zero are called *leaves*. If T is a spanning out-tree, i.e. $V(T) = V(D)$, then T is called an *out-branching* of D. Given a digraph D, the *Minimum Leaf Out-Branching* problem (*MinLOB*) is the problem of finding an out-branching with the minimum possible number of leaves in D. Denote this minimum by $\ell_{\min}(D)$. When D has no out-branching, we write $\ell_{\min}(D) = 0$. Notice that not every digraph D has an out-branching. It is not difficult to see that D has an out-branching (i.e., $\ell_{\min}(D) > 0$) if and only if D has just one strongly connected component without incomming arcs [5]. Since the last condition can be checked in linear time [5], we may often assume that $\ell_{\min}(D) > 0$.

We first study MinLOB restricted to acyclic digraphs (abbreviated *MinLOB-DAG*). MinLOB-DAG was considered in US patent [10], where its application to the area of database systems was described. Demers and Downing [10] also suggested a heuristic approach to MinLOB-DAG. However no argument or assertion has been made to provide the validity of their approach and to investigate its

R. Fleischer and J. Xu (Eds.): AAIM 2008, LNCS 5034, pp. 235–246, 2008.
© Springer-Verlag Berlin Heidelberg 2008

computational complexity. Using another approach, we give a simple proof in Section 2 that MinLOB-DAG can be solved in polynomial time.

Since MinLOB generalizes the hamiltonian directed path problem, MinLOB is NP-hard. In this paper, we consider three parameterizations of MinLOB and show that two of them are NP-complete for every value of the parameter, but the third one is fixed-parameter tractable. The parameterized problems and related results are given in Sections 3 and 4. Further research is discussed in Section 5.

We recall some basic notions of parameterized complexity here, for a more in-depth treatment of the topic we refer the reader to [9,13,21].

A parameterized problem Π can be considered as a set of pairs (I, k) where I is the *problem instance* and k (usually an integer) is the *parameter*. Π is called *fixed-parameter tractable (FPT)* if membership of (I, k) in Π can be decided in time $O(f(k)|I|^c)$, where $|I|$ is the size of I, $f(k)$ is a computable function, and c is a constant independent from k and I. Let Π and Π' be parameterized problems with parameters k and k', respectively. An *fpt-reduction R from Π to Π'* is a many-to-one transformation from Π to Π', such that (i) $(I, k) \in \Pi$ if and only if $(I', k') \in \Pi'$ with $|I'| \leq g(k)$ for a fixed computable function g and (ii) R is of complexity $O(f(k)|I|^c)$. A *reduction to problem kernel* (or *kernelization*) is an fpt-reduction R from a parameterized problem Π to itself. In kernelization, an instance (I, k) is reduced to another instance (I', k'), which is called the *problem kernel*; $|I'|$ is the *size* of the kernel.

It is easy to see that a decidable parameterized problem is FPT if and only if it admits a kernelization (cf. [13,21]); however, the problem kernels obtained by this general result have impractically large size. Therefore, one tries to develop kernelizations that yield problem kernels of smaller size. The survey of Guo and Niedermeier [14] on kernelization lists some problem for which polynomial size kernels and exponential size kernels were obtained. Notice that if a kernelization can be done in time $O(n^{O(1)} + f(k))$, then we can obtain so-called an *additive FPT* algorithms, i.e., an algorithm of running time $O(n^{O(1)} + g(k))$, where $f(k)$ and $g(k)$ are independent of n, which is often significantly faster than its 'multiplicative' counterpart.

All digraphs in this paper are finite with no loops or parallel arcs. We use terminology and notation of [5]; in particular, for a digraph D, $V(D)$ and $A(D)$ denote its vertex and arc sets. The symbols n and m will denote the number of vertices and arcs in the digraph under consideration.

2 MinLOB-DAG

Let D be an acyclic digraph. We may assume that D has a unique vertex r of in-degree 0 as otherwise D has no out-branchings. Let $V = V(D)$ and $V' = \{v' : v \in V\}$. Let us define a bipartite graph B of D with partite sets X and X' as follows: $X = V$, $X' = V' \setminus \{r'\}$ and $E(B) = \{xy' : x \in X, y' \in X', xy \in A(D)\}$.

Consider the following algorithm for finding a minimum leaf out-branching T in an input acyclic digraph D. The algorithm outputs T if it exists and 'NO', otherwise.

MINLEAF

1. **if** the number of vertices with in-degree 0 equals 1 **then**
   $r \leftarrow$ the vertex of in-degree 0 **else** return 'NO'
2. construct the bipartite graph B of D
3. find a maximum matching M in B
4. $M^* \leftarrow M$
5. **for** all $y' \in X'$ not covered by M **do**
   $M^* \leftarrow M^* \cup \{$an arbitrary edge incident with $y'\}$
6. $A(T) \leftarrow \emptyset$
7. **for** all $xy' \in M^*$ **do** $A(T) \leftarrow A(T) \cup \{xy\}$
8. return T

Theorem 1. *Let D be an acyclic digraph. Then MINLEAF returns a minimum leaf out-branching if one exists, or returns 'NO' otherwise in time $O(m + n^{1.5}\sqrt{m/\log n})$.*

Proof. We start with proving the validity of the algorithm. Observe that an acyclic digraph has an out-branching if and only if there exists only one vertex of in-degree zero. Hence Step 1 returns 'NO' precisely when $\ell_{\min}(D) = 0$.

Let M be the maximum matching obtained in Step 2, let $V(M)$ be the set of vertices of B covered by M, and let $Z = X \setminus V(M)$ and $Z' = X' \setminus V(M)$.

First we claim that Z is the set of the leaves of T, the out-branching of D obtained in the end of Step 7. Consider the edge set M^* obtained at the end of Step 5. First observe that for each vertex $y' \in Z'$, there exists an edge of $E(B)$ which is incident with y' since r is the only vertex of in-degree zero and thus no vertex of Z' is isolated. Moreover, all neighbors of y' are covered by M due to the maximality of M. It follows that $M^* \supseteq M$ covers all vertices of X' and leaves Z uncovered. Notice that r is covered by M. Indeed there exists a vertex u such that r is the only in-neighbor of u in D. Hence if r was not covered by M then u' would not be covered by M either, which means we could extend M by ru', a contradiction.

Consider T which has been obtained in the end of Step 7. Clearly $d_T^-(v) = 1$ for all $v \in V(D) \setminus \{r\}$ due to the construction of M^*. Moreover D does not have a cycle, which means that T is connected and thus is an out-branching. Finally no vertex of Z has an out-neighbor in T while all the other vertices have an out-neighbor. Now the claim holds.

Conversely, whenever there exists a minimum leaf out-branching T of D with the leaf set Z, we can build a matching in B which covers exactly $X \setminus Z$ among the vertices of X. Indeed, simply reverse the process of building an out-branching T from M^* described at Step 7. If some vertex $x \in X$ has more than one neighbor in X', eliminate all but one edge incident with x.

Secondly we claim that T obtained in MINLEAF(D) is of minimum number of leaves. Suppose to the contrary that the the attained out-branching T is not a minimum leaf out-branching of D. Then a minimum leaf out-branching can be used to produce a matching of B that covers more vertices of X than M

does using the argument in the preceding paragraph, a contradiction. Hence MINLEAF(D) returns a min leaf out-branching T at Step 8.

Finally we analyze the computational complexity of MINLEAF(D). Each step of MINLEAF(D) takes at most $O(m)$ time except for Step 3. The computation time required to perform Step 3 is the same as that of solving the maximum cardinality matching problem on a bipartite graph. The last problem can be solved in time $O(|V(B)|^{1.5}\sqrt{|E(B)|/\log|V(B)|})$ [4]. Hence, the algorithm requires at most $O(m + n^{1.5}\sqrt{m/\log n})$ time.   □

3   Parameterizations of MinLOB

The following is a natural way to parameterize MinLOB.

MinLOB Parameterized Naturally (MinLOB-PN)
Instance: A digraph D.
Parameter: A positive integer k.
Question: Is $\ell_{\min}(D) \leq k$?

Clearly, this problem is NP-complete already for $k = 1$ as for $k = 1$ MinLOB-PN is equivalent to the hamiltonian directed path problem. Let v be an arbitrary vertex of D. Transform D into a new digraph D_k by adding k vertices $v_1, v_2, \ldots, v_k$ together with the arcs $vv_1, vv_2, \ldots, vv_k$. Observe that D has a hamiltonian directed path terminating at v if and only if $\ell_{\min}(D_k) \leq k$. Since the problem is NP-complete of checking whether a digraph has a hamiltonian directed path terminating at a prescribed vertex, we conclude that MinLOB-PN is NP-complete for every fixed k.

Clearly, $\ell_{\min}(D) \leq n - 1$ for every digraph D of order $n > 1$. Consider a different parameterizations of MinLOB.

MinLOB Parameterized Below Guaranteed Value (MinLOB-PBGV)
Instance: A digraph D of order n with $\ell_{\min}(D) > 0$.
Parameter: A positive integer k.
Question: Is $\ell_{\min}(D) \leq n - k$?
Solution: An out-branching B of D with at most $n - k$ leaves or the answer 'NO' to the above question.

Note that we consider MinLOB-PBGV as a search problem, not just as a decision problem. In the next section we will prove that MinLOB-PBGV is fixed-parameter tractable. We will find a problem kernel of order $O(k \cdot 2^k)$ and construct an additive FPT algorithm of running time $O(2^{O(k \log k)} + n^3)$. To obtain our results we use notions and properties of vertex cover and tree decomposition of underlying graphs and Las Vergnas' theorem on digraphs.

The parametrization MinLOB-PBGV is of the type *below a guaranteed value*. Parameterizations above/below a guaranteed value were first considered by Mahajan and Raman [20] for the problems Max-SAT and Max-Cut; such parameterizations have lately gained much attention, cf. [11,15,16,17,21] (it worth noting that Heggernes, Paul, Telle, and Villanger [17] recently solved the longstanding minimum interval completion problem, which is a parametrization above guaranteed value). For directed graphs there have been only a couple of results on problems parameterized above/below a guaranteed value, see [6,12].

Let us denote by $K_{1,p-1}$ the *star digraph* of order p, i.e., the digraph with vertices $1, 2, \ldots, p$ and arcs $12, 13, \ldots, 1p$. Our success with MinLOB-PBGV may lead us to considering the following stronger (than MinLOB-PBGV) parameterizations of MinLOB.

MinLOB Parameterized Strongly Below Guaranteed Value (MinLOB-PSBGV)

Instance: A digraph D of order n with $\ell_{\min}(D) > 0$.
Parameter: An integer $k \geq 2$.
Question: Is $\ell_{\min}(D) \leq n/k$?

Unfortunately, MinLOB-PSBGV is NP-complete for every fixed $k \geq 2$. To prove this consider a digraph D of order n and a digraph H obtained from D by adding to it the star digraph $K_{1,p-1}$ on $p = \lfloor n/(k-1) \rfloor$ vertices $(V(D) \cap V(K_{1,p-1}) = \emptyset)$ and appending an arc from vertex 1 of $K_{1,p-1}$ to an arbitrary vertex y of D. Observe that $\ell_{\min}(H) = p - 1 + \ell_{min}(D, y)$, where $\ell_{min}(D, y)$ is the minimum possible number of leaves in an out-branching rooted at y, and that $\frac{1}{k}|V(H)| = p + \epsilon$, where $0 \leq \epsilon < 1$. Thus, $\ell_{\min}(H) \leq \frac{1}{k}|V(H)|$ if and only if $\ell_{min}(D, y) = 1$. Hence, the hamiltonian directed path problem with fixed initial vertex (vertex y in D) can be reduced to MinLOB-PSBGV for every fixed $k \geq 2$ and, therefore, MinLOB-PSBGV is NP-complete for every $k \geq 2$.

4   Solving MinLOB-PBGV

The *underlying graph* $UG(D)$ of a digraph D is obtained from D by omitting all orientation of arcs and by deleting one edge from each resulting pair of parallel edges. For a digraph D, let $\alpha(D)$ denote the independence number of $UG(D)$.

Theorem 2 (Las Vergnas[19]). *If a digraph D has an out-branching, then* $\ell_{\min}(D) \leq \alpha(D)$.

For an out-branching B of D, let $L(B)$ denote the set of leaves of B. We will prove the following claim which implies the theorem:

Claim 1. *Let B be an out-branching of D with more than $\alpha(D)$ leaves. Then D contains an out-branching B' such that $L(B')$ is a proper subset of $L(B)$.*

Proof. We will prove this claim by induction on the number n of vertices in D. For $n \leq 2$ the result holds; thus, we may assume that $n \geq 3$ and consider an out-branching B of D with $|L(B)| > \alpha(D)$. Clearly, D has an arc xy such that x, y are leaves of B. If the in-neighbor p of y in B is of out-degree at least 2, then $L(B') \subset L(B)$, where $B' = B + xy - py$. So, we may assume that $d_B^+(p) = 1$. Observe $\alpha(D - y) \leq \alpha(D) < |L(B)| = |L(B - y)|$. Hence by the induction hypothesis, $D - y$ has an out-branching B'' such that $L(B'') \subset L(B - y)$. Notice that $L(B - y) = L(B) \cup \{p\} \setminus \{y\}$. If $p \in L(B'')$, then observe that $L(B'' + py) \subset L(B)$. Otherwise, $L(B'' + xy) \subseteq L(B) \setminus \{x\} \subset L(B)$. $\qquad\square$

A *vertex cover* of D is a vertex cover of $UG(D)$.

Lemma 1. *Let D be a digraph of order n with $\ell_{\min}(D) > 0$. In time $O(1.28^k + n^3)$, we can find either an out-branching of D with at most $n - k$ leaves or a vertex cover of D of size less than k.*

Proof. It is well-known that $\alpha(D) + \beta(D) = n$, where $\beta(D)$ is the minimum size of a vertex cover of D. First we can use the vertex cover algorithm of [8] to find a vertex cover of size less than k in time $O(1.2745^k k^4 + kn)$. If no such vertex cover exists, we have $\alpha(D) \leq n - k$ and by Theorem 2, D contains an out-branching B such that $|L(B)| \leq n - k$. To find such an out-branching we can use the procedure LEAFRED described below, which is just an algorithmic version of the proof of Claim 1.

The procedure LEAFRED(D, B) takes as an input a digraph D with $\beta(D) \geq k$ and an out-branching B of D and finds a new out-branching with leaves at most $\alpha(D)$. We may assume that the input out-branching B is a DFS tree. We denote by $p(y)$ the parent of a leaf vertex y in the given out-branching. The correctness of LEAFRED(D, B) follows from the proof of Claim 1 and it is not hard to see that its time complexity is $O(n^3)$.

LEAFRED(D, B)
improve $\leftarrow$ true
while improve $=$ true **do**   {
    while there is an arc xy such that x and y are leaves and $d^+(p(y)) \geq 2$
      do $B \leftarrow B + xy - p(y)y$
    if there is an arc xy such that x and y are leaves of B **then** {
        $B'' \leftarrow$ LEAFRED$(D - y, B - y)$
        if $p(y) \in L(B'')$ **then** $B' \leftarrow B'' + p(y)y$
        else $B' \leftarrow B'' + xy$ }
    else $B' \leftarrow B$
    if $|L(B')| < |L(B)|$ **then** $B \leftarrow B'$ **else** improve $\leftarrow$ false   }
    return B $\square$

It follows from Lemma 1 that there is an FPT algorithm that given an instance (D, k) of the MinLOB-PBGV problem either returns a solution or specifies a vertex cover of D of size less than k. We are going to show that, in the latter case, there is a possibility of kernelization. Let U be a vertex cover of D. Let $S \subseteq U$, $u \in U$ (we allow $S = \emptyset$). We denote by $V(S, u)$ a subset of vertices w of $V(D) \setminus U$ such that $(u, w) \in A(D)$ and $N^+(w) = S$. Let $NList$ be the set of all non-empty items $V(S, u)$. Now we create a set V' according to the following algorithm called **KERNEL**.

1. $V' \leftarrow \emptyset$
2. let all the elements of $NList$ be unmarked
3. **while** $NList$ has at least one unmarked item $V(S, u)$ **do**
       if $|V(S, u) \setminus V'| \leq 2|U|$ **then**
           $V' \leftarrow V' \cup V(S, u)$ **else**
           $V' \leftarrow V' \cup W$, where W is an arbitrary subset of $2|U|$ vertices of $V(S, u) \setminus$
           V'
       endif
       mark $V(S, u)$
     endwhile
4. $V' \leftarrow V' \cup U$
5. **if** D has a vertex v^* with in-degree 0 **then** $V' \leftarrow V' \cup \{v^*\}$
6. return V'

Let D' be the subgraph of D induced by V'. The following lemma claims that D' can serve as a kernel of D with respect to the MinLOB-PBGV problem.

Lemma 2. (D, k) *is a 'YES' instance of the MinLOB-PBGV problem if and only if* (D', k) *is.*

Proof. In order to prove this lemma it is more convenient to think of the MinLOB-PBGV problem as a problem of constructing an out-branching with at least k non-leaf vertices rather than at most $n - k$ leaf vertices.

Assume that (D', k) is a 'YES' instance of the MinLOB-PBGV problem and let B' be an out-branching of D' having at least k non-leaf vertices. By definition of V', every vertex w of $V(D) \setminus V'$ has at least one in-neighbor $p(w)$ which belongs to $U \subseteq V'$. Add each such vertex w to B' together with arc $(p(w), w)$. Clearly, the resulting graph B is an out-branching of D whose set of non-leaf-vertices is a superset of the set of non-leaf vertices of B'. Thus, B has at least k non-leaf vertices which shows that (D, k) is a 'YES' instance of the MinLOB-PBGV problem.

Assume now that (D, k) is a 'YES' instance of the MinLOB-PBGV problem and let B be an out-branching of D having at least k non-leaf vertices. Let B^* be the subgraph of B induced by a set of vertices V^* defined as follows.

1. V^* contains all vertices of U and all the non-leaf vertices of B.
2. Let u be a non-leaf vertex such that the set $X(u)$ of children of u which are leaf vertices of B and belong to $V(D) \setminus U$ is non-empty. Then V^* contains exactly one vertex of $X(u)$.

Since all the non-leaf vertices of B belong to V^*, B^* is an out-tree. In addition, observe that every non-leaf vertex of B remains a non-leaf vertex in B^*. Indeed, consider an arbitrary non-leaf vertex u. If at least one child v of u is a non-leaf vertex itself or $v \in U$ then v is included in V^* according to the first item of definition of V^*. Otherwise, all the children of u are leaf vertices which belong to $V(D) \setminus U$. According to the second item of the definition of V^* at least one such child belongs to V^*. It follows that B^* is an out-tree with at least k non-leaf vertices.

Let $v_1, \ldots, v_t$ be the vertices of V^* enumerated in some arbitrary order. *Following this order*, we associate with each v_i a vertex $u_i \in V'$ according to the following procedure. If $v_i \in V'$ then $u_i = v_i$. Otherwise observe that the in-degree of v_i is at least one. Associate with v_i a vertex w such that w is the parent of v_i in B^* if v_i is a non-root and an arbitrary in-neighbor of v_i in D otherwise. Let u_i be an arbitrary vertex of $V(N^+(v_i), w) \cap V'$ which is not equal to u_j for any $j < i$.

Statement 1. *The procedure of construction of $u_1, \ldots, u_t$ is sound in the sense that if $v_i \notin V'$, the procedure always finds an available vertex for u_i.*

Proof of Statement 1. Observe that for $v_i \notin V'$, we have $|V(N^+(v_i), w) \cap V'| \geq 2|U|$ by the description of the kernelization algorithm. Since u_i is chosen among vertices in $V(N^+(v_i), w) \cap V'$ for each $v_i \notin V'$, it suffices to show that the number of vertices of $V(D) \setminus U$ among $u_1, \ldots, u_t$ does not exceed $2|U|$. Indeed, by construction of $u_1, \ldots, u_t$, we have $v_i \in U$ if and only if $u_i \in U$. Therefore, we may equivalently show that $|(V(D) \setminus U) \cap V^*| \leq 2|U|$.

The vertices of $V(D) \setminus U$ in V^* can be either non-leaf or leaf vertices. Each of the non-leaf vertices has a child in U and, of course, no two vertices share a child. Therefore the number of these vertices is at most U. On the other hand, by construction of V^*, each non-leaf vertex of B^* has at most one child which is a leaf vertex of $V(D) \setminus U$. Taking into account that the parent of a vertex of $V(D) \setminus U$ may be only a vertex of U, it follows that the number of vertices of the second category is also at most $|U|$ and the overall number of vertices of $V(D) \setminus U$ in B^* is at most $2|U|$ as required.   □

Statement 2. *Let (v_i, v_j) be an arc of B^*. Then $(u_i, u_j) \in A(D')$.*

Proof of Statement 2. The statement is clearly true if $v_i = u_i$ and $v_j = u_j$. If $v_i \neq u_i$ but $v_j = u_j$, the statement follows because by selection of u_i, u_i has the same out-neighborhood as v_i. If $v_i = u_i$ but $v_j \neq u_j$ then the statement follows because by selection of u_j, the parent of v_j in B^* is an in-neighbor of u_j in D and hence in D'. Finally the case where $v_i \neq u_i$ and $v_j \neq u_j$ cannot happen because, by definition of a vertex cover, there is no arc between two vertices of $V(D) \setminus U$.   □

It follows from the combination of statements that graph D' has a subgraph B_1 whose set of vertices is $u_1, \ldots, u_t$ and which is isomorphic to B^*. Observe that any vertex of $V' \setminus V(B_1)$ has an in-neighbor among the vertices of B_1. Indeed, any vertex w of $V' \setminus V(B_1)$ belongs to $V(D) \setminus U$. Hence all the in-neighbors of w in D belong to U and thus to $V(B_1)$. Consequently, if w does not have in-neighbors in $V(B_1)$, the in-degree of w in D is 0. It follows that w is the root vertex of B and hence the root vertex of B^*. By construction of $u_1, \ldots, u_t$, we have w is necessarily one of u_i-s, a contradiction. Therefore for each vertex w of $V' \setminus V(B_1)$, we can select an in-neighbor $p(w) \in V(B_1)$ of w in D'. Add vertex w and arc $(p(w), w)$ to B_1 for each $w \in V' \setminus V(B_1)$, and let B_2 be the resulting digraph. It is not hard to see that B_2 is an out-branching of D' with at least k non-leaf vertices, which shows that (D', k) is a 'YES' instance of the MinLOB-PBGV problem.   □

Combining Lemmas 1 and 2 we obtain the following theorem.

Theorem 3. *The MinLOB-PBGV problem is FPT. In particular, there is an* $O(1.28^k + n^3)$ *time algorithm which given an instance* (D, k) *of the MinLOB-PBGV problem, either produces a solution or reduces the instance* (D, k) *to an instance* (D', k) *where* $|V(D')| = O(k \cdot 2^k)$.

Proof. It follows from Lemma 1 that there is an $O(1.28^k + n^3)$ algorithm which finds either vertex cover U of D of size at most $k - 1$ or produces an out-branching with at most $n - k$ leaves. Assume that the algorithm has found a vertex cover U such that $|U| < k$. Consider the transformation from (D, k) to (D', k) performed by the algorithm KERNEL *given* the vertex cover U. Lemma 2 proves that (D, k) is a 'YES' instance if and only if (D', k) is. We will show that this transformation takes $O(n^3)$ time and $|V(D')| = O(k \cdot 2^k)$ and that will complete the proof.

For $S \subseteq U$, let $V_S = \{x \in V(D) \setminus U : N^+(x) = S\}$. Going through all vertices of D one by one and comparing their out-neighborhoods, we can find all sets V_S in time $O(n^3)$. Now for each fixed $u \in U$ we can construct all non-empty sets $V(S, u)$ in time $O(n^2)$. Thus, all non-empty sets $V(S, u)$ can be found in time $O(n^3)$. Given the whole list $NList$, the construction of V' requires $O(n^3)$.

Let us now compute the number of vertices of D'. In terms of $|U|$, the number of elements of sets $V(S, u)$ which belong to V' is at most $2|U| \cdot 2^{|U|}$. In addition at most $|U| + 1$ vertices are added to V' at the end of KERNEL. Thus the number of vertices of V' is $O(|U| \cdot 2^{|U|}) = O(k \cdot 2^k)$ since $|U| < k$.   □

Thus we have shown that the MinLOB-PBGV problem has a kernel of order proportional to $k \cdot 2^k$. Now we are going to clarify how we explore this kernel in order to get the desired out-branching. A straightforward exploration of all possible out-branchings (using, e.g., the main algorithm of [18]) is not a good choice because the number of different out-branchings may be up to p^{p-1}, where $p = |V(D')| = (k \cdot 2^k)$. Indeed, by the famous Kelly's formula the number of spanning trees in the complete graph K_p on p vertices equals p^{p-2}. In the complete digraph on p vertices, one can get p out-branchings from each spanning tree of K_p by assigning a vertex to be the root.

In order to achieve a better running time we provide an alternative way of showing the fixed-parameter tractability of the MinLOB-PBGV problem based on the notion of *tree decomposition*.

A *tree decomposition* of an (undirected) graph G is a pair (X, U) where U is a tree whose vertices we will call *nodes* and $X = \{X_i : i \in V(U)\}$ is a collection of subsets of $V(G)$ (called *bags*) such that

1. $\bigcup_{i \in V(U)} X_i = V(G)$,
2. for each edge $\{v, w\} \in E(G)$, there is an $i \in V(U)$ such that $v, w \in X_i$, and
3. for each $v \in V(G)$ the set of nodes $\{i : v \in X_i\}$ form a subtree of U.

The *width* of a tree decomposition $(\{X_i : i \in V(U)\}, U)$ equals $\max_{i \in V(U)} \{|X_i| - 1\}$. The *treewidth* of a graph G is the minimum width over all tree decompositions of G. We use the notation $\mathrm{tw}(G)$ to denote the treewidth of a graph G.

By a *tree decomposition of a digraph* D we will mean a tree decomposition of the underlying graph $UG(D)$. Also, $\mathrm{tw}(D) = \mathrm{tw}(UG(D))$.

Theorem 4. *There is an algorithm that, given an instance (D, k) of the MinLOB-PBGV problem, either finds a solution or establishes a tree decomposition of D of width at most k.*

Proof. By Lemma 1, there is an algorithm which either finds a solution or specifies a vertex cover C of D of size at most k. Let $I = \{v_1, \ldots, v_s\} = V(D) \backslash C$. Consider a star U with nodes $x_0, x_1, \ldots, x_s$ and edges $x_0 x_1, x_0 x_2, \ldots, x_0 x_s$. Let $X_0 = C$ and $X_i = X_0 \cup \{v_i\}$ for $i = 1, 2, \ldots, s$ and let X_j be the bag corresponding to x_j for every $j = 0, 1, \ldots, s$. Observe that $(\{X_0, X_1, \ldots, X_s\}, U)$ is a tree decomposition of D and its width is at most k. $\square$

Theorem 4 shows that an instance (D, k) of the MinLOB-PBGV problem can be reduced to another instance with treewidth $O(k)$. Using standard dynamic programming techniques we can solve this instance in time $2^{O(k \log k)} n^{O(1)}$. On the first glance it seems that this running time makes the above kernelization redundant. However, although the $O(k \cdot 2^k)$ kernel is not polynomial, it is much smaller than $2^{O(k \log k)}$. Therefore if we first find the kernel and then establish the tree decomposition, the resulting dynamic programming algorithm will run in time $2^{O(k \log k)} + n^{O(1)}$ *without* changing the constant at $k \log k$. More precisely, Theorem 3 and Theorem 4 imply the following corollary.

Corollary 1. *Let D a digraph of order n. Suppose that a tree-decomposition of D of width k is specified. Suppose also that given this tree-decomposition the MinLOB-PBGV problem can be solved in time $2^{ck \log k} n^{O(1)}$. Then for any instance (D, k), the MinLOB-PBGV problem can be solved in time $O(2^{ck \log k + dk} + 1.28^k + n^3)$, where d is a constant.*

Proof. The additional dk at the exponent follows from replacing $n^{O(1)}$ by $(k2^k)^{O(1)}$. To obtain a vertex cover of size at most k, if one exists, takes $O(1.28^k + kn)$ time. The kernelization can be done in $O(n^3)$ time by Theorem 3. $\square$

The above results imply the following:

Theorem 5. *The MinLOB-PBGV problem can be solved by an additive FPT algorithm of running time $O(2^{O(k \log k)} + n^3)$.*

5   Discussion and Further Research

We have proved that MinLOB-PBGV is FPT. It would be interesting to check whether MinLOB-PBGV admits significantly more efficient FPT algorithms, i.e., algorithms of complexity $O(c^k n^{O(1)})$, where c is a constant. The same question is of interest for the following related problem, which is the natural parametrization of the Maximum Leaf Out-Branching problem.

MaxLOB Parameterized Naturally (MaxLOB-PN)
Instance: A digraph D.
Parameter: A positive integer k.
Question: Does D have an out-branching with at least k leaves?

Alon et al. [1,2] proved that this problem is FPT for several special classes of digraphs such as strongly connected digraphs and acyclic digraphs, and Bonsma and Dorn [7] proved that the problem is FPT. Note that in the three papers, MaxLOB-PN algorithms are of running time $O(2^{k(\log k)^{O(1)}} \cdot n^{O(1)})$.

Interestingly, MaxLOB-PN remains NP-complete even when the given digraph D is acyclic [3], which is in a clear contrast with MinLOB-PBGV unless P=NP.

Acknowledgements. Research of Gutin and Kim was supported in part by an EPSRC grant. Part of the paper was written when Razgon was vising Department of Computer Science, Royal Holloway, University of London. The research of Razgon at the Department of Computer Science, University College Cork was supported by Science Foundation Ireland Grant 05/IN/I886.

References

1. Alon, N., Fomin, F., Gutin, G., Krivelevich, M., Saurabh, S.: Parameterized Algorithms for Directed Maximum Leaf Problems. In: Arge, L., Cachin, C., Jurdziński, T., Tarlecki, A. (eds.) ICALP 2007. LNCS, vol. 4596, pp. 352–362. Springer, Heidelberg (2007)
2. Alon, N., Fomin, F., Gutin, G., Krivelevich, M., Saurabh, S.: Better Algorithms and Bounds for Directed Maximum Leaf Problems. In: Arvind, V., Prasad, S. (eds.) FSTTCS 2007. LNCS, vol. 4855, Springer, Heidelberg (2007)
3. Alon, N., Fomin, F.V., Gutin, G., Krivelevich, M., Saurabh, S.: Spanning directed trees with many leaves. Report arXiv: 0803.0701 (2008)
4. Alt, H., Blum, N., Melhorn, K., Paul, M.: Computing of maximum cardinality matching in a bipartite graph in time $O(n^{1.5}\sqrt{m/\log n})$. Inf. Proc. Letters 37, 237–240 (1991)
5. Bang-Jensen, J., Gutin, G.: Digraphs: Theory, Algorithms and Applications. Springer, Heidelberg (2000), www.cs.rhul.ac.uk/books/dbook/
6. Bang-Jensen, J., Yeo, A.: The minimum spanning strong subdigraph problem is fixed parameter tractable. Discrete Applied Math. (to appear)
7. Bonsma, P.S., Dorn, F.: An FPT Algorithm for Directed Spanning k-Leaf. Preprint 046-2007, Combinatorial Optimization & Graph Algorithms Group, TU Berlin (November 2007) (preprint 046-2007)
8. Chandran, L.S., Grandoni, F.: Refined memorization for vertex cover. Inform. Proc. Letters 93, 125–131 (2005)
9. Downey, R.G., Fellows, M.R.: Parameterized Complexity. Springer, Heidelberg (1999)

10. Demers, A., Downing, A.: Minimum leaf spanning tree. US Patent no. 6,105,018 (August 2000)
11. Fernau, H.: Parameterized Algorithmics: A Graph-theoretic Approach. Habilitation thesis, U. Tübingen (2005)
12. Fernau, H.: Parameterized Algorithmics for Linear Arrangement Problems(manscript, July 2005)
13. Flum, J., Grohe, M.: Parameterized Complexity Theory. Springer, Heidelberg (2006)
14. Guo, J., Niedermeier, R.: Invitation to Data Reduction and Problem Kernelization. ACM SIGACT News 38, 31–45 (2007)
15. Gutin, G., Rafiey, A., Szeider, S., Yeo, A.: The Linear Arrangement Problem Parameterized Above Guaranteed Value. Theory of Computing Systems 41, 521–538 (2007)
16. Gutin, G., Szeider, S., Yeo, A.: Fixed-Parameter Complexity of Minimum Profile Problems. Algorithmica (to appear)
17. Heggernes, P., Paul, C., Telle, J.A., Villanger, Y.: Interval completion with few edges. In: Proc. STOC 2007 - 39th ACM Symposium on Theory of Computing, pp. 374–381 (2007)
18. Kapoor, S., Ramesh, H.: An Algorithm for Enumerating All Spanning Trees of a Directed Graph. Algorithmica 27, 120–130 (2000)
19. Las Vergnas, M.: Sur les arborescences dans un graphe orienté. Discrete Math. 15, 27–29 (1976)
20. Mahajan, M., Raman, V.: Parameterizing above guaranteed values: MaxSat and MaxCut. J. Algorithms 31, 335–354 (1999)
21. Niedermeier, R.: Invitation to Fixed-Parameter Algorithms. Oxford University Press, Oxford (2006)

Graphs and Path Equilibria

Stéphane Le Roux*

École normale supérieure de Lyon, Université de Lyon, LIP, CNRS, INRIA, UCBL

Abstract. The quest for optimal/stable paths in graphs concerns a few practical or theoretical areas. Taking part in the quest, this paper adopts an abstract, general, equilibrium-oriented approach: it uses (quasi-arbitrary) arc-labelled digraphs, and assumes little about the structure of the sought paths and the definition of equilibrium, *i.e.* optimality/stability. The paper gives both a sufficient condition and a necessary condition for equilibrium existence for every "graph", pinpoints the difference between these conditions, and shows coincidence when optimality relates to a total order. These results are applied to network routing.

Keywords: Labelled directed graph, path, preference, equilibrium, optimisation, strict weak order, sufficient condition, necessary condition, induction.

1 Introduction

This paper summarises a 40-page document [4] that presents detailed proofs and more examples. This paper provides an abstract formalism that enables generic proofs, yet accurate results, about path equilibria in graphs. Other approaches to optimisation in graphs may be found in [3], for instance.

This paper introduces *dalographs*, *i.e.* finite, arc-labelled, directed graphs with non-zero outdegree *wlog* (without loss of generality). In this paper, paths are infinite *wlog*. Note that the outdegree constraint ensures existence of infinite paths, starting from any node. This uniformity facilitates an "algebraic" approach. Also, paths are non-self-crossing: their shapes resemble 6 or 0 but not 8 or A. This is often desirable, but may yield a loss of generality in some areas.

In this formalism, a path induces an ultimately periodic sequence of labels (of arcs that are involved in the path). An arbitrary binary relation over such sequences is assumed and named *preference*. Using the preference to compare the sequences induced by two paths *starting from the same node*, one can compare these paths. If a path has no successor wrt the new relation/preference, it is deemed maximal. A *strategy* is an object built on a dalograph. It amounts to every node's choosing an outgoing arc. Hence from any node, a strategy induces a path. Finally, an *equilibrium* is a strategy inducing maximal paths at any node.

This paper gives a sufficient condition on the preference for equilibrium existence in any dalograph, and examples of relations complying with it: lexicographic and Pareto ordering. The condition involves *strict weak orders*, which

---

* http://perso.ens-lyon.fr/stephane.le.roux/. Now working at INRIA-Microsoft Research. I thank Pierre Lescanne for his comments on the draft of this paper.

R. Fleischer and J. Xu (Eds.): AAIM 2008, LNCS 5034, pp. 247–258, 2008.

are discussed *e.g.* in [1]. To prove sufficiency, design the *seeking-forward* function that expects a node and returns a path starting from the node: Informally, pick a (candidate) path that is preference-wise maximal at the given node; follow it until its current suffix, which is also a path, is not maximal among the suffixes that form a path when appended to the current prefix; pick a suffix that is maximal among these possible suffixes, and append it to the current prefix to form a new candidate path, and so on till there is only one possible suffix. This process terminates due to the non-self-crossing constraint, and it yields a *hereditary maximal path*: a path that is maximal at all of its nodes. Equilibrium existence is proved by induction on the number of arcs in the dalograph, as follows:

- Compute a hereditary maximal path in the dalograph.
- Remove the arcs of the dalograph that the path ignored while visiting adjacent nodes. This yields a smaller dalograph.
- Compute an equilibrium on this smaller dalograph and add the ignored arcs back. This yields an equilibrium for the original dalograph.

Conversely, this paper gives a necessary condition on the preference for equilibrium existence in every dalograph. The condition involves binary relation closures that are actually related to the sufficient condition: e.g. sufficiency involves preference's transitivity, and necessity involves the preference's transitive closure. These closures preserve equilibrium existence: if a preference ensures equilibrium existence, so do its closures. By the notion of *simple closure*, the combination of the above-mentioned closures also preserves equilibrium existence, which gives a non-trivial necessary condition. This paper also provides an example of a relation that does not meet the requirements of this condition.

An example shows that the necessary condition is not sufficient in general, although it is sufficient when preference is a total order. However, it is unclear whether or not the sufficient condition is necessary, and what a practical necessary *and* sufficient condition would be. As an application to networking, one derives a sufficient condition on routing policies for stable routing to exist. This condition is also necessary when the policy is a total order, as suggested above.

Section 2 defines dalographs and their equilibria; section 3 discusses strict weak orders; section 4 proves a sufficient condition for equilibrium existence and gives examples and an application to network routing; section 5 defines simple closures, proves a necessary condition for equilibrium existence, and gives an example; section 6 compares the sufficient condition and the necessary condition, and presents a further application to network routing in the total order case.

Unless said otherwise, universal quantifiers are usually omitted in front of formal statements: *e.g.* a claim of the form $P(x, y)$ should be read $\forall x, y, \ P(x, y)$. Also, the notation $P(x) \triangleq Q(x)$ means that P is defined as coinciding with Q.

2 Dalographs and Equilibria

This section defines dalographs, walks, paths, and derives a notion of equilibrium from a notion of maximality for these paths. The more general will be those dalographs and paths, the more general will be the derived notion of equilibrium.

Definition 1 (Dalograph). *A dalograph is a finite directed graph whose arcs are labelled and whose nodes have non-zero outdegree.*

In the definition above, both labelling arcs instead of nodes and demanding that all nodes have non-zero outdegree do not yield a loss of generality, as justified in [4]. The dalograph below has four (squared) nodes and five labels.

$$\square \overset{a_5}{\underset{a_4}{\rightleftarrows}} \square \;-\; a_3 \rightarrow \square \;-\; a_1 \rightarrow \square \;\; a_2 \circlearrowleft$$

Definition 2 (Walks and paths). *Walks in a digraph are defined as follows.*

- *The empty word is a walk, and o is a walk for any node o of the digraph.*
- *If $o_0 \ldots o_n$ is a walk, if o does not occur in $o_0, \ldots, o_{n-1}$, and if oo_0 is an arc of the digraph, then $oo_0 \ldots o_n$ is also a walk.*

Note that if a node occurs twice in a walk, it occurs only twice and it occurs at the end of the walk which is said to be looping. *A looping walk $uovo$ (u and v are walks, o is a node) induces a* path, *i.e. the ultimately periodic sequence $u(ov)^\omega$. Given a walk x and a path Γ such that $x\Gamma$ is also a path, Γ is said to be a* continuation *of x. Note that every walk in a dalograph has a continuation, and that a looping walk $uovo$ has a unique continuation $(vo)^\omega$.*

In the definition above, paths being infinite does not yield a loss of generality, as justified in [4]. A non-looping walk u may be represented by the left-hand picture below; a looping walk $uovo$ may be represented by the right-hand picture.

$$\rule{2em}{0.4pt}\; u \twoheadrightarrow \qquad \rule{2em}{0.4pt}\; uovo \;\rule{1em}{0.4pt}\!{\dashv}$$

The usual induction principle for walks goes from the empty walk to bigger walks, along the inductive definition of walks. Here, walks are bounded since dalographs are finite. This enables an alternative induction principle for walks.

Lemma 3 (Nibbling induction principle for walks). *Let P be a predicate on walks in a dalograph g. Assume that for every walk x, "$P(xo)$ for all walks xo" implies "$P(x)$" (where o is a node). Then $P(x)$ for every walk x.*

A strategy is an object built on a dalograph by choosing an outgoing arc at each node. In the strategy below, the choices are represented by double lines.

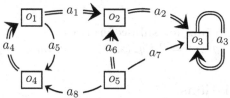

Definition 4 (Strategy). *Given a dalograph g, a strategy s on g is a pair (g, c), where c is a function from the nodes of g to themselves, and such that for all nodes o, the pair $(o, c(o))$ is an arc of g.*

Let o be a node of some strategy $s = (g, c)$. Following the choices of s starting from $c(o)$ defines (co-inductively) the path $p(s, o) \triangleq c(o) \cdot p(s, c(o))$. For instance in the picture above, $p(s, o_4) = o_1 o_2 o_3^\omega$. In turn, any path Γ *induces* a (ultimately periodic) sequence of labels $seq(\Gamma)$. For instance above, $seq(o_5 o_2 o_3^\omega) = a_6 a_2 a_3^\omega$. These sequences may be compared through a *preference* binary relation $\prec$. Moreover, if paths Γ_1 and Γ_2 start from the same node and induce sequences γ_1 and γ_2 with $\gamma_1 \prec \gamma_2$, one writes $\Gamma_1 \prec \Gamma_2$ by abuse of notation.

The following definition captures the notion of maximality (with respect to a preference) of a path among the continuations of a given walk. In addition, this paper needs to discuss paths all of whose subpaths are maximal from their starting points. The notion of hereditary maximality captures this idea.

Definition 5 (Maximal continuation and hereditary maximal path). *If $xo\Gamma$ is a path in a dalograph g, and if $o\Gamma \not\prec o\Gamma'$ for all paths $xo\Gamma'$ in g, one writes $m_{g,\prec}(xo, \Gamma)$. Let Γ be a path. If $m_{g,\prec}(o, \Gamma')$ for any decomposition $\Gamma = xo\Gamma'$ where xo is non-looping, one writes $hm_{g,\prec}(\Gamma)$. One forgets the index $(g, \prec)$ when there is no ambiguity.*

A strategy on a dalograph is a local equilibrium at a node if it induces a hereditary maximal path at this node. A global equilibrium is intended to be a strategy inducing a maximal path at every node, so it can be defined as a strategy that is a local equilibrium at every node.

Definition 6 (Local and global equilibrium)
$$LEq_\prec(s, o) \triangleq m_{g,\prec}(o, p(s, o)) \quad and \quad GEq_\prec(s) \triangleq \forall o \in g, \, LEq_\prec(s, o).$$

In the rest of this paper, "equilibrium" may mean "global equilibrium". In the strategy below, arcs are labelled with natural numbers and $\prec$ is the lexicographic extension of the usual order to infinite sequences of natural numbers. So, the strategy is a local equilibrium for node o' but not for node o. Then, lemma 7 says that subpreference preserves (local) equilibrium.

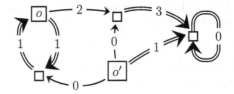

Lemma 7 (Equilibrium for subpreference)
$$\prec \, \subseteq \, \prec' \quad \Rightarrow \quad (LEq_{\prec'}(s, o) \Rightarrow LEq_\prec(s, o) \quad \wedge \quad GEq_{\prec'}(s, o) \Rightarrow GEq_\prec(s, o)).$$

3   Binary Relations

This section rephrases the notion of strict weak order (studied in [1]) and introduces a few predicates on binary relations over sequences. It also proves a few properties. As usual, a binary relation $\prec$ is irreflexive if it complies with the formula $\alpha \not\prec \alpha$. More generally, it is asymmetric if $\alpha \prec \beta \Rightarrow \beta \not\prec \alpha$. In addition, the corresponding incomparability relation is defined by $\alpha \sharp \beta \overset{\Delta}{=} \alpha \not\prec \beta \wedge \beta \not\prec \alpha$.

Definition 8 (Strict weak order). *A strict weak order is an asymmetric relation whose negation is transitive.*

The following definition and lemma characterise strict weak orders.

Definition 9 (Imitation). *Let $\prec$ be an asymmetric relation. In this case, the following two formulas are equivalent. If they hold, $\prec$ is said to be imitating.*
$(\alpha \sharp \beta \wedge \gamma \prec \alpha) \Rightarrow \gamma \prec \beta$ *and* $(\alpha \sharp \beta \wedge \alpha \prec \gamma) \Rightarrow \beta \prec \gamma$.

Lemma 10. *Let $\prec$ be a binary relation, and let $\sharp$ be the corresponding noncomparability relation. The following three propositions are equivalent.*

1. $\prec$ *is a strict weak order.*
2. $\prec$ *is transitive and $\sharp$ is an equivalence relation.*
3. $\prec$ *is imitating and has no cycle of length 2 or 3.*

The picture below is meant to give an intuition of what a strict weak order is. The circles represent equivalence classes of $\sharp$. Here we have $\gamma \prec \alpha, \beta, \delta$ and $\gamma, \alpha, \beta \prec \delta$ and $\alpha \sharp \beta$. Informally, a strict weak order looks like a knotted rope.

Let $\prec$ be a binary relation over infinite sequences. Let W be a non-empty finite set of non-empty finite sequences, with at most one sequence of length 1. If for all u in W there is v in W such that $u\alpha \prec v\beta$, one writes $W\alpha \prec W\beta$. Five predicates on $\prec$ are defined below. First, E-prefix means preservation by prefix elimination. Second, Gen-E-prefix generalises E-prefix. Third, A-transitivity generalises transitivity. Fourth, assume that $\prec$ is transitive and preserved by prefix addition. If $\alpha \prec u\alpha$ then $u\alpha \prec u^2\alpha$ and $\alpha \prec u\alpha \prec \cdots \prec u^n\alpha \prec \ldots$, and one might see u^ω as a possible upper bound for this chain. Subcontinuity captures this informal thought. Fifth, alt-subcontinuity gives an alternative (slightly more complex) definition for subcontinuity. A lemma follows.

Definition 11. *A relation $\prec$ over sequences is described with the left-hand words when complying with the right-hand formulas, where sequences are non-empty.*

$E - prefix$	$u\alpha \prec u\beta \Rightarrow \alpha \prec \beta$
$Gen - E - prefix$	$W\alpha \prec W\beta \Rightarrow \alpha \prec \beta$
$A - transitive$	$(\alpha \prec \beta \wedge u\beta \prec \gamma) \Rightarrow u\alpha \prec \gamma$
$Subcontinuous$	$\alpha \prec u\alpha \Rightarrow u^\omega \not\prec \alpha$
$Alt - subcontinuous$	$\alpha \prec t\beta \Rightarrow (v\alpha \prec \beta \vee \alpha \prec (tv)^\omega)$

Lemma 12. *A E-prefix strict weak order is A-transitive, and a E-prefix subcontinuous strict weak order is alt-subcontinuous.*

4   Equilibrium Existence, Examples, and Application

This section proves a sufficient condition for equilibrium existence in any dalograph, it gives examples of relations meeting the requirements of the sufficient condition, and it applies the result to network routing.

The notion of semi-hereditary maximality lies and makes the connection between maximality and and hereditary maximality. A lemma follows its definition.

Definition 13 (Semi-hereditary maximal path). *Let x be a non-empty walk of continuation Γ. If $m(xy, \Gamma')$ for all decompositions $\Gamma = y\Gamma'$ where xy is a walk, one writes $shm(x, \Gamma)$.*

Lemma 14. $shm(x, \Gamma) \Rightarrow m(x, \Gamma)$ and $m(x, o\Gamma) \Rightarrow shm(xo, \Gamma) \Rightarrow shm(x, o\Gamma)$.

Given a non-empty walk, the function below expects and returns a continuation of the walk. Informally: If the walk is looping, return its sole continuation. Otherwise, follow the given continuation until its current suffix is not a maximal continuation of the new walk composed of the given walk and the current prefix of the given continuation. At that point, feed the function with the new walk and any of its maximal continuation. Given u a non-empty finite sequence, u^f (resp. u^l) represents the first (resp. last) element of u.

Definition 15 (Seeking-forward function). *Let $\prec$ be an acyclic preference and g a dalograph. The function F expects a non-empty walk in g and a continuation of this walk. More specifically, $F(x, \Gamma)$ is defined recursively along lemma 3.*

- *If x is a looping walk of (unique) continuation Γ, let $F(x, \Gamma) \triangleq \Gamma$.*
- *If x is not a looping walk then case split as follows.*
  1. *If $m(x, \Gamma)$ then $F(x, \Gamma) \triangleq oF(xo, \Gamma')$, where $\Gamma = o\Gamma'$.*
  2. *If $\neg m(x, \Gamma)$ then $F(x, \Gamma) \triangleq oF(xo, \Gamma')$ for some $o\Gamma'$ such that $m(x, o\Gamma')$ and $x^l\Gamma \prec x^l o\Gamma'$.*

The properties of lemma 16 were proved in the same order they are stated. Their combination helps prove lemma 17 on sufficiency for equilibrium existence.

Lemma 16. *Let g be a dalograph, $\prec$ an irreflexive and A-transitive preference, u a non-empty suffix of x, and xo a walk. The following formulas hold.*

$$uF(x, \Gamma) \prec u^f\Delta \Rightarrow u\Gamma \prec u^f\Delta$$
$$F(x, F(x, \Gamma)) = F(x, \Gamma)$$
$$F(x, o\Gamma) = o\Gamma \Rightarrow F(xo, \Gamma) = \Gamma$$

Moreover, if $\prec$ is E-prefix, then $F(x, \Gamma) = \Gamma \Rightarrow shm(x, \Gamma)$. If in addition $\not\prec$ is alt-subcontinuous, then $shm(o, \Gamma) \Rightarrow hm(o\Gamma)$.

Theorem 17. *If a preference is included in a subcontinuous E-prefix strict weak order, then any dalograph has a global equilibrium with respect to the preference.*

Proof. By lemma 7, assume *wlog* that the preference $\prec$ is a subcontinuous E-prefix strict weak order. If all outdegrees equal 1, we are done by irreflexivity since there is only one possible strategy. Now assume that there is a node o with at least two outgoing arcs, and proceed by induction on the number of arcs in the dalograph. If it has one arc only, we are done by the previous case. Now assume that the claim holds for dalographs with n or less arcs, and consider g with $n+1$ arcs. By lemma 12 and assumption, $\prec$ is irreflexive, A-transitive, E-prefix, and alt-subcontinuous. By lemma 16 there is a hereditary maximal path Γ starting from the node o. Let $u(av)^\omega$ be the induced sequences of labels, depicted below where u and v are finite (possibly empty) sequences of labels and a is a label.

$$o \,\text{---}\, u \twoheadrightarrow \overleftarrow{} \overset{a}{\underset{v}{\rightrightarrows}}$$

Build g' from g by removing the arcs dismissed by the choices along Γ. There is at least one such arc since Γ starts at o. Below, g is to the left; g' to the right.

g' has n or less arcs, so by induction hypothesis, it has an equilibrium s', which is depicted below. Double lines represent the choices (one per node).

By construction of g' from g, there is only one possible path from o in g', namely Γ, so s' induces Γ at o. Build a new strategy s (related to g) from s', by adding back the arcs that were removed when defining g', as shown below.

The rest of the proof shows that s is an equilibrium. Since Γ is hereditary maximal, s is a local equilibrium for any node on Γ. Let o' be a node outside Γ, and let Γ' be the path induced by s (and s') from o'. Let a path Γ^3 start from o'. If Γ^3 does not involve any arc dismissed by Γ, then Γ^3 is also valid in g', so it is not greater than Γ' since s' is an equilibrium. If Γ^3 involves such an arc, then the situation looks like the picture below, where $\Gamma = v\Gamma''$ and $\Gamma^3 = u\Delta$.

By hereditary maximality, $\Gamma'' \not\prec \Delta$, so $u\Gamma'' \not\prec u\Delta$ by E-prefix. The path $u\Gamma''$ is also valid in g', so $\Gamma' \not\prec u\Gamma''$ since s' is an equilibrium. So $\Gamma' \not\prec u\Delta$ by transitivity of the negation. So Γ' is also maximal in g, and s is an equilibrium. $\quad\square$

Below, two examples of relations that are (included in) subcontinuous E-prefix strict weak orders. Other examples involving limit sets are detailed in [4].

Definition 18 (Lexicographic extension). *Let* $(A, \prec)$ *be a strict weak order. The lexicographic extension of* $\prec$ *is defined for infinite sequences over A.*

$$\frac{a \prec b}{a\alpha \prec^{lex} b\beta} \qquad\qquad \frac{\alpha \prec^{lex} \beta \qquad a \sharp b}{a\alpha \prec^{lex} b\beta}$$

For instance $0 \cdot 1^\omega <^{lex} 1 \cdot 0^\omega$ and $(0 \cdot 3)^\omega <^{lex} 1^\omega <^{lex} (3 \cdot 0)^\omega$, where $<$ is the usual order over the naturals. The Pareto extension (smaller than lexicographic) also allows comparing infinite vectors. A vector is greater than another one if it is not smaller for all components and if it is greater for some component.

Definition 19 (Pareto extension). *Let* $(A, \prec)$ *be a strict weak order. The Pareto extension of* $\prec$ *is defined for infinite sequences of elements of A.*

$$\alpha \prec^P \beta \overset{\Delta}{=} \quad \forall n \in \mathbb{N}, \beta(n) \not\prec \alpha(n) \quad \wedge \quad \exists n \in \mathbb{N}, \alpha(n) \prec \beta(n)$$

For instance $0 \cdot 1^\omega \not<^P 1 \cdot 0^\omega$ but $(0 \cdot 1)^\omega <^P 1^\omega <^P (1 \cdot 3)^\omega$.

Theorem 20. *(By theorem 17) A dalograph over elements of a strict weak order has a global equilibrium with respect to the lexicographic/Pareto extension.*

The following defines routing problems and states a condition for every routing problem to have a routing equilibrium. The condition is decidable for routing problems with decidable policy, which ought to be the case in the real world. This approach is different from [2] but may complement it nicely.

Definition 21 (Routing problem). *A* routing policy *is a binary relation over finite words over labels. A* routing problem *is a finite digraph whose arcs are labelled with the labels. In addition, only one node, the* target, *has outdegree zero. The target is reachable from any node via some walk. A* routing strategy *maps every node (but the target) to one of its outgoing arcs. A strategy is a* routing equilibrium *if at any node, the induced path leads to the target, and if for any node, no strategy induces better (wrt the routing policy) such path.*

Lemma 22. *If a routing policy is (included in) a E-prefix strict weak order* $\prec^r$ *such that* $v \not\prec^r uv$ *for all* v *and* u, *every routing problem has an equilibrium.*

Proof. A routing problem is translated into a dalograph: add a dummy node dn, add arcs from the target and dn to dn, and add dummy labels dl on both arcs. Call L the labels but the dummy label, and define the following preference.

$$\frac{u \prec^r v}{udl^\omega \prec vdl^\omega} \qquad\qquad \frac{\alpha \in L^\omega \qquad u \in L^*}{\alpha \prec udl^\omega}$$

Like $\prec^r$, $\prec$ is E-prefix. By $v \not\prec^r uv$ assumption, $\prec$ is subcontinuous. By theorem 17, the dalograph has an equilibrium, which yields a routing equilibrium. $\square$

5   Preservation of Equilibrium Existence

This section defines the notion of simple closure, states a few properties (all of which are detailed in [4]), and shows that the combination of the A-transitive and gen-E-prefix closures preserves equilibrium existence. An example follows.

Lemma 23. *Let f_0 to f_n be functions of type $A \to A$. Let Q be a predicate on A. Assume that each f_k preserves Q. Then for all x in A and all w words on the f_k, we have $Q(x) \Rightarrow Q(w(x))$.*

In the next definition and lemma, relations have arity r and X is a vector $(X_1, \ldots, X_r)$ of dimension r. Below, a class of closures (on relations) is defined together with a union operation. A preservation property follows.

Definition 24 (Simple closure). *Let f be an operator on relations. If f is inductively defined with the first inference rule below and some rules having the same form as the second inference rule below, then f is a simple closure.*

$$\frac{R(X)}{f(R)(X)} \qquad \frac{K(X, \{X^i\}_{i \in C}) \qquad \wedge_{i \in C} f(R)(X^i)}{f(R)(X)}$$

Let f (resp. g) be a simple closure that is defined by the induction rules F_1 to F_n (resp. G_1 to G_m). Then $f + g$, the rule union of f and g, is defined by the F_i and the G_j, and it is also a simple closure.

Lemma 25. *Let Q be a predicate on relations that is preserved by the simple closures f_0 to f_n and subrelation, i.e. $R \subset R' \Rightarrow Q(R') \Rightarrow Q(R)$. Then for all relations R of finite subdomain S, $Q(R)$ implies $Q(\Sigma_k f_k(R) \mid_S)$.*

The A-transitive (resp. gen-E-prefix) closure of a binary relation is its smallest A-transitive (resp. gen-E-prefix) superrelation. Lemma 27 is proved by rule induction and graph merging. (Gen-E-prefix closure enjoys a similar property.)

Definition 26 (A-transitive closure and gen-E-prefix closure)

$$\frac{\alpha \prec \beta}{\alpha \prec^{st} \beta} \qquad \frac{\alpha \prec^{st} \beta \qquad u\beta \prec^{st} \gamma}{u\alpha \prec^{st} \gamma} \qquad \frac{\alpha \prec \beta}{\alpha \prec^{sep} \beta} \qquad \frac{W\alpha \prec^{sep} W\beta}{\alpha \prec^{sep} \beta}$$

Lemma 27. *Let $\prec$ be a preference, and let α and β be such that $\alpha \prec^{st} \beta$. There exists a dalograph g with the following properties.*

- *The dalograph g has the following shape (dashed lines represent and delimit the rest of the dalograph).*

- *The path inducing β is not branching after the top node.*
- *Any equilibrium for g involves the path inducing β.*

The following lemma states that under some conditions (similar to the conclusions of lemma 27), superrelation preserves equilibrium existence.

Lemma 28. *Let $\prec$ and $\prec'$ be two preferences. Assume that $\prec$ is included in $\prec'$ and that for any $\alpha \prec' \beta$, there exists a dalograph g with the following properties.*

- *The dalograph below is a subgraph of the dalograph g.*

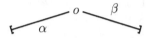

- *The path inducing β is not branching after the top node o.*
- *Any $\prec$-equilibrium for g involves the path inducing β.*

In this case, if all dalographs have $\prec$-equilibria, then they have $\prec'$-equilibria.

The combination closure of a binary relation is its smallest superrelation that is A-transitive and gen-E-prefix. The sought property follows.

Definition 29 (Combination closure)

$$\frac{\alpha \prec \beta}{\alpha \prec^c \beta} \qquad \frac{\alpha \prec^c \beta \quad u\beta \prec^c \gamma}{u\alpha \prec^c \gamma} \qquad \frac{W\alpha \prec^c W\beta}{\alpha \prec^c \beta}$$

Theorem 30. *If all dalographs have $\prec$-equilibria, then they have $\prec^c$-equilibria.*

The relation that is defined below does not guarantee equilibrium existence.

Definition 31. *Let $(E, \prec)$ be a strict weak order. A total order on the $\sharp$-equivalence classes is defined as follows: $A^\sharp \prec B^\sharp \triangleq \exists x \in A^\sharp, \exists y \in B^\sharp, x \prec y$. The max (resp. min) of a finite subset A of E is the maximal (resp. minimal) $\sharp$-class intersecting A. The max-min order is defined on finite subsets of E.*

$$A \prec^{Mm} B \triangleq max(A) \prec max(B) \vee (max(A) = max(B) \wedge min(A) \prec min(B))$$

For instance $\{1,2,3\} <^{Mm} \{0,4\}$ and $\{0,2,3\} <^{Mm} \{1,3\}$. The max-min set order is defined below and proved not to guarantee equilibrium existence.

Definition 32 (Max-min set order). *Let $(E, \prec)$ be a strict weak order. For α an infinite sequence over E, let S_α be the set of all elements occurring in α.*

$$\alpha \prec^{Mms} \beta \triangleq S_\alpha \prec^{Mm} S_\beta$$

Lemma 33. *There exists a natural-labelled dalograph with no $<^{Mms}$-equilibrium.*

Proof. $2(02)^\omega <^{Mms} 21^\omega$ and $1^\omega <^{Mms} (02)^\omega$. Therefore any E-prefix and transitive relation including $<^{Mms}$ is not irreflexive. Conclude by theorem 30. $\square$

6   Sufficient Condition and Necessary Condition

Theorem 34 gathers the main results of the paper, and rewrites them to enable comparison between the sufficient and the necessary conditions. These coincide when the preference is a strict total order, as stated by corollary 35.

Theorem 34

The preference $\prec$ is included in some $\prec'$.
The preference $\prec'$ is a E-prefix and subcontinuous strict weak order.
$$\Updownarrow$$
The preference $\prec$ is included in some $\prec'$.
The preference $\prec'$ is E-prefix, subcontinuous, transitive, and irreflexive.
The non comparability relation $\not\parallel'$ is transitive.
$$\Downarrow$$
Every dalograph has a $\prec$-equilibrium.
$$\Downarrow$$
The preference $\prec$ is included in some $\prec'$.
The preference $\prec'$ is (gen-) E-prefix, (A-) transitive, and irreflexive.

Corollary 35. *Let a preference be a strict total order. All dalographs have equilibria iff the preference is E-prefix and subcontinuous.*

In general, the necessary condition is not a sufficient condition, as shown by the following example. (More examples in [4].) Let $\prec$ be defined as followed.

$$u_1 y_1 \beta_2 \prec v_1 x_1 \alpha_1 \qquad u_2 y_2 \beta_1 \prec v_2 x_2 \alpha_2$$
$$v_1 x_2 \alpha_2 \prec u_1 \alpha_1 \qquad v_2 x_1 \alpha_1 \prec u_2 \alpha_2$$
$$v_1 x_1 y_1 \beta_2 \prec u_1 \alpha_1 \qquad v_2 x_2 y_2 \beta_1 \prec u_2 \alpha_2$$
$$v_1 x_2 y_2 \beta_2 \prec u_1 \alpha_1 \qquad v_2 x_1 y_1 \beta_1 \prec u_2 \alpha_2$$

The preference $\prec$ complies with the necessary condition but the dalograph below has no $\prec$-equilibrium. Indeed, the node o_1 "wants" to follow a path leading to α_1 or β_1, while the node o_2 "wants" to follow a path leading to α_2 or β_2.

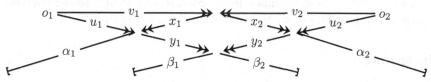

In the total order case, the condition for equilibrium in dalographs yields a necessary and sufficient condition for routing equilibrium existence in routing problems. Necessity invokes constructive arguments that are similar to the ones used for the necessary condition in dalographs. However, the proof is simple enough so that just doing it is simpler than invoking a big theorem.

Theorem 36. *Let $\prec^r$ be a total order routing policy. Every routing problem has an equilibrium iff $\prec^r$ is E-prefix and $uv \prec^r v$ for all v and non-empty u.*

Proof. Right-to-left: by lemma 22. Left-to-right: by contraposition, assume that either $\prec^r$ is not E-prefix or there exists u and v such that $v \prec^r uv$. First case, $\prec^r$ is not E-prefix. So there exists u, v and w such that $uv \prec^r uw$ and $w \prec^r v$. So the left-hand routing problem below has no routing equilibrium.

Second case, there exists u and v such that $v \prec^r uv$. The right-hand routing problem above has no routing equilibrium. □

7 Conclusion and Open Questions

This paper defines dalographs, an abstract and general formalism for path equilibria in graph-like systems. It shows a sufficient condition on preferences for equilibrium existence in all dalographs. The proof defines a function recursively along the nibbling principle (from big walks to small walks), and uses an induction on the number of arcs in a dalograph. A necessary condition on preferences is also given via equilibrium-preservation properties proved by rule induction on simple closures and by a few cut-and-paste on dalographs. The necessary condition is not sufficient in general, but these conditions coincide when the preference is a strict total order (this corollary has a direct proof given in [4]). However, whether or not the sufficient condition is also necessary is an open question. A few examples show the usefulness of the two conditions in general, and the theoretical results are applied to a network routing problem that is also addressed in [2]. How the two approaches complement each other is still unclear.

This paper is also useful to another respect: many other formalisms that are different from dalographs also use a notion of preference. In some of those, preferences may be seen as total orders without serious loss of generality: any preference guaranteeing equilibrium existence is included in some total order with the same guarantee that is preserved by subrelation. In this case, considering only total orders somehow accounts for all preferences. However in dalographs, there might be a guaranteeing preference each of whose linear extension gives no guarantee. In this case, assuming total ordering may yield a (non-recoverable) loss of generality. This discussion is continued yet left open in [4].

References

1. Fishburn, P.: Intransitive indifference in preference theory: A survey. Operations Research 18, 207–228 (1970)
2. Griffin, T., Sobrinho, J.: Metarouting. In: SIGCOMM 2005: Proceedings of the 2005 conference on Applications, technologies, architectures, and protocols for computer communications, pp. 1–12. ACM Press, New York (2005)
3. Lawler, E.: Combinatorial Optimization: Networks and Matroids. Dover (2001)
4. Le Roux, S.: Graphs and path equilibria. Research report, INRIA (2007), http://hal.inria.fr/inria-00195379/fr/

No l Grid-Points in Spaces of Small Dimension

Hanno Lefmann

Fakultät für Informatik, TU Chemnitz, D-09107 Chemnitz, Germany
lefmann@informatik.tu-chemnitz.de

Abstract. Motivated by a question raised by Pór and Wood in connection with compact embeddings of graphs in $\mathbb{Z}^d$, we investigate generalizations of the no-three-in-line-problem. For several pairs (k, l) we give algorithmic lower, and upper bounds on the largest sizes of subsets S of grid-points from the d-dimensional $T \times \cdots \times T$-grid, where no l distinct grid-points of S are contained in a k-dimensional affine or linear subspace.

1 Introduction

The No-Three-in-Line problem, which has been raised originally by Dudeney [7], asks for the maximum number of grid-points, which can be chosen from the $T \times T$-grid in $\mathbb{Z}^2$, i.e., from the set $\{0, \ldots, T-1\} \times \{0, \ldots, T-1\}$, such that no three distinct points are on a line, see [5,16]. It has been proved by Erdös [10] that this maximum number is $\Theta(T)$. The lower bound follows by considering for primes T the grid-points $(x, x^2 \bmod T)$, $x = 0, \ldots, T-1$. For the upper bound observe that each horizontal line may contain at most two grid-points. For constructons of (near-)optimal solutions for small values of T see [11,12].

Cohen, Eades, Lin and Ruskey [6] investigated embeddings of graphs into $\mathbb{Z}^3$ such that distinct edges (represented by segments) do not cross each other in a point distinct from the endpoints. One is interested in compact representations of graphs, in particular, in the minimum volume of an axis-aligned bounding box in $\mathbb{Z}^3$, which contains the drawing. The endpoints of crossing edges in a drawing of a graph are coplanar. In connection with this, it has been proved in [6] that there is a set of $\Omega(T)$ grid-points in the $T \times T \times T$-grid, which does not contain four distinct coplanar grid-points, and up to a constant factor this lower bound is best possible. Thus, the minimum volume of an axis-aligned bounding box for a crossing-free drawing of the complete graph K_n on n vertices in $\mathbb{Z}^3$ is $\Theta(n^3)$, see [6] and compare [17].

Pór and Wood [18] considered embeddings of graphs in $\mathbb{Z}^3$, where the line segments, which represent the edges, do not cross any vertex distinct from its endpoints. Then n grid-points in $\mathbb{Z}^3$ yield a crossing-free drawing of K_n, if no three of the n grid-points are on a line. Pór and Wood proved in [18] that there are $\Omega(T^2)$ grid-points in the $T \times T \times T$-grid with no three collinear points, by considering the set of all triples $(x, y, (x^2 + y^2) \bmod T)$, $x, y \in \{0, \ldots, T-1\}$, for T a prime with $T \equiv 3 \bmod 4$. This gives the upper bound $O(n^{3/2})$ on the minimum volume of a bounding box of a drawing of K_n in $\mathbb{Z}^3$.

R. Fleischer and J. Xu (Eds.): AAIM 2008, LNCS 5034, pp. 259–270, 2008.
© Springer-Verlag Berlin Heidelberg 2008

For higher dimensions, Pór and Wood [18] raised the question of determining vol (n, d, k), which is defined as the minimum volume of an axis-aligned bounding box for embeddings of n points in $\mathbb{Z}^d$, such that no $(k + 2)$ points are contained in a k-dimensional affine space. By a partition of such a bounding box into vol $(n, d, k)^{(d-k)/d}$ many affine k-dimensional subspaces they obtained vol $(n, d, k) \geq (n/(k + 1))^{d/(d-k)}$. Known from [6] is that vol $(n, 3, 2) = \Theta(n^3)$, and in general vol $(n, d, d - 1) = \Theta(n^d)$.

A closely related problem – finding in the d-dimensional $T \times \cdots \times T$-grid large sets of grid-points with distinct slopes (or distances) – has been considered in [9,15] and has applications to the problem of measuring distances by using radar or sonar signals, see [14], compare also [20] for avoiding collinear triples in connection with perturbation techniques.

Motivated by the question of Pór and Wood, we consider here for fixed integers $d, k, l \geq 1$ lower and upper bounds on the function $f_d(l, k, T)$, which is defined as the maximum number of grid-points in the d-dimensional $T \times \cdots \times T$-grid, such that no l distinct grid-points are contained in a k-dimensional affine subspace. A lower bound of $f_d(k + 2, k, T) = \Omega(T^\beta)$ yields immediately the upper bound vol $(n, d, k) = O(n^{d/\beta})$ on the minimum volume of a bounding box. In the following the focus is on the investigation of the growth of the function $f_d(l, k, T)$ rather than on vol (n, d, l). In Section 2 we investigate constructive lower bounds for $f_d(l, T) := f_d(l, 1, T)$, i.e., no l distinct points are on a line. We obtain $f_d(l, T) = \Theta(l \cdot T^{d-1})$ for $l \geq d + 1$. For $l \leq d$ we prove $f_d(l, T) = \Omega(T^{d(l-2)/(l-1)} \cdot poly(\log T))$. We also give a simple but useful construction for obtaining lower bounds on $f_D(l, T)$ from lower bounds on $f_d(l, T)$, $d < D$. In Section 3 we give new upper bounds on $f_d(k + 2, k, T)$ for integers $k \geq 1$, in particular $f_d(k + 2, k, T) = O(T^{2d/(k+2)})$ for k even. We also consider distributions of grid-points, where no l points are contained in a k-dimensional linear subspace and disprove a suggested order of the corresponding function $f_d^{lin}(l, k, T)$, see [4,5]. Moreover, in connection with a question of Füredi [13] we show for any finite set $S \subset \mathbb{R}^2$, which does not contain l collinear points, a lower bound on the largest size of a subset $S' \subseteq S$, where S' does not contain k collinear points, $3 \leq k < l$. All of our arguments for proving lower bounds are of a probabilistic nature, however, they easily can be made constructive in polynomial time by using derandomization arguments. In Section 4 we finish with some concluding remarks.

2   No l Collinear Points

For integers d, l, T with $d \geq 2$ and $3 \leq l \leq T$ let $f_d(l, T)$ denote the largest size of a subset S of grid-points in the d-dimensional $T \times \cdots \times T$-grid, such that no l distinct points of S are collinear. By monotonicity we have $f_d(l+1, T) \geq f_d(l, T)$. Well-known is the following upper bound on $f_d(l, T)$:

Proposition 1. *For integers d, l, T with $d \geq 2$ and $3 \leq l \leq T$ it is*

$$f_d(l, T) \leq (l - 1) \cdot T^{d-1}. \tag{1}$$

Proof. Let S be a subset of grid-points in the d-dimensional $T \times \cdots \times T$-grid, such that no l distinct points of S are collinear. Partition the set of grid-points in the $T \times \cdots \times T$-grid into T^{d-1} lines, where each line has the form $(a_1, \ldots, a_i, x, a_{i+2}, \ldots, a_d)$ for fixed $a_1, \ldots, a_i, a_{i+2}, \ldots, a_d \in \{0, \ldots, T-1\}$. Each line contains at most $(l-1)$ grid-points from S, hence $|S| \leq (l-1) \cdot T^{d-1}$. $\square$

Next we give lower bounds on $f_d(l, T)$ for arbitrary integers $l \geq 3$.

Theorem 1. *For fixed integers $d \geq 2$ there exists a constant $c > 0$ such that for all integers l, T with $3 \leq l \leq T$ it is*

$$f_d(l, T) \geq \begin{cases} c \cdot l \cdot T^{d-1} & \text{if} \quad l \geq d+1 \\ c \cdot T^{d\frac{l-2}{l-1}} & \text{if} \quad l \leq d. \end{cases}$$

The lower bound on $f_d(d+1, T)$ in Theorem 1 is by Brass and Knauer [4]. They observed for primes T and integers q that the set of all grid-points $(x_1, \ldots, x_d)$ with $x_1 + x_2^2 + \cdots + x_d^d \equiv q \mod T$ in the d-dimensional $T \times \cdots \times T$-grid contains at most d collinear points, thus $f_d(d+1, T) = \Omega(T^{d-1})$. Hence, by Proposition 1 we have $f_d(d+1, T) = \Theta(T^{d-1})$.

As mentioned in the introduction, Pór and Wood [18] obtained $f_3(3, T) = \Omega(T^2)$, which is bigger than the lower bound in Theorem 1, compare also the remarks following Lemma 1. However, Theorem 1 holds for all pairs (d, l) for fixed d, and the lower bounds match up to constant factors the upper bounds (1) for every $l \geq d+1$.

Before proving Theorem 1, we introduce some notation.

For integers $a_1, \ldots, a_d$, which are not all equal to 0, let $gcd\,(a_1, \ldots, a_d) > 0$ denote the *greatest common divisor* of $a_1, \ldots, a_d$.

Let $P = (p_1, \ldots, p_d)$ and $Q = (q_1, \ldots, q_d)$ be distinct grid-points in the d-dimensional $T \times \cdots \times T$-grid. Let PQ denote the segment between the grid-points P and Q, including the endpoints P and Q. The segment PQ contains exactly $(gcd\,(p_1 - q_1, \ldots, p_d - q_d) + 1)$ grid-points.

A *hypergraph* $\mathcal{G}$ is given by a pair $(V, \mathcal{E})$, where V is its vertex-set and $\mathcal{E} \subseteq \mathcal{P}(V)$ is its edge-set. A subset $I \subseteq V$ of the vertex-set V is called *independent*, if I does not contain any edges from $\mathcal{E}$, i.e., $E \not\subseteq I$ for each edge $E \in \mathcal{E}$. The largest size of an independent set in $\mathcal{G}$ is the *independence number* $\alpha(\mathcal{G})$. A 2-cycle in $\mathcal{G} = (V, \mathcal{E})$ is a pair $\{E, E'\}$ of distinct edges $E, E' \in \mathcal{E}$ with $|E \cap E'| \geq 2$. A 2-cycle $\{E, E'\}$ is called $(2, j)$-cycle if $|E \cap E'| = j$. A hypergraph $\mathcal{G}$ without any 2-cycles is called *linear*. A hypergraph $\mathcal{G} = (V, \mathcal{E})$ is called l-*uniform*, if each edge $E \in \mathcal{E}$ contains exactly l vertices.

In our arguments we use Túran's theorem for hypergraphs, see [19]:

Theorem 2. *Let $\mathcal{G} = (V, \mathcal{E}_l)$ be an l-uniform hypergraph on $|V| = N$ vertices with average-degree $t^{l-1} := l \cdot |\mathcal{E}_l|/N \geq 1$.*

Then, the independence number $\alpha(\mathcal{G})$ of $\mathcal{G}$ fulfills:

$$\alpha(\mathcal{G}) \geq \frac{l-1}{l} \cdot \frac{N}{t}. \tag{2}$$

An independent set $I \subseteq V$ with $|I| \geq ((l-1)/l) \cdot (N/t)$ can be found in time $O(N + |\mathcal{E}_l|)$.

Next we prove Theorem 1:

Proof. We only have to consider the case $l \neq d+1$. Form an l-uniform hypergraph $\mathcal{G} = (V, \mathcal{E}_l)$ with vertex-set V, which consists of all T^d grid-points in the d-dimensional $T \times \cdots \times T$-grid. For distinct grid-points $P_1, \ldots, P_l$ let $\{P_1, \ldots, P_l\} \in \mathcal{E}_l$ be an edge iff $P_1, \ldots, P_l$ are collinear. We want to find a large independent set $I \subseteq V$ in $\mathcal{G}$, as I yields a subset of grid-points, where no l distinct grid-points are on a line.

We give an upper bound on the size $|\mathcal{E}_l|$ of the edge-set. Let $P_1, \ldots, P_l$ be distinct, collinear grid-points in the $T \times \cdots \times T$-grid, where w.o.l.g. $P_2, \ldots, P_{l-1}$ are contained in the segment $P_1 P_l$. There are T^d choices for the grid-point $P_1 = (p_{1,1}, \ldots, p_{1,d})$. Any d-tuple $(s_1, \ldots, s_d) \in \{-T+1, -T+2, \ldots, T-1\}^d$ fixes at most one grid-point $P_l = (p_{1,1} + s_1, \ldots, p_{1,d} + s_d)$ in the $T \times \cdots \times T$-grid. By symmetry, which we take into account by a factor of 2^d, we may assume that $s_1, \ldots, s_d \geq 0$. Given the grid-points P_1 and P_l with $P_l - P_1 = (s_1, \ldots, s_d) \neq (0, \ldots, 0)$, on the segment $P_1 P_l$ there are less than $\binom{gcd(s_1, \ldots, s_d)}{l-2}$ choices for the grid-points $P_2, \ldots, P_{l-1}$. With the inequality $\binom{N}{k} \leq ((e \cdot N)/k)^k$ we obtain

$$|\mathcal{E}_l| \leq 2^d \cdot T^d \cdot \sum_{s_1=0}^{T-1} \cdots \sum_{s_d=0}^{T-1} \binom{gcd(s_1, \ldots, s_d)}{l-2}$$

$$\leq 2^d \cdot T^d \cdot \sum_{s_1=0}^{T-1} \cdots \sum_{s_d=0}^{T-1} \left(\frac{e \cdot gcd(s_1, \ldots, s_d)}{l-2} \right)^{l-2}. \qquad (3)$$

For a given divisor $g \in \{1, \ldots, T-1\}$ there are $\lceil T/g \rceil \leq 2 \cdot T/g$ integers $x \in \{0, \ldots, T-1\}$ which are divisible by g, hence (3) becomes

$$|\mathcal{E}_l| \leq (2 \cdot T)^d \cdot \sum_{g=1}^{T} \left(\frac{2 \cdot T}{g} \right)^d \cdot \left(\frac{e \cdot g}{l-2} \right)^{l-2} \leq (2 \cdot T)^{2d} \cdot \left(\frac{9}{l} \right)^{l-2} \cdot \sum_{g=1}^{T} g^{l-d-2} \qquad (4)$$

The sum $\sum_{g=1}^{T} g^{l-d-2}$ is $O(T^{l-d-1}/l)$ for $l \geq d+2$, and $O(\log T)$ for $l = d+1$, and $O(1)$ for $l \leq d$. Thus, by (4) for fixed $d \geq 2$ for a constant $c > 0$ we infer

$$|\mathcal{E}_l| \leq \begin{cases} c \cdot 9^{l-1} \cdot \dfrac{T^{l+d-1}}{l^{l-1}} & \text{if} \quad l \geq d+2 \\ c \cdot T^{2d} & \text{if} \quad l \leq d. \end{cases}$$

Hence, the average-degree $t^{l-1} = l \cdot |\mathcal{E}_l|/T^d$ of $\mathcal{G}$ fulfills for a constant $c' > 0$:

$$t \leq \begin{cases} c' \cdot \dfrac{T}{l} & \text{if} \quad l \geq d+2 \\ c' \cdot T^{\frac{d}{l-1}} & \text{if} \quad l \leq d. \end{cases}$$

By Theorem 2 we can find in time $O(T^d + |\mathcal{E}_l|)$ an independent set $I \subseteq V$ with

$$|I| \geq \begin{cases} \frac{1}{2 \cdot c'} \cdot l \cdot T^{d-1} & \text{if} \quad l \geq d+2 \\[2ex] \frac{1}{2 \cdot c'} \cdot T^{d\frac{l-2}{l-1}} & \text{if} \quad l \leq d. \end{cases}$$

The grid-points, which correspond to the vertices of the independent set I satisfy, that no l distinct points are collinear. $\qquad\square$

To improve the results from Theorem 1 for fixed $l \leq d$ by a logarithmic factor, we use the following result of Ajtai, Komlos, Pintz, Spencer and Szemeredi [1] in a version arising from work in [3] and [8].

Theorem 3. *Let $l \geq 3$ be a fixed integer. Let $\mathcal{G} = (V, \mathcal{E}_l)$ be an l-uniform, linear hypergraph on $|V| = N$ vertices with average-degree $t^{l-1} = l \cdot |\mathcal{E}_l|/N$.*
Then the independence number $\alpha(\mathcal{G})$ of $\mathcal{G}$ satisfies for a constant $C_l > 0$:

$$\alpha(\mathcal{G}) \geq C_l \cdot \frac{N}{t} \cdot (\log t)^{\frac{1}{l-1}}. \tag{5}$$

An independent set $I \subseteq V$ with $|I| = \Omega((N/t) \cdot (\log t)^{1/(l-1)})$ can be found in polynomial time.

Theorem 4. *Let $d, l \geq 2$ be fixed integers with $l \leq d$. Then there exists a constant $c > 0$ such that for all integers $T \geq 1$:*

$$f_d(l, T) \geq c \cdot T^{d\frac{l-2}{l-1}} \cdot (\log T)^{\frac{1}{l-1}}. \tag{6}$$

Proof. We form a non-uniform hypergraph $\mathcal{G} = (V, \mathcal{E}_l \cup \mathcal{E}_{l+1})$. The vertex-set consists of all T^d grid-point from the d-dimensional $T \times \cdots \times T$-grid and for $m = l, l+1$ and distinct grid-points $P_1, \ldots, P_m$ it is $\{P_1, \ldots, P_m\} \in \mathcal{E}_m$ iff $P_1, \ldots, P_m$ are collinear. By the proof of Theorem 1 (see remarks following inequality (4)) for $l \leq d$ we have for constants $c_1, c_2 > 0$ that

$$|\mathcal{E}_l| \leq c_1 \cdot T^{2d} \quad \text{and} \quad |\mathcal{E}_{l+1}| \leq c_2 \cdot T^{2d} \cdot \log T. \tag{7}$$

For $\varepsilon := d/l^2$ we select with probability $p := T^\varepsilon/T^{d/(l-1)}$ uniformly at random and independently of each other grid-points from the d-dimensional $T \times \cdots \times T$-grid. Let V^* be the random set of chosen grid-points, let $\mathcal{E}_m^* := \mathcal{E}_m \cap [V^*]^m$, and let $E(|V^*|)$, $E(|\mathcal{E}_m^*|)$, $m = l, l+1$, be their expected sizes. We infer with (7):

$$E(|V^*|) = p \cdot T^d = T^{\varepsilon + d(l-2)/(l-1)}$$
$$E(|\mathcal{E}_l^*|) = p^l \cdot |\mathcal{E}_l| \leq p^l \cdot c_1 \cdot T^{2d} \leq c_1 \cdot T^{\varepsilon l + d(l-2)/(l-1)}$$
$$E(|\mathcal{E}_{l+1}^*|) = p^{l+1} \cdot |\mathcal{E}_{l+1}| \leq p^{l+1} \cdot c_2 \cdot T^{2d} \cdot \log T \leq c_2 \cdot T^{\varepsilon(l+1) + d(l-3)/(l-1)} \cdot \log T.$$

By Markov's and Chernoff's inequalities (this argument can be easily derandomized in time polynomial in T by using the method of conditional probabilities) there exists a subset $V^* \subseteq V$ of grid-points such that

$$|V^*| = (1 - o(1)) \cdot T^{\varepsilon + d(l-2)/(l-1)} \tag{8}$$
$$|\mathcal{E}_l^*| \leq 3 \cdot c_1 \cdot T^{\varepsilon l + d(l-2)/(l-1)} \tag{9}$$
$$|\mathcal{E}_{l+1}^*| \leq 3 \cdot c_2 \cdot T^{\varepsilon(l+1) + d(l-3)/(l-1)} \cdot \log T. \tag{10}$$

By (8) and (10) it is $T^{\varepsilon(l+1)+d(l-3)/(l-1)} \cdot \log T = o(T^{\varepsilon+d(l-2)/(l-1)})$ for $\varepsilon = d/l^2$, thus

$$|\mathcal{E}_{l+1}^*| = o(|V^*|). \tag{11}$$

Let $\mathcal{G}^* = (V^*, \mathcal{E}_l^* \cup \mathcal{E}_{l+1}^*)$ be the induced subhypergraph of $\mathcal{G}$ on V^*. We delete one vertex from each edge $E \in \mathcal{E}_{l+1}^*$. For distinct edges $E, E' \in \mathcal{E}$ with $|E \cap E'| \geq 2$, all grid-points in $E \cup E'$ are collinear, as two distinct grid-points determine a unique line. Thus, we have destroyed all 2-cycles in $\mathcal{G}^*$. Let V^{**} be the set of remaining vertices, where $|V^{**}| = (1-o(1)) \cdot |V^*| \geq |V^*|/2$ by (11). The induced, uniform subhypergraph $\mathcal{G}^{**} = (V^{**}, \mathcal{E}_l^{**})$ of $\mathcal{G}$ on V^{**} with $\mathcal{E}_l^{**} := \mathcal{E}_l^{**} \cap [V^{**}]^l$ is linear. We infer with (8) and (9):

$$|V^{**}| \geq (1/2) \cdot T^{\varepsilon+d(l-2)/(l-1)} \quad \text{and} \quad |\mathcal{E}_l^{**}| \leq 3 \cdot c_1 \cdot T^{\varepsilon l+d(l-2)/(l-1)}, \tag{12}$$

hence the average-degree t^{l-1} of $\mathcal{G}^{**}$ satisfies:

$$t^{l-1} = \frac{l \cdot |\mathcal{E}_l^{**}|}{|V^{**}|} \leq 6 \cdot c_1 \cdot l \cdot T^{\varepsilon(l-1)} := t_0^{l-1}. \tag{13}$$

Since $\mathcal{G}^{**}$ is linear, we can apply Thereom 3 and we infer with (12) and (13) for the independence number $\alpha(\mathcal{G})$ for constants $C_l, C_l' > 0$:

$$\alpha(\mathcal{G}) \geq \alpha(\mathcal{G}^{**}) \geq C_l \cdot \frac{|V^{**}|}{t} \cdot (\log t)^{1/(l-1)} \geq C_l \cdot \frac{|V^{**}|}{t_0} \cdot (\log t_0)^{1/(l-1)} \geq$$

$$\geq C_l \cdot \frac{(1/2) \cdot T^{\varepsilon+d(l-2)/(l-1)}}{(6 \cdot c_1 \cdot l)^{1/(l-1)} \cdot T^\varepsilon} \cdot \left(\log \left((6 \cdot c_1 \cdot l)^{1/(l-1)} \cdot T^\varepsilon\right)\right)^{1/(l-1)}$$

$$\geq C_l' \cdot T^{d(l-2)/(l-1)} \cdot (\log T)^{1/(l-1)},$$

and by Theorem 3 such an independent set can be constructed in time polynomial in T. This shows $f_d(l, T) = \Omega(T^{d(l-2)/(l-1)} \cdot (\log T)^{1/(l-1)})$. □

Some of the few known lower bounds on $f_d(l, T)$ have been shown for small dimension $d \geq 2$. Having a good construction for small d might also give a good construction for higher dimensions as the following observation shows.

Lemma 1. *For integers $d, g, h, l \geq 2$ it is*

$$f_{g+h}(l, T) \geq f_g(l, T) \cdot f_h(l, T).$$

By Lemma 1 we infer by induction that $f_{gd}(l, T) \geq (f_d(l, T))^g$ for integers $g \geq 1$.

Proof. Let S and S' be subsets of a g- and h-dimensional $T \times \cdots \times T$-grid with $|S| = f_g(l, T)$ and $|S'| = f_h(l, T)$, respectively, where S and S' do not contain any l distinct points on a line. Let $S^* := S \times S'$ be the Cartesian product of S and S' in the $(g+h)$-dimensional $T \times \cdots \times T$-grid, i.e., $|S^*| = f_g(l, T) \cdot f_h(l, T)$. Assume for contradiction, that the set S^* contains l distinct collinear grid-points, say $s_i = s + \lambda_i \cdot v$, $i = 1, \ldots, l$, with pairwise distinct $\lambda_1, \ldots, \lambda_l$, where addition is component-wise. Restricting to the first g coordinates, i.e., to the points $s_i(\leq g) = s(\leq g) + \lambda_i \cdot v(\leq g) \in S$ in dimension g, by choice of the set S these points cannot be all distinct, and for some $i \neq j$ we have $s_i(\leq g) = s_j(\leq g)$, thus $(\lambda_i - \lambda_j) \cdot v(\leq g) = (0, \ldots, 0)$, and hence $v(\leq g) = (0, \ldots, 0)$. Restricting to the last h coordinates, i.e., to the points $s_i(\geq g+1) = s(\geq g+1) + \lambda_i \cdot v(\geq g+1) \in S'$ in dimension h yields $v(\geq g+1) = (0, \ldots, 0)$, hence $v = v(\leq g) \times v(\geq g+1) = (0, \ldots, 0)$, which is a contradiction. $\quad\square$

By using $f_3(3, T) = \Omega(T^2)$ from [18], and $f_2(3, T) = \Omega(T)$ we infer with Lemma 1 that that $f_d(3, T) = \Omega(T^{\lceil (2d-2)/3 \rceil})$, which is bigger than the lower bound $f_d(3, T) = \Omega(T^{d/2})$ from Theorem 1. By Lemma 1 we infer with Theorem 1 for $l = d + 1$ that $f_{gd}(d+1, T) = \Omega(T^{g(d-1)})$, which yields asymptotically the same lower bound as in Theorem 1.

However, for $l \leq d$ Theorem 4 and Lemma 1 gives $f_{gd}(l, T) = \Omega(T^{gd(l-2)/(l-1)} \cdot (\log T)^{g/(l-1)})$, which is better than the direct application of Theorem 4, and in general we have:

Corollary 1. *For fixed integers $d, l \geq 3$ with $l \leq d$ and any integers $T \geq 1$ it is*

$$f_d(l, T) = \Omega(T^{d(l-2)/(l-1)} \cdot (\log T)^{\lceil d/l \rceil/(l-1)}).$$

3 No $(k+2)$ Points in Affine k-Space or Linear $(k+1)$-Space

Here we consider higher dimensional versions of Theorem 1. For $l \geq k+2$, let $f_d(l, k, T)$ denote the maximum number of grid-points in the d-dimensional $T \times \cdots \times T$-grid, such that no l distinct grid-points are contained in a k-dimensional affine subspace of $\mathbb{R}^d$. We have by monotonicity $f_d(l+1, k, T) \geq f_d(l, k, T)$.

The d-dimensional $T \times \cdots \times T$-grid can be partitioned into T^{d-k} many k-dimensional affine spaces, namely for fixed $a_1, \ldots, a_{d-k} \in \{0, \ldots, T-1\}$, into the k-dimensional affine spaces given by all points $(a_1, \ldots, a_{d-k}, x_{d-k+1}, \ldots, x_d)$, hence it follows

$$f_d(l, k, T) \leq (l-1) \cdot T^{d-k}. \tag{14}$$

For $k = d - 1$ and fixed $l \geq d + 1$ the upper bound (14) is asymptotically sharp, namely for primes T the set of points $(x \bmod T, x^2 \bmod T, \ldots, x^d \bmod T)$, $x = 0, \ldots, T-1$, on the modular moment-curve meets every $(d-1)$-dimensional affine space in at most d points, compare [5,20], thus

$$f_d(l, d-1, T) = \Theta(T). \tag{15}$$

We can improve on the upper bound (14) for certain pairs (l, k) as follows.

Lemma 2. *Let $d, k \geq 1$ with $k \leq d-1$ be fixed integers. Then, for some constant $c > 0$ it is:*

$$f_d(k+2, k, T) \leq c \cdot T^{\frac{d}{\lceil (k+1)/2 \rceil}} \tag{16}$$

For even $k \geq 2$, the upper bound (16) on $f_d(k+2, k, T)$ is smaller than (14) for $k < d-2$, and the exponents of T are equal for $k = d-2$ in both bounds, and (14) is smaller than (16) only for $k = d-1$. For odd $k \geq 1$, (16) is smaller than (14) for $(d-1)/2 - \sqrt{(d-1)^2/4 - d} \leq k \leq (d-1)/2 + \sqrt{(d-1)^2/4 - d}$.

Proof. Let $k \geq 2$ be even and set $g := k/2$. Let S be a subset of the d-dimensional $T \times \cdots \times T$-grid, where no $(k+2)$ distinct points in S are contained in a k-dimensional affine subspace, w.l.o.g. $|S| \geq k+2$. Consider the set S_{g+1} of all $(g+1)$-term sums of distinct elements from S with addition component-wise:

$$S_{g+1} := \{s_1 + \cdots + s_{g+1} \mid s_1, \ldots, s_{g+1} \in S \text{ are pairwise distinct}\}.$$

We claim that for distinct grid-points $s_1, \ldots, s_{g+1} \in S$ and distinct $t_1, \ldots, t_{g+1} \in S$ with $\{s_1, \ldots, s_{g+1}\} \neq \{t_1, \ldots, t_{g+1}\}$ it is

$$s_1 + \cdots + s_{g+1} \neq t_1 + \cdots + t_{g+1}. \tag{17}$$

Otherwise, we have $s_1 + \cdots + s_{g+1} = t_1 + \cdots + t_{g+1}$ for some distinct grid-points $s_1, \ldots, s_{g+1} \in S$ and distinct $t_1, \ldots, t_{g+1} \in S$. Assume that for some integer $j \geq 1$ it is $s_i = t_i$, $i = 0, \ldots, j-1$, and that $s_j, \ldots, s_{g+1}, t_j, \ldots, t_{g+1}$ are pairwise distinct grid-points. Then, it is $s_j + \cdots + s_{g+1} = t_j + \cdots + t_{g+1}$, hence we have found $2 \cdot (g+2-j) = k+4-2 \cdot j$ distinct grid-points in S, which are contained in a $(k+2-2 \cdot j)$-dimensional affine space. Adding $2 \cdot j$ further distinct grid-points from S to $s_j, \ldots, s_{g+1}, t_j, \ldots, t_{g+1}$ yields $(k+2)$ grid-points in the set S, which are contained in a k-dimensional affine space, a contradiction.

By (17) we infer $|S_{g+1}| = \binom{|S|}{g+1}$, and all points in S_{g+1} are contained in a $((g+1) \cdot T) \times \cdots \times ((g+1) \cdot T)$-grid, thus we obtain

$$\binom{|S|}{g+1} = |S_{g+1}| \leq ((g+1) \cdot T)^d,$$

and with $k = 2 \cdot g$ for a constant $c > 0$ we have $|S_{g+1}| \leq c \cdot T^{2d/(k+2)}$, hence $f_d(k+2, k, T) \leq c \cdot T^{2d/(k+2)}$.

Let $k \geq 1$ be odd. If a subset S of the d-dimensional $T \times \cdots \times T$-grid does not contain $(k+2)$ distinct grid-points, which are contained in a k-dimensional affine subspace, then S also does not contain $(k+1)$ distinct points, which are contained in a $(k-1)$-dimensional affine subspace. With the already proved upper bound for even values we infer for $k \geq 1$ odd for a constant $c > 0$ that $f_d(k+2, k, T) \leq f_d(k+1, k-1, T) \leq c \cdot T^{2d/(k+1)}$. □

Concerning lower bounds, Brass and Knauer proved in [4] for fixed integers $d, k, l \geq 2$ by a random selection of grid-points from the $T \times \cdots \times T$-grid that

$$f_d(l, k, T) = \Omega(T^{d-k-(d(k+1)/(l-1))}). \tag{18}$$

Then (18) guarantees only $f_d(l, k, T) = \Omega(T)$ for $l - 1 \geq d(k+1)/(d-k-1)$ for $k \leq d-2$. One can improve (18) a little by using a (slightly) different argument:

Lemma 3. *For fixed integers $d, k, l \geq 1$ with $k \leq d-1$ and $l \geq k+2$ and integers $T \geq 1$ it is:*

$$f_d(l, k, T) = \Omega(T^{d-k-(k(d+1)/(l-1))}).$$ (19)

Notice that (19) is bigger than (18) for $k < d$.

Proof. Form a l-uniform hypergraph $\mathcal{G} = (V, \mathcal{E}_l)$ with vertex-set V consisting of all T^d grid-points from the d-dimensional $T \times \cdots \times T$-grid. For grid-points $P_1, \ldots, P_l \in V$ let $\{P_1, \ldots, P_l\} \in \mathcal{E}_l$ iff $P_1, \ldots, P_l$ are contained in a k-dimensional affine subspace. We want to get a large independent set in $\mathcal{G}$. Each k-dimensional affine subspace contains at most T^k grid-points from the d-dimensional $T \times \cdots \times T$-grid. The number of k-dimensional affine subspaces, which intersect the d-dimensional $T \times \cdots \times T$-grid in at least $(k+1)$ grid-points, is at most $\binom{T^d}{k+1}$. We infer for a constant $c > 0$

$$|\mathcal{E}| \leq \binom{T^k}{l} \cdot \binom{T^d}{k+1} \leq c \cdot T^{kl+d(k+1)},$$

hence the average-degree t^{l-1} of $\mathcal{G}$ fulfills for some constant $c' > 0$:

$$t^{l-1} = \frac{l \cdot |\mathcal{E}_l|}{|V|} \leq \frac{l \cdot c \cdot T^{kl+d(k+1)}}{T^d} \leq c' \cdot T^{k(d+l)}.$$ (20)

By Theorem 2 and (20) we can find in time polynomial in T an independent set $I \subseteq V$ in $\mathcal{G}$, such that for a constant $c'' > 0$ it is

$$|I| \geq \frac{l-1}{l} \cdot \frac{T^d}{c'^{1/(l-1)} \cdot T^{k(d+l)/(l-1)}} \geq c'' \cdot T^{d-k-(k(d+1)/(l-1))}. \qquad \square$$

Some more appropiate representations of low-dimensional subspaces would possibly give a better counting of the bad objects than that in the proof of Lemma 3, if one wants to apply a random selection process as given above.

Next we consider *linear* subspaces. Let $f_d^{lin}(l, k, T)$ denote the maximum number of grid-points in the d-dimensional $T \times \cdots \times T$-grid, such that no l grid-points are contained in a k-dimensional *linear* subspace. From number theory it is known [5] that for fixed $d \geq 2$ it is $f_d^{lin}(2, 1, T) = \Theta(T^d)$. Bárány, Harcos, Pach and Tardos proved in [2] that $f_d^{lin}(d, d-1, T) = \Theta(T^{d/(d-1)})$ for fixed $d \geq 2$. Based on this, Brass and Krauer [4] conjectured (stated as a problem in [5]) that

$$f_d^{lin}(k+1, k, T) = \Theta(T^{\frac{(d-k)d}{d-1}}). \quad (?)$$ (21)

However, one can show the following:

Lemma 4. *For fixed integers d, k with $1 \leq k \leq d-1$ there exists a constant $c > 0$, such that for every integer $T \geq 1$ it is*

$$f_d^{lin}(k+1, k, T) \leq c \cdot T^{\frac{d}{\lceil k/2 \rceil}}.$$ (22)

For odd k the upper bound (22) is asymptotically smaller than the conjectured growth of $f_d^{lin}(k+1, k, T)$ in (21) for $1 < k < d-2$ with equality for $k = d-2$. Similarly, for even k the upper bound (22) is smaller than in (21) for $d/2 - \sqrt{d^2/4 - 2d + 2} < k < d/2 + \sqrt{d^2/4 - 2d + 2}$. Hence, (21) does not hold for several values of k, d.

Proof. The proof is similar to that of Lemma 2, therefore we only sketch it. Let $k \geq 1$ be an odd integer and set $g := (k+1)/2$. Let S be a subset of the d-dimensional $T \times \cdots \times T$-grid, where no $(k+1)$ distinct points from S are contained in a k-dimensional linear subspace, w.l.o.g. $|S| \geq k + 1$. Let

$$S_g := \{s_1 + \cdots + s_g \mid s_1, \ldots, s_g \in S \text{ are pairwise distinct}\}.$$

As in the proof of Lemma 2, for distinct grid-points $s_1, \ldots, s_g \in S$ and distinct $t_1, \ldots, t_g \in S$ with $\{s_1, \ldots, s_g\} \neq \{t_1, \ldots, t_g\}$ it is $s_1 + \cdots + s_g \neq t_1 + \cdots + t_g$, as otherwise we can find $(k+1)$ distinct grid-points in S, which are contained in a k-dimensional linear subspace, a contradiction, hence $|S_g| = \binom{|S|}{g} \leq (g \cdot T)^d$, and we have $f_d^{lin}(k+1, k, T) = O(T^{2d/(k+1)})$ for odd $k \geq 1$.

For even $k \geq 2$ we infer as in the proof of Lemma 2 that for a constant $c > 0$ it is $f_d^{lin}(k+1, k, T) \leq f_d^{lin}(k, k-1, T) \leq c \cdot T^{2d/k}$.   □

Related here is a problem, which has been investigated by Füredi in [13]. He considered finite sets $S \subset \mathbb{R}^2$ of points, which for fixed $l \geq 3$ do not contain l collinear points. He asks for the largest size $\alpha_k(S)$ of a subset of S, which does not contain any k points on a line, where $k < l$. He proved in [13] that $\alpha_k(S) = \Omega(|S|^{(k-2)/(k-1)})$. As asked for in [13], this lower bound on $\alpha_k(S)$ can be improved to $\alpha_k(S) = \Omega(|S|^{(k-2)/(k-1)} \cdot (\log |S|)^{1/(k-1)})$ as the following shows.

Theorem 5. *Let $d, k, l \geq 2$ be fixed integers with $3 \leq k < l$. Let $S \subset \mathbb{R}^d$ be a finite set with $|S| = N$, where S does not contain l collinear points.*

Then, one can find in time polynomial in N a subset $S' \subseteq S$, such that S' does not contain k collinear points, and

$$|S'| = \Omega(N^{\frac{k-2}{k-1}} \cdot (\log N)^{\frac{1}{k-1}}). \tag{23}$$

Proof. We construct a hypergraph $\mathcal{G} = (S, \mathcal{E}_k)$ with vertex-set S. For any k distinct points $P_1, \ldots, P_k \in S$ let $\{P_1, \ldots, P_k\} \in \mathcal{E}_k$ iff $P_1, \ldots, P_k$ are collinear. We want to find a large independent set in $\mathcal{G}$. The set S with $|S| = N$ generates at most $\binom{N}{2}$ lines, as two distinct points determine uniquely a line. Each line contains at most $(l-1)$ points from S, hence on each line the number of k-element sets of collinear points is at most $\binom{l-1}{k}$, and we infer for a constant $c > 0$:

$$|\mathcal{E}_k| \leq \binom{N}{2} \cdot \binom{l-1}{k} \leq c \cdot N^2. \tag{24}$$

Next we give upper bounds on the numbers $s_{2,j}(\mathcal{G})$ of $(2, j)$-cycles in $\mathcal{G}$, $j = 2, \ldots, k-1$. For a $(2, j)$-cycle $\{E, E'\}$ in $\mathcal{G}$ all points in $E \cup E'$ are collinear.

Thus we have $s_{2,j}(\mathcal{G}) = 0$ for $j \leq 2 \cdot k - l$, as the set S does not contain l collinear points. For $j > 2 \cdot k - l$ we obtain as in (24) for some constant $c' > 0$:

$$s_{2,j}(\mathcal{G}) \leq \binom{N}{2} \cdot \binom{l-1}{2 \cdot k - j} \leq c' \cdot N^2. \tag{25}$$

For $\varepsilon := 1/k^2$, we select uniformly at random and independently of each other points from S with probability $p := N^\varepsilon/N^{1/(k-1)}$. Let $S^* \subseteq S$ be the random set of chosen points and let $\mathcal{G}^* = (S^*, \mathcal{E}_k^*)$ with $\mathcal{E}_k^* := \mathcal{E}_k \cap [S^*]^k$ be the induced subhypergraph of $\mathcal{G}$ on S^*. The expected numbers satisfy $E[|S^*|] = p \cdot |S| = p \cdot N$, and $E[|\mathcal{E}_k^*|] = p^k \cdot |\mathcal{E}_k|$ and $E[s_{2,j}(\mathcal{G}^*)] = p^{2k-j} \cdot s_{2,j}(\mathcal{G})$, $j = 2, \ldots, k-1$. By Markov's and Chernoff's inequality and with (24) and (25) there exists an induced subhypergraph $\mathcal{G}^* = (S^*, \mathcal{E}_k^*)$ of $\mathcal{G}$ such that

$$|S^*| \geq p \cdot N/2 = N^{\frac{k-2}{k-1}+\varepsilon}/2 \tag{26}$$

$$|\mathcal{E}_k^*| \leq 3 \cdot p^k \cdot |\mathcal{E}_k| \leq 3 \cdot c \cdot N^{\frac{k-2}{k-1}+\varepsilon k} \tag{27}$$

$$s_{2,j}(\mathcal{G}^*) \leq 3 \cdot p^{2k-j} \cdot s_{2,j}(\mathcal{G}) \leq 3 \cdot c' \cdot N^{\frac{j-2}{k-1}+\varepsilon(2k-j)}. \tag{28}$$

By (26) and (28) it is $N^{\frac{j-2}{k-1}+\varepsilon(2k-j)} = o(N^{\frac{k-2}{k-1}+\varepsilon})$, $j = 2, \ldots, k-1$, for $\varepsilon \leq 1/(2 \cdot k^2)$, thus

$$s_{2,j}(\mathcal{G}^*) = o(|S^*|). \tag{29}$$

Discard one vertex from each $(2,j)$-cycle in $\mathcal{G}^*$, $j = 2, \ldots, k-1$. By (29) we obtain a subset $S^{**} \subseteq S^*$ with $|S^{**}| = (1 - o(1)) \cdot |S^*| \geq |S^*|/2$. The induced subhypergraph $\mathcal{G}^{**} = (S^{**}, \mathcal{E}_k^{**})$ of $\mathcal{G}^*$ on S^{**} with $\mathcal{E}_k^{**} := \mathcal{E}_k^* \cap [S^{**}]^k$ is linear. With $|\mathcal{E}_k^{**}| \leq |\mathcal{E}_k^*|$ and (27) we infer for the average-degree $(t^{**})^{k-1}$ of $\mathcal{G}^{**}$:

$$(t^{**})^{k-1} := \frac{k \cdot |\mathcal{E}_k^{**}|}{|V^{**}|} \leq 6 \cdot k \cdot c \cdot \frac{N^{\frac{k-2}{k-1}+\varepsilon k}}{N^{\frac{k-2}{k-1}+\varepsilon}} = 6 \cdot k \cdot c \cdot N^{\varepsilon(k-1)} =: (t_0^{**})^{k-1} \tag{30}$$

By Theorem 3 with (30) one can find in time polynomial in N an independent set $I \subseteq S^{**}$, such that, as $\varepsilon > 0$ is fixed, for constants $C_k, C_k' > 0$ we have

$$|I| \geq C_k \cdot \frac{|S^{**}|}{t^{**}} \cdot (\log t^{**})^{\frac{1}{k-1}} \geq C_k \cdot \frac{|S^{**}|}{t_0^{**}} \cdot (\log t_0^{**})^{\frac{1}{k-1}} \geq$$

$$\geq C_k \frac{(1/2) \cdot N^{\frac{k-2}{k-1}+\varepsilon}}{(6 \cdot k \cdot c)^{\frac{1}{k-1}} \cdot N^\varepsilon} \cdot \left(\log \left((6 \cdot k \cdot c)^{\frac{1}{k-1}} \cdot N^\varepsilon \right) \right)^{\frac{1}{k-1}}$$

$$\geq C_k' \cdot N^{\frac{k-2}{k-1}} \cdot (\log N)^{\frac{1}{k-1}}.$$

The set I does not contain k distinct collinear points. $\square$

4   Concluding Remarks

We have considered extensions of the no-three-in-line-problem, motivated by questions on the embeddings of graphs in $\mathbb{Z}^d$. Very desirable would be constructive lower, and upper bounds on the function $f_d(l, k, T)$, as already for the case

$k = 2$ and arbitrary fixed d the known lower bounds are quite weak. Towards this, a better counting of the incidences of given sets of grid-points and affine or linear subspaces seems to be necessary.

References

1. Ajtai, M., Komlós, J., Pintz, J., Spencer, J., Szemerédi, E.: Extremal Uncrowded Hypergraphs. J. Comb. Theory Ser. A 32, 321–335 (1982)
2. Bárány, I., Harcos, G., Pach, J., Tardos, G.: Covering Lattice Points by Subspaces. Per. Math. Hung. 43, 93–103 (2001)
3. Bertram–Kretzberg, C., Lefmann, H.: The Algorithmic Aspects of Uncrowded Hypergraphs. SIAM J. on Comput. 29, 201–230 (1999)
4. Brass, P., Knauer, C.: On Counting Point-Hyperplane Incidences. Comp. Geo. 25, 13–20 (2003)
5. Brass, P., Moser, W., Pach, J.: Research Problems in Discrete Geometry, pp. 417–421. Springer, Heidelberg (2005)
6. Cohen, R.-F., Eades, P., Lin, T., Ruskey, F.: Three-Dimensional Graph Drawing. Algorithmica 17, 199–208 (1996)
7. Dudeney, H.E.: Amusements in Mathematics, Nelson, London, pp. 94–222 (1917)
8. Duke, R.A., Lefmann, H., Rödl, V.: On Uncrowded Hypergraphs. Rand. Struct. & Algorithms 6, 209–212 (1995)
9. Erdős, P., Graham, R.L., Rusza, I.,, H.: Bounds for Arrays of Dots with Distinct Slopes or Lengths. Combinatorica 12, 39–44 (1992)
10. Erdös, P., Roth, K.-F.: On a Problem of Heilbronn. J. London Math. Soc. 26, 198–204 (1951)
11. Flammenkamp, A.: Progress in the No-three-in-a-Line-Problem. J. Comb. Theory, Ser. A 60, 305–311 (1992)
12. Flammenkamp, A.: Progress in the No-three-in-a-Line-Problem. II. J. Comb. Theory, Ser. A 81, 108–113 (1998)
13. Füredi, Z.: Maximal Independent Subsets in Steiner Systems and in Planar Sets. SIAM J. Disc. Math. 4, 196–199 (1991)
14. Golomb, S.W.: Construction of Signals with Favourable Correlation Properties. Surveys in Combinatorics, London Mathematical Society LNS 166, 1–39 (1991)
15. Golomb, S.W., Taylor, H.: Two-dimensional Synchronization Patterns for Minimum Ambiguity. IEEE Transactions Information Theory IT-28, 600–604 (1982)
16. Guy, R.: Unsolved Problems in Number Theory, 2nd edn. Springer, Heidelberg (1994)
17. Pach, J., Thiele, T., Tóth, G.: Three Dimensional Grid Drawings of Graphs. In: Chazelle, B., Goodman, J.E., Pollack, R. (eds.) Advances in Discrete and Computational Geometry. Contemporary Math. 223, pp. 251–255. AMS (1999)
18. Pór, A., Wood, D.W.: No-Three-in-Line-in-3D. Algorithmica 47, 481–488 (2007)
19. Spencer, J.: Turán's Theorem for k-Graphs. Disc. Math. 2, 183–186 (1972)
20. Thiele, T.: Geometric Selection Problems and Hypergraphs, PhD thesis, FU Berlin (1995)

The Secret Santa Problem

Leo Liberti[1] and Franco Raimondi[2]

[1] LIX, École Polytechnique, F-91128 Palaiseau, France
`liberti@lix.polytechnique.fr`
[2] Dept. of Computer Science, University College London, UK
`f.raimondi@cs.ucl.ac.uk`

Abstract. Consider a digraph where the vertices represent people and an arc (i, j) represents the possibility of i giving a gift to j. The basic question we pose is whether there is an anonymity-preserving "gift assignment" such that each person makes and receives exactly one gift, and such that no person i can infer the remaining gift assignments from the fact that i is assigned to give a gift to j. We formalize this problem as a graph property involving vertex disjoint circuit covers, give a polynomial algorithm to decide this property for any given graph and provide a computational validation of the algorithm.

1 Introduction

The problem we deal with is well described by the following Wikipedia [14] entry:

> Secret Santa is a Christmas ritual involving a group of people exchanging anonymous gifts. Participants names are placed in a hat and each person draws a name for whom they are to buy a gift. Presents are then exchanged anonymously. There is usually a gift giving occasion, where all the presents are placed on a table, with the name of the receiver, but not the giver.

We assume that additional constraints may exist in the definition of the problem. For instance, it may be required that self-gifts and gifts between certain pairs of participants should be avoided. These constraints are enforced on a graph model: participants are represented by vertices and the possibility of a participant giving a gift to another participant is represented by an arc between the corresponding vertices.

Previous academic work on the Secret Santa problem is scarce. A secure protocol for the distributed solution of the Secret Santa problem is proposed in [11], with the corresponding implementation being described in [12]. Some published works in social sciences exist [3]. A scholarly discussion ensued in 1999-2001 in the *Mathematical Gazette* [7,9,1] focussing on the probability of picking a gift assignment without mutual gifts. This is extended in [8] to deal with more constraints on pairs of people that cannot exchange gifts, and in [13] to include at least a cyclic assignment of given length.

R. Fleischer and J. Xu (Eds.): AAIM 2008, LNCS 5034, pp. 271–279, 2008.

We use a digraph to model arbitrary constraints on the possibility of people exchanging gifts and propose a formalization of the Secret Santa problem as a decision problem on digraphs. Our main result is that the problem of determining whether anonymous gift assignments are possible on a given graph is in **P**. An investigation of anonymous communication channels along the lines of [2] provides further application-driven motivation for studying the Secret Santa problem.

An instance of the Secret Santa Problem is a connected digraph $G = (V, A)$ where V is the set of the participants, and $(i, j) \in A$ if participant i is allowed to make a gift to participant j. If symmetry is assumed (i.e., if we assume that if i can buy a gift for j then j can do the same for i) then A contains pairs of opposing arcs (i, j) and (j, i). In what follows, given a vertex $i \in V$, we let $\delta^+(i) = \{j \in V \mid (i, j) \in A\}$ and $\delta^-(i) = \{j \in V \mid (j, i) \in A\}$. An istance of the Secret Santa problem has a solution if for each person $i \in V$ there exists another assigned person $j \in V$ such that $(i, j) \in A$ such that i makes a gift to j (e.g. if $V| = \{1, 2\}$ and $A = \{(1, 2)\}$ there is no solution, for 2 has no assigned person). We model solutions as follows.

Definition 1.1. *A Vertex Disjoint Circuit Cover (VDCC) for $G = (V, A)$ is a subset $S \subseteq A$ of arcs of G such that: (a) for each $v \in V$ there is a unique $u \in V$, called the predecessor of v and denoted by $\pi_S(v)$, such that $(u, v) \in S$; (b) for each $v \in V$ there is a unique $u \in V$, called the successor of v and denoted by $\sigma_S(v)$, such that $(v, u) \in S$. We denote by $\mathcal{C}$ the set of all VDCCs in G.*

Let $x_{ij} \geq 0$ be real non-negative continuous variables for all $(i, j) \in A$, and consider the equations:

$$\forall i \in V \quad \sum_{j \in \delta^+(i)} x_{ij} = 1 \tag{1}$$

$$\forall i \in V \quad \sum_{j \in \delta^-(i)} x_{ji} = 1. \tag{2}$$

The support of any mapping $x^* : A \to \mathbb{R}^+$ satisfying (1)-(2) defines a VDCC (this follows by total unimodularity of the constraint matrix of (1)-(2)). Assuming G admits at least a VDCC, it is easy to see that x^* defines a permutation on $\{1, \ldots, |V|\}$ and therefore an assignment of maximum size on the undirected (bipartite) graph subjacent to the bipartite digraph $B = (U_1, U_2, A')$ where $U_1 = U_2 = V$ and A' are the same arcs as in A such that their heads are in U_1 and their tails in U_2. The best method for finding such an assignment is reported in [10] (Thm. 16.5) as $O(\nu(B)^{\frac{1}{2}}|A|)$, where $\nu(B)$ is the maximum size of a matching in B. Since in our case $\nu(G) = |V|$, we obtain a total (polynomial) worst-case complexity of $O(|V|^{\frac{1}{2}}|E|)$ for solving the VDCC.

Since gifts must be exchanged anonymously, not all VDCCs are acceptable: e.g. when $V = \{1, 2\}$ and $A = \{(1, 2), (2, 1)\}$, there is a unique VDCC given by $(x_{12}, x_{21}) = (1, 1)$, so each person knows that the other person will make them a gift. Informally, we define a graph G as a Secret Santa graph if it admits at least a VDCC ensuring gift anonymity; i.e., if knowing a (donor,receiver) pair in the VDCC does not uniquely identify any other (donor,receiver) pair. Such a VDCC is an "acceptable" solution (precise definitions are given in Defn. 2.1).

The rest of the paper is organised as follows: in Section 2 we formalize the Secret Santa problem and discuss a few basic properties. In Section 3 we give a polynomial-time algorithm for deciding whether a given graph has the Secret Santa property or not. In Section 4 we discuss some computational results. Section 5 concludes the paper.

2 Characterisation of Anonymity and Basic Results

Given a connected digraph $G = (V, A)$, let $n = |V|$ and $m = |A|$. We aim to characterize the set $\mathcal{S} \subseteq \mathcal{C}$ of "acceptable" solutions (i.e. anonymity-preserving VDCCs) formally: Secret Santa graphs are those for which $|\mathcal{S}| > 0$. The anonymity requirement on VDCCs reflects the notion of *ignorance* in epistemic logic [4], and is translated in graph-theoretical terms in the following definition.

Definition 2.1. *A graph G is a Secret Santa graph (SESAN) if there exists a VDCC S for G such that for each pair of distinct arcs $a, b \in S \cap A$, there is another VDCC T for G with $a \in T$ and $b \notin T$. The set*

$$\mathcal{V}(S) = \{S\} \cup \{T_{ab} \in \mathcal{C} \mid a \in S \cap T_{ab} \wedge b \in S \smallsetminus T_{ab}\} \tag{3}$$

is a verification family for G, and S is a witness. Elements of a verification family are called acceptable solutions.

By the definition of SESAN, even if a participant knows his/her own gift assignment a, he/she does not gain any knowledge with respect to any other gift assignment b. We define $\mathcal{S}$ as the union of all verification families for G.

Proposition 2.2. *If the graph $G = (V, A)$ contains a vertex v such that $|\delta^-(v)| = 1$ or $|\delta^+(v)| = 1$ then G is not SESAN.*

Proof. Let $\delta^-(v) = \{(u)\}$ (the proof for $\delta^+(v)$ is the same). All VDCCs S will necessarily contain (u, v), thereby contradicting Defn. 2.1. $\qquad \square$

The converse to Prop. 2.2 is of course false: the complete digraph over three vertices is an example of a graph where $|\delta^-(i)| = |\delta^+(i)| = 2$ for each $i \in \{1, 2, 3\}$, but since there are only two possible VDCCs, G is not SESAN.

Lemma 2.3. *A verification family for $G = (V, A)$ contains at most $\tau(V) = |V|(|V| - 1) + 1$ VDCCs.*

Proof. This follows trivially from Eq. (3), as apart from S there is at most one VDCC for each pair of distinct arcs in S (the fact that $|S| = |V|$ follows trivially from (1)-(2)). $\qquad \square$

We formally define the Secret Santa problem as follows.

Definition 2.4. SECRET SANTA PROBLEM *(SESANP). Given a graph $G = (V, A)$, decide whether it is SESAN.*

Notice that the SESANP asks for the existence of particular subgraphs (the VDCCs) whose added condition (anonymity) requires checking against $O(m^2)$ similar subgraphs. A naive approach of finding an arbitrary VDCC and then checking over all pairs of arcs whether it is acceptable might yield the answer NO without proving that the graph is not SESAN, for a different initial choice might have yielded a different answer. In order to make this approach work, one would need a complete enumeration of an exponentially large set (that of all VDCCs), suggesting that SESANP might be **NP**-complete. It turns out, however, that SESANP is in **P** (see Sect. 3).

2.1   Some Examples

Consider the directed graph obtained from K_4 (in Fig. 1, left) by replacing edge edge with two antiparallel arcs. The second graph on from the left of Fig 1 is a possible witness; a verification family is displayed on the right, therefore the graph is SESAN.

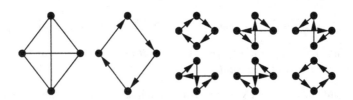

Fig. 1. K_4, a witness and its verification family

Consider now the graph in Figure 2. This graph is SESAN: it is sufficient to take two equal solutions for K_4, as in Figure 1, to guarantee an acceptable solution. However, not all VDCCs for this graph are acceptable solutions. Indeed, on the right hand side of Figure 2, the arc (i, j) does not guarantee anonymity for the arc (j, i), because there is no other VDCC in which (i, j) appears, and (j, i) does not. In this case the set $\mathcal{S}$ is a proper subset of $\mathcal{C}$.

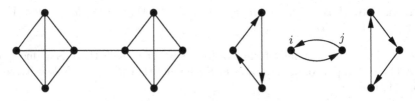

Fig. 2. An instance with two copies of K_4 and a non-anonymous VDCC

3   A Polynomial Time Algorithm

Algorithm 1 is a polynomial time algorithm for deciding whether a given graph is SESAN. The algorithm is based on the following observation: let T be a VDCC that does not guarantee anonymity. This implies that there exists an arc $a \in T$ such that, for some arc $b \in T$, all the VDCCs of G containing a also contain b. The key observation here is that *all* VDCCs of G containing a cannot satisfy the anonymity requirement (because of the necessary presence of b). The algorithm incrementally builds a set α of arcs that are not permitted and uses this set as additional constraints when looking for possible VDCCs. If no VDCC can be found satisfying the additional constraints given by α the graph is not SESAN.

Let P be the constraint satisfaction problem (1)-(2) such that $x \geq 0$. For $\alpha \subseteq A$ define P^α as P with the added constraints $x_{ij} = 0$ for each $(i,j) \in \alpha$. For given $(i,j),(k,l) \in A$ define P^α_{ijkl} as P^α with the added constraints $x_{ij} = 1$ and $x_{kl} = 0$. Recall $\mathcal{C}$ is the set of all VDCCs and $\mathcal{S}$ is the set of all acceptable solutions. Let $\mathcal{C}^\alpha$ (resp. $\mathcal{S}^\alpha$) be the set of all VDCCs (resp. acceptable solutions) not containing the arcs in α and let $\mathcal{C}^\alpha_{ijkl}$ (resp. $\mathcal{S}^\alpha_{ijkl}$) be the subset of $\mathcal{C}^\alpha$ (resp. $\mathcal{S}^\alpha$) containing (i,j) but not (k,l).

Lemma 3.1. *For any $\alpha \subseteq A$ and distinct $(i,j),(k,l) \in A$, if P^α_{ijkl} is infeasible then no acceptable solution in $\mathcal{S}^\alpha$ contains the arc (i,j).*

Proof. Since P^α_{ijkl} is infeasible, $\forall T \in \mathcal{C}^\alpha\ ((i,j) \in T \rightarrow (k,l) \in T)$, hence $\forall T \in \mathcal{C}^\alpha\ ((i,j) \in T \rightarrow T \notin \mathcal{S}^\alpha)$. This implies that $\forall S \in \mathcal{S}^\alpha\ ((i,j) \notin S)$.

Theorem 3.2. *Alg. 1 correctly solves the SESANP.*

Proof. By Lemma 3.1 and Line 13 in Alg. 1, no arc in α is contained in an acceptable solution. First assume G is SESAN and suppose Alg. 1 fails. This happens either when P^α is infeasible at Line 3 or when $|\alpha| > |A| - |V|$ at Line 2. The former case implies that $\mathcal{C}^\alpha = \emptyset$ and hence $\mathcal{S}^\alpha = \emptyset$, which means that all the acceptable solution must have an arc in α, a contradiction with the construction of α in Lines 11-12. The latter case would imply an acceptable solution with fewer than $|V|$ arcs, again a contradiction as $|T| = |V|$ for all $T \in \mathcal{C}$. Therefore the algorithm terminates with an acceptable solution. Assume now that G is not SESAN. Then for each $\alpha \subseteq A$ (and hence also the sets α generated during the algorithm run) there exist distinct $(i,j),(k,l) \in A$ such that P^α_{ijkl} is infeasible, i.e. $\mathcal{C}^\alpha_{ijkl} = \emptyset$. By Lines 11-12 and 3, the only possibility for $|\alpha|$ not to increase monotonically at each outer iteration is for P^α to be infeasible. Since $|\alpha|$ is bounded above by $|A| - |V|$, in either case the algorithm terminates with IsSesan = FALSE.

Note that P, P^α and P^α_{ijkl} are simply instances of the VDCC problem (i.e. the problem of determining whether a given graph has a VDCC) on graphs that are modifications of the original digraph G given by the forced absence of the arcs in α and (k,l) and by the forced presence of the arc (i,j): when these instances are infeasible, the maximum matching mentioned in Sect. 1 has size strictly smaller

---

Algorithm 1. Polynomial algorithm for solving the SESANP.

Require: $G = (V, A)$.
Ensure: Whether G is SESAN or not.
1: Let $\alpha = \emptyset$, ExitLoop = FALSE, IsSesan = FALSE
2: **while** $|\alpha| \le |A| - |V|$ and ExitLoop = FALSE **do**
3:    **if** P^α is feasible **then**
4:        Let S be a solution to P^α
5:        Let ExitLoop = TRUE
6:    **else**
7:        Let IsSesan = FALSE
8:        exit While loop
9:    **end if**
10:    **for all** $(i,j), (k,l) \in S : (i,j) \ne (k,l)$ **do**
11:        **if** P^α_{ijkl} is infeasible **then**
12:            Let $\alpha \leftarrow \alpha \cup \{(i,j)\}$
13:            Let ExitLoop = FALSE
14:            exit For loop
15:        **end if**
16:    **end for**
17:    **if** ExitLoop = TRUE **then**
18:        Let IsSesan = TRUE
19:    **end if**
20: **end while**
21: **if** IsSesan = TRUE **then**
22:    G is SESAN
23: **else**
24:    G is not SESAN
25: **end if**

---

than $|V|$. Solving $P, P^\alpha, P^\alpha_{ijkl}$ has the same worst-case polynomial complexity as finding a VDCC in G, namely $O(n^{\frac{1}{2}}m)$.

Lemma 3.3. *Alg. 1 has worst case $O(n^{\frac{5}{2}}m^2)$ time complexity.*

Proof. An n^2 term arises because of the internal loop on the distinct arcs in S (Line 10), as $|S| = |V|$. An m term arises because of the external loop (Line 2), and because $|\alpha|$ increases at each outer iteration (Line 13) unless the algorithm terminates. The remaining $n^{\frac{1}{2}}m$ term refers to the solution of each P^α_{ijkl} problem in Line 11.

Corollary 3.4 *SESANP is in* **P**.

4   Computational Results

We tested Alg. 1 on a class of randomly generated graph instances. As the main target application of the SESANP is in communication protocols, communication

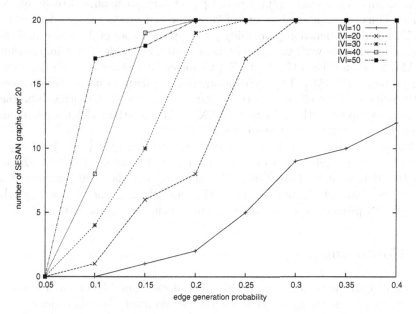

Fig. 3. Proportion of SESAN random graphs with p ranging in $[0.05, 0.4]$

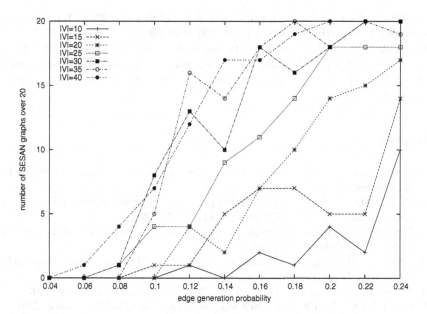

Fig. 4. Proportion of SESAN random graphs with p ranging in $[0.04, 0.24]$

between any two agents (gifts between participants) is assumed to be bidirectional. Thus, we generated groups of 20 undirected random graphs with vertex set V and edge generation probability p for various values of $|V|$ and p, and then replaced each edge with two antiparallel directed arcs. Alg. 1 was implemented in AMPL [5] and the ILOG CPLEX 10.1 solver [6] was deployed on the VDCC sub-problems $P^\alpha, P^\alpha_{ijkl}$. This yields a practical algorithm that is nonpolynomial in the worst case but efficient on the average case, as we solved each sub-problem using the simplex method. Using CPLEX's barrier method yields a polynomial algorithm but is practically less efficient.

The plot in Fig. 3 refers to $|V| \in \{10i \mid 1 \leq i \leq 5\}$ and $p \in \{0.05i \mid 1 \leq i \leq 8\}$. The plot in Fig. 4 refers to $|V| \in \{10 + 5i \mid 0 \leq i \leq 6\}$ and $p \in \{0.04 + 0.02i \mid 0 \leq i \leq 10\}$. It took around 4h of user CPU time to solve the 2340 instances on an Intel Core Duo 1.2GHz and 1.5GB RAM running Linux. The plots suggest that the SESAN property is correlated to graph density and graph size.

5   Conclusion

We formalized the Secret Santa problem as a decision problem related to finding subgraphs of a given graph with a particular structure (vertex-disjoint circuit covers) subject to an anonymity condition, and proved that it is in **P**. We provided an $O(|V|^{\frac{5}{2}}|A|^2)$ polynomial algorithm and a limited computational study thereof.

Future work will focus on a generalized decision problem: given a graph, a particular graph structure and a particular anonymity property, are there families of subgraphs with the given structure that are anonymous according to the given property? A practical interest is attached, for example, to path-structured subgraphs in the study of networks providing anonymity of the source and/or intermediate vertices.

References

1. Boyd, A.V., Ridley, J.N.: The return of Secret Santa. Mathematical Gazette 85(503), 307–311 (2001)
2. Chaum, D.: The dining cryptographers problem: unconditional sender and recipient untraceability. Journal of Cryptology 1(1), 65–75 (1988)
3. Duncan, B.: Secret Santa reveals the secret side of giving. Technical Report 0601, University of Colorado at Denver (2006)
4. Fagin, R., Halpern, J.Y., Moses, Y., Vardi, M.Y.: Reasoning about Knowledge. MIT Press, Cambridge (1995)
5. Fourer, R., Gay, D.: The AMPL Book. Duxbury Press, Pacific Grove (2002)
6. ILOG. ILOG CPLEX 10.1 User's Manual. ILOG S.A., Gentilly, France (2006)
7. McGuire, K.M., Mackiw, G., Morrell, C.H.: The Secret Santa problem. Mathematical Gazette 83(498), 467–472 (1999)
8. Penrice, S.: Derangements, permanents and christmas presents. American Mathematical Monthly 98(7), 617–620 (1991)

9. Pinkham, R.: The Secret Santa problem revisited. Mathematical Gazette 85(502), 96–97 (2001)

10. Schrijver, A.: Combinatorial Optimization: Polyhedra and Efficiency. Springer, Berlin (2003)

11. Tel, G.: Cryptografie, Beveiliging van de digitale maatschappij. Addison-Wesley, Reading (2002)

12. Verelst, J.: Secure computing and distributed solutions to the Secret Santa problem. Master's thesis, Computer Science Dept., University of Utrecht (2003)

13. White, M.: The Secret Santa problem. Rose-Hulman Institute of Technology Undergraduate Math Journal 7(1) (paper 5) (2006)

14. Secret Santa. Wikipedia (2006), http://en.wikipedia.org/wiki/Secret_Santa

Finding Optimal Refueling Policies in Transportation Networks

Shieu-Hong Lin

Biola University, La Mirada CA 90639, USA
shieu-hong.lin@biola.edu

Abstract. We study the combinatorial properties of optimal refueling policies, which specify the transportation paths and the refueling operations along the paths to minimize the total transportation costs between vertices. The insight into the structure of optimal refueling policies leads to an elegant reduction of the problem of finding optimal refueling policies into the classical shortest path problem, which ends in simple and more efficient algorithms for finding optimal refueling policies.

1 Introduction

A vehicle refueling policy is a path in a transportation network together with the series of refueling operations when passing through the vertices on the path to reach a destination vertex from a starting vertex in the network while always maintaining the fuel level between a lower limit L and an upper limit U throughout the entire process. An optimal refueling policy minimizes the total fuel cost to reach the destination vertex given an initial fuel level and a required minimal final fuel level. Different from the shortest path problem, fuel prices at the vertices must be considered in addition to the distances between vertices, and there are situations in which optimal refueling policies involve non-simple paths. Optimal refueling polices can be determined by solving mixed integer programs [6]. Lin et al. [4] gives a linear-time algorithm for determining an optimal refueling policy given a fixed path and relate the problem to the bounded-inventory economic lot-sizing problems [1] [5]. Given a general transportation network of n vertices and the all-pairs shortest-distance information, Khuller et al. [3] provides algorithms for (i) finding all-pairs optimal refueling policies given the initial fuel levels at the vertices in $O(n^4)$ time in general or in $O(n^3 k * \min(k, \log n))$ time if the vehicle is constrained to use at most k refueling stops respectively and (ii) finding an optimal refueling policy for a given pair of vertices and an initial fuel level in $O(n^3)$ time in general or in $O(n^2 k \log n)$ time with the k-stop constraint. Both of Lin and Khuller et al. assume the required minimal final fuel level reaching the destination to be the same as the lower fuel limit L.

In this paper, we first analyze the combinatorial structure of optimal $[L, L]$ refueling policies, which are refueling policies required to begin with a fuel level at the lower limit L and ending with a fuel level at least at the lower limit L. We prove that finding optimal $[L, L]$ vehicle refueling policies in a transportation

R. Fleischer and J. Xu (Eds.): AAIM 2008, LNCS 5034, pp. 280–291, 2008.

network G is equivalent to finding shortest paths in an optimal transition graph derived from G, which is essentially a finite automaton modelling all possible optimal transitions between the upper and the lower limits of fuel level between vertices together with a distance measure representing the optimal transition costs. This leads to simple algorithms that can determine optimal $[L, L]$ refueling policies for all pairs of vertices in a network of n vertices in $O(n^3)$ time or in $O(n^3 \log k)$ time with the additional k-stop constraint. With the all-pairs optimal (k-stop-bounded) $[L, L]$ vehicle refueling policies determined, we can then (i) determine all-pairs optimal (k-stop-bounded) refueling policies given various initial fuel levels at the vertices and ending with a fuel level at least at the lower limit L in $O(n^3)$ time, and (ii) determine an optimal refueling policy given a pair of vertices, an initial fuel level, and a required minimal final fuel level in $O(n^2)$ time.

2   The Optimal Refueling Policy Problems

Consider a vehicle with a fixed fuel tank capacity operates in a transportation network delivering commodities between pairs of locations in the network. The vehicle needs to refuel at fuel stations to maintain a minimum level of fuel in the fuel tank all the time. Given a pair of locations s and t together with an initial fuel level starting at s and a required fuel level when arriving at t, we would like to determine an optimal refueling policy specifying a path from s to t, the fuel stations on the path to stop for refueling, and the amounts of fuel to add in the fuel stations to minimizes the total refueling cost.

Definition 1. (Vehicle-network instances). An vehicle-network instance is a six-tuple $\langle L, U, V, A, X, P \rangle$ where

- $L, U \in \mathbb{Z}^+ \cup \{0\}$, $0 \leq L < U$, L is the minimum fuel level the vehicle needs to maintain all the time while the minimum fuel level U is simply the full tank capacity of the vehicle,
- V and A together form a directed graph $G = (V, A)$ modelling a transportation network where V is the set of vertices representing points of interest in the network including locations of suppliers, customers, cities, and fuel stations in the transportation network, while A is the set of directed edges with each directed edge (u, v) in A representing a transportation link from vertex u to vertex v that requires no more than $U - L$ amount of fuel,
- $X : V \times V \to [0, U - L] \cup \{\infty\}$ is the fuel consumption function where $X(u, v)$ equals the amount of fuel consumed to reach vertex v from vertex u for $(u, v) \in A$, $X(u, u) = 0$ for $u \in V$, and $X(u, v) = \infty$ for $(u, v) \notin A$ when $u \neq v$, and in the following of the paper we use $\hat{X}(u, v)$ to denote the shortest distance from u to v with X as the underlying distance measure over the directed graph $G = (V, A)$, and
- $P : V \to \mathbb{Z}^+ \cup \{0, \infty\}$ is the fuel price function where $P(u)$ denotes the unit fuel price charged in vertex u and $P(u) = \infty$ if u does not have a fuel station.

Definition 2. (Vehicle refueling policies). Given a vehicle-network instance $I = \langle L, U, V, A, X, P \rangle$, a $[l_s, l_t]$ refueling policy from vertex s to vertex t is a sequence of vertex-and-refueling-amount pairs $\pi_{s \to t}^{l_s, l_t} = \langle (v_0, f_0), \ldots, (v_m, f_m) \rangle$ where $m \geq 0$, $\forall i \in [0, m]$ $v_i \in V$, $v_0 = s$, $v_m = t$, f_i, l_s and l_t are integers, $L \leq l_s, l_t \leq U$, and $0 \leq f_i \leq U - L$ such that

- $\forall i \in [0, m-1]$ $(v_i, v_{i+1}) \in A$ and $\langle v_0, \ldots, v_m \rangle$ is a path from s to t,
- $\forall i \in [0, m]$, $FE_\pi(v_i) =$
  $l_s + \sum_{1 \leq j < i} f_j - \sum_{1 \leq j < i} X(v_j, v_{j+1}) \geq L$,
- $\forall i \in [0, m]$, $FL_\pi(v_i) =$
  $l_s + \sum_{1 \leq j \leq i} f_j - \sum_{1 \leq j < i} X(v_j, v_{j+1}) \leq U$, and
- $FL_\pi(v_m) =$
  $l_s + \sum_{1 \leq j \leq m} f_j - \sum_{1 \leq j < m} X(v_j, v_{j+1}) \geq l_t$

where l_s and l_t are the initial fuel level and the required minimal final fuel level at vertices s and t respectively, f_i is the amount of fuel to purchase at vertex v_i, $FE_\pi(v_i)$ denotes the fuel level when entering the ith vertex v_i on the path, and $FL_\pi(v_i)$ denotes the fuel level when leaving v_i or when finally ending at v_m when $i = m$.

Remark: A $[l_s, l_t]$ refueling policy from vertex s to vertex t represents a feasible refueling solution along an operational path $\langle v_0, \ldots, v_m \rangle$ of length m, allowing the vehicle to progressively move from vertex v_i to vertex v_{i+1} for $0 \leq i < m$, starting with an initial fuel level l_s at the starting vertex $s = v_0$, stopping at vertex v_i to purchase f_i units of fuel if f_i is not zero, ending in the destination vertex $t = v_m$ with a final fuel level of l_t or more in the end. The refueling policy ensures in the process it never reaches a fuel level lower than the lower limit L or higher than upper limit U by keeping $FE_\pi(v_i)$, the fuel level when entering the ith vertex v_i on the operation path, to at least L and keeping $FL_\pi(v_i)$, the fuel level when leaving v_i (or when settling down at v_m when $i = m$) to at most U. Note that $\langle v_0, \ldots, v_m \rangle$ is not necessarily a simple path since the vehicle may repeatedly leave a vertex v_i to refuel at other vertices with lower fuel prices and come back to v_i later with a higher fuel level needed to proceed with remainder of the path. A graphical example is shown in the appendix.

Definition 3. (Operational costs, refueling stops, and policy sets). For a $[l_s, l_t]$ refueling policy $\pi_{s \to t}^{l_s, l_t} = \langle (v_0, f_0), \ldots, (v_m, f_m) \rangle$ from vertex s to vertex t, we define the operational cost of the policy as $Cost(\pi_{s \to t}^{l_s, l_t}) = \sum_{1 \leq i \leq m} P(v_i) * f_i$. We say v_i is a refueling vertex in $\pi_{s \to t}^{l_s, l_t}$ if and only if $f_i > 0$ and denote the set of refueling vertices from which the vehicle must stop to purchase fuel as $Refuelings(\pi_{s \to t}^{l_s, l_t}) = \{f_i | f_i > 0\}$. For a vehicle-network instance $I = \langle L, U, V, A, X, P \rangle$, we denote the set of all $[l_s, l_t]$ refueling policies from vertex s to vertex t as $\prod_{s \to t}^{l_s, l_t}$, and denote the set of all $[l_s, l_t]$ refueling policies from vertex s to vertex t using at most k refueling stops as $\prod(k)_{s \to t}^{l_s, l_t} = \{\pi | \pi \in \prod_{s \to t}^{l_s, l_t}, |Refuelings(\pi)| \leq k\}$.

Definition 4. (The optimal refueling policy problems) Given a vehicle-network instance $I = \langle L, U, V, A, X, P \rangle$, a $[l_s, l_t]$ refueling policy $\pi_{s \to t}^{l_s, l_t}$ from vertex s to vertex t is an optimal $[l_s, l_t]$ (k-stop-bounded) refueling policy from vertex s to vertex t if and only if $Cost(\pi_{s \to t}^{l_s, l_t}) \leq Cost(\pi)$ for every π in $\prod_{s \to t}^{l_s, l_t}$ (for every π in $\prod(k)_{s \to t}^{l_s, l_t}$). **(i)** The computational task of the single-pair $[l_s, l_t]$ optimal (k-stop-bounded) refueling policy problem is to determine an optimal $[l_s, l_t]$ (k-stop-bounded) refueling policy $\pi_{s \to t}^{l_s, l_t}$ from vertex s to vertex t given a pair of vertices s and t in V and the fuel levels $l_s, l_t \in [L, U]$. **(ii)** Given the initial fuel levels at the vertices, the computational task of the all-pairs $[*, L]$ optimal (k-stop-bounded) refueling policy problem is to determine the optimal (k-stop-bounded) refueling policies $\pi_{s \to t}^{l_s, L}$ where l_s is the given initial fuel level at vertex s for all pairs of vertices s and t ($s \neq t$) in V **(iii)** The computational task of the all-pairs $[L, L]$ optimal (k-stop-bounded) refueling policy problem is to determine the optimal $[L, L]$ (k-stop-bounded) refueling policies $\pi_{s \to t}^{L, L}$ for all pairs of vertices s and t ($s \neq t$) in V.

Remark: Later in the paper, we show that solutions to the optimal $[L, L]$ (k-stop-bounded) refueling policy problem can be used to efficiently determine solutions to the other optimal refueling policy problems. The k-stop-bounded versions of the problems arise in practice since each refueling stop takes time and at times the number of stops must be bounded above to ensure timely arrival. The all-pairs $[*, L]$ optimal (k-stop-bounded) refueling policy problem is the one studied by Khuller et al. [3].

3 Combinatorial Properties of Optimal $[L, L]$ Refueling Policies

In this section, we explore the combinatorial properties of optimal $[L, L]$ refueling policies and define the related terminology. In the next section, based on these combinatorial properties we develop polynomial-time algorithms for solving optimal refueling policy problems. Note that the definitions in the following all implicitly assume a vehicle-network instance $I = \langle L, U, V, A, X, P \rangle$ in the context and we use $\hat{X}(u, v)$ to denote the shortest distance from u to v with X as the underlying distance measure over the directed graph $G = (V, A)$ modelling the transportation network.

Definition 5. (Stop sequences and $\rho(\pi)$). The stop sequence $Stops(\pi)$ of a refueling policy $\pi = \langle (v_0, f_0), \ldots, (v_m, f_m) \rangle$ where $m \geq 1$ is the maximal subsequence of the operational path $\langle v_0, \ldots, v_m \rangle$ that contains the starting vertex $s = v_0$, the destination vertex $t = v_m$, and all the refueling vertices in between. We define $\rho(\pi) = |Refuelings(\pi) - \{s, t\}|$ and denote the stop sequence of the refueling policy π as $Stops(\pi) = \langle v_{i_0}, v_{i_1}, \ldots, v_{i_{\rho(\pi)}}, v_{i_{\rho(\pi)+1}} \rangle$ where $i_0 = 0$, $v_{i_0} = v_0 = s$, $i_{\rho(\pi)+1} = m$, $v_{i_{\rho(\pi)+1}} = v_m = t$, and $\{v_{i_1}, \ldots, v_{i_{\rho(\pi)}}\} = Refuelings(\pi) - \{s, t\}$ with $0 = i_0 < i_1 < \ldots < i_{\rho(\pi)} < i_{\rho(\pi)+1} = m$ when $\rho(\pi) > 0$.

Definition 6. (The shortest-subpath property). Given a path $\langle v_0, \ldots, v_m \rangle$, we refer to $\langle v_j, \ldots, v_{j'} \rangle$ where $0 \leq j < j' \leq m$ as the subpath from v_j to $v_{j'}$ within the path. Given a refueling policy $\pi = \langle (v_0, f_0), \ldots, (v_m, f_m) \rangle$ with the stop sequence $Stops(\pi) = \langle v_{i_0}, v_{i_1}, \ldots, v_{i_{\rho(\pi)}}, v_{i_{\rho(\pi)+1}} \rangle$ for a vehicle-network instance $I = \langle L, U, V, A, X, P \rangle$, we say π satisfies the shortest-subpath property if and only if for every k in $[0, \rho(\pi)]$, the subpath $\langle v_{i_k}, v_{i_k+1}, \ldots, v_{i_{k+1}} \rangle$ from v_{i_k} to $v_{i_{k+1}}$ within the operational path $\langle v_0, \ldots, v_m \rangle$ is a shortest path from vertex v_{i_k} to $v_{i_{k+1}}$ in the directed graph $G = (V, A)$ with the fuel consumption fuction X regarded as the distance measure.

Lemma 1. (The shortest-subpath property and optimality). Given a $[L, L]$ refueling policy $\pi_{s \to t}^{L,L}$ where $s \neq t$, if $\pi_{s \to t}^{L,L}$ does not satisfy the shortest-subpath property, then there exists a $[L, L]$ refueling policy $\widehat{\pi}_{s \to t}^{L,L}$ satisfying the shortest-subpath property such that $Cost(\widehat{\pi}_{s \to t}^{L,L}) \leq Cost(\pi_{s \to t}^{L,L})$ and $\widehat{\pi}_{s \to t}^{L,L}$ has the same stop sequences $Stops(\widehat{\pi}_{s \to t}^{L,L}) = Stops(\pi_{s \to t}^{L,L})$.

Proof Sketch. Since no fuel is purchased between two refueling stops, the fuel prices at the vertices visited in between do not affect the total cost the refueling policy and it will always reduce cost by following a shortest path to reach from one refueling stop to the next refueling stop.

Definition 7. (The LU adjacency property). Given a refueling policy π with the stop sequence $Stops(\pi) = \langle v_{i_0}, v_{i_1}, \ldots, v_{i_{\rho(\pi)}}, v_{i_{\rho(\pi)+1}} \rangle$, we say the refueling policy π satisfies the LU adjacency property if and only if for every k in $[0, \rho(\pi)]$, either $FL_\pi(v_{i_k}) = U$ or $FE_\pi(v_{i_{k+1}}) = L$, i.e. either the tank is full when leaving v_{i_k} or the tank is at the minimum fuel level required when arriving at the next refueling stop $v_{i_{k+1}}$.

Lemma 2. (The LU adjacency property and optimality). Given a $[L, L]$ refueling policy $\pi_{s \to t}^{L,L}$ where $s \neq t$, if $\pi_{s \to t}^{L,L}$ does not satisfy the LU adjacency property, then there exists a $[L, L]$ refueling policy $\widehat{\pi}_{s \to t}^{L,L}$ satisfying the LU adjacency property such that $Cost(\widehat{\pi}_{s \to t}^{L,L}) \leq Cost(\pi_{s \to t}^{L,L})$, $\widehat{\pi}_{s \to t}^{L,L}$ has no more refueling vertices than $\pi_{s \to t}^{L,L}$, and $Stops(\widehat{\pi}_{s \to t}^{L,L})$ is a subsequence of $Stops(\pi_{s \to t}^{L,L})$.

Proof Sketch. Consider two adjacent refueling stops v_{i_k} and $v_{i_{k+1}}$. If $FL_\pi(v_{i_k}) \neq U$ and $FE_\pi(v_{i_{k+1}}) \neq L$, we can keep increasing the fuel purchased at the stop with the lower fuel price and decreasing the fuel purchased at the other stop without affecting the refueling decisions at other stops until either $FL_\pi(v_{i_k}) = U$ or $FE_\pi(v_{i_{k+1}}) = L$ or one of the two refueling stops can be eliminated from the stop sequence to reach a new reduced case with fewer refueling stops. Lemma 2 then follows by induction on the number of stops in the stop sequences.

Theorem 1. (Properties of optimal $[L, L]$ refueling policies). Every optimal $[L, L]$ (k-stop-bounded) refueling policy satisfies the shortest-subpath property and the LU adjacency property.

Proof sketch. It follows from Lemma 1 and Lemma 2.

Definition 8. (*LU* **states,** *LU* **successors,** *LU* **transitions, and transition costs**). Given a $[L, L]$ refueling policy $\pi = \pi_{s \to t}^{L,L} = \langle (v_0, f_0), \ldots, (v_m, f_m) \rangle$ satisfying the LU adjacency property and $s \neq t$ with the stop sequence $Stops(\pi) = \langle v_{i_0}, v_{i_1}, \ldots, v_{i_{\rho(\pi)}}, v_{i_{\rho(\pi)+1}} \rangle$, we say (v_{i_k}, L) is an L state of π if $FE_\pi(v_{i_k}) = L$ and (v_{i_k}, U) is a U state of π if $FL_\pi(v_{i_k}) = U$. We refer to the L states and U states in π together as the *LU* states of π. We define a *LU* successor relation and four types of state transitions and their costs among *LU* states of π and refer to the transitions as the *LU* transitions of π as follows.

(i) Given an L state(v_{i_k}, L), (v_{i_k}, U) is an *LU* successor of (v_{i_k}, L) and $\langle (v_{i_k}, L), (v_{i_k}, U) \rangle$ is an $[L, U]$ transition in π with the cost $TC_\pi((v_{i_k}, L), (v_{i_k}, U)) = f_{i_k} * P(v_{i_k})$ if and only if $FE_\pi(v_{i_k}) = L$ and $FL_\pi(v_{i_k}) = U$.

(ii) Given an L state(v_{i_k}, L), $(v_{i_{k+1}}, L)$ is an *LU* successor of (v_{i_k}, L) and $\langle (v_{i_k}, L), (v_{i_{k+1}}, L) \rangle$ is a $[L, L]$ transition in π with the cost $TC_\pi((v_{i_k}, L), (v_{i_{k+1}}, L)) = f_{i_k} * P(v_{i_k})$ if and only if $FE_\pi(v_{i_k}) = L$ and $FE_\pi(v_{i_{k+1}}) = L$.

(iii) Given a U state (v_{i_k}, U), $(v_{i_{k+1}}, U)$ is an *LU* successor of (v_{i_k}, U) and $\langle (v_{i_k}, U), (v_{i_{k+1}}, U) \rangle$ is a $[U, U]$ transition in π with the cost $TC_\pi((v_{i_k}, U), (v_{i_{k+1}}, U)) = f_{i_{k+1}} * P(v_{i_{k+1}})$ if and only if $FL_\pi(v_{i_k}) = U$ and $FL_\pi(v_{i_{k+1}}) = U$.

(iv) Given a U state (v_{i_k}, U), $(v_{i_{k+2}}, L)$ is an *LU* successor of (v_{i_k}, U) and $\langle (v_{i_k}, U), (v_{i_{k+2}}, L) \rangle$ is a $[U, L]$ transition in π with the cost $TC_\pi((v_{i_k}, U), (v_{i_{k+2}}, L)) = f_{i_{k+1}} * P(v_{i_{k+1}})$ if and only if $FL_\pi(v_{i_k}) = U$, and $FE_\pi(v_{i_{k+2}}) = L$. We refer to vertex $v_{i_{k+1}}$ above as the medium vertex of the $[U, L]$ transition.

Definition 9. (*LU* **transition sequences**). Given a $[L, L]$ refueling policy $\pi = \pi_{s \to t}^{L,L} = \langle (v_0, f_0), \ldots, (v_m, f_m) \rangle$ satisfying the LU adjacency property and $s \neq t$ with the stop sequence $Stops(\pi) = \langle v_{i_0}, v_{i_1}, \ldots, v_{i_{\rho(\pi)}}, v_{i_{\rho(\pi)+1}} \rangle$, an *LU* state transition sequence of π is a representation of *LU* transitions from (s, L) to (t, L) as a sequence $\mathbf{LU}_\pi = \langle (v_{j_0}, l_0), \ldots, (v_{j_{\rho(\pi)+1}}, l_{\rho(\pi)+1}) \rangle$ of length $\rho(\pi) + 2$ where $l_i \in \{L, U\}$ for i in $[0, \rho(\pi)+1]$, $j_0 = 0$, $v_{j_0} = v_0 = s$, $j_{\rho(\pi)+1} = m$, $v_{j_{\rho(\pi)+1}} = v_m = t$, $l_0 = L$, $l_{\rho(\pi)+1} = L$, $(v_{j_{k+1}}, l_{k+1})$ is the successor state of (v_{j_k}, l_k) for k in $[0, \rho(\pi)]$, and $0 = j_0 \leq j_1 \leq \cdots \leq j_{\rho(\pi)} < j_{\rho(\pi)+1} = m$ when $\rho(\pi) > 0$. We define the cost of an *LU* state transition sequence $\mathbf{LU}_\pi = \langle (v_{j_0}, l_0), \ldots, (v_{j_{\rho(\pi)+1}}, l_{\rho(\pi)+1}) \rangle$ as $Cost(\mathbf{LU}_\pi) = \sum_{0 \leq j \leq \rho(\pi)} TC_\pi((v_{j_k}, l_k), (v_{j_{k+1}}, l_{k+1}))$, i.e. the summation of the *LU* state transition costs between adjacent *LU* states in $\mathbf{LU}_\pi$.

Lemma 3. (**Equality of the total** *LU* **transition cost and the total refueling cost**). Given a $[L, L]$ refueling policy $\pi = \pi_{s \to t}^{L,L} = \langle (v_0, f_0), \ldots, (v_m, f_m) \rangle$ satisfying the LU adjacency property and $s \neq t$, the cost of every *LU* state transition sequence $\mathbf{LU}_\pi$ of π always equals the cost of the refuelling policy π, i.e. $Cost(\mathbf{LU}_\pi) = Cost(\pi)$, and the number of refueling stops in π equals the length of the transition sequence minus one.

Proof Sketch. Each *LU* transition involves exactly one refueling operation and the transition cost is exactly the cost of the refueling operation. Therefore the number of refueling stops in π equals the length of the transition sequence minus one. The sum of transition costs of the transition sequence therefore always equals the sum of the refueling costs, i.e. the total operation cost of the refueling policy. This is true even if in degenerate cases there may be multiple ways of interpreting π into LU transition sequences.

Definition 10. (Optimal LU transition paths). For the four types of LU transitions respectively, we define the following four types of optimal LU transition paths in the directed graph $G = (V, A)$:

- a path p is an optimal transition path for an $[L, U]$ transition $\langle (v_i, L), (v_i, U) \rangle$ if and only if p is the self loop $\langle v_i, v_i \rangle$,
- a path p is an optimal transition path for an $[L, L]$ transition $\langle (v_i, L), (v_j, L) \rangle$ where $v_i \neq v_j$ if and only if p is a shortest path from v_i to v_j in $G = (V, A)$,
- a path p is an optimal transition path for an $[U, U]$ transition $\langle (v_i, U), (v_j, U) \rangle$ where $v_i \neq v_j$ if and only if p is a shortest path from v_i to v_j in $G = (V, A)$, and
- a path p is an optimal transition path for an $[U, L]$ transition $\langle (v_i, U), (v_j, L) \rangle$ where $v_i \neq v_j$ if and only if $p = \langle v_i, \ldots, v_k, \ldots v_j \rangle$ where $v_i \neq v_k \neq v_j$, the subpath $\langle v_i, \ldots, v_k \rangle$ in p is a shortest path from v_i to v_k in $G = (V, A)$, the subpath $\langle v_k, \ldots, v_j \rangle$ in p is a shortest path from v_k to v_j in $G = (V, A)$, and $v_k = \arg\min_v (\hat{X}(v_i, v) + \hat{X}(v, v_j) - (U - L)) * P(v)$ over every vertex v in V where $\hat{X}(v_i, v) \leq (U - L))$, $\hat{X}(v, v_j) \leq (U - L))$, and $\hat{X}(v_i, v) + \hat{X}(v, v_j) > (U - L)$. We refer to vertex v_k above as the medium vertex of the optimal transition path for the $[U, L]$ transition.

Definition 11. (The optimal LU transition property). Given an $[L, L]$ refueling policy $\pi = \langle (v_0, f_0), \ldots, (v_m, f_m) \rangle$, we say π satisfies the optimal LU transition property if and only if (i) π satisfies both the shortest subpath property and the LU adjacency property and (ii) for every LU transition sequence $\boldsymbol{LU}_\pi = \langle (v_{j_0}, l_0), \ldots, (v_{j_{\rho(\pi)+1}}, l_{\rho(\pi)+1}) \rangle$ of π and for every k in $[0, \rho(\pi)]$, the subpath $\langle v_{j_k}, v_{1+j_k} \ldots, v_{j_{k+1}} \rangle$ from v_{j_k} to v_{j_k+1} within the operational path $\langle v_0, , \ldots, v_m \rangle$ is an optimal LU transition path in $G = (V, A)$.

Theorem 2. (More on the properties of optimal $[L, L]$ refueling policies). Every optimal $[L, L]$ (k-stop-bounded) refueling policy satisfies the optimal LU transition property.

Proof Sketch. By Theorem 1, every optimal $[L, L]$ (k-stop-bounded) refueling policy must satisfy both the shortest subpath property and the LU adjacency property. It follows then the transition paths (i.e. the sequences of vertices visited between two adjacent refueling stops) for $[L, L]$ transitions and $[U, U]$ must be the shortest paths between the adjacent refueling vertices. By Lemma 3, every optimal $[L, L]$ (k-stop-bounded) refueling policy must also have the minimal-cost $[U, L]$ transition between a pair of adjacent U state and an L state in the policy, which ensures that for an $[U, L]$ transition, the transition path (i.e. the sequences of vertices visited between two adjacent pairs of adjacent refueling stops) must have the optimal transition structure required by the optimal LU transition property.

Corollary 1. (The optimal LU transition costs). For every optimal $[L, L]$ (k-stop-bounded) refueling policy $\pi = \langle (v_0, f_0), \ldots, (v_m, f_m) \rangle$, the transition costs of LU transitions in π always have the following optimal-transition-cost structure:

- if $\langle (v_i, L), (v_i, U) \rangle$ is a $[L, U]$ transition in π, then the transition cost is
  $TC_\pi((v_i, L), (v_i, U)) = f_i * P(v_i) = (U - L) * P(v_i)$,
- if $\langle (v_i, L), (v_j, L) \rangle$ is a $[L, L]$ transition in π, then the transition cost is
  $TC_\pi((v_i, L), (v_j, L)) = f_i * P(v_i) = \hat{X}(v_i, v_j) * P(v_i)$,
- if $\langle (v_i, U), (v_j, U) \rangle$ is a $[U, U]$ transition in π, then the transition cost is
  $TC_\pi((v_i, U), (v_j, U)) = f_j * P(v_j) = \hat{X}(v_i, v_j) * P(v_j)$, and
- if $\langle (v_i, U), (v_j, L) \rangle$ is a $[U, L]$ transition in π, then the transition cost is
  $TC_\pi((v_i, U), (v_j, L)) = f_k * P(v_k) = (\hat{X}(v_i, v_k) + \hat{X}(v_k, v_j) - (U - L)) * P(v_k)$
  where v_k is the medium vertex of an optimal transition path for the $[U, L]$ transition $\langle (v_i, U), (v_j, L) \rangle$.

Proof Sketch. The optimal transition costs above follow from Theorem 2, the definition of optimal LU transition paths in Definition 10, and the definition of LU transitions in Definition 8.

Theorem 3. (Reduction of finding optimal $[L, L]$ refueling policies into finding optimal LU transition sequences). Finding an optimal $[L, L]$ (k-stop-bounded) refueling policy from vertex s to vertex t is equivalent to finding a minimum-cost LU transition sequence (of no more than k transitions) from the LU state (s, L) to the LU state (t, L), in which the LU transition costs have the optimal-transition-cost structure depicted in Corollary 1.

Proof Sketch. It follows from Lemma 3 and Corollary 2.

Lemma 4. (Compact representation of optimal $[L, L]$ refueling policies). Given the information of all-pairs shortest paths in the network, for each pair of vertices s and t we can compactly represent an optimal $[L, L]$ (k-stop-bounded) refueling policy $\pi = \pi_{s \to t}^{L,L} = \langle (v_0, f_0), \ldots, (v_m, f_m) \rangle$ as the refueling stop sequence $Stops(\pi) = \langle v_{i_0}, v_{i_1}, \ldots, v_{i_{\rho(\pi)}}, v_{i_{\rho(\pi)+1}} \rangle$ together with $\langle f_{i_0}, f_{i_1}, \ldots, f_{i_{\rho(\pi)}} \rangle$, the sequence of refueling amounts at v_{i_k}'s for $0 \leq k \leq \rho(\pi)$.

Proof Sketch. Since no refueling occurs between two refueling stops, it is redundant to record the refueling information of $f_i = 0$ for every vertex v_i that is not a refueling stop. And because every optimal $[L, L]$ (k-stop-bounded) refueling policy satisfies the shortest subpath property, it is redundant to record the vertices between two adjacent refueling stops since we can use any shortest path between them and still maintain the optimality of the policy.

Theorem 4. (Determining an $[L, L]$ refueling policy from an LU transition sequence). Given an LU transition sequence $\boldsymbol{LU}$ of length m of an optimal $[L, L]$ (k-stop-bounded) refueling policy, Algorithm 1 below can determine in $O(m)$ time the compact representation of an $[L, L]$ refueling policy π that satisfies the optimal LU transition property and has an LU transition sequence $\boldsymbol{LU}_\pi = \boldsymbol{LU}$. If two different $[L, L]$ refueling policies both satisfy the optimal LU transition property and both have LU as a LU transition sequence, then they both have the same total refueling cost.

Proof sketch. It takes linear time since it is a linear scan through the transition sequence and it takes constant time to process each LU transition. The way the LU transitions are processed from Theorem 2, the definition of optimal LU transition paths in Definition 10, the optimal LU transition costs described in Corollary 2, and the definition of LU transitions in Definition 8.

Algorithm 1. Transforming an optimal LU transition sequence into
a compact representation of the corresponding optimal $[L, L]$ refueling policy
as a sequence of refueling stops and a sequence of refueling amounts.
Input: a vehicle-network instance $\langle L, U, V, A, X, P \rangle$,
an LU transition sequence LU of an optimal $[L, L]$ (k-stop-bounded)
refueling policy π, and for all pairs of vertices v_i and v_j, the information
of the shortest distance $\hat{X}(v_i, v_j)$ and the medium vertex of an optimal $[U, L]$
transition path from v_i to v_j.
Output: the compact representation of the corresponding optimal policy π as
the sequence of refueling stops $Stops(\pi) = \langle v_{i_0}, v_{i_1}, \ldots, v_{i_{\rho(\pi)}}, v_{i_{\rho(\pi)+1}} \rangle$, and
the corresponding refueling amount sequence $F_\pi = \langle f_{i_0}, f_{i_1}, \ldots, f_{i_{\rho(\pi)}} \rangle$.
Steps:
1. Start with empty sequences $Stops(\pi) = \langle \rangle$, $F_\pi = \langle \rangle$
2. Examine every pair of adjacent LU states in the LU transition sequence LU
   from the first pair to the last pair:
{
   if the pair is an $[L, U]$ transition $\langle (v_i, L), (v_i, U) \rangle$,
      append v_i into $Stops(\pi)$ and append the refueling amount $U - L$ into F_π;
   if the pair is an $[L, L]$ transition $\langle (v_i, L), (v_j, L) \rangle$,
      append v_i into $Stops(\pi)$ and
      append the refueling amount $\hat{X}(v_i, v_j) * P(v_i)$ into F_π;
   if the pair is an $[U, U]$ transition $\langle (v_i, U), (v_j, U) \rangle$,
      append v_j into $Stops(\pi)$ and
      append the refueling amount $\hat{X}(v_i, v_j) * P(v_j)$ into F_π;
   if the pair is an $[U, L]$ transition $\langle (v_i, U), (v_j, L) \rangle$ nd v_k is
      a a medium vertex of an optimal $[U, L]$ transition path from v_i to v_j,
      append v_k into $Stops(\pi)$ and append the refueling amount
      $(\hat{X}(v_i, v_k) + \hat{X}(v_k, v_j) - (U - L)) * P(v_k)$ into F_π.
}

4   Solving the Optimal Refueling Policy Problems

Based on these combinatorial properties explored in the previous section, in the following we show that finding optimal $[L, L]$ vehicle refueling policies in a transportation network G is equivalent to finding shortest paths in an optimal LU transition graph derived from G, which is essentially a finite automaton modelling all possible optimal LU transitions between vertices together with a distance measure representing the optimal LU transition costs. This leads to simple and efficient polynomial-time algorithms for solving the all-pairs $[L, L]$ (k-stop-bounded) optimal refueling policy problem. We then show ways to efficiently solve other optimal refueling policy problems based on the solutions to the all-pairs optimal (k-stop-bounded) $[L, L]$ vehicle refueling problem.

Definition 12. (The optimal LU transition graphs) The optimal LU transition graph for a vehicle-network instance $I = \langle L, U, V, A, X, P \rangle$ is a directed graph $G_I^{LU} = (V_I^{LU}, A_I^{LU})$ with a distance measure $X_I^{LU} : V_I^{LU} \times V_I^{LU} \rightarrow \mathbb{Z}^+ \cup \{0, \infty\}$ where

- the vertex set $V_I^{LU} = \{(v, L) | v \in V\} \cup \{(v, U) | v \in V\}$ is the set of all LU states,
- the set of directed edges $A_I^{LU} = \{\langle (v, L), (v, U) \rangle | v \in V\} \cup$
  $\{\langle (v, L), (v', L) \rangle | v, v' \in V, \hat{X}(v, v') \leq U - L\} \cup$
  $\{\langle (v, U), (v', U) \rangle | v, v' \in V, \hat{X}(v, v') \leq U - L\} \cup$
  $\{\langle (v, U), (v', L) \rangle | v, v' \in V, \hat{X}(v, v') > U - L,$
  $\exists v'' \in V, \hat{X}(v, v'') \leq U - L, \hat{X}(v'', v') \leq U - L\}$
  is the set of all possible LU transitions between LU states, and
- the distance metric function X_I^{LU} encodes information of the optimal LU transition costs described in Corollary 1:
  (i) $X_I^{LU}(y, y) = 0$ for $y \in V_I^{LU}$,
  (ii) $X_I^{LU}(y, z) = \infty$ for $(y, z) \notin A_I^{LU}$ when $y \neq z$,
  (iii) $X_I^{LU}((v, L), (v, U)) = (U - L) * P(v)$ for
  an optimal $[L, U]$ transition $\langle (v, L), (v, U) \rangle \in A_I^{LU}$,
  (iv) $X_I^{LU}((v, L), (v', L)) = \hat{X}(v, v') * P(v)$ for
  an optimal $[L, L]$ transition $\langle (v, L), (v', L) \rangle \in A_I^{LU}$,
  (v) $X_I^{LU}((v, U), (v', U)) = \hat{X}(v, v') * P(v)$ for
  an optimal $[U, U]$ transition $\langle (v, U), (v', U) \rangle \in A_I^{LU}$, and
  (vi) $X_I^{LU}((v, U), (v', L)) = (\hat{X}(v, v'') + \hat{X}(v'', v') - (U - L)) * P(v'')$ where
  $v'' \in V$ is the medium vertex in an optimal transition path for the $[U, L]$
  transition $\langle (v, U), (v', L) \rangle \in A_I^{LU}$.

Lemma 5. (Complexity of constructing the optimal LU transition graphs). Given a vehicle-network instance $I = \langle L, U, V, A, X, P \rangle$, the optimal LU transition graph $G_I^{LU} = (V_I^{LU}, A_I^{LU})$, the optimal LU transition paths, the medium vertices on optimal $[U, L]$ transition paths, and the distance measure X_I^{LU} can all be determined in $O(n^3)$ time where $n = |V|$ is the number of vertices in the network.

Proof Sketch. Applying the Floyd-Warshall algorithm for all-pairs shortest paths, we can determine the shortest distance $\hat{X}(u, v)$ between any pair of vertices in the transporation network $G = (V, A)$ in $O(n^3)$ time. The optimal LU transition paths are simply shortest paths in the cases of $[L, L]$, $[L, U]$, and $[U, U]$ transitions. In the case of $[U, L]$ transitions, the optimal LU transition paths and the medium vertices can be determined by examining all possible concatenations of two shortest paths with a common medium vertex as described in

Definition 10, which takes $O(n^3)$ time time all together for the n^2 pairs of $[U, L]$ transitions. The optimal transition costs as the metric X_I^{LU} can be determined in $O(n^3)$ time similarly as described in Corollary 1.

Theorem 5. (Reduction to the all-pairs shortest path problem). Given a vehicle-network instance $I = \langle L, U, V, A, X, P \rangle$, $\pi = \pi_{s \to t}^{L,L}$ is an optimal $[L, L]$ (k-stop-bounded) refueling policy from s to t if and only if an LU transition sequence $\boldsymbol{LU}_\pi$ of π is a shortest path from (s, L) to (t, L) (of length k or less) in the optimal LU transition graph G_I^{LU}. Finding all-pairs $[L, L]$ (k-stop-bounded) optimal refueling policies for a vehicle-network instance $I = \langle L, U, V, A, X, P \rangle$ is equivalent to finding all-pairs shortest paths (of length up to k) in the optimal LU transition graph G_I^{LU}.

Proof sketch. According to Theorem 2 and the definition of the optimal LU transition graph, each LU transition sequence $\boldsymbol{LU}_\pi$ of each $[L, L]$ refueling policy $\pi = \pi_{s \to t}^{L,L}$, $s \neq t$ that satisfies the optimal LU transition property uniquely represents a path $p = \boldsymbol{LU}_\pi = \langle (s, L), \ldots (t, L) \rangle$ from (s, L) to (t, L) in the optimal LU transition graph G_I^{LU}. By Lemma 3, a shortest path from (s, L) to (t, L) (of length k or less) in the optimal LU transition graph G_I^{LU} is also an LU transition sequence of an optimal $[L, L]$ (k-stop-bounded) refueling policy from s to t.

Corollary 2. (Complexity of the all-pairs optimal $[L, L]$ refueling policy problem). Algorithm 2 below can solve the all-pairs optimal $[L, L]$ refueling policy problem in $O(n^3)$ time and solve the all-pairs $[L, L]$ optimal k-stop-bounded refueling policy problem in $O(n^3 \log k)$ time where n is the number of vertices in the transportation network.

Proof Sketch. It follows from Theorem 5, Lemma 5, and the fact that the Floyd-Warshall algorithm takes $O(n^3)$ time while the distance-matrix repeated-squaring algorithm takes $O(n^3 \log k)$ time [2].

Algorithm 2. All-pairs $[L, L]$ optimal (k-stop-bounded) refueling policies.
Input: a vehicle-network instance $I = \langle L, U, V, A, X, P \rangle$, and
        optionally a bound k on the maximal number of refueling stops allowed.
Output: compact representations of all-pairs optimal $[L, L]$ refueling policies
        or all-pairs optimal $[L, L]$ k-stop-bounded refueling policies if k is given
        as sequences of refueling stops and the corresponding refueling amounts.
Steps:
1. Determine the optimal LU transition graph G_I^{LU}, the medium vertices of optimal $[U, L]$ transition paths, and the distance measure X_I^{LU}.
2. If k is not given, determine all-pairs shortest paths in G_I^{LU}
        using the Floyd-Warshall algorithm [2],
    otherwise determine all-pairs shortest paths of length up to k
        using the distance-matrix repeated-squaring algorithm [2].
3. View each path found in step 2 above as an LU transition sequence $\boldsymbol{LU}$ and apply algorithm 1 to determine the corresponding optimal $[L, L]$ (k-stop-bounded) refueling policy π.

Corollary 3. (Reduction of the all-pairs $[*, L]$ optimal refueling policy problem). The solution to an instance of the all-pairs $[*, L]$ optimal refueling policy problem can be determined in $O(n^3)$ time given the solution to the all-pairs $[L, L]$ optimal refueling policy problem on the same instance.

Proof Sketch. To reduce it to the case where all vertices have the same initial fuel level L, for each vertex s that has an initial fuel level $l_s > L$ we add an additional virtual vertex s' that has a unit fuel price of zero, an initial fuel level of L, and can only directly reach s by consuming $U - l_s$ units of fuel. An optimal $[L, L]$ policy from s' then corresponds to an optimal $[l_s, L]$ policy from s. For each vertex s (out of n vertices), it then takes $O(n^2)$ additional time to update and LU transition graph to find the $[l_s, L]$ policies from s to other vertices.

Corollary 4. (Reduction of the single-pair optimal refueling policy problem). The solution to an instance of the single-pair optimal $[l_s, l_t]$ optimal refueling policy problem can be determined in $O(n^2)$ time given the solution to the all-pairs $[L, L]$ optimal refueling policy problem on the same instance.

Proof Sketch. To reduce it to the case where all vertices have both the initial fuel level and the final minimum fuel level equal to L, we can add an additional virtual vertex s' for s as described in the proof sketch for Corollary 3, and for t add an additional virtual vertex t' that is only directly reachable from t by consuming l_t units of fuel. An optimal $[L, L]$ policy from s' to t' then corresponds to an optimal $[l_s, l_t]$ policy from s to t. It then takes $O(n^2)$ additional time to update and LU transition graph to find an optimal $[l_s, l_t]$ policy from s to t.

References

1. Atamtürk, A., Küçükyavuz, S.: Lot sizing with inventory bounds and fixed costs: polyhedral study and computation. Operations Research 53, 711–730 (2005)
2. Corman, T.H., Leiserson, C.E.: Introduction to Algorithms, 2nd edn. MIT Press, Cambridge (2001)
3. Khuller, S., Malekian, A., Mestre, J.: To fill or not to fill: the gas station problem. In: Proceedings of European Symposia on Algorithms (2007)
4. Lin, S.-H., Gertsch, N., Russell, J.: A linear-time algorithm for finding optimal refueling policies. Operations Research Letters 35(3), 290–296 (2007)
5. Love, S.F.: Bounded production and inventory models with piecewise concave costs. Management Science 20, 313–318 (1973)
6. Suzuki, Y.: A generic model for motor-carrier fuel optimization. Working paper, College of Business, Iowa State University

Scale Free Interval Graphs

Naoto Miyoshi[1], Takeya Shigezumi[1], Ryuhei Uehara[2],
and Osamu Watanabe[1]

[1] Tokyo Institute of Technology, Tokyo 152-8552, Japan
[2] JAIST, Ishikawa 923-1292, Japan

Abstract. Scale free graphs have attracted attention by their non-uniform structure that can be used as a model for various social and physical networks. In this paper, we propose a natural and simple random model for generating scale free interval graphs. The model generates a set of intervals randomly, which defines a random interval graph. The main advantage of the model is its simpleness. The structure/properties of the generated graphs are analyzable by relatively simple probabilistic and/or combinatorial arguments, which is different from the most of the other models for which we need to approximate the processes by certain differential equations. We indeed show that the distribution of degrees follows power law, and it achieves large cluster coefficient.

Keywords: scale free graph, small world network, interval graphs.

1 Introduction

Since early works by Watts & Strogatz [8] and Barabási & Albert [2], small world networks are the focus of recent interest because of their potential as models for the interaction networks of complex systems in real world [1,7]. There are three major properties that a small world network and/or a scale free network has (see, e.g., [4]): (SF) the node connectivities follow a scale free power law distribution, (CC) two neighbors of a node are also connected by an edge with high probability, and (SW) any two nodes are connected by a short path through a very few nodes called hubs.

Up to now, many models have been proposed and their properties have been investigated. Aside from few deterministic models, most of the randomized models are based on some dynamic *recursive* construction of random graphs. Thus, the analysis of certain properties of the obtained graphs becomes rather complicated, and it is not so easy to see the combinatorial structure of the obtained graphs. Typically, for example, in order to obtain a formula for the distribution of degrees (for showing the property (SF) mentioned above), one has to approximate the process by some differential equations and solve them. Therefore, although many random graph models have been proposed, we think that it is yet important to introduce some random graph model that can be easier to analyze by somewhat standard probabilistic/combinatorial methods. This is important in particular for designing and analyzing algorithms for scale free networks.

R. Fleischer and J. Xu (Eds.): AAIM 2008, LNCS 5034, pp. 292–303, 2008.

In this paper, we propose a simple random model for generating scale free interval graphs. Interval graphs have many applications from scheduling to bioinformatics. A graph $G = (V, E)$ is an interval graph if and only if G has an interval representation $\mathcal{I}$ such that each vertex v corresponds to an interval I_v and two vertices u and v are adjacent in G if and only if corresponding intervals I_u and I_v share a common interval on $\mathcal{I}$. For defining a random interval graph model, we introduce a way to randomly generate an interval representation $\mathcal{I}$; some standard random process is used for choosing intervals' starting points, and a power law distribution is used for determining intervals' lengths. This model has the following intuitive reasoning: Each interval is regarded as a period of existence, i.e., life, of some object (or creature), and relationships are created between these objects who have an overlap of lives. A power law distribution of a lifespan is derived from the simple rule "longer intervals tend to survive yet longer" (since experience is the best teacher).

Technically we consider a random model for generating interval representations. For combinatorial analysis, it is easier to assume that all intervals start at integer points and their lengths are integers. Thus, we adapt *the immigration and death process* for randomly choosing intervals' starting points as integers; this model has been studied well in the queuing theory as the infinite server model. We use a power law distribution on integers for determining lengths of generated intervals.

Although our interval model is defined as a random process, it is also possible to consider random interval distributions in a static way. For example (under the condition that n intervals are generated in a given period) we may assume that the starting points of these intervals are uniformly distributed in the period. Thus, the probabilistic/combinatorial structure of the model gets more clear, and we may be able to use various techniques for analyzing the obtained graphs. In fact, by relatively standard methods, we show that the obtained random interval graphs satisfy two properties of the scale free networks, namely, (SF) and (CC).

2 Preliminaries

We first introduce the notions for a (undirected) graph $G = (V, E)$ of which each *edge* $e = \{u, v\}$ in $E \subseteq V^2$ has no ordering. We only consider simple graphs without multiedges and self loops. The *neighborhood* of a vertex v in V is the set $N(v) = \{u \in V \mid \{u, v\} \in E\}$, and the *degree* of v is $|N(v)|$ denoted by $\deg_G(v)$. The subscript G can be omitted if no confusion can arise. We sometimes denote by $v \sim u$ if $u \in N(v)$. A sequence of distinct vertices $v_1, v_2, \ldots, v_\ell$ is a *path*, denoted by $(v_1, v_2, \ldots, v_\ell)$, if $\{v_j, v_{j+1}\} \in E$ for each $1 \leq j < \ell$. The *length* of a path is the number of edges on the path. For two vertices u and v, the *distance* of the vertices, denoted by $d(u, v)$, is the minimum length of the paths joining u and v. We define $d(u, v) = \infty$ if u is not reachable to v. The graph G is *connected* if $d(u, v) < \infty$ for each pair of vertices.

A graph (V, E) with $V = \{v_1, v_2, \ldots, v_n\}$ is an *interval graph* if there is a finite set of intervals $\mathcal{I} = \{I_{v_1}, I_{v_2}, \ldots, I_{v_n}\}$ on the real line such that $\{v_i, v_j\} \in E$ if and only if $I_{v_i} \cap I_{v_j} \neq \emptyset$ for each i and j with $0 < i, j \leq n$. We call the set $\mathcal{I}$ of intervals an *interval representation* of the graph. For each interval I, we denote by $R(I)$ and $L(I)$ the right and left endpoints of the interval, respectively (hence we have $L(I) \leq R(I)$ and $I = [L(I), R(I)]$). For any interval representation $\mathcal{I}$ and a point p, $\xi[p]$ denotes the set of intervals that contain the point p. We denote by $I_{v_i} \sim I_{v_j}$ if $I_{v_i} \cap I_{v_j} \neq \emptyset$, which means same as $v_i \sim v_j$ for an interval graph, and denote the length of an interval I by $|I|$.

In this paper, we focus on a discrete interval model. In this model, each interval I has two integer endpoints $L(I)$ and $R(I)$, and each interval is closed interval with minimum length 0. This model seems the most natural and simple one. However, sometimes, it is (intuitively) better to assume that the minimum length of an interval is 1. In this case, we may use another (but equivalent) interval model that consists of open intervals of length at least one. In the following, we use $[i..j]$ to denote the set of integers $\{i, i+1, \ldots, j\}$.

2.1   Scale Free Graph

Many social networks can be modeled as a scale free graph such that the degrees of the graph follow a scale free power law distribution [4]. More precisely, given a random distribution on some family of graphs, we consider the following condition for a random graph under this distribution: (SF) the probability that a vertex v has $\deg(v) = k$ is proportional to $k^{-\gamma}$ for some positive constant γ. We call such a random graph (more precisely, a random graph distribution) satisfying this condition *scale free*. Two other properties are required for the notion of *small world*. The first one is about "clustering coefficient", which characterizes the probability that two neighbors of a node are adjacent. The second one is the average (or longest) distance between any pair of vertices in the graph. In this paper we consider the first property and leave the second one for our future topic.

We explain a condition for the small world property on the clustering coefficient. For a vertex $v \in V$, *clustering coefficient of v*, denoted by $\mathsf{CC}(v)$, is defined by:

$$\frac{|\{\{u, w\} \in E \mid u, w \in N(v)\}|}{\binom{\deg(v)}{2}}$$

The *clustering coefficient of $G = (V, E)$*, denoted by $\mathsf{CC}(G)$, is defined by the arithmetical mean of the clustering coefficient of v in V. By definition, we immediately have the following:

$$\mathsf{CC}(G) = \frac{1}{|V|} \sum_{v \in V} \mathsf{CC}(v) = \frac{1}{|V|} \left(\sum_{v \in V} \sum_{u, w \in V \setminus \{v\}, u \neq w} \Pr[u \sim w \mid u \sim v \text{ and } w \sim v] \right).$$

As a desired property of small world graphs, for a given random distribution on some family of graphs, the following condition has been proposed: (CC) for some constant $c > 0$, $\mathsf{CC}(G)$ under the distribution is larger than c.

2.2   Probability Distributions

Our random interval graph model is defined based on a random interval generation model, a way of generating intervals randomly. To determine each interval's starting point, we use some random processes studied in the queuing theory; on the other hand, we use power law distribution for determining the length of each interval. Here we recall basic distributions and their important properties.

We use the Poisson distribution for specifying the distribution of intervals' starting points.   For specifying the distribution of intervals' lengths, we use a power law distribution.

We say that a random variable L on non-negative integers follows a *discrete power law distribution* with parameter α (which we denote $\mathcal{P}(\alpha)$) if it satisfies the following.

$$\Pr[L = k] = \frac{1}{\zeta(\alpha)}(k+1)^{-\alpha}, \quad (k \geq 0)$$

where $\zeta(\alpha) = \sum_{i=1}^{\infty} i^{-\alpha}$ is the Riemann's zeta function. Here we note the following property for this random variable L following $\mathcal{P}(\alpha)$.

$$p_k = \Pr[L \geq k+1 \mid L \geq k] = \frac{\zeta(\alpha, k+2)}{\zeta(\alpha, k+1)} \tag{1}$$

where $\zeta(\alpha, n) = \sum_{i=n}^{\infty} i^{-\alpha}$ is the generalized zeta function. This probability, say p_k, increases as k increasing. This gives the simple rule as mentioned in Section 1; "longer intervals tend to survive yet longer".

3   New Model of Scale Free Interval Graphs

We here present the random generation model of interval graphs. We use the immigration and death process [3] to generate an interval representation. The immigration and death process is one of the waiting queue model such as the customers arrive independently of other customers and there exist infinite number of gates for service.

In our model, we set a clock $T = 1, 2, \ldots$ and put intervals on each time using the Poisson distribution. The algorithm for our model, *put-intervals* (λ, α, n), is shown in Algorithm 1. In this algorithm, the variable T is the clock for the arriving time, t_T holds the number of intervals that begin at time T, and the subprocedure **Poisson**(λ) is a random procedure returning an integer according to Poisson(λ). To decide the length of an interval, we use the conditional probability p_l given in Equation (1), which means that if an interval I has the current length l, the final length of this interval, L, is longer than $l + 1$ with probability p_l. We call sometimes p_l as the probability of survive at length l. Note that the distribution of the length of intervals are $\mathcal{P}(\alpha)$.

Actually, this is an approximated approach to the model below. Consider a coin such that lands on heads with probability p. Flip the coin m times at each time step T and if the coin lands on head, we put an interval starting at time T. The number of heads on time T follows $B(m, p)$, and we can approximate it

Algorithm 1. *Put-intervals* (λ, α, n)

```
input  : Parameters λ, α, and n.
output: A set of intervals I.
begin
    T = 1, i = 1, I = φ;
    while i ≤ n do
        t_T = Poisson(λ);
        start t_T intervals from T, i.e.,
        for j = i to i + t_T − 1 do
            set L(I_j) = T;
            let the initial length l_j = 0;
        end
        i = i + t_T;
        for each interval I_i' exists at T do
            decide it will be dead or alive at T + 1 with probability p_{l_i'};
            if alive then
                l_i' = l_i' + 1;
            end
            else
                let the right endpoint of the interval R(I_i') = T;
                add I_i' = [L(I_i'), R(I_i')] = [T − l_i', T] to I;
            end
        end
        proceed the clock T to T + 1;
    end
    output I.
end
```

with the Poisson distribution, Poisson(λ), if m tends to infinity as the expected number of heads, $pm = \lambda$, remains fixed.

We here consider values of $\zeta(\alpha)$ and how it is related intervals' lifespan for a typical value of the parameter α.

Example 1. It has been usually claimed that typical scale-free networks satisfy (SF) with $\alpha = 2.1 \simeq 2.8$. Since our later analysis shows that the smaller α gives the smaller clustering coefficient in our model, we consider $\alpha = 2.1$ for our example. Then, since $\zeta(2.1) \simeq 1.560$, we have *on average* $n_0 \simeq 0.641n$, $n_1 \simeq 0.150n$, and $n_2 \simeq 0.064n$, where n_i denotes the number of vertices such that corresponding interval has length i, and n denotes the number of vertices.

4   Scale Free Property

In this section, we will show that our scale free interval graph has the degree sequence following a power law distribution.

To consider the degree of a vertex, let us define $\xi(T)$ for time T and $A(I)$ for an interval I. $\xi(T)$ is the number of intervals which exist on time T in our

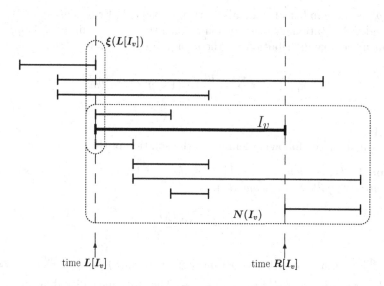

Fig. 1. An example of the degree of v: $\deg(v) = 9$. There are 6 intervals on time $L[I_v]$, 6 intervals are put on time $[L[I_v]..R[I_v]]$ and 3 intervals starts at time $L[I_v]$.

algorithm 1. $A(I)$ represents the number of intervals whose left endpoints are put on $[L[I]..R[I]]$. It is easy to see that the degree of a vertex v is the sum of $\xi(L[I_v])$ and $A(I_v)$ minus $t_{L[I_v]}$. (See figure 1). $\xi(L[I_v]) - t_{L[I_v]}$ means the number of intervals which exist on time $L[I_v]$ and started before $L[I_v]$. We will analyze the stationary distribution $\pi(k)$ of $\xi(T)$ (i.e., $\pi(k) = \lim_{T \to \infty} \Pr[\xi(T) = k]$) and $A(I)$. First, we will show that $\xi(T) - t_T$ follows Poisson($\lambda \frac{\zeta(\alpha-1)-1}{\zeta(\alpha)}$) in the steady state. Second, we will show that $\Pr[A(I) = k]$ follows a power law distribution for large k. Third, we conclude with the fact that a power law distribution dominates the Poisson distribution for large degrees. In the rest of this section, we use the $f(x) \sim g(x)$ notation to approximate $f(x)$ by $g(x)$. Precisely, "$f(x) \sim g(x)$ as $x \to \infty$" stands for "$\lim_{x \to \infty} f(x)/g(x) = 1$".

Consider the time T of the procedure *put-intervals*. Some intervals exist and each of them has their current length ≥ 0. Since the length of an interval follows $\mathcal{P}(\alpha)$, the probability of survive depends on the current length of the interval. Let the p_i (for $i \geq 0$) be the probability such that an interval whose current length is i at time T will survive at time $T+1$. Since we consider that the length of an interval follows $\mathcal{P}(\alpha)$, p_i is derived from equation (1).

Let ρ_i^T be the number of intervals which are alive and have current length i at time T. As the time T will proceed, ρ_{i+1}^{T+1} is depends only on ρ_i^T because some of intervals of ρ_i^T will survive at time $T+1$ with probability p_i and others ends at time T. From this observation, we obtain this formula for $i \geq 0$:

$$\Pr[\rho_{i+1}^{T+1} = k] = \sum_{m=k}^{\infty} \binom{m}{k} p_i^k (1-p_i)^{m-k} \Pr[\rho_i^T = m]. \tag{2}$$

Since ρ_0^T is the number of intervals starting at time T, $\Pr[\rho_0^T = k] = e^{-\lambda}\frac{\lambda^k}{k!}$. Let us consider the stationary distribution π_i such that $\pi_i(k) = \lim_{T\to\infty} \Pr[\rho_i^T = k]$. For the stationary distribution π_i, the equation (2) becomes

$$\pi_{i+1}(k) = \sum_{m=k}^{\infty} \binom{m}{k} p_i^k (1 - p_i)^{m-k} \pi_i(m) \tag{3}$$

and $\pi_0(k) = e^{-\lambda}\frac{\lambda^k}{k!}$.

We will show the following lemma as the solution of the equation (3).

Lemma 1. *Let us denote* $P_i = \prod_{j=0}^{i-1} p_j$ *for* $i \geq 1$ *and* $P_0 = 1$. *The stationary distribution* π_i *follows* $Poisson(\lambda P_i)$;

$$\pi_i(k) = e^{-\lambda P_i}\frac{(\lambda P_i)^k}{k!}.$$

Proof. The proof is done by induction. For $i = 0$, $\pi_0(k) = e^{-\lambda P_0}\frac{(\lambda P_0)^k}{k!}$. Assume it holds for $i \leq k$, i.e., $\pi_i(k) = e^{-\lambda P_i}\frac{(\lambda P_i)^k}{k!}$. The stationary distribution is:

$$\pi_{i+1}(k) = \sum_{m=k}^{\infty} \frac{m!}{(m-k)!k!} p_i^k (1 - p_i)^{m-k} e^{-\lambda P_i}\frac{(\lambda P_i)^m}{m!}$$

$$= e^{-\lambda P_i}\frac{(\lambda p_i P_i)^k}{k!} \sum_{m'=0}^{\infty} \frac{\{\lambda P_i(1 - p_i)\}^{m'}}{m'!} = e^{-\lambda P_i}\frac{(\lambda P_{i+1})^k}{k!} e^{\lambda P_i(1-p_i)}\frac{(\lambda P_{i+1})^k}{k!}.$$

Note that we used $p_i P_i = P_{i+1}$ in the above.   □

We now can obtain the stationary distribution of $\xi(T)$, however, what we want to know is the distribution of $\xi(T) - t_T$ since the degree of a vertex is $(\xi(L[I]) - t_{L[I]}) + A(I)$. we conclude the first part of the degree analysis with the following lemma.

Lemma 2. *Let* $\xi(T)$ *be the number of intervals which are alive at time* T *and* t_T *be the number of intervals starting at time* T. *Then* $\xi(T) - t_T$ *follows* $Poisson(\lambda\frac{\zeta(\alpha-1)-1}{\zeta(\alpha)})$ *in the steady state.*

Proof. Since π_is are independent and follow $Poisson(\lambda P_i)$, the sum $\xi(T) - \pi_0 = \sum_{i=1}^{\infty} \pi_i$ also follows the Poisson distribution with parameter $\sum_{i=1}^{\infty} \lambda P_i = \lambda\frac{\zeta(\alpha-1)-1}{\zeta(\alpha)}$.   □

Second, we will show that $A(I)$ follows a power law distribution. Recall that for any T_0 and $T \geq 0$, the number of intervals which start on time $[T_0..T_0+T]$ follows the Poisson distribution with parameter $\lambda(T+1)$. Let us suppose that the interval I has length l. The number of intervals starting on $[L[I]..R[I]](= [L[I]..L[I]+l])$ is;

$$\Pr[A(I) = k \mid |I| = l] = e^{-\lambda(l+1)}\frac{\{\lambda(l+1)\}^k}{k!} \tag{4}$$

Since the length of an interval follows $\mathcal{P}(\alpha)$,

$$\Pr[A(I) = k] = \sum_{l=0}^{\infty} e^{-\lambda(l+1)} \frac{\{\lambda(l+1)\}^k}{k!} \Pr[|I| = l]$$

$$= \sum_{l=1}^{\infty} e^{-\lambda l} \frac{(\lambda l)^k}{k!} \frac{1}{\zeta(\alpha)} l^{-\alpha} = \frac{\lambda^k}{\zeta(\alpha)k!} \sum_{l=1}^{\infty} e^{-\lambda l} l^{k-\alpha} = \frac{\lambda^k}{\zeta(\alpha)k!} \sum_{l=0}^{\infty} e^{-\lambda l} l^{k-\alpha}$$

For the last formula , we show the following lemma.

Lemma 3.

$$\frac{\lambda^k}{k!} \sum_{l=0}^{\infty} l^{k-\alpha} e^{-\lambda l} \sim \lambda^{\alpha-1} k^{-\alpha} \quad as \ k \to \infty.$$

Proof. Let $f(k) = \frac{\lambda^k}{k!} \sum_{l=0}^{\infty} l^{k-\alpha} e^{-\lambda l}$ and suppose $k > \alpha$. We then have

$$f(k) \geq \frac{\lambda^k}{k!} \int_0^{\infty} x^{k-\alpha} e^{-\lambda x} \, dx - \frac{\lambda^{\alpha}}{k!} (k-\alpha)^{k-\alpha} e^{-(k-\alpha)},$$

$$f(k) \leq \frac{\lambda^k}{k!} \int_0^{\infty} x^{k-\alpha} e^{-\lambda x} \, dx + \frac{\lambda^{\alpha}}{k!} (k-\alpha)^{k-\alpha} e^{-(k-\alpha)}.$$

Let the integral part $g(k) = \frac{\lambda^k}{k!} \int_0^{\infty} x^{k-\alpha} e^{-\lambda x} \, dx$ and the rest $h(k) = \frac{\lambda^{\alpha}}{k!} (k - \alpha)^{k-\alpha} e^{-(k-\alpha)}$. We show that $g(k) \sim \lambda^{\alpha-1} k^{-\alpha}$ as $k \to \infty$ and $h(k) = o(k^{-\alpha})$ as $k \to \infty$. Changing the variable to $u = \lambda x$, we have

$$g(k) = \frac{\lambda^{\alpha-1}}{k!} \int_0^{\infty} u^{k-\alpha} e^{-u} \, du = \frac{\lambda^{\alpha-1}}{k!} \Gamma(k - \alpha + 1),$$

where Γ denotes the Gamma function $\Gamma(s) = \int_0^{\infty} u^{s-1} e^{-u} \, du$. We use the following properties of Gamma function;

(i) $\Gamma(s+1) = s\, \Gamma(s)$,

(ii) $\Gamma(s) = \lim_{k \to \infty} \frac{k!\, k^s}{s\,(s+1)\cdots(s+k)} \quad (s \neq 0, -1, -2, \cdots)$.

Then, we have for $\alpha \neq 0, 1, 2, \ldots$,

$$g(k) \times k^{\alpha} = \lambda^{\alpha-1} \frac{(k-\alpha)\cdots(1-\alpha)(-\alpha)}{k!\, k^{-\alpha}} \Gamma(-\alpha) \to \lambda^{\alpha-1} \quad as \ k \to \infty.$$

When α is a nonnegative integer, we have

$$g(k) = \lambda^{\alpha-1} \frac{(k-\alpha)!}{k!} \sim \lambda^{\alpha-1} k^{-\alpha} \quad as \ k \to \infty.$$

Finally, for the term $h(k)$, applying Stirling's formula $k! \sim \sqrt{2\pi}\, k^{k+1/2} e^{-k}$ as $k \to \infty$, we have

$$h(k) \times k^{\alpha} = \lambda^{\alpha} \frac{k^k e^{-(k-\alpha)}}{k!} \left(1 - \frac{\alpha}{k}\right)^{k-\alpha} \sim \frac{\lambda^{\alpha}}{\sqrt{2\pi}} k^{-1/2} \to 0 \quad as \ k \to \infty.$$

$\square$

Applying Lemma 3 , we obtain

$$\Pr[A(I) = k] \sim \frac{1}{\zeta(\alpha)} \lambda^{\alpha-1} k^{-\alpha}.$$

Third, we will show that we can neglect the effect of $\xi(L[I]) - t_{L[I]}$ if the degree of v is large enough. We present the following lemma.

Lemma 4. *Let $\bar{F}$ and $\bar{G}$ be the tail probability of a power law distribution with parameter α and the Poisson distribution with parameter $\lambda\mu$, respectively. In precise, using the constant c, $\bar{F}(k) = c\sum_{i=k}^{\infty} i^{-\alpha}$ and $\bar{G}(k) = \sum_{i=k}^{\infty} e^{-\lambda\mu} \frac{(\lambda\mu)^k}{k!}$. Then we have;*

$$\frac{\bar{G}(k)}{\bar{F}(k)} \to 0 \quad as \ k \to \infty.$$

Proof. Let us recall that $k! \geq \left(\frac{k}{3}\right)^k$ and if $\lim_{x\to\infty} f(x) = 0$ and $\lim_{x\to\infty} g(x) = 0$, then

$$\lim_{x\to\infty} \frac{g(x)}{f(x)} = \lim_{x\to\infty} \frac{\frac{d}{dx}g(x)}{\frac{d}{dx}f(x)}.$$

$$\lim_{k\to\infty} \frac{\bar{G}(k)}{\bar{F}(k)} = \lim_{k\to\infty} \frac{\sum_{i=k}^{\infty} e^{-\lambda\mu} \frac{(\lambda\mu)^i}{i!}}{c\sum_{i=k}^{\infty} i^{-\alpha}} < \lim_{k\to\infty} \frac{e^{-\lambda\mu} \sum_{i=k}^{\infty} \left(\frac{3\lambda\mu}{i}\right)^i}{c\sum_{i=k}^{\infty} i^{-\alpha}}$$

$$< \lim_{k\to\infty} \frac{e^{-\lambda\mu} \int_k^{\infty} \left(\frac{3\lambda\mu}{x}\right)^x dx}{c\int_k^{\infty} x^{-\alpha}dx} = \frac{e^{-\lambda\mu}}{c} \lim_{x\to\infty} \frac{\left(\frac{3\lambda\mu}{x}\right)^x}{x^{-\alpha}} = 0$$

$\square$

By Lemma 4 with $\mu = \frac{\zeta(\alpha-1)-1}{\zeta(\alpha)}$, for sufficiently large degree vertices, $A(I_v)$ dominates $\xi[L[I_v]] - t_{L[I_v]}$ on the degree distribution. We now conclude with the following result.

Theorem 1. *A scale free interval graph generated according to our discrete model has the degree sequence following $\mathcal{P}(\alpha)$ for large degrees.*

5   Clustering Coefficient

In this section, we will show the constant lower bound of the expected value of the clustering coefficient. First, we show that there are many *short* intervals and, second, the expected value of the clustering coefficient of those short intervals are large. For given $G = (V, E)$, we partition V into $V_0, V_1, \ldots$ such that V_i contains vertices corresponding to intervals of length i. Let n_i be the number of vertices in V_i, and n the number of vertices in V. Our goal is to show how to compute a lower bound of the clustering coefficient of a scale free interval graph since it depends on the distribution of n_i. Typically, we have the following lower bound.

Example 2. By Example 1, we assume that the expected values of n_0, n_1, and n_2 are $0.641n$, $0.150n$, and $0.064n$, respectively. Then the expected value of the clustering coefficient of G, $\mathsf{CC}(G)$, is at least 0.7713.

We have the key lemma is the following, which is independent of the degree distribution:

Lemma 5. *Let I_u, I_v, I_w be any three intervals placed randomly. We assume that the positions of the intervals are independent, and the universal interval is long enough. Then, $\Pr[I_u \sim I_w \mid I_u \sim I_v \text{ and } I_w \sim I_v] = \dfrac{l_u l_v + l_v l_w + l_w l_u + l_u + l_v + l_w + 1}{(l_u + l_v + 1)(l_w + l_v + 1)}$.*

Proof. To simplify, we shift the whole intervals and fix $L(I_v) = 0$ and $R(I_v) = l_v$. Then $R(I_u)$ takes i in $[0..l_v + l_u]$ with conditional probability $\frac{1}{l_u + l_v + 1}$ given $I_u \sim I_v$. Similarly, $R(I_w)$ takes j in $[0..l_v + l_w]$ with conditional probability $\frac{1}{l_w + l_v + 1}$ given $I_w \sim I_v$.

We first assume that $l_w \leq l_v \leq l_u$. Then, for each i in $[0..l_v + l_u]$, we have the following cases; $|I_u \cap I_v| = i$ for $0 \leq i < l_v$, $|I_u \cap I_v| = l_v$ for $l_v \leq i \leq l_u$, and $|I_u \cap I_v| = l_u + l_v - i$ for $l_u < i \leq l_u + l_v$. That is, we have $l_v + l_u + 1$ different cases that I_u intersects with I_v. Moreover by the useful property of the Poisson distribution, we can treat each of them occurring with the same probability [6]. Now, we turn to consider the cases that I_w intersects with $I_u \cap I_v$. The number of cases that I_w intersects with I_v is $l_v + l_w + 1$. Among them, when $0 \leq i < l_v$, I_w intersects with $I_u \cap I_v$ for each $j = R(I_w)$ with j in $[0..i + l_w]$. If $l_v \leq i \leq l_u$, I_w always intersects with $I_u \cap I_v = I_v$. The case $l_u < i \leq l_u + l_v$ is symmetric. Hence, taking average, we have

$$\Pr[I_u \sim I_w \mid I_u \sim I_v \text{ and } I_w \sim I_v]$$

$$= \frac{1}{l_v + l_w + 1}\left(2\sum_{i=0}^{l_v - 1} \frac{i + l_u + 1}{l_v + l_u + 1} + (l_w - l_v + 1) \right)$$

$$= \frac{l_u l_v + l_v l_w + l_w l_u + l_u + l_v + l_w + 1}{(l_u + l_v + 1)(l_w + l_v + 1)}.$$

In the other two cases ($l_v < l_w, l_u$, and $l_w, l_u < l_v$), we can analyze in a similar way, and obtain equations which imply the same results. $\square$

Hereafter, we denote by $f(l_v; l_u, l_w) = \frac{l_u l_v + l_v l_w + l_w l_u + l_u + l_v + l_w + 1}{(l_u + l_v + 1)(l_w + l_v + 1)}$. It is easy to check that for any fixed positive integer l_v, $f(l_v; l_u, l_w)$ is a nondecreasing function for l_u and l_w. We also note that $f(0; l_u, l_w) = 1$ for any l_u and l_w, which means that any two intervals I_u and I_w intersecting with I_v of length 0, I_u and I_w share a common interval I_v, which is a point.

Now, we turn to the computation of the lower bound of the expected value of $\mathsf{CC}(G)$. We denote by $\mathsf{CC}(V_i)$ the expected value of the clustering coefficient of a vertex in V_i. Then we have $\mathsf{CC}(G) = \frac{1}{n}\sum_{i=0,1,\dots} n_i \mathsf{CC}(V_i)$. In this section, our goal is to give a good lower bound of $\mathsf{CC}(V_i)$. In our model, first few V_is are influential. Hence we can give a good lower bound by analyzing them.

Lemma 6. *We have* $CC(V_0) = 1$, *and* $CC(V_1) > (63n^2 - 9n_0^2 - n_1^2 - 18n_0n - 6n_1n - 6n_0n_1 - 183n + 51n_0 + 15n_1 + 112)/(72(n - 2)(n - 1))$.

Proof. Due to the limit of space, we omitted the proof. For the complete proof, see [5]. $\qquad\qquad\qquad\qquad\qquad\qquad\qquad\qquad\qquad\qquad\qquad\qquad\qquad\qquad\qquad$ □

By the equation of Lemma 6, we have a lower bound of $CC(G)$ for fixed α. For example, letting $n_0 = 0.641n$ and $n_1 = 0.150n$ (see Example 1), we have $CC(G) = \frac{1}{n}\sum_{i=0,1,\ldots} n_i CC(V_i) > \frac{1}{n}\sum_{i=0,1} n_i CC(V_i) = \frac{1}{n}(n_0 + \frac{46.2647n^2 - 148.059n + 112}{72(n-1)(n-2)}n_1) = \frac{1}{n}(0.641n + \frac{452n^2 - 1395n + 1008}{648(n-1)(n-2)}0.150n) = \frac{0.737385(n-1.99726)(n-1.02891)}{(n-1)(n-2)} \simeq 0.7374$. In Lemma 6, we only consider three sets V_0, V_1, and $V \setminus (V_0 \cup V_1)$. We can repeat the idea once more, and obtain a better lower bound. The computations in the proof are straightforward and tediously, and hence omitted.

Lemma 7. *We have* $CC(V_1) > (1656n^2 - 324n_0^2 - 64n_1^2 - 9n_2^2 - 432n_0n - 192n_1n - 72n_2n - 288n_1n_0 - 108n_2n_0 - 48n_2n_1 - 4776n + 1476n_0 + 576n_1 + 201n_2 + 2800)/(1800(n-1)(n-2))$ *and* $CC(V_2) > (500n^2 - 100n_0^2 - 25n_1^2 - 4n_2^2 - 200n_0n - 100n_1n - 40n_2n - 100n_1n_0 - 40n_2n_0 - 20n_2n_1 - 1460n + 540n_0 + 245n_1 + 92n_2 + 921)/(600(n-1)(n-2))$.

Using the equations of Lemma 7, we have a better lower bound of $CC(G)$ for fixed α. For example, letting $n_0 = 0.641n$, $n_1 = 0.150n$, and $n_2 = 0.064n$, we have $CC(G) = \frac{1}{n}\sum_{i=0,1,\ldots} n_i CC(V_i) > \frac{1}{n}\sum_{i=0,1,2} n_i CC(V_i) = \frac{1}{n}\left(n_0 + \frac{1178.5n^2 - 3730.62n + 2800}{1800(n-1)(n-2)}n_1 + \frac{301.125n^2 - 1071.22n + 921}{600(n-1)(n-2)}n_2\right) = \frac{0.771328(n-1.99648)(n-1.04782)}{(n-1)(n-2)} \simeq 0.7713$.

6   Concluding Remarks

In this paper, we have proposed the scale free interval graph model, and analyzed that it has power law degree distribution and large clustering coefficient. Actually, we had considered the time-continuous model which is almost the same as the time-discrete model introduced in Section 3. For the time-continuous model, we also showed the following results.

Theorem 2. *A scale free interval graph generated according to our time-continuous model has the degree sequence following* $\mathcal{P}(\alpha)$ *for large degrees.*

Theorem 3. *For* $\alpha = 2.1$ *(same as Example 1), A scale free interval graph generated according to our time-continuous model has the expected clustering coefficient of G is at least 0.159. Note that it is independent of the size of the graph.*

However, our model seems to not satisfy the third property (SW). The property (SW) is that any two nodes are joined by short path, which is estimated by average or longest distance between any two nodes in G. Our experimental results showed that the average distance and the diameter of the graph are both linear in n. We leave for future works that proposing scale free interval graph model that has the property (SW).

Acknowledgment

This research was supported in part by JSPS Global COE program "Computationism as a Foundation for the Sciences".

References

1. Barabási, A.: Linked: The New Science of Networks. Perseus Books Group (2002)
2. Barabási, A., Albert, R.: Emergence of Scaling in Random Networks. Science 286(5439), 509–512 (1999)
3. Cox, D.R., Isham, V.: Point Processes. Chapman & Hall, Boca Raton (1980)
4. Newman, M.: The structure and function of complex networks. SIAM Review 45, 167–256 (2003)
5. Miyoshi, N., Shigezumi, T., Uehara, R., Watanabe, O.: Scale Free Interval Graphs. Dept. of Math. and Comp. Sciences Tokyo Institute of Technology Research Reports (Series C), series C-255 (2008),
   http://www.is.titech.ac.jp/research/research-report/C/C-255.pdf
6. Takács, L.: Introduction to the Theory of Queues. Oxford University Press, Oxford (1962)
7. Watts, D.J.: Small Worlds: The Dynamics of Networks Between Order and Randomness. Princeton University Press, Princeton (2004)
8. Watts, D.J., Strogatz, D.H.: Collective Dynamics of 'Small-World' Networks. Nature 393, 440–442 (1998)
9. Wolff, R.W.: Poisson Arrivals See Time Averages. Operations Research 30, 223–231 (1982)

On Representation of Planar Graphs by Segments

Sadish Sadasivam and Huaming Zhang*

Computer Science Department
University of Alabama in Huntsville
Huntsville, AL, 35899, USA
{ssadasiv,hzhang}@cs.uah.edu

Abstract. In this paper, we introduce Vertex-face contact representation (VFCR for short) for 2-connected plane multigraphs. We present a simple linear time algorithm for constructing a VFCR for 2-connected plane graphs. Our algorithm only uses an st-orientation for G and its corresponding st-orientation for the dual graph of G. We also show that one kind of vertex-vertex contact representation (VVCR) for 2-connected bipartite planar graphs introduced by Fraysseix et al. [2,3] can be easily obtained by applying our algorithm. In general, our algorithm produces a more compact representation than their algorithm.

Then we investigate st-orientations for 3-connected planar graphs. We prove that a 3-connected planar graph G with n vertices and f faces, has an st-orientation with the length of its longest directed path $\leq \frac{2n}{3} + 2\lceil \sqrt{n/3} \rceil + 5$. This implies that such a graph G admits a VFCR in a grid with non-trivial size bound. This non-trivial size bound also applies to the vertex-vertex contact representation [2,3] for a large class of 2-connected bipartite planar graphs.

1 Introduction

Geometric representation of planar graphs has drawn extensive attention from computer science community. There are many kinds of geometric representations of planar graphs. In [2,3], Fraysseix et al. introduced *vertex-vertex contact representation* (VVCR for short) for 2-connected bipartite planar graphs. In this representation, each vertex is represented by either a vertical or a horizontal segment so that no two of them have a common interior point, and two segments have a point in common if and only if the corresponding vertices are adjacent in G.

Consider VVCR for 2-connected bipartite planar graphs. Without loss of generality, we can assume that the given bipartite planar graph is maximal, i.e., all its facial cycles are quadrangles. Let $G = (V_1 \cup V_2, E)$ be such a 2-connected maximal bipartite planar graph, where V_1 and V_2 are disjoint vertex subsets and both are independent vertex subsets. In [2,3], Fraysseix et al. presented a linear time algorithm to construct a VVCR for G in a grid of size $W \times H$, where

* This research is supported in part by NSF grant CCF-0728830.

R. Fleischer and J. Xu (Eds.): AAIM 2008, LNCS 5034, pp. 304–315, 2008.
© Springer-Verlag Berlin Heidelberg 2008

the width W is always $(|V_1| - 1)$, and the height H is always $(|V_2| - 1)$. (Of course, here the roles played by width and height are symmetric, i.e., they are interchangeable.)

We introduce the following geometric representation for plane graphs (a plane graph is a planar graph with a fixed planar embedding):

Definition 1. *Let G be a 2-connected plane multigraph. A* vertex-face contact representation *(VFCR for short) of G is a representation such that: a) Each vertex is drawn as a horizontal line segment (called* vertex segment*), each face is drawn as a vertical line segment (called* face segment*). (b) No two segments have a common interior point. A vertex segment and a face segment have a point in common if and only if their corresponding vertex and face are adjacent in G.*

Let G be a 2-connected plane multigraph. The *angular graph* of G is obtained through the following steps: (1) add one vertex for each face (including exterior face), call it *face node*; (2) add an edge between each face node and all its neighboring vertices in G; (3) delete all the original edges in G. The angular graph of G is denoted by $\mathcal{A}(G)$. Obviously, $\mathcal{A}(G)$ is a 2-connected maximal bipartite plane graph. The mapping from G to $\mathcal{A}(G)$ can be easily done in linear time. On the other hand, let G be a 2-connected maximal bipartite plane graph with n vertices. G is 2-colorable and we color its vertices by black and white. Each face f of G is adjacent to two black vertices. We connect the two black vertices in f by an edge within f. The collection of these edges is a multigraph G_b on black vertices. G_w can be defined similarly on the set of white vertices of G. Both G_w and G_b are 2-connected plane multigraphs. The transformation from G to G_b and G_w can also be done in linear time.

Simple observations give the following technical lemma.

Lemma 1. *1. Let G be a 2-connected plane multigraph. A VFCR of G is a VVCR of $\mathcal{A}(G)$.*

2. Let G be a 2-connected maximal bipartite plane graph. A VVCR of G is a VFCR of G_b and G_w. (Subject to a 90-degree rotation if necessary. This is because we insist that the vertex segments are placed horizontally in a VFCR.)

Applying above lemma and the existence of VVCR for 2-connected maximal bipartite plane graphs, we actually prove that every 2-connected plane multigraph $G = (V, E)$ has a VFCR and it can be constructed in linear time. The size of the VFCR for G is always $(|F| - 1) \times (|V| - 1)$, where $|F|$ is the number of faces in G, $|V|$ is the number of vertices of G.

In this paper, we consider the problem the other way around. Namely, we consider the problem from VFCR first. We present a linear time algorithm for constructing VFCR for a 2-connected plane multigraph. Our algorithm has the following advantages: (1) The algorithm is simple and it only uses an st-orientation $\mathcal{O}$ of G and its corresponding st-orientation $\mathcal{O}^*$ of the dual graph of G. The coordinates of the segments carry clear combinatorial meanings. The grid size is $\text{length}(\mathcal{O}^*) \times \text{length}(\mathcal{O})$, where $\text{length}(\mathcal{O})$ ($\text{length}(\mathcal{O}^*)$, respectively) is the length of the longest directed path in $\mathcal{O}$ ($\mathcal{O}^*$, respectively). (2) Although the grid size bound remains to be the trivial bound $(|F|-1) \times (|V|-1)$, it in general produces VFCR (and hence VVCR) in a smaller grid size for 2-connected plane graphs in practice. Actually, Papamanthou and Tollis [8] introduced a practical algorithm on controlling the length of the longest directed paths in st-orientations and they conjectured that for a 2-connected planar graph, there exists an st-orientation of G such that $(\text{length}(\mathcal{O}) + \text{length}(\mathcal{O}^*)) \leq 2|V|$.

For a 3-connected planar graph, all its planar embeddings are derived from its unique spherical embedding. Therefore, its faces and vertices maintain the same adjacency relations in all its different planar embeddings. Hence, it is meaningful to define VFCR for 3-connected planar graphs. Let G be a 3-connected planar graph with n vertices and f faces. We present a linear time algorithm to construct an st-orientation of G such that either $\text{length}(\mathcal{O}) \leq \frac{2n}{3} + 2\lceil \sqrt{n/3} \rceil + 5$, or $\text{length}(\mathcal{O}^*) \leq \frac{2f}{3} + 2\lceil \sqrt{f/3} \rceil + 5$. This bound result is of independent interest. In addition, it guarantees a non-trivial grid size bound for VFCR for 3-connected planar graphs and hence it guarantees a non-trivial grid size bound for VVCR for a large class of bipartite planar graphs, i.e., the class of angular graphs of 3-connected planar graphs.

2   Preliminaries

A graph $G = (V, E)$ is a *multigraph* if G allows more than one edge between its any two vertices. A graph G is a *planar graph* if G admits a planar embedding without edge crossings except possibly at their common end vertices. A *plane graph* G is a planar graph with a fixed planar embedding. G is called a *directed graph* (digraph for short) if each edge of G is assigned a direction. We abbreviate the words "counterclockwise" and "clockwise" as ccw and cw respectively.

The *dual graph* $G^* = (V^*, E^*)$ of a plane graph G is defined as follows: For each face F of G, G^* has a node v_F. For each edge e in G, G^* has an edge $e^* = (v_{F_1}, v_{F_2})$ where F_1 and F_2 are the two faces of G with e on their common boundaries. e^* is called the *dual edge* of e. For each vertex $v \in V$, the dual face of v in G^* is denoted by v^*.

An *orientation* $\mathcal{O}$ of a multigraph G is a digraph obtained from G by assigning a direction to each edge of G. We will use G to denote both the resulting digraph and the underlying undirected graph unless otherwise specified. (Its meaning will be clear from the context.) The length of the longest directed path in $\mathcal{O}$ is denoted by $\text{length}(\mathcal{O})$.

For a 2-connected plane multigraph G and an exterior edge (s, t), an orientation of G is called an *st-orientation* if the resulting digraph is acyclic with s as the only source and t as the only sink.

Given an st-orientation $\mathcal{O}$ of 2-connected plane multigraph G, consider the dual graph G^* of G. For each $e \in G$, we direct its dual edge e^* from the face on the left of e to the face on the right of e when we walk on e along its direction in $\mathcal{O}$. We then reverse the direction of $(s, t)^*$. It was shown in [9,10] that this orientation is an st-orientation of G^* with $(s, t)^*$ as the distinguished exterior

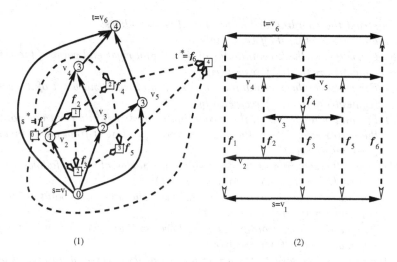

(1)   (2)

Fig. 1. (1) An st-orientation $\mathcal{O}$ of a graph G and its corresponding st-orientation $\mathcal{O}^*$ of G^*. (2) A VFCR for G.

edge. We will denote this orientation of G^* by $\mathcal{O}^*$ and call it the *corresponding st-orientation* of $\mathcal{O}$. For each vertex v in G, the length of the longest directed path from s to v in $\mathcal{O}$ will be denoted by $\alpha(v)$. Similarly, for each node f in G^*, $\alpha(f)$ is the length of the longest directed paths from s^* to f in $\mathcal{O}^*$. Note that $\alpha(s) = 0$, $\alpha(s^*) = 0$, $\alpha(t) = \text{length}(\mathcal{O})$, and $\alpha(t^*) = \text{length}(\mathcal{O}^*)$. For example, Figure 1 (a) gives an example of an st-orientation of a 2-connected plane graph G and its corresponding st-orientation of G^*. The edges of G are drawn in solid lines, while the edges of G^* are drawn in dashed lines. The number inside the circle for each vertex v is $\alpha(v)$. The number inside the square for each face f is $\alpha(f)$. Obviously, $\text{length}(\mathcal{O}) = 4$ and $\text{length}(\mathcal{O}^*) = 4$.

3   VFCR for 2-Connected Plane Graphs

Let G be a 2-connected plane multigraph. Using an st-orientation of G and its corresponding st-orientation of G^*, with some modification to the existing *visibility representation* algorithm from [9,10], we have the following algorithm for constructing a VFCR for G.

Algorithm 1. Vertex-face contact representation
Input: A 2-connected plane multigraph $G = (V, E)$.
Output: A VFCR for G.

1. *Select an exterior edge $(s, t) \in E$. Compute an st-orientation $\mathcal{O}$ for G, where s is the source, t is the sink.*
2. *Construct its dual graph G^* and its dual orientation $\mathcal{O}^*$.*
3. *For each vertex $v \in G$ and $f \in G^*$, compute $\alpha(v)$ and $\alpha(f)$ respectively.*

4. *Construct the coordinates for vertices of G as follows:*
   (a) *For each vertex, its y coordinate is $\alpha(v)$.*
   (b) *For vertex s, the x coordinates of its left and right end point of the vertex segment are 0, $\alpha(t^*)$ respectively.*
   (c) *For vertex t, the x coordinates of its left and right end point of the vertex segment are 0, $\alpha(t^*)$ respectively.*
   (d) *For any other vertex v, the x coordinates of its left and right end point of the vertex segment are the smallest and biggest $\alpha(f)$ from all its incident faces respectively.*
5. *Construct the coordinates for faces of G, i.e. nodes of G^* as follows:*
   (a) *For each face f, its x coordinate is $\alpha(f)$.*
   (b) *For face s^*, the y coordinates of its lower and upper end point of the face segment are 0, $\alpha(t)$ respectively.*
   (c) *For face t^*, the y coordinates of its lower and upper end point of the face segment are 0, $\alpha(t)$ respectively.*
   (d) *For any other face f, the y coordinates of its lower and upper end point of the face segment are the smallest and biggest $\alpha(v)$ from all its incident vertices respectively.*

The grid size of the VFCR constructed by Algorithm 1 is length($\mathcal{O}^*$) × length($\mathcal{O}$). It is bounded by $(|F| - 1) \times (|V| - 1)$. For a particular VFCR constructed by Algorithm 1, its compactness only depends on the choice of the st-orientation $\mathcal{O}$. For a practical algorithm on controlling length($\mathcal{O}$) and length($\mathcal{O}^*$), we refer the readers to [8].

Figure 1 (2) produces a VFCR for G in Figure 1 (1), by using $\mathcal{O}$ of G in Figure 1 (1) in Algorithm 1. The grid size produced by Algorithm 1 for G in Figure 1 (2) is 4 × 4. Consider the angular graph $\mathcal{A}(G)$ for G. It has 6 black vertices and 6 white vertices. Applying Lemma 1 (1), the VFCR for G is also a VVCR for $\mathcal{A}(G)$. If we had used the algorithm presented in [2,3] to construct a VVCR for $\mathcal{A}(G)$, the grid size would be $(6 - 1) \times (6 - 1) = 5 \times 5$. Hence, in general, Algorithm 1 produces more compact representations. We have the following theorem for the correctness of Algorithm 1.

Theorem 1. *1. Let G be a 2-connected plane multigraph. Algorithm 1 produces a VFCR for G in linear time.*

2. Let G be a 2-connected maximal bipartite plane graph, then applying Algorithm 1 to either G_w or G_b produces a VVCR for G in linear time.

4 st-Orientations of 3-Connected Planar Graphs and Their Applications in VFCR

Let G be a 3-connected planar graph. G has a unique embedding in the sphere. After choosing a particular face in the spherical embedding of G to be the exterior face, the unique spherical embedding leads to a particular planar embedding of G. In all its planar embeddings, vertices and faces of G preserve exactly the same adjacency relations. Therefore, VFCR constructed for any particular planar embedding of G (which is a plane graph) can be viewed as a VFCR for the original planar graph G. Therefore, it is meaningful to define VFCR for 3-connected planar graphs.

In this section, we present an algorithm to construct an st-orientation $\mathcal{O}$ for a proper planar embedding for 3-connected planar graphs such that length($\mathcal{O}$) is non-trivially bounded. Note that, this result naturally yields a non-trivial grid size bound for VFCR for 3-connected planar graphs.

Next we introduce the concept and properties of realizer for 3-connected plane graphs [1,5,6].

Definition 2. Let $G = (V, E)$ be a 3-connected plane graph. v_1, v_2, v_n be three vertices on its exterior face in ccw order. A *realizer* of G is a triplet of rooted directed spanning trees $\{T_1, T_2, T_n\}$ of G with the following properties:

1. For $i \in \{1, 2, n\}$, the root of T_i is v_i, the edges of G are directed from children to parent in T_i.
2. Each edge of G is contained in at least one and at most two spanning trees. If an edge of G is contained in two spanning trees, then it has different directions in the two trees.
3. For each vertex v of G, when v is not the root of the tree, it has exactly one edge leaving v in each of T_1, T_2, T_n. The ccw order of the edges incident to v is: leaving in T_1, entering in T_n, leaving in T_2, entering in T_1, leaving in T_n, and entering in T_2 (Figure 2 shows two examples of edge directions around a vertex v). Each entering block could be empty. An edge with two opposite directions is considered twice. The first and the last incoming edges are possibly coincident with the outgoing edges.
4. For $i \in \{1, 2, n\}$, all the incoming edges of the root v_i belong to T_i.

We color the edges in T_1 by blue, T_2 by green, and T_n by red. According to the definition of realizer, each edge of G is assigned one or two colors, and is said to be *1-colored* or *2-colored*, respectively.

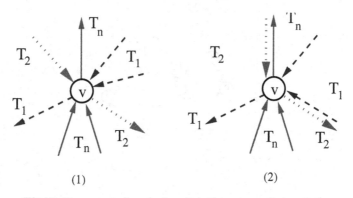

(1)   (2)

Fig. 2. Two examples of edge directions around a vertex v

For each vertex v of G and $i \in \{1, 2, n\}$, $p_i(v)$ denotes the path in T_i from v to the root v_i of T_i. A subpath of $p_i(v)$ between the end vertex v and an ancestor u of v in T_i is denoted by $p_i(v, u)$. The subpath of the external face of G with end vertices v_1 and v_2 and not containing v_n is denoted by $ext(v_1, v_2)$. The subpaths $ext(v_2, v_n)$ and $ext(v_n, v_1)$ are defined similarly. The properties of realizer have been studied extensively in [1,5,6]. We summarize them in the following lemma.

Lemma 2. *Let $G = (V, E)$ be a 3-connected plane graph, v_1, v_2, v_n be its three exterior vertices in ccw. Then G admits a realizer $\mathcal{R} = (T_1, T_2, T_n)$, where vertex v_i is the root of T_i for $i \in \{1, 2, n\}$. The realizer can be constructed in linear time. It satisfies the following properties:*

1. *For each vertex v of G, $p_1(v)$, $p_2(v)$ and $p_n(v)$ have only the vertex v in common.*
2. *For vertices v_1, v_2, v_n the following hold: $p_1(v_2) = p_2(v_1) = ext(v_1, v_2)$; $p_2(v_n) = p_n(v_2) = ext(v_2, v_n)$; $p_n(v_1) = p_1(v_n) = ext(v_n, v_1)$.*

Figure 3 shows an example of a 3-connected plane graph and one of its realizers. Simple observation also verifies the following property for 3-connected plane graph G with a realizer $\mathcal{R} = (T_1, T_2, T_n)$:

Property 1. Let v be a vertex other than v_1, v_2, v_n of G.

1. All ancestors of v in T_1 (T_2, T_n respectively) constitute a nonempty set and they appear before v in the ccw postordering of the vertices of G with respect to T_n (T_1, T_2 respectively).
2. All ancestors of v in T_2 (T_n, T_1 respectively) constitute a nonempty set and they appear before v in the cw postordering of the vertices of G with respect to T_n (T_1, T_2 respectively).

An *ordering* $\mathcal{O}$ of a set consisting of elements $a_1, a_2, \ldots, a_k$ is written as $\mathcal{O} = < a_1, a_2, \ldots, a_k >$. For two elements a_i and a_j, if a_i appears before a_j in $\mathcal{O}$, we write $a_i \prec_\mathcal{O} a_j$. The concatenation of two ordered lists $\mathcal{O}_1$ and $\mathcal{O}_2$ is written as $\mathcal{O}_1\mathcal{O}_2$.

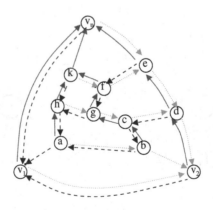

Fig. 3. A 3-connected plane graph and one of its realizer

Let $T = (V, E)$ be a rooted tree drawn in the plane. A *balanced partition* [11] of T is the partition of V into three ordered subsets A, B, C such that: Let a_i be the ith vertex of T in counterclockwise postordering, and b_i be the ith vertex of T in clockwise postordering. Marking the vertices of T in the order $a_1, b_1, a_2, b_2, \ldots, a_i, b_i, \ldots$, continue this process as long as the next pair of the vertices a_{i+1}, b_{i+1} have not been marked. This process stops when either $a_{k+1} = b_{k+1}$ or b_{k+1} is already marked. This vertex is called the *merge vertex* of T. When the marking process stops, the un-marked vertices of T form a single path from the merge vertex a_{k+1} to the root of T. This path is called the *leftover* path of T. Then $A = < a_1, a_2, \ldots, a_k >$, $B = < b_1, b_2, \ldots, b_k >$, and C is the leftover path ordered from the merge vertex to the root of T. We have the following lemma.

Lemma 3. *Let G be a 3-connected plane graph with n vertices. G has at most 5 exterior vertices $v_1, v_2, \cdots, v_n$ in ccw order. (Therefore, the ellipsis here could mean 0, 1, or 2 vertices.) Let $\mathcal{R} = \{T_1, T_2, T_n\}$ be a realizer of G. T_i is rooted at v_i. For each $i \in \{1, 2, n\}$, let (A_i, B_i, C_i) be the balanced partition of T_i. Then there exists $i \in \{1, 2, n\}$ such that $|C_i| \leq n/3 + 4$.*

In a balanced partition, the subgraph induced by vertices in $A_i \cup B_i$ for $i \in \{1, 2, n\}$ defines a *ladder graph* of order k. Ladder graph is defined as follows [11]:

Definition 3. *A ladder graph of order k is a spanning subgraph of the following plane graph $L = (V_L, E_L)$. The vertex set V_L is partitioned into two ordered lists $A = < a_1, a_2, \ldots, a_k >$ and $B = < b_1, b_2, \ldots, b_k >$. $E_L = E_A \cup E_B \cup E_{cross}$, where: $E_A = \{(a_i, a_{i+1}) | 1 \leq i < k\}$; $E_B = \{(b_i, b_{i+1}) | 1 \leq i < k\}$; and E_{cross} consists of edges between a vertex $a_i \in A$ and a vertex $b_j \in B$. The edges in E_{cross} are called cross edges of L.*

An ordering $\mathcal{O}$ of the vertices of a ladder graph $L = (A \cup B, E_L)$ is *consistent with respect to L* if for any $i < j$, $a_i \prec_{\mathcal{O}} a_j$ and $b_i \prec_{\mathcal{O}} b_j$.

The following lemma was proved in [11]:

Lemma 4. *Let $L = (A \cup B, E_L)$ be a ladder graph of order k. Then L has a consistent ordering $\mathcal{O}$ such that length($\mathcal{O}$) $\leq k + 2\lceil\sqrt{k}\rceil - 1$. $\mathcal{O}$ can be constructed in linear time.*

Next, we show how to use T_1, T_2, T_n to construct an *st*-orientation $\mathcal{O}$ of G such that length($\mathcal{O}$) is non-trivially bounded.

Algorithm 2. An *st*-orientation for 3-connected plane graphs
Input: A 3-connected plane graph $G = (V, E)$ with n vertices and at most 5 exterior vertices. $v_1, v_2, \cdots, v_n$ be its exterior vertices in ccw order. (The ellipsis here could mean 0, 1, or 2 vertices.)
Output: An st-orientation $\mathcal{O}$ of G.

1. *Construct a realizer $\mathcal{R} = \{T_1, T_2, T_n\}$ for G. T_i is rooted at v_i for $i \in \{1, 2, n\}$.*

2. *Construct the balanced partition* $\{A_i, B_i, C_i\}$ *for* T_1, T_2, T_n.
3. *Select a* T_i *from* $i \in \{1, 2, n\}$ *with* $|C_i| \leq n/3 + 4$, *there are three cases:*
   Case 1: $i = n$.
   (a) *Order the set* $\{v_1, v_2\}$ *as* $< v_1, v_2 >$. *Denote it by* $\mathcal{O}_1$.
   (b) *Construct a consistent ordering for the ladder graph induced by the vertex set* $(A_n - \{v_1\}) \cup (B_n - \{v_2\})$, *where the order of this ladder graph is* $(k_n - 1) = (|A_n| - 1)$. *Denote this ordering by* $\mathcal{O}_2$.
   (c) *Order the vertices in* C_n *from the merge vertex towards* v_n. *Denote this ordering by* $\mathcal{O}_3$
   (d) *Concatenate* $\mathcal{O}_1 \mathcal{O}_2 \mathcal{O}_3$ *to be* $\mathcal{O}$.
   Case 2: $i = 2$.
   (a) *Order the set* $\{v_1, v_n\}$ *as* $< v_1, v_n >$. *Denote it by* $\mathcal{O}_1$
   (b) *Construct a consistent ordering for the ladder graph induced by the vertex set* $(A_2 - \{v_n\}) \cup (B_2 - \{v_1\})$, *where the order of this ladder graph is* $(k_2 - 1) = (|A_2| - 1)$. *Denote this ordering by* $\mathcal{O}_2$.
   (c) *Order the vertices in* C_2 *from the merge vertex towards* v_2. *Denote this ordering by* $\mathcal{O}_3$
   (d) *Concatenate* $\mathcal{O}_1 \mathcal{O}_2 \mathcal{O}_3$ *to be* $\mathcal{O}$.
   Case 3: $i = 1$.
   (a) *Order the vertices in* $ext(v_2, v_n)$ *from* v_2 *towards* v_n. *Denote this ordering by* $\mathcal{O}_0$. *Note that, it has at most 4 vertices.*
   (b) *Remove the vertices on* $ext(v_2, v_n)$ *from* A_1, B_1. *Consider the tree* $T_1' = T_1 - ext(v_2, v_n)$. T_1' *has a partition* $\{A_1' = A_1 - ext(v_2, v_n), B_1' = B_1 - ext(v_2, v_n), C_1\}$. *This partition is not necessarily balanced.* $ext(v_2, v_n)$ *has at most 4 vertices.* $v_2 \in A_2$, *and* $v_n \in B_2$. *Therefore* $|A_1'|$ *differs from* $|B_1'|$ *by at most 2. Remove at most 2 vertices from the lower ends from one of* A_1' *and* B_1' *to balance the partition. Denote this partition by* $\{A_1'', B_1'', C_1\}$. *Order the removed 1 or 2 vertices consistently with their orders in* A_1 *or* B_1. *Denote this ordering by* $\mathcal{O}_1$. *Note that* $|A_1''| = |B_1''|$ *and they differ from* $|A_1| = |B_1|$ *by at most 3. Therefore, the new ladder graph induced by* $A_1'' \cup B_1''$ *has order at least* $(k_1 - 3) = (|A_1| - 3)$. *(If* G *were a 3-connected plane graph with more exterior vertices, then the ladder graph induced by* $A_1'' \cup B_1''$ *could have smaller order, which could cause problem for controlling the length of our intended st-orientation. So we require* G *to have at most 5 exterior vertices.)*
   (c) *Construct a consistent numbering for the ladder graph induced by* $A_1'' \cup B_1''$. *Denote this ordering by* $\mathcal{O}_2$.
   (d) *Order the vertices in* C_1 *form the merge vertex towards* v_1. *Denote this ordering by* $\mathcal{O}_3$
   (e) *Concatenate* $\mathcal{O}_0 \mathcal{O}_1 \mathcal{O}_2 \mathcal{O}_3$ *to be* $\mathcal{O}$.

Next lemma proves the ordering $\mathcal{O}$ produced above by Algorithm 2 is indeed an st-orientation and its length($\mathcal{O}$) is non-trivially bounded.

Lemma 5. *Let* G *be a 3-connected plane graph* $G = (V, E)$ *with* n *vertices and at most 5 exterior vertices. Then* $\mathcal{O}$ *constructed by Algorithm 2 is an st-orientation of* G *with* length($\mathcal{O}$) $\leq \frac{2n}{3} + 2\lceil \sqrt{n/3} \rceil + 5$.

Proof. According to Algorithm 2, we have three different cases to consider. We will only prove for the case $i = n$. The other two cases for $i = 1, i = 2$ can be similarly proved. We have the following observations:

1. v_1 is the first in $\mathcal{O}$. v_2 is the second in $\mathcal{O}$. v_n is the last in $\mathcal{O}$.
2. For $v \neq v_n$, its parent in T_n is a neighbor of v in G and it is after v in $\mathcal{O}$.
3. For $v = v_2$, its neighbor v_1 is before v_2 in $\mathcal{O}$.
4. For $v \notin \{v_1, v_2\}$, if $v \in A_n$, then according to Property 1 (1), its parent in T_1 is before v in $\mathcal{O}$. Similarly, if $v \in B_n$, then its parent in T_2 is before v in $\mathcal{O}$. If $v \in C_n$, then its parents in both T_1 and T_2 are before v in $\mathcal{O}$.

Therefore, $\mathcal{O}$ is indeed an st-orientation of G. v_1 is the single source. v_n is the single sink.

No matter which T_i we use to construct $\mathcal{O}$, there is a sub-ordering $\mathcal{O}_2$ using a ladder graph in $\mathcal{O}$. Denote its order by k. Note that we choose $i \in \{1, 2, n\}$ so that $|C_i| \leq n/3 + 4$. $2|A_i| + |C_i| = n$, $i \in \{1, 2, n\}$, and $k \geq (|A_i| - 3)$. Therefore, $k \geq |A_i| - 3 \geq \frac{n - (n/3 + 4)}{2} - 3 \geq (n/3 - 5)$.

A ladder graph of order k has $2k$ vertices. By Lemma 4, a consistent ordering constructed for the ladder graph can pick at most $k + 2\lceil \sqrt{k} \rceil - 1 + 1$ vetices. (The last $+1$ term comes from the fact that the number of vertices on a path is one more than its length.) Therefore, $\mathcal{O}_2$ constructed for the ladder graph of order k bypasses at least $2k - (k + 2\lceil \sqrt{k} \rceil) = k - 2\lceil \sqrt{k} \rceil$ vertices. $k - 2\lceil \sqrt{k} \rceil$ is almost a monotonically increasing function except that from $k = m^2$ to $k = m^2 + 1$, it decreases by 1. Then at $k = m^2 + 2$, its value increases by 1 again. Given this observation and the fact $k \geq n/3 - 5$, $\mathcal{O}_2$ bypasses at least $(n/3 - 5) - 2\lceil \sqrt{n/3 - 5} \rceil - 1 \geq (n/3 - 5) - 2\lceil \sqrt{n/3} \rceil - 1$. (The last -1 term is because that the function $k - 2\lceil \sqrt{k} \rceil$ might decrease by 1 at certain places.) Therefore $length(\mathcal{O}) \leq n - (n/3 - 5 - 2\lceil \sqrt{n/3} \rceil - 1) - 1 = \frac{2n}{3} + 2\lceil \sqrt{n/3} \rceil + 5$. (The subtraction of last 1 is because that the length of a path is one less than the number of the vertices on it.)

Symmetrically, we can start from the 3-connected dual graph G^* and construct an st-orientation for it and then its dual st-orientation for G. Therefore, we have the following main theorem. Its proof is obvious and hence omitted.

Theorem 2. *Let G be a 3-connected planar graph with n vertices and f faces.*

1. *G admits a VFCR on a grid with size bounded by $(f-1) \times (\frac{2n}{3} + 2\lceil \sqrt{n/3} \rceil + 5)$, which is constructable in linear time.*
2. *G admits a VFCR on a grid with size bounded by $(\frac{2f}{3} + 2\lceil \sqrt{f/3} \rceil + 5) \times (n-1)$, which is constructable in linear time.*

Observe that for the above grid size bound, either the width bound is trivial, or the height bound is trivial. It would be nice if we could have both height and width bounds to be non-trivial. Next, we present a 3-connected planar graph G with n vertices and $f = n$ faces, such that for any st-orientation $\mathcal{O}$ of G, either its $length(\mathcal{O}) = (n - 1)$, or $length(\mathcal{O}^*) = (f - 1)$. Therefore, it is not possible to have both non-trivial grid size bounds at least by using our algorithm. We conjecture that this is true for all VFCR algorithms.

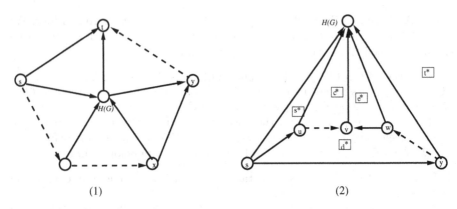

Fig. 4. A wheel graph of order n. Dashed lines represent a path with possibly many vertices on it. (1) $\mathcal{H}(G)$ is an interior vertex in the planar embedding, (2) $\mathcal{H}(G)$ is an exterior vertex in the planar embedding.

Definition 4. *A wheel graph W_n of order n is a graph containing a cycle of order $(n-1)$, and for which every graph vertex in the cycle is connected to one other graph vertex (which is called the hub of G and denoted by $\mathcal{H}(G)$).*

Figure 4 (1) gives a wheel graph with n vertices. Note that, a wheel graph is 3-connected planar. Its dual is isomorphic to itself. Since all the non-hub vertices are symmetric, wheel graph essentially has two different planar embeddings. See Figure 4 (1) and (2). In the planar embedding of Figure 4 (1), $\mathcal{H}(G)$ is an interior vertex. On the other hand, in the planar embedding of Figure 4 (2), $\mathcal{H}(G)$ is an exterior vertex.

Theorem 3. *Let G be the wheel graph with n vertices. Then for any st-orientation $\mathcal{O}$ for G, either* length($\mathcal{O}$) *is* $(n-1)$, *or* length($\mathcal{O}^*$) *is* $(f-1)$.

Proof. We have three different cases to consider.

Case 1: $s \neq \mathcal{H}(G)$ and $t \neq \mathcal{H}(G)$. See Figure 4 (1). According to the property of st-orientation, the exterior face has two cycles, one is (s, t). The other is the directed path $(s, \cdots, x, y, \cdots, t)$. Consider $\mathcal{H}(G)$ in $\mathcal{O}$, all its incoming edges are consecutively around $\mathcal{H}(G)$. So do all its outgoing edges. Let $(x, \mathcal{H}(G))$ be the last incoming edge for $\mathcal{H}(G)$ in ccw direction, $(y, \mathcal{H}(G))$ be the first outgoing edges for $\mathcal{H}(G)$ in ccw direction. Then (x, y) is an edge on the path $(s, \cdots, x, y, \cdots, t)$. Therefore, the path $(s, \cdots, x, \mathcal{H}(G), y, \cdots, t)$ is a directed path in $\mathcal{O}$ and its length is $(n-1)$.

Case 2: $t = \mathcal{H}(G)$. See Figure 4 (2). The faces of G are indicated by squares. Consider the face d^*. Its facial cycle is decomposed into two directed paths. Both paths start from s. Suppose v is the sink for the face d^*. Consider the dual st-orientation $\mathcal{O}^*$ for G^*, its source is s^*. Its sink is t^*. Its longest directed path is $(s^*, \cdots, c^*, d^*, e^*, \cdots, t^*)$. It picks up every face of G^*. Therefore, length($\mathcal{O}^*$) is $(f-1) = (n-1)$.

Case 3: $s = \mathcal{H}(G)$. The proof is similar to Case 2.

5 Conclusion

In this paper, we introduce VFCR for 2-connected plane multigraphs. We present a linear time algorithm for constructing VFCR for 2-connected plane multigraphs. For 3-connected planar graphs, we present a non-trivial bound on the length of its longest directed path for its st-orientations. This non-trivial bound yields non-trivial grid size bound for VFCR for 3-connected planar graphs. The relation between VFCR and VVCR is also presented.

References

1. Di Battista, G., Tamassia, R., Vismara, L.: Output-sensitive Reporting of Disjoint Paths. Algorithmica 23(4), 302–340 (1999)
2. Fraysseix, H.D., de Mendez, P.O., Pach, J.: Representation of planar graphs by segments. Intuitive Geometry 63, 109–117 (1991)
3. Fraysseix, H.D., de Mendez, P.O., Pach, J.: A Left-First Search Algorithm for Planar Graphs. Discrete & Computational Geometry 13, 459–468 (1995)
4. Lempel, A., Even, S., Cederbaum, I.: An algorithm for planarity testing of graphs. In: Theory of Graphs Proc. of an International Symposium, Rome, July 1966, pp. 215–232 (1966)
5. Felsner, S.: Convex drawings of Planar Graphs and the Order Dimension of 3-Polytopes. Order 18, 19–37 (2001)
6. Miura, K., Azuma, M., Nishizeki, T.: Canonical decomposition, realizer, Schnyder Labelling and orderly spanning trees of plane graphs. International Journal of Foundations of Computer Science 16, 117–141 (2005)
7. Ossona de Mendez, P.: Orientations bipolaires, PhD thesis, Ecole des Hautes Etudes en Sciences Sociales, Paris (1994)
8. Papamanthou, C., Tollis, I.G.: Applications of parameterized st-orientations in graph drawings. In: Healy, P., Nikolov, N.S. (eds.) GD 2005. LNCS, vol. 3843, pp. 355–367. Springer, Heidelberg (2006)
9. Rosenstiehl, P., Tarjan, R.E.: Rectilinear planar layouts and bipolar orientations of planar graphs. Discrete Comput. Geom. 1, 343–353 (1986)
10. Tamassia, R., Tollis, I.G.: An unified approach to visibility representations of planar graphs. Discrete Comput. Geom. 1, 321–341 (1986)
11. He, X., Zhang, H.: Nearly Optimal Visibility Representations of Plane Graphs. In: Bugliesi, M., Preneel, B., Sassone, V., Wegener, I. (eds.) ICALP 2006. LNCS, vol. 4051, pp. 407–418. Springer, Heidelberg (2006)

An Optimal On-Line Algorithm for Preemptive Scheduling on Two Uniform Machines in the ℓ_p Norm

Tianping Shuai[1,*], Donglei Du[2,**], and Xiaoyue Jiang[3]

[1] School of Sciences, Beijing University of Posts and Telecommunication
shuaitp@gmail.com
[2] Faculty of Business Administration, University of New Brunswick
ddu@unb.ca
[3] College of Engineering, Louisiana State University
jiang@lsu.edu

Abstract. One of the basic and fundamental problems in scheduling is to minimize the machine completion time vector in the ℓ_p norm (a direct extension of the l_∞ norm: the makespan) on uniform parallel machines. We concentrate on the on-line and preemptive version of this problem where jobs arrive one by one over a list and are allowed to be preempted. We present a best possible deterministic on-line scheduling algorithm along with a matching lower bound when there are two machines, generalizing existing results for the identical machines scheduling problem in the literature. The main difficulty in the design of the algorithm and the analysis of the resultant competitive ratio as well as the proof of the lower bound is that the competitive ratio is only known to be the root of some equation systems, which admits no analytic solution—a distinct feature from most existing literature on competitive analysis. As a consequence, we develop some new ideas to tackle this difficulty. Specifically we need to exploit the properties of the equations system that defines the competitive ratio.

Keywords: On-line algorithm; Scheduling; Preemption; ℓ_p norm.

1 Introduction

We investigate the problem of scheduling a set of jobs, arriving one by one over a list, on two uniform parallel machines M_1 and M_2 with speeds $s_1 = 1$ and $s_2 = s > 1$, respectively. If a job with size p_j is assigned to machine M_i, then p_j/s_i time units are required to process this job. Machines are available at time zero. Whenever a job is revealed, it is allocated with one or several non-overlapping time slots with preemption being allowed, to possibly different

* This work was done while the author was visiting the Faculty of Business Administration, University of New Brunswick. Supported in part by NSFC grant 10726058.
** Supported in part by NSERC grant 283103 and URF, FDF at UNB.

R. Fleischer and J. Xu (Eds.): AAIM 2008, LNCS 5034, pp. 316–327, 2008.

machines. Such an allocation is called a (feasible) schedule. The objective is to find a schedule which minimizes the ℓ_p norm of the machines' completion times, that is, $\sqrt[p]{L_1^p + L_2^p}$, where L_1 and L_2 are the completion times of the last job on M_1 and M_2, respectively. We denote this problem as $Q2|pmtn, on - line|\ell_p$. The off-line version of problem, where the full information (number of jobs and sizes of jobs) on all the jobs is known in advance, can be denoted as $Q2|pmtn|\ell_p$ using the three-field notation in [12].

The makespan, the most studied objective function in scheduling, is just a special ℓ_p norm—that is—the ℓ_∞. Scheduling in the *general* ℓ_p norm has also been widely studied in the literature [1], [2], [3], [4], [5], [6], [7], [8], [10], [11], [13], [14], [15], and the ℓ_p norm of the completion time vector is one of the basic and fundamental objectives investigated in scheduling theory. While the makespan only characterizes the latest completion time among all machines, the general ℓ_p norm is more appropriate when we are interested in the average behavior of the machine completion times rather than the worst-case scenario. A particular application of ℓ_p norm scheduling in disk storage allocation problem is illustrated in [7].

The main focus of this work is on the competitive analysis of the on-line scheduling problem introduced in the beginning. We will measure the quality of an *on-line algorithm ON*, one for the on-line scheduling problem, by its *competitive ratio* r_{ON}, which is defined to be the superimum of ratio C_{ON}/C_{OFF} over all problem instances, where C_{ON} and C_{OFF} denote respectively the ℓ_p norm of machine completion time vector of the on-line schedule constructed by ON and that of the corresponding (off-line) optimal schedule.

Related Work. Scheduling in the ℓ_p norm has been investigated from both on-line (and semi-online) [2], [7], [8], [14], [13], [15] and offline [1], [3], [4], [5], [6], [10], [11] points of view. But we only review the results related to the on-line case as it is the main focus of this work. Readers who are interested in the off-line case are referred to the references listed here and further pointers therein.

The most related paper to the current work is [8], where a best possible on-line algorithm is devised for $P2|pmtn, on - line|\ell_p$. The next related result is [14], in which the non-preemptive problem $P2|on - line|\ell_p$ is investigated. Finally, [16] (independently [9]) develops best possible on-line algorithms for $Q2|pmtn|C_{\max}$, the ℓ_∞ norm. Moreover, several semi-online models of $P2||\ell_p$, and $Q2||\ell_p$ are studied (both preemptive and non-preemptive versions) in [7], [13], [15].

Our Results. The main contribution of this work is to generalize the result of [8] to the uniform machines case. Specifically, we devise a best possible deterministic on-line algorithms for $Q2|pmtn, on - line|\ell_p$ along with a matching lower bound. The main difficulty in the design of the algorithm and the analysis of the competitive ratio as well as the proof of the lower bound is that the competitive ratio is only known to be the root of some equation systems, which admits no analytic solution. We follow a similar (but more involved) approach developed in [8] to tackle this difficulty. Specifically we need to exploit the properties of the equations system that defines the competitive ratio. It is well-known that machine scheduling problems on two uniform machines are unique as their resolutions tend to be complicated and often generate novel ideas that are usually useful in attacking scheduling problem of more than two machines. Therefore, the investigation of such problems is of important theoretical significance.

The rest of the paper is organized as follows. After some preliminary results in Section 2, we present an on-line algorithm algorithm and prove its competition ratio in Section 3. Then we show a matching lower bound in Section 4.

2   Preliminaries and Notations

We first prepare us with some preliminary results, which are summarized in the following lemmas. Throughout this paper, for simplicity, *increasing* always means nondecreasing, and *decreasing* always means nonincreasing.

In the problem $Q2|pmtn, on-line|\ell_p$, assume $p > 1$ and machine M_1 has speed 1 and machine M_2 has speed $s > 1$. Define the following functions and quantities.

$$\hat{s} := s^{\frac{1}{p-1}},$$

$$\hat{t} := \frac{1}{1+\hat{s}^p},$$

$$\mu_1 := 1 + \frac{(\hat{s}-1)^{p/(p-1)}}{\hat{s}^{p+1}\left(\hat{s}-\hat{t}^{p-1}\right)^{1/(p-1)}} \in \left(1, 1+\frac{1}{\hat{s}^p}\right),$$

$$\bar{t}(\mu) := \frac{\mu - \hat{s}}{\mu(1+\hat{s}) - \hat{s}},$$

$$\varphi(\mu, t) := \left(\frac{\mu}{\mu-1}\right)^{p-1}\left(1 - \frac{1-t}{\hat{s}}\right)^p + \frac{(1-t)^p}{\hat{s}} - s^p\mu^{p-1}, \mu \neq 1,$$

$$\phi(\mu, t) := (st)^p + (1-t)^p - s^p(\mu\hat{t})^{p-1},$$

$$H(\mu, t) := \hat{s}\varphi(\mu, t) - \phi(\mu, t).$$

The following properties for $\bar{t}$, φ, ϕ and H will be useful hereafter.

Lemma 1

1. *Function $\bar{t}(\mu)$ is increasing for $\mu > \hat{s}/(1+\hat{s})$, implying that $\bar{t}(\mu) \leq \hat{t}$ for $1 < \mu \leq \hat{s}^p/(s-1)$.*
2. *For any fixed $\mu \neq 1$, function $\varphi(\mu, t)$ is decreasing for $t \leq \bar{t}(\mu)$, increasing for $t > \bar{t}(\mu)$, and achieves its minimum value at $\bar{t}(\mu)$.*
3. *For any fixed $\mu \neq 1$, function $\phi(\mu, t)$ is decreasing for $t \leq \hat{t}$, increasing for $t > \hat{t}$, and achieves its minimum value at $t = \hat{t}$.*
4. *Function $H(\mu, \hat{t})$ is decreasing for $\mu > 1$.*

Proof. (1), (2) and (3) being easy by examining the derivative, we only prove (4). The following calculation shows that the derivative $H'(\mu, \hat{t}) \leq 0$ for $u > 1$, implying that $H(\mu, \hat{t})$ is decreasing for $\mu > 1$:

$$H'(\mu, \hat{t}) = (p-1)\mu^{p-2}\left\{-\frac{\hat{s}}{(\mu-1)^p}\left((1+\hat{s}^p-s)\hat{t}\right)^p - s^p\left(\hat{s}-\hat{t}^{p-1}\right)\right\} \leq 0.$$

The following lemma is useful in the proof of Lemma 3.

Lemma 2. *The following equation has a unique root $\hat{\mu}$ within the interval $(1, 1 + 1/\hat{s}^p)$.*

$$H(\mu, \hat{t}) = 0.$$

Proof. The desired result follows if we could show that $H(\mu, \hat{t})$ has opposite signs on the two boundaries of $(1, 1 + 1/\hat{s}^p)$ because $H(\mu, \hat{t})$ is monotonic by Lemma 1(4).

First of all, there exists a positive number $\varepsilon \ll 1$, such that the value at the left boundary of $(1, 1 + 1/\hat{s}^p)$:

$$H(1 + \varepsilon, \hat{t}) = \left(\frac{1 + \varepsilon}{\varepsilon}\right)^{p-1} \left((1 + \hat{s}^p - s)\hat{t}\right)^p \hat{s} - (s\hat{t})^p - s^p(1 + \varepsilon)^{p-1}\left(\hat{s} - \hat{t}^{p-1}\right) > 0.$$

Second of all, we show that the value $H(1 + 1/\hat{s}^p, \hat{t}) \le 0$ at the right boundary of $(1, 1 + 1/\hat{s}^p)$. Let

$$H(1 + 1/\hat{s}^p, \hat{t}) = \hat{s}\hat{t}W(s),$$

where

$$W(s) := (1 + \hat{s}^p - s)^p + s^{1/(1-p)} + s - (1 + \hat{s}^p)^p - \frac{s^p\hat{t}^{p-1}}{\hat{s}}.$$

Note that $W(s)$ is decreasing by checking the derivative:

$$W'(s) = \frac{p^2}{p-1}\hat{s}\{(1 + \hat{s}^p - s)^{p-1} - (1 + \hat{s}^p)^{p-1}\} + \{1 - p(1 + \hat{s}^p - s)^{p-1}\}$$
$$+ \frac{p}{p-1}\hat{s}^{-p}(s^p\hat{t}^{p-1} - 1) + ps^p\left((1 + \hat{s}^p)^{-1} - s^{-p}\right)\hat{t}^{p-1} \le 0.$$

So, for $s > 1$, $W(s) \le W(1) = 3 - 2^p - 2^{1-p} \le 0$ for all $p > 1$, implying that $H(1 + \hat{s}^{-p}, \hat{t}) \le 0$.

Define two intervals $T = [0, \hat{t}]$ and $U = [\mu_1, \hat{\mu}]$, where $\hat{\mu}$ is the unique root defined in Lemma 2.

Lemma 3. *The following equations system has a unique solution (μ, t) such that $t \in T$ and $\mu \in U$.*

$$\begin{cases} \varphi(\mu, t) = 0, \\ \phi(\mu, t) = 0. \end{cases}$$

Proof. First, we show that there exists a unique positive solution $t(\mu) \in T$ for any given $\mu \in U$. Consider the equation $H(\mu, t) = 0$, or equivalently,

$$\hat{s}\left(\frac{\mu}{\mu - 1}\right)^{p-1}\left(1 - \frac{1}{\hat{s}}(1 - t)\right)^p - (st)^p - s^p\mu^{p-1}\left(\hat{s} - \hat{t}^{p-1}\right) = 0 \qquad (1)$$

A direct calculation reveals that the derivative $H'_t(\mu, t) \ge 0$ for $\mu \in U$ and $t \in T$, implying $H(\mu, t)$ is an increasing function of $t \in T$. Consequently, the desired result follows if we could show that $H(\mu, t)$ has opposite signs on the two boundaries of $(0, \hat{t})$. (i) $t = 0$: Note that $H(\mu, 0)/\mu^{p-1} = (\mu - 1)^{1-p}(\hat{s} - 1)^p/s - s^p(\hat{s} - (1 + \hat{s}^p)^{1-p})$ is an decreasing function for $\mu \geq 1$ and $H(\mu_1, 0)/\mu_1^{p-1} = 0$, implying that $H(\mu, 0)/\mu_1^{p-1} \leq H(\mu_1, 0)/\mu_1^{p-1} \leq 0$ for any $\mu \geq \mu_1$. (ii). $t = \hat{t}$: $H(\mu, \hat{t}) \geq 0$ for any $\mu \leq \hat{\mu}$ by Lemma 1(4) and Lemma 2.

Next, we show the important property that $t(\mu)$ is an increasing function of $\mu \in U$, which will be needed to prove the existence and uniqueness of solution $\mu \in U$. Implicitly differentiating Equation (1) yields

$$t'(\mu) = \frac{(p-1)\mu^{p-2}\left\{s^p\left(\hat{s} - \hat{t}^{p-1}\right) + \frac{\hat{s}}{(\mu-1)^p}\left(1 - \frac{1-t}{\hat{s}}\right)^p\right\}}{p\left(\left(\frac{\mu}{\mu-1}\right)^{p-1}\left(1 - \frac{1-t}{\hat{s}}\right)^{p-1} - s^p t^{p-1}\right)}$$

The numerator being non-negative is evidential for $t \in T$ and $\mu \in U$ by noting that $1 > (1 - t)/s$. Denote $D_1(t) = \mu(\hat{s} - 1 + t)$ and $D_2(t) = \hat{s}^{p+1}(\mu - 1)t$. Then the denominator is given by

$$\frac{p}{s(\mu-1)^{p-1}}\left(D_1(t)^{p-1} - D_2(t)^{p-1}\right).$$

Note that $D(t) := D_1(t) - D_2(t)$ is a linear function of t for any fixed μ. Moreover, (i) $D(0) = \mu(\hat{s} - 1) > 0$. (ii) $\mu \leq \hat{\mu} \leq 1 + 1/\hat{s}^p \leq \hat{s}^p/(s - 1)$ implies that $D(\hat{t}) = (\hat{s}(\mu(1 - s) + \hat{s}^p))/(1 + \hat{s}^p) > 0$. Consequently, $D(t) > 0$ for all $t \in T$, implying the denominator is positive. Therefore $t'(\mu) \geq 0$, implying $t(\mu)$ is an increasing function of $\mu \in U$.

Finally, we are ready to show that there exists a unique solution $u \in U$. Note that, for any $\mu \in U$, $t(\mu)$ being increasing implies that $t(\mu) \leq t(\hat{\mu}) = \hat{t}$ by the definitions of $\hat{\mu}$ and $\hat{t}$. Define the following function based on $\varphi(\mu, t)$:

$$\xi(\mu) := \hat{s}\varphi(\mu, t) = \hat{s}\left(\frac{\mu}{\mu-1}\right)^{p-1}\left(1 - \frac{1}{\hat{s}}(1 - t)\right)^p + (1 - t)^p - s^{p+1/(p-1)}\mu^{p-1}$$

$$\stackrel{t=t(\mu)}{=} (st)^p + (1 - t)^p - s^p(\mu\hat{t})^{p-1} = \phi(\mu, t)$$

Lemma 1(3) and $t(\mu)$ being increasing implies that $(st)^p + (1 - t)^p$ is decreasing for $t \in T$. Consequently $\xi(\mu) = \phi(\mu, t(\mu))$ is decreasing for $\mu \in U$. Note that $t(\mu_1) = 0$ and $t(\hat{\mu}) = \hat{t}$. Therefore, (i) $\mu_1 \leq 1 + 1/\hat{s}^p$ implies that $\xi(\mu_1) = 1 - (s^p\mu_1^{p-1})/((1 + \hat{s}^p)^{p-1}) > 0$. (ii) $\hat{\mu} > 1$ and $t(\hat{\mu}) = \hat{t}$ imply that

$$\xi(\hat{\mu}) = \left(\frac{s}{1 + 1/\hat{s}^p}\right)^p + \left(\frac{\hat{s}^p}{1 + 1/\hat{s}^p}\right)^p - \frac{s^p\mu^{p-1}}{(1 + 1/\hat{s}^p)^{p-1}} = \frac{s^p(1 - \hat{\mu}^{p-1})}{(1 + 1/\hat{s}^p)^{p-1}} < 0.$$

Now we can conclude that there is a unique $\mu \in U$ since $\xi(\mu)$ is non-increasing for $\mu \in U$ with opposite signs on the two boundaries.

From now on, denote (μ_*, t_*) to be the unique solution of the system (1)-(2), and let $\alpha = \mu_*^{p-1}$. From the proof of Lemma 3, it is clear that $\mu_* \geq \mu_1 > 1$.

The following result is an immediate consequence of Lemmas 1 and 3.

Lemma 4. *Assume $t \in T$.*

1. $\varphi(\mu_*, t) = 0$ *if and only if* $t = t_*$; $\varphi(\mu_*, t) > 0$ *if and only if* $t > t_*$; $\varphi(\mu_*, t) < 0$ *if and only if* $t < t_*$;
2. $\phi(\mu_*, t) = 0$ *if and only if* $t = t_*$; $\phi(\mu_*, t) > 0$ *if and only if* $t < t_*$; $\phi(\mu_*, t) < 0$ *if and only if* $t > t_*$.

Proof. (1) We consider two separate cases.

Case 1. If $\mu_* < \hat{s}$, then $\bar{t}(\mu^*) < 0$. Therefore $\varphi(\mu_*, t)$ is an increasing function for $t \geq \bar{t}(\mu^*)$ by Lemma 1(2). Consequently, the desired result follows as t_* is the unique root of $\varphi(\mu_*, t)$ in $t \in T$ by Lemma 3.

Case 2. If $\mu_* \geq \hat{s}$, then $\bar{t}(\mu^*) \geq 0$. Note that $\mu_* - 1 \geq \hat{s} - 1$ implies that

$$\varphi(\mu_*, 0) \leq \left(\frac{\mu_*}{\hat{s}-1}\right)^{p-1} \left(1 - \frac{1}{\hat{s}}\right)^p + \frac{1}{\hat{s}} - s^p \mu_*^{p-1}$$

$$= \mu_*^{p-1}\left(\frac{1}{s} - s^p\right) + \frac{1}{\hat{s}}\left(1 - \left(\frac{\mu_*}{\hat{s}}\right)^{p-1}\right) < 0.$$

Therefore $\varphi(\mu_*, t) < 0$ for $t \in [0, \bar{t}(\mu^*)]$ and it is increasing for $t > \bar{t}(\mu^*)$ by Lemma 1(2). Moreover, Lemma 1(1) and $\mu_* \leq \hat{\mu} \leq 1 + 1/\hat{s}^p \leq \hat{s}^p/(s-1)$ implies that $\bar{t}(\mu^*) \leq \hat{t}$ and hence $\bar{t}(\mu^*) \in T$. Consequently, the desired result follows as t_* is the unique root of $\varphi(\mu_*, t)$ in $t \in T$ by Lemma 3.

(2) Note that $\phi(\mu_*, 0) = 1 + s^p - s^p(\mu_*\hat{t})^{p-1} > 0$ and $\phi(\mu_*, t)$ is decreasing for $t \in T$ by Lemma 1(3). Consequently, the desired result follows as t_* is the unique root of $\varphi(\mu_*, t)$ in $t \in T$ by Lemma 3.

3   Upper Bound

We devise an on-line algorithm and show its competitive ratio to be $\sqrt[p]{\alpha}$ in this section. In the next section (Section 4), we shall prove that no deterministic on-line algorithm can achieve a competitive ratio smaller than $\sqrt[p]{\alpha}$.

For simplicity, we will identify a job with its length whenever there is no confusion. We introduce some notations first. In any on-line algorithm, after scheduling some jobs, let L_i ($i = 1, 2$) be the machine load on machine M_i prior to the new incoming job x. Thus $L = L_1 + sL_2$ is the total workload before x, and the objective value $C = \sqrt[p]{L_1^p + L_2^p}$. Denote C_{opt} to be the optimal off-line objective prior to x, then for any job z prior to x,

$$C_{opt} \geq \sqrt[p]{\left(\max\left\{\frac{z}{s}, \hat{s}\hat{t}L\right\}\right)^p + \left(\min\left\{L - z, L\hat{t}\right\}\right)^p}$$

Define the following functions and quantities:

$$f(x) = \left(\frac{1}{s}(L_1 + x) + \left(1 - \frac{1}{\hat{s}}\right)L_2\right)^p + \frac{L_2^p}{\hat{s}} - \alpha\left(\left(\frac{x}{s}\right)^p + L^p\right),$$

$$g(x,y) = (L_1 + y)^p + \left(L_2 + \frac{x-y}{s}\right)^p - \alpha(L+x)^p \hat{t}^{p-1},$$

$$x_1 = \frac{L_1 + s(1 - 1/\hat{s})L_2}{\mu_* - 1},$$

$$x_2 = \frac{\mu_* L - (1 + \hat{s}^p)L_1}{1 + \hat{s}^p - \mu_*},$$

$$x_3 = \frac{\mu_* \hat{s} L - L_2(1 - 1/\hat{s}^p)(1 + \hat{s}^p)^{p-1}}{(1 + \hat{s}^p)^{p-1}/s - \mu_* \hat{s}}.$$

The following simple facts will be used shortly.

Lemma 5

1. *Function $f(x)$ is increasing within the interval $x \in [0, x_1]$, and decreasing within the interval $x \in [x_1, +\infty]$. Therefore f achieves its maximum at $x = x_1$,*
2. *Function $g(x,x) = (L_1 + x)^p + L_2^p - \alpha(L+x)^p \hat{t}^{p-1}$ is decreasing within the interval $x \in [0, x_2]$, and increasing within the interval $x \in [x_2, +\infty]$. Therefore $g(x,x)$ achieves its minimum at $x = x_2$,*
3. *Function $g(x, L_2/\hat{s} - L_1) = L_2^p/\hat{s}^p + (L_2 + (x + L_1 - L_2/\hat{s})/s)^p - \alpha(L+x)^p \hat{t}^{p-1}$ is decreasing within the interval $x \in [0, x_3]$, and increasing within the interval $x \in [x_3, +\infty]$. Therefore $g(x, L_2/\hat{s} - L_1)$ achieves its minimum at $x = x_3$,*
4. *Function $g(\hat{s}^p L, L_2/\hat{s} - L_1) = f(\hat{s}^p L)$.*
5. *$x_2 \geq L_2/\hat{s} - L_1$.*

Proof. (1), (2) and (3) being easy to prove by checking the derivative and (4) being easy to verify. We only prove (5). Note that

$$x_2 - \left(\frac{L_2}{\hat{s}} - L_1\right) = \left(\frac{s\mu_*}{1 + \hat{s}^p - \mu_*} - \frac{1}{\hat{s}}\right) L_2$$

$$\geq \left(\frac{s}{1 + \hat{s}^p - 1} - \frac{1}{\hat{s}}\right) L_2 = 0,$$

where the first inequality follows because the first term therein $s\mu/(1 + \hat{s}^p - \mu)$ is an increasing function of μ and $\mu^* \geq 1$.

The main algorithm is presented below.

Algorithm H: When a new job x arrives, do the following:
Step 1. Solve the following mathematical program with a single variable y:

$$\max \ y$$

$$\left(\frac{\mu_*}{\mu_* - 1}\right)^{p-1} \left(\frac{L_1 + y}{s} + \left(1 - \frac{1}{\hat{s}}\right)\left(L_2 + \frac{x-y}{s}\right)\right)^p + \frac{(L_2 + (x-y)/s)^p}{\hat{s}} \leq \alpha(x+L)^p \quad (2)$$

$$0 \leq y \leq x \quad (3)$$

$$0 \leq y \leq \frac{1}{\hat{s}} L_2 - L_1 \quad (4)$$

Step 2. Let y^* be an optimal solution. Assign y^* and $x - y^*$ to M_1 and M_2 respectively.

Evidently, algorithm H maintains the following invariant at any state:

$$\left(\frac{\mu_*}{\mu_* - 1}\right)^{p-1} \left(\frac{L_1}{s} + \left(1 - \frac{1}{\hat{s}}\right) L_2\right)^p + \frac{L_2^p}{\hat{s}} \leq \alpha L^p.$$

Moreover, note that the mathematical program with the single variable y in Step 1 of the algorithm H is a convex program, which can be solved efficiently, implying the running time of Algorithm H is solvable in polynomial time. However, we will not address more on time complexity, since the primary focus in competitive analysis is the identification of competitive ratio.

Theorem 1. *The competitive ratio of Algorithm H is no more than $\sqrt[p]{\alpha}$.*

Proof. First, we show that the feasible region defined by (2)-(4) is non-empty, implying the correctness of the algorithm. When there is no job initially, $y = 0$ is obviously a feasible solution. Inductively, assume 0 were a feasible solution at some state of the algorithm after scheduling a certain number of jobs. Let the current machine load be L_i ($i = 1, 2$) on M_i. Consider the next incoming job $x \geq 0$. Then 0 obviously satisfies constraint (3) and also satisfies (2) by the inductive hypothesis. Moreover, $y = 0$ satisfies constraint (4) because the machine load on M_2 is $L_2 + x/s$ and the load on M_1 did not change. Therefore $(L_2 + x/s)/\hat{s} - L_1 \geq L_2/\hat{s} - L_1 \geq 0$, where the last inequality follows from the inductive hypothesis.

Next, we show that our algorithm is $\sqrt[p]{\alpha}$–competitive. This will be done by induction on the number of jobs. It is trivial when there is no job scheduled. Suppose the algorithm H is $\sqrt[p]{\alpha}$–competitive after scheduling some jobs. Let x be the next job and we show that the algorithm H is still $\sqrt[p]{\alpha}$–competitive after scheduling x.

The rest of the argument will be based on the following idea. We are done if we can show either (i) $f(x) \leq 0$ when $x \geq \hat{s}^p L$, as $\sqrt[p]{(x/s)^p + L^p}$ is obviously a lower bound of the optimal off-line value in this case; or (ii), $g(x, y^*) \leq 0$ as $(L + x)^p \hat{t}^{p-1}$ is always a lower bound for the optimal off-line value. We consider two separate cases.

Case 1. Equation (2) is tight, i.e., satisfied as equality, in the optimal solution y^*. Obviously we can assume that $L + x > 0$ to exclude the trivial case. Let $\lambda = (L_1 + y^*)/(L + x)$, then $\lambda \leq \hat{t}$ by the constraints (3)-(4). Moreover, the constraint (2) becomes

$$\frac{1}{s^p} \left\{ \left(\frac{\mu_*}{\mu_* - 1}\right)^{p-1} \left(1 - \frac{1}{\hat{s}}(1 - \lambda)\right)^p + \frac{1}{\hat{s}}(1 - \lambda)^p \right\} = \alpha = \mu_*^{p-1},$$

which is equivalent to $\varphi(\mu_*, \lambda) = 0$. By the definition of (μ_*, t_*), and the proof of Lemma 4, we know $\lambda = t_*$ is the only root of $\varphi(\mu_*, t) = 0$. Therefore

$$s^p \frac{g(x, y*)}{(L + x)^p} = s^p \lambda^p + (1 - \lambda)^p - \alpha s^p \hat{t}^{p-1} = s^p t_*^p + (1 - t_*)^p - \mu_*^{p-1} s^p \hat{t}^{p-1}$$

$$= \phi(\mu_*, t_*) = 0$$

Case 2. Equation (2) is not tight in y^*. Then we have either $y^* = x \leq L_2/\hat{s} - L_1$ or $y^* = L_2/\hat{s} - L_1 \leq x$ due to the optimality of y^*, which leads to two subcases.

Case 2.1. $y^* = x \leq L_2/\hat{s} - L_1$.

Consider $g(x, y^*) = g(x, x) = (L_1 + x)^p + L_2^p - \alpha(L + x)^p/(1 + \hat{s}^p)^{p-1}$. By Lemma 5(2), $g(x, x)$ achieves its minimum at $x = x_2$, and it is decreasing within the interval $x \in [0, x_2]$. Moreover, $x_2 \geq L_2/\hat{s} - L_1$ by Lemma 5(5). Therefore $g(x, x)$ can only achieve its maximum value at the left boundary 0. Hence $g(x, x) \leq \max_{0 \leq x \leq L_2/\hat{s} - L_1} g(x, x) = g(0, 0) \leq 0$, where the last inequality follows from the induction hypothesis.

Case 2.2. $y^* = L_2/\hat{s} - L_1 \leq x$.

1. If $x \geq \hat{s}^p L$, consider $f(x) = (L_1/s + (1 - 1/\hat{s})L_2 + x/s)^p + L_2^p/\hat{s} - \alpha((x/s)^p + L^p)$. By Lemma 5(1), $f(x)$ achieves its maximum at $x = x_1$. So

$$f(x) \leq \max_{x \geq 0} f(x) = f(x_1)$$

$$= \left(\frac{\mu_*}{\mu_* - 1}\right)^p \left(\frac{1}{s}L_1 + \left(1 - \frac{1}{\hat{s}}\right)L_2\right)^p + \frac{L_2^p}{\hat{s}}$$

$$- \left(\frac{\mu_*^{p-1}}{(\mu_* - 1)^p} \left(\frac{L_1 + s(1 - \frac{1}{\hat{s}})L_2}{s}\right)^p + \alpha L^p\right)$$

$$= \left(\frac{\mu_*}{\mu_* - 1}\right)^{p-1} \left(\frac{1}{s}L_1 + \left(1 - \frac{1}{\hat{s}}\right)L_2\right)^p + \frac{L_2^p}{\hat{s}} - \alpha L^p \leq 0$$

where the last inequality follows from the induction hypothesis.

2. If $x \leq \hat{s}^p L$, consider

$$g(x, y^*) = g\left(x, \frac{L_2}{\hat{s}} - L_1\right)$$

$$= \frac{1}{\hat{s}^p}L_2^p + \left(L_2 + \frac{x + L_1 - \frac{1}{\hat{s}}L_2}{s}\right)^p - \alpha(L + x)^p \hat{t}^{p-1}$$

By Lemma 5(3), $g(x, L_2/\hat{s} - L_1)$ achieves its minimum at $x = x_3$, and it is decreasing within the interval $x \in [0, x_3]$. Therefore $g(x, L_2/\hat{s} - L_1)$ can only achieve its maximum value at the boundaries.

$$g\left(x, \frac{L_2}{\hat{s}} - L_1\right) \leq \max_{L_2/\hat{s} - L_1 \leq x \leq \hat{s}^p L} g\left(x, \frac{L_2}{\hat{s}} - L_1\right)$$

$$= \max\left\{g\left(\frac{L_2}{\hat{s}} - L_1, \frac{L_2}{\hat{s}} - L_1\right), g\left(\hat{s}^p L, \frac{L_2}{\hat{s}} - L_1\right)\right\}$$

Note that the first term in the last quantity $g(L_2/\hat{s} - L_1, L_2/\hat{s} - L_1) \leq g(0, 0) \leq 0$ by Lemma 5(2) and the second term $g(\hat{s}^p L, L_2/\hat{s} - L_1) = f(\hat{s}^p L) \leq \max_{x \geq 0} f(x) = f(x_1) \leq 0$ by Lemma 5(4) and Lemma 5(1). Therefore $g(x, y^*) \leq 0$.

This completes the proof of the theorem.

4   Lower Bound

We show that the upper bound proved in the previous section is actually best possible.

Theorem 2. *No deterministic on-line algorithm has a competitive ratio smaller than $\sqrt[p]{\alpha}$.*

Proof. Consider the sequence of jobs with size 1, $\hat{s}^p$, w, where w will be specified shortly. We only need to consider those schedules with no simultaneous idle time. Suppose there exists an on-line algorithm A with competitive ratio less than $\sqrt[p]{\alpha}$. Without loss of generality, algorithm A schedules the first job on M_2. Let $0 \le x \le 1$ be the amount assigned to M_1 after the second job. The optimal off-line objective value now is $1 + \hat{s}^p$. Note that $x^p + (\hat{s} + (1-x)/s)^p < \alpha(1 + \hat{s}^p)$ implies

$$\frac{(1+\hat{s}^p)^p}{s^p}\phi(\mu_*, x\hat{t}) = x^p + \left(\hat{s} + \frac{1-x}{s}\right)^p - \alpha(1+\hat{s}^p) < 0,$$

which again implies that $x\hat{t} > t_*$. We shall now prove a contradictory claim that $x\hat{t} \le t_*$ and hence completes the proof.

Let

$$w = \frac{1 + \hat{s}^p - \hat{s} - (1-x)/s}{\mu_* - 1}.$$

First, we show that $w \ge \hat{s}^p(1 + \hat{s}^p)$. Let $t' = s(\hat{s}^p(\mu_* - 1) - 1) + 1$. If $t_* \ge t'$, then $x\hat{t} > t_*$ implies that

$$w = \frac{1 + \hat{s}^p - \hat{s} - \frac{1-x}{s}}{\mu_* - 1} \ge \frac{(1 + \hat{s}^p)\left(1 - \frac{1-t_*}{s}\right)}{\mu_* - 1}$$

$$\ge \frac{(1 + \hat{s}^p)(1 - \frac{1}{s} + (s(\hat{s}^p(\mu_* - 1) - 1) + 1)/s}{\mu_* - 1} = \hat{s}^p(1 + \hat{s}^p).$$

So, we only need to prove $t_* \ge t'$. Clearly, this is true when $t' \le 0$. Therefore, we assume that $t' \ge 0$. Note that

$$\varphi(\mu_*, t') = \left(\frac{\mu_*}{\mu_* - 1}\right)^{p-1}\left(1 + \frac{s}{\hat{s}}(\hat{s}^p(\mu_* - 1) - 1)\right)^p + \frac{(s(1 - \hat{s}^p(\mu_* - 1)))^p}{\hat{s}} - \alpha s^p$$

$$= s^p \mu_*^{p-1}\left\{\left(\frac{1}{s} + \frac{1}{\hat{s}}(\hat{s}^p(\mu_* - 1) - 1)\right)\left(s - \frac{\hat{s}^{p-2} - 1}{s(\mu_* - 1)}\right)^{p-1}\right.$$

$$\left. + \frac{(1 - \hat{s}^p(\mu_* - 1))^p}{\hat{s}\mu_*^{p-1}} - 1\right\}$$

$$\le s^p \mu_*^{p-1}\left\{\left(\left(\frac{1}{s} + \frac{1}{\hat{s}}(\hat{s}^p(\mu_* - 1) - 1)\right) + \frac{(1 - \hat{s}^p(\mu_* - 1))}{\hat{s}}\right) \times\right.$$

$$\left. \max\left\{\left(s - \frac{\hat{s}^{p-2} - 1}{s(\mu_* - 1)}\right)^{p-1}, \left(\frac{(1 - \hat{s}^p(\mu_* - 1))}{\mu_*}\right)^{p-1}\right\} - 1\right\}$$

$$= s^p \mu_*^{p-1} \left\{ \frac{1}{s} \max \left\{ \left(s - \frac{\hat{s}^{p-2} - 1}{s(\mu_* - 1)} \right)^{p-1}, \left(\frac{(1 - \hat{s}^p(\mu_* - 1))}{\mu_*} \right)^{p-1} \right\} - 1 \right\} \leq 0.$$

The last inequality is true since $t' \geq 0$ and $\mu_* \leq 1 + 1/\hat{s}^p$. By lemma 4, we have $t_* \geq t'$.

Next, $w \geq \hat{s}^p(1 + \hat{s}^p)$ implies that the optimal off-line value is $\sqrt[p]{(w/s)^p + (1 + \hat{s}^p)^p}$, while the best objective value that the on-line algorithm A can get, by put $\hat{s} + (1 - x)/s - x$ of w on M_1 and the rest on M_2, is

$$\sqrt[p]{(\hat{s} + (1 - x)/s)^p + \left(\hat{s} + \frac{1 + w - \hat{s} - (1 - x)/s}{s} \right)^p}.$$

Then

$$\sqrt[p]{(\hat{s} + (1 - x)/s)^p + \left(\hat{s} + \frac{1 + w - \hat{s} - (1 - x)/s}{s} \right)^p} < \sqrt[p]{\alpha((w/s)^p + (1 + \hat{s}^p)^p)},$$

or equivalently

$$(\hat{s} + (1 - x)/s)^p + \left(\frac{\mu_*}{\mu_* - 1} \right)^{p-1} \left(\frac{\hat{s}^p + 1 - \hat{s} - (1 - x)/s}{s} \right)^p - \mu_*^{p-1}(1 + \hat{s}^p)^p < 0.$$

So

$$\varphi(\mu_*, x\hat{t})$$
$$= (s\hat{t})^p \left\{ \frac{1}{\hat{s}} \left(\hat{s} + \frac{1-x}{s} \right)^p - \mu_*^{p-1}(1 + \hat{s}^p)^p + \left(\frac{\mu_*}{\mu_* - 1} \right)^{p-1} \left(\frac{\hat{s}^{p+1} + \hat{s} - 1 - \hat{s}^p + x}{\hat{s}^p} \right)^p \right\}$$
$$< (s\hat{t})^p \left\{ \left(\hat{s} + \frac{1-x}{s} \right)^p - \mu_*^{p-1}(1 + \hat{s}^p)^p + \left(\frac{\mu_*}{\mu_* - 1} \right)^{p-1} \left(\frac{\hat{s}^p + 1 - \hat{s} - (1-x)/s}{s} \right)^p \right\} < 0.$$

Hence $\varphi(\mu_*, x\hat{t}) < 0$. This implies that $x\hat{t} < t_*$ by lemma 4, which is an obvious contradiction to the previous claim that $x\hat{t} > t_*$.

References

1. Alon, N., Azar, Y., Woeginger, G.J., Yadid, T.: Approximation schemes for scheduling. In: SODA, pp. 493–500 (1997)
2. Avidor, A., Azar, Y., Sgall, J.: Ancient and new algorithms for load balancing in the ℓ_p norm. Algorithmica 29, 422–441 (2001)
3. Azar, Y., Epstein, A.: Convex programming for scheduling unrelated parallel machines. In: STOC, pp. 331–337 (2005)
4. Azar, Y., Epstein, A., Epstein, L.: Load balancing of temporary tasks in the ℓ_p norm. Theoretical Computer Science 361(2-3), 314–328 (2006)
5. Azar, Y., Epstein, L., Richter, Y., Woeginger, G.J.: All-norm approximation algorithms. In: Penttonen, M., Schmidt, E.M. (eds.) SWAT 2002. LNCS, vol. 2368, pp. 288–297. Springer, Heidelberg (2002)

6. Azar, Y., Taub, S.: All-Norm Approximation for Scheduling on Identical Machines. In: Hagerup, T., Katajainen, J. (eds.) SWAT 2004. LNCS, vol. 3111, pp. 298–310. Springer, Heidelberg (2004)
7. Chandra, A.K., Wong, C.K.: Worst-case analysis of a placement algorithm related to storage allocation. SIAM Journal on Computing 1, 249–263 (1975)
8. Du, D.-L., Jiang, X., Zhang, G.: Optimal preemptive online scheduling to minimize lp norm on two processors. Journal of Manufacturing and Management Optimization 1(3), 345–351 (2005)
9. Epstein, L., Noga, J., Seiden, S., Sgall, J., Woeginger, G.J.: Randomized online scheduling on two uniform machines. Journal of Scheduling 4(2), 71–92 (2001)
10. Epstein, L., Tassa, T.: Optimal Preemptive Scheduling for General Target Functions. Journal of Computer and System Sciences 72(1), 132–162 (2006)
11. Kumar, V.S.A., Marathe, M.V., Parthasarathy, S., Srinivasan, A.: Approximation Algorithms for Scheduling on Multiple Machines. In: FOCS, pp. 254–263 (2005)
12. Lawler, E.L., Lenstra, J.K., Rinnooy Kan, A.H.G., Shmoys, D.B.: Sequencing and scheduling: Algorithms and complexity. In: Graves, S.C., Rinnooy Kan, A.H.G., Zipkin, P.H. (eds.) Logistics of Production and Inventory, pp. 445–522. North-Holland, Amsterdam (1993)
13. Lin, L.: Semi-online scheduling algorithm under the ℓ_p norm on two identical machines. Journal of Zhejiang University (Science Edition) 34(2), 148–151 (2007) (in chinese)
14. Lin, L., Tan, Z.Y., He, Y.: Deterministic and randomized scheduling problems under the ℓ_p norm on two identical machines. Journal of Zhejiang University Science 6(1), 20–26 (2005)
15. Tan, Z., He, Y., Epstein, L.: Optimal On-line Algorithms for the Uniform Machine Scheduling Problem with Ordinal Data. Information and Computation 196(1), 57–70 (2005)
16. Wen, J., Du, D.-L.: Preemptive on-line scheduling for two uniform processors. Operations Research Letters 23, 113–116 (1998)

An Optimal Strategy for Online Non-uniform Length Order Scheduling⋆

Feifeng Zheng[1,*], E. Zhang[2], Yinfeng Xu[1,3], and Xiaoping Wu[1]

[1] School of Management, Xi'an JiaoTong University,
Xi'an, Shaanxi, 710049, China
[2] School of Information Management and Engineering,
Shanghai University of Finance and Economics,
Shanghai, 200433, China
[3] The State Key Lab for Manufacturing Systems Engineering,
Xi'an, Shaanxi, 710049, China
zhengff@mail.xjtu.edu.cn

Abstract. This paper will study an online non-uniform length order scheduling problem. For the case where online strategies have the knowledge of Δ beforehand, which is the ratio between the longest and shortest length of order, Ting [3] proved an upper bound of $(\frac{6\Delta}{\log \Delta} + O(\Delta^{5/6}))$ and Zheng et al. [2] proved a matching lower bound. This work will consider the scenario where online strategies do not have the knowledge of Δ at the beginning. Our main work is a $(\frac{6\Delta}{\log \Delta} + O(\Delta^{5/6}))$-competitive optimal strategy, extending the result of Ting [3] to a more general scenery.

Keywords: Scheduling, Online Strategy, Competitive Ratio.

1 Introduction

Since there are many dynamic and unpredictable factors in modern manufacturing business, including production order scheduling and processing operations, many authors adopt online theory to describe and investigate the scenario, and online scheduling has caught much interest among extensive scheduling literature in recent decades. In online scheduling, there is a manufacturer who may accept or decline orders that arrive one by one over time. Each order will stay in the system to be satisfied after arrival until it expires, i.e., its deadline cannot be met at the time even started at once. The manufacturer will gain a profit from each completed order. He may also abort an order on running to start a new one in favor of larger profit, and the aborted order has to be started again from the beginning to be satisfied. That is, we consider the preemption-restart online model to maximize the total profit of completed orders.

⋆ Corresponding author. Tel.: +86-29-8266-5034; fax: +86-29-8266-5543. This work was supported by NSF of China under Grants 70525004, 70702030, 70602031, 70121001 and 60736027, and Doctoral Fund of Ministry of Education of China 20070698053.

R. Fleischer and J. Xu (Eds.): AAIM 2008, LNCS 5034, pp. 328–336, 2008.

In most literature, it is usually assumed that the shortest and longest order is of one unit and Δ units of length respectively. Most literature studied the case where the knowledge of Δ is known to online strategies at the first beginning. Fung et al. [1] studied an online broadcast problem. Translated into our terminology, they proved a $\Delta + 2\sqrt{\Delta} + 2$-competitive strategy ACE (Another Completes Earlier). Zheng et al. [2] present a lower bound of $\Omega(\Delta/\ln\Delta)$. Ting [3] proved a matching upper bound of $O(\Delta/\log\Delta)$, and left it open whether there exist $O(\Delta/\log\Delta)$-competitive online strategies without the knowledge of Δ. In the above studies, online strategies make use of Δ. Kim et al. [4] presented a 5-competitive greedy strategy GD for the case of unit length of order, i.e., $\Delta = 1$. GD makes an abortion if a newly arrival order has weight α times that of the order being processed, otherwise GD continues the current service. Zheng et al. [5] proved that GD is $4\Delta + 1$-competitive with $\alpha = 2$ for the case where $\Delta \geq 1$. GD is treated as an online strategy that acts without the knowledge of Δ and it performs poorly for the case where $\Delta > 1$. For the other case where Δ is unknown beforehand, it is harder for online strategies to act efficiently, and to the best of our knowledge there is little result considering to maximize profits. We believe that it is a reasonable case in real business since the manufacturer usually can not foresee the exact information of future orders, including order length. So, the work will focus on the performance of online strategies in the latter case and give a positive answer to the problem proposed in Ting [3].

We describe our problem formally as follows. There is one manufacturer who processes orders arriving over time. Each order J has four attributes, namely, $a(J)$: the arrival time, $p(J)$: the processing time or the length, $w(J)$: the profit or weight to be obtained only if J is finished and $d(J)$: the deadline by which the order has to be completed to be satisfied. $p(J)$, $w(J)$ and $d(J)$ become known on the arrival of J, i.e., at time $a(J)$. $1 \leq p(J) \leq \Delta$ where Δ is assumed w.l.o.g. to be a natural number. The goal is to maximize the total profit of completed orders within a time period.

1.1 Related Work

One related direction is the scenario where $p(J)$ is unknown to online strategies until order J has been completed. The scenery is called *non-clairvoyant scheduling* and was first investigated by Motwani et al [7]. They aimed to minimize the total flow time, presenting a lower bound of $\Omega(n^{1/3})$ for deterministic non-clairvoyant strategies and of $\Omega(\log n)$ for randomized ones where n is the number of orders, respectively. Kalyanasundaram and Pruhs [8] proposed a $O(\log n \log \log n)$-competitive randomized strategy RMLF against a adaptive adversary. Becchetti and Leonardi [9] further showed that the RMLF strategy is in fact $O(\log n)$-competitive, matching the lower bound.

Another quite related line is the scenario where the length of order is selected from a finite set of real numbers but not an arbitrary number within $[1, \Delta]$. Lipton and Tomkins [10] studied the scenario to maximize resource utilization. One of their results is an upper bound of 2 for non-preemption strategies in the case where $p(J)$ is either 1 or Δ and $d(J) = a(J)+p(J)$. Goldwasser [11] extended

Lipton and Tomkins's work and investigated the case where orders have slack time equal to $k \geq 0$ times of order length, i.e., $d(J) - a(J) = (k+1)p(J)$. They proved a matching upper and lower bound of $2 + \frac{\lceil \Delta \rceil - 1}{\Delta}$ when $\frac{1}{\Delta} \leq k < 1$, and a matching bound of $1 + \frac{\lceil \Delta \rceil}{\Delta}$ when $1 \leq k < \Delta$ (please refer to [11] for details).

1.2   Our Results

In this work, we will study the preemption-restart online model to maximize the total profit of completed orders. An online deterministic strategy that does not make use of Δ is proposed and proved to be $(\frac{6\Delta}{\log \Delta} + O(\Delta^{5/6}))$-competitive when $\Delta > 20$, and $(\frac{\sqrt{20}}{\sqrt{20}-1}\Delta + 6.8)$-competitive when $1 < \Delta \leq 20$ respectively. Since the lower bound of $\Omega(\Delta / \ln \Delta)$ in [2] applies to the case without the knowledge of Δ, the online strategy is optimal when $\Delta > 20$.

The rest of the paper is organized as follows. Section 2 gives the definition of competitive ratio, and presents the online deterministic strategy. We propose several fundamental properties for the strategy in Section 3 and prove its optimal competitiveness in Section 4. Section 5 concludes the work.

2   Definitions and Online Strategy Description

2.1   Competitive Ratio Definition

To gauge the performance of an on-line strategy $\mathcal{A}$, the competitive ratio analysis (refer to Borodin and El-yaniv, 1998 [6]) is often used. Denote by $\Gamma_{\mathcal{A}}(I)$ and $\Gamma^*(I)$ the schedules produced by $\mathcal{A}$ and by an optimal offline strategy OPT on an order input set I respectively, and by $|\Gamma_{\mathcal{A}}(I)|$ and $|\Gamma^*(I)|$ the total profit of completed orders in $\Gamma_{\mathcal{A}}(I)$ and $\Gamma^*(I)$ respectively. The competitive ratio of $\mathcal{A}$ is then defined as $r_{\mathcal{A}} = \sup_I \frac{|\Gamma^*(I)|}{|\Gamma_{\mathcal{A}}(I)|}$.

2.2   An Online Strategy

For the model without preemption penalties, Ting [3] proposed the BALANCE strategy applying the knowledge of Δ. In this section, we will put forward the Dynamic Preemption Criteria (abbr. DPC) strategy that does not make use of the knowledge of Δ. Similar to BALANCE, DPC makes two kinds of preemptions: the *good-profit preemption* that would increase the supposed profit of processing, and the *short-length preemption* that would reduce the completion time but not increase the supposed profit of processing.

Before describing DPC strategy, we give some preliminary definitions. Given an arbitrary order input list Γ, let $\sigma = (J_0, \ldots, J_m)$ be a *preempting chain* produced by DPC such that J_i is preempted by J_{i+1} for $0 \leq i < m$ and J_m is completed by DPC. J_0 preempts none, that is, σ cannot be backward extended. If $m = 0$, then σ consists of a single order to be completed by DPC. If J_i preempts J_{i-1} by good-profit preemption, J_i is called a *gp* order, otherwise if

J_i preempts J_{i-1} by short-length preemption, it is a sp order. J_0 is a gp order since it preempts none. Suppose that J_i is a gp order, and J_j is a sp order for $j = i+1, \ldots, k$, we call J_i the *nearest preceding gp* order of J_j for $j = i, \ldots, k$. Note that J_i itself is the nearest preceding gp order of J_i. Let δ_t be the maximum (ratio of) length of order among those that have arrived by time t, at which DPC gets to know δ_t. Let $\lambda_t = 3\delta_t/\log \delta_t$. Denote by $s(J_i)$ the time at which DPC starts J_i. DPC is formally described in below.

Strategy DPC. The strategy is triggered either when it completes an order or when there comes a new one. After completing an order, DPC will start another one with the largest profit among those having arrived but not been satisfied yet. Otherwise it keeps idle if there is no such an order. Suppose that during DPC is processing order J whose nearest preceding gp order is J', there comes a new order R at time $t \in (s(J), s(J) + p(J))$. DPC will abort J to start R immediately if one of the following two conditions is satisfied.

C1. $w(R) \geq \sqrt{\max\{20, \delta_{s(J')}\}} w(J')$.

C2. The supposed completion time of R, $t_c = a(R) + p(R)$, is strictly earlier than $s(J) + p(J)$, and $w(R) \geq \begin{cases} \left(\dfrac{2^{d(s(J'), t_c)}}{(\delta_t)^{1/3}} \right) w(J'), & \delta_{s(J')} > 20; \\ w(J'), & otherwise. \end{cases}$ where

$d(s(J'), t_c) = \lfloor \frac{t_c - s(J')}{\lambda_{s(J')}} \rfloor$.

Note that $w(R)$ is compared with $w(J')$ but not $w(J)$ in both C1 and C2 conditions. If R preempts J by C1 condition, then R is a gp order, otherwise if R preempts J by C2 condition, then R is a sp order and $a(R) + p(R) < s(J) + p(J) \leq s(J') + p(J')$.

3   Basic Properties

By construction of DPC, all the orders in $\sigma = (J_0, \ldots, J_m)$ except J_0 are started by DPC on their arrival. So, $s(J_i) = a(J_i)$ holds for $0 < i \leq m$. For J_0, it may or may not be started by DPC on its arrival.

Property 1. In an arbitrary preempting chain $\sigma = (J_0, \ldots, J_m)$ produced by DPC, $p(J_i) \leq \delta_{s(J_i)} \leq \delta_{s(J_{i+1})}$ for $0 \leq i \leq m - 1$.

Proof. By the definition of δ_t, it is a non-decreasing function in time t. First, $\delta_{s(J_i)} \leq \delta_{s(J_{i+1})}$ due to $s(J_i) < s(J_{i+1})$. Since J_i shall arrive on or before $s(J_i)$, $p(J_i)$ is a candidate for $\delta_{s(J_i)}$, implying that $p(J_i) \leq \delta_{s(J_i)}$.   □

Assume that there are $n + 1$ ($n \geq 0$) gp orders in $\sigma = (J_0, \ldots, J_m)$. We will divide σ into $n + 1$ segments $\sigma = (\tau_0, \ldots, \tau_n)$ such that in each segment $\tau_i = (R_{i,0}, R_{i,1}, \ldots, R_{i,k_i})$ ($0 \leq i \leq n$), the first order $R_{i,0}$ is a gp order and the other k_i orders are sp orders. If $k_i = 0$, it means that there is no sp orders in τ_i. For $0 \leq i < n$, the last order in τ_i will be preempted by $R_{i+1,0}$ by C1 condition. For τ_n, $R_{n,k_n} = J_m$.

Property 2. In an arbitrary preempting chain $\sigma = (\tau_0, \ldots, \tau_n)$ produced by DPC, we have that $s(R_{i+1,0}) - s(R_{i,0}) < p(R_{i,0})$ for $0 \leq i < n$.

Proof. By the definition of segment τ_i, $R_{i,j}$ $(1 \leq j \leq k_i)$ is a *sp* order. So, $s(R_{i,k_i}) + p(R_{i,k_i}) < s(R_{i,0}) + p(R_{i,0})$. On the other hand, $s(R_{i+1,0}) < s(R_{i,k_i}) + p(R_{i,k_i})$ since R_{i,k_i} is preempted by $R_{i+1,0}$ at time $s(R_{i+1,0})$. So, $s(R_{i+1,0}) - s(R_{i,0}) < p(R_{i,0})$. □

Let $|\sigma|$ be the total profit obtained by DPC in σ. Let $|\sigma^*|$ and $|\tau_i^*|$ be the total profit of orders that are started by OPT during DPC is serving σ and τ_i respectively. Then, $|\sigma^*| = \sum_{i=0}^{n} |\tau_i^*|$. We will first bound $|\tau_i^*|$ with related to $w(R_{i,0})$. For notational convenience, denote by a_i, w_i and p_i the arrival time, weight and processing length of order $R_{i,0}$ respectively, and by s_i the time DPC starts $R_{i,0}$. For $1 \leq j \leq k_i$, denote by $a_{i,j}, w_{i,j}, p_{i,j}$ and $s_{i,j}$ those of $R_{i,j}$ respectively in the rest of Section 2. For segment τ_i $(0 \leq i < n)$, we have the following lemma.

Lemma 1. *For $0 \leq i < n$, if $\delta_{s_i} \leq 20$, then*

$$|\tau_i^*| < (\delta_{s_i} + \sqrt{20})w_i,$$

otherwise if $\delta_{s_i} > 20$, then

$$|\tau_i^*| \leq (\frac{6\delta_{s_i}}{\log \delta_{s_i}} + \sqrt{\delta_{s_i}})w_i.$$

Proof. Case 1. $\delta_{s_i} \leq 20$. First, τ_i starts and ends at time s_i and s_{i+1} respectively, and $s_{i+1} - s_i < p_i$ by Property 2. Moreover, $p_i \leq \delta_{s_i}$ by Property 1. OPT starts at most $\lfloor s_{i+1} - s_i \rfloor \leq \lfloor p_i \rfloor \leq \delta_{s_i}$ *sp* orders with weight less than w_i followed by one *gp* order with weight less than $\sqrt{20}w_i$. So, $|\tau_i^*| < (\delta_{s_i} + \sqrt{20})w_i$ in this case.

Case 2. $\delta_{s_i} > 20$. Since the values of δ_{s_i} and λ_{s_i} are fixed during DPC is serving τ_i, by the same reasoning as in the proof of Lemma 3 in [3], we have that

$$|\tau_i^*| \leq \left(\frac{\lambda_{s_i} \cdot 2^{d(s_i, s_{i+1})+1}}{(\delta_{s_i})^{1/3}} + \sqrt{\delta_{s_i}} \right) w_i$$

Together with Properties 1 and 2, we have $2^{d(s_i, s_{i+1})} \leq 2^{\lfloor \frac{p_i}{\lambda_{s_i}} \rfloor} \leq 2^{\lfloor \frac{\delta_{s_i}}{\lambda_{s_i}} \rfloor} \leq (\delta_{s_i})^{1/3}$ for $0 \leq i < n$. Thus,

$$|\tau_i^*| \leq (\frac{6\delta_{s_i}}{\log \delta_{s_i}} + \sqrt{\delta_{s_i}})w_i.$$

The lemma is proved.   □

For segment τ_n, the last order is $R_{n,k_n} = J_m$ to be completed by DPC. OPT will start its last order in τ_n strictly earlier than $s(J_m) + p(J_m) = s_{n,k_n} + p_{n,k_n}$. $|\tau_n|$ can be upper bounded in below by the same reasoning as in the proof of Lemma 3 in [3].

$$|\tau_n^*| \leq \left(\frac{\lambda_{s_n} \cdot 2^{d(s_n, s_{n,k_n} + p_{n,k_n})+1}}{(\delta_{s_n})^{1/3}} + \sqrt{\delta_{s_n}} \right) w_n.$$

The following lemma is fundamental in competitive analysis later on.

Lemma 2. *Given an arbitrary* $i \geq 0$, $f(x, i) = (1 + \frac{6\sqrt{x}}{\log x})(1 + \frac{1}{\sqrt{x}} + \frac{1}{(\sqrt{x})^2} + \ldots + \frac{1}{(\sqrt{x})^i}) = (1 + \frac{6\sqrt{x}}{\log x})\frac{\sqrt{x}}{\sqrt{x}-1}(1 - (\frac{1}{\sqrt{x}})^{i+1})$ *is an increase function of* x *when* $x > 20$.

Proof. First, $f(x, +\infty) = (1 + \frac{6\sqrt{x}}{\log x})\frac{\sqrt{x}}{\sqrt{x}-1}$ increases in x since its first order derivative to x is positive when $x > 20$. Moreover, $1 - (\frac{1}{\sqrt{x}})^{i+1}$ increases in x for $i \geq 0$. Hence, $f(x, i)$ increases in x. The lemma follows.   □

4   Competitive Analysis

Theorem 1. *In online non-uniform length order scheduling, DPC is* $(\frac{6\Delta}{\log \Delta} + O(\Delta^{5/6}))$-*competitive when* $20 < \Delta$ *and* $(\frac{\sqrt{20}}{\sqrt{20}-1}\Delta + 6.8)$-*competitive when* $1 < \Delta \leq 20$, *respectively.*

Proof. Let Γ be the schedule produced by DPC for a given order input list. Assume w.l.o.g. that DPC completes η (≥ 1) orders in Γ. We can divide Γ into η preempting chains so that in each preempting chain only the last order is completed by DPC. Since the total profit of both OPT and DPC in Γ is the simple summation of their respective total profit in each preempting chain, to upper bound the ratio between the profit of OPT and that of DPC in Γ, it suffices to bound the ratio in an arbitrary preempting chain $\sigma = (\tau_0, \ldots, \tau_n)$, where τ_i ($0 \leq i \leq n$) are segments defined before. For DPC, $|\sigma| = w_m$ by the definition of σ. Assume that $\delta_{s_k} \leq 20$ and $\delta_t > 20$ for some $t \in (s_k, s_{k+1}]$, where $0 \leq k < n$. We will discuss two cases where $0 \leq k \leq n - 1$ and where $k = n$.

Case 1. $0 \leq k \leq n-1$. Let $r_i = \sqrt{\max\{20, \delta_{s_i}\}}$. Since $R_{i,0}$ preempts $R_{i-1,k_{i-1}}$ by C1 condition, $w_i \leq w_{i+1}/r_i \leq \ldots \leq w_n/\prod_{j=i}^{n-1} r_j$.

$$
\begin{aligned}
|\sigma^*| - |\tau_n^*| &= \sum_{i=0}^{n-1} |\tau_i^*| \\
&\leq \sum_{i=0}^{k} (\delta_{s_k} + \sqrt{20})w_i + \sum_{i=k+1}^{n-1} (\frac{6\delta_{s_i}}{\log \delta_{s_i}} + \sqrt{\delta_{s_i}})w_i \\
&\leq \sum_{i=0}^{k} \frac{(\delta_{s_k} + \sqrt{20})w_n}{(\sqrt{20})^{n-i}} + \sum_{i=k+1}^{n-1} (\frac{6\delta_{s_i}}{\log \delta_{s_i}} + \sqrt{\delta_{s_i}})\frac{w_n}{\prod_{j=i}^{n-1} r_j} \\
&< \frac{(\delta_{s_k} + \sqrt{20})w_n}{(\sqrt{20} - 1)(\sqrt{20})^{n-k-1}} + \sum_{i=k+1}^{n-1} (\frac{6\delta_{s_i}}{\log \delta_{s_i}} + \sqrt{\delta_{s_i}})\frac{w_n}{\prod_{j=i}^{n-1} r_j}
\end{aligned}
$$

where the first inequality holds due to Lemma 1, and the second inequality holds since $\sqrt{20} \leq r_i$ and then $w_i \leq w_n/(\sqrt{20})^{n-i}$ for $0 \leq i \leq k$. For the first summation item on the righthand side of the above third inequality, $\frac{(\delta_{s_k}+\sqrt{20})w_n}{(\sqrt{20}-1)(\sqrt{20})^{n-k-1}} \leq \frac{(20+\sqrt{20})w_n}{(\sqrt{20}-1)(\sqrt{20})^{n-k-1}} = O(1)w_n$ for $0 \leq k \leq n-1$. In below we will bound the second summation item. Note that $r_i = \sqrt{\delta_{s_i}}$ for $k+1 \leq i < n$. Combining $\delta_{s_i} \leq \delta_{s_{i+1}}$ with Lemma 2,

$$\sum_{i=k+1}^{n-1} \left(\frac{6\delta_{s_i}}{\log \delta_{s_i}} + \sqrt{\delta_{s_i}} \right) \frac{w_n}{\prod_{j=i}^{n-1} r_j}$$

$$\leq \left(\frac{6\sqrt{\delta_{s_{k+1}}}}{\log \delta_{s_{k+1}}} + 1 \right) \frac{w_n}{\prod_{j=k+2}^{n-1} r_j} + \sum_{i=k+2}^{n-1} \left(\frac{6\delta_{s_i}}{\log \delta_{s_i}} + \sqrt{\delta_{s_i}} \right) \frac{w_n}{\prod_{j=i}^{n-1} r_j}$$

$$\leq \left(\frac{6\sqrt{\delta_{s_{k+2}}}}{\log \delta_{s_{k+2}}} + 1 \right) \left(1 + \frac{1}{\sqrt{\delta_{s_{k+2}}}} \right) \frac{w_n}{\prod_{j=k+3}^{n-1} r_j} + \sum_{i=k+3}^{n-1} \left(\frac{6\delta_{s_i}}{\log \delta_{s_i}} + \sqrt{\delta_{s_i}} \right) \frac{w_n}{\prod_{j=i}^{n-1} r_j}$$

$$\leq \dots$$

$$\leq \left(\frac{6\sqrt{\delta_{s_{n-1}}}}{\log \delta_{s_{n-1}}} + 1 \right) \left(1 + \frac{1}{\sqrt{\delta_{s_{n-1}}}} + \dots + \frac{1}{(\sqrt{\delta_{s_{n-1}}})^{n-k-2}} \right) w_n$$

$$\leq \left(\frac{6\sqrt{\Delta}}{\log \Delta} + 1 \right) \frac{\sqrt{\Delta}}{\sqrt{\Delta} - 1} \left(1 - \frac{1}{(\sqrt{\Delta})^{n-k-1}} \right) w_n$$

where the last inequality holds due to $\delta_{s_{n-1}} \leq \delta_{s_n} \leq \Delta$ and Lemma 2. The above inequality is less than $\left(\frac{6\sqrt{\Delta}}{\log \Delta} + 1 \right) \frac{\sqrt{\Delta}}{\sqrt{\Delta} - 1} w_n$ if $k < n - 1$, and otherwise it is equal to 0 if $k = n - 1$. Thus,

$$|\sigma^*| - |\tau_n^*| \leq \left(\frac{6\sqrt{\Delta}}{\log \Delta} + 1 \right) \frac{\sqrt{\Delta}}{\sqrt{\Delta} - 1} w_n + O(1) w_n.$$

We already upper bound $|\tau_n^*|$ by $\left(\frac{\lambda_{s_n} \cdot 2^{d(s_n, s_n, k_n + p_{n, k_n}) + 1}}{(\delta_{s_n})^{1/3}} + \sqrt{\delta_{s_n}} \right) w_n$. Thus, the ratio between what OPT and DPC gains in σ can be bounded in two subcases according to whether τ_n consists of a single order, i.e., $k_n = 0$.

Case 1.1. $k_n = 0$ and then DPC completes $R_{n,0}$, implying that $|\sigma| = w_n$.

$$|\tau_n^*| \leq \left(\frac{\lambda_{s_n} \cdot 2^{d(s_n, s_n + p_n) + 1}}{(\delta_{s_n})^{1/3}} + \sqrt{\delta_{s_n}} \right) w_n$$

$$\leq \left(\frac{6\delta_{s_n}}{\log \delta_{s_n}} + \sqrt{\delta_{s_n}} \right) w_n$$

$$\leq \left(\frac{6\Delta}{\log \Delta} + \sqrt{\Delta} \right) w_n.$$

OPT may still satisfy $R_{n,0}$ after DPC does if the order has sufficiently large deadline, and then OPT gains at most $|\sigma^*| + w_n$ in σ. Hence, $(|\sigma^*| - |\tau_n^*|)/|\sigma| = O(\frac{\sqrt{\Delta}}{\log \Delta})$, and then $(|\sigma^*| + w_n)/|\sigma| \leq \frac{6\Delta}{\log \Delta} + O(\sqrt{\Delta})$ in this case.

Case 1.2. $k_n > 0$. In this case, $|\sigma| = w_{n,k_n} \geq \frac{2^{d(s_n, s_n, k_n + p_{n, k_n})}}{(\delta_{s_n})^{1/3}} w_n$, which is larger than $\frac{w_n}{(\delta_{s_n})^{1/3}} > \frac{w_n}{\Delta^{1/3}}$ due to $d(s_n, s_{n,k_n} + p_{n,k_n}) > 0$. So, $(|\sigma^*| - |\tau_n|)/|\sigma| = O(\frac{\Delta^{5/6}}{\log \Delta})$. And

$$|\tau_n^*| \leq \left(\frac{\lambda_{s_n} \cdot 2^{d(s_n, s_n, k_n + p_{n, k_n}) + 1}}{(\delta_{s_n})^{1/3}} + \sqrt{\delta_{s_n}} \right) w_n.$$

Similarly, OPT may satisfy $R_{n,0}$ with profit w_{n,k_n} after DPC does. With $\delta_{s_n} \leq \Delta$, we have that $|\tau_n^*|/|\sigma| \leq 2\lambda_{s_n} + \sqrt{\delta_{s_n}}\Delta^{1/3} \leq \frac{6\Delta}{\log \Delta} + O(\Delta^{5/6})$ and then $(|\sigma^*| + w_{n,k_n})/|\sigma| \leq \frac{6\Delta}{\log \Delta} + O(\Delta^{5/6})$ in this case.

Case 2. $k = n$ and then $\delta_{s_n} \leq 20$. In this case, $w_i \leq w_n/(\sqrt{20})^{n-i}$ for $0 \leq i \leq n$. DPC will complete R_{n,k_n} with weight w_{n,k_n}, which is at least w_n by construction of DPC. That is, $|\sigma| \geq w_n$. For OPT, $|\sigma^*|$ is bounded in below by Lemma 1.

$$|\sigma^*| = \sum_{i=0}^{n} |\tau_i^*| \leq \sum_{i=0}^{n} (\delta_{s_k} + \sqrt{20})w_i \leq \sum_{i=0}^{n} \frac{(\delta_{s_k} + \sqrt{20})}{(\sqrt{20})^{n-i}} w_n < \frac{\sqrt{20}(\delta_{s_k} + \sqrt{20})}{(\sqrt{20} - 1)} w_n.$$

OPT may satisfy R_{n,k_n} after DPC does and thus

$$(|\sigma^*| + w_{n,k_n})/|\sigma| \leq \frac{\sqrt{20}(\delta_{s_k} + \sqrt{20})}{(\sqrt{20} - 1)} + 1 < \frac{\sqrt{20}}{(\sqrt{20} - 1)}\delta_{s_k} + 6.8.$$

If $20 < \Delta$, $(|\sigma^*| + w_{n,k_n})/|\sigma| < \frac{20\sqrt{20}}{(\sqrt{20}-1)} + 6.8 = o(\Delta)$ by $\delta_{s_k} \leq 20$, otherwise if $\Delta \leq 20$, then $(|\sigma^*| + w_{n,k_n})/|\sigma| < \frac{\sqrt{20}}{(\sqrt{20}-1)}\Delta + 6.8$ due to $\delta_{s_k} \leq \Delta$.
The theorem follows.   □

5   Conclusion

This paper discussed an online order scheduling scenery where online strategies does not have the knowledge of Δ at all beforehand. We mainly present a $(\frac{6\Delta}{\log \Delta} + O(\Delta^{5/6}))$-competitive optimal strategy DPC, extending the previous conclusion in [3] from the case with the knowledge of Δ to the one without the knowledge. It is interesting that whether randomization can help to break this bound.

References

1. Fung, S.P.Y., Chin, F.Y.L., Poon, C.K.: Laxity helps in broadcast scheduling. In: Proceedings of 11th Italian Conference on Theoretical Computer Science, Siena, Italy, pp. 251–264 (2005)
2. Zheng, F.F., Fung, S.P.Y., Chan, W.T., Chin, F.Y.L., Poon, C.K., Wong, P.W.H.: Improved On-line Broadcast Scheduling with Deadlines. In: Proceedings of the 12th Annual International Computing and Combinatorics Conference, Taipei, Taiwan, pp. 320–329 (2006)
3. Ting, H.F.: A near optimal scheduler for on-demand data broadcasts. In: 6th Italian Conference on Algorithms and Complexity, Rome, Italy, pp. 163–174 (2006)
4. Kim, J.H., Chwa, K.Y.: Scheduling broadcasts with deadlines. In: Proceedings of 9th Italian Conference on Theoretical Computer Science, Big Sky, MT, USA, pp. 415–424 (2003)
5. Zheng, F.F., Dai, W.Q., Xiao, P., Zhao, Y.: Competitive Strategies for On-line Production Order Disposal Problem. In: 1st International Conference on Algorithmic Applications In Management, Xi'an, China, pp. 46–54 (2005)

6. Borodin, A., El-yaniv, R.: Online computation and competitive analysis. Cambridge University Press, England (1998)
7. Motwani, R., Phillips, S., Torng, E.: Nonclairvoyant Scheduling. Theoretical Computer Science 130(1), 17–47 (1994)
8. Kalyanasundaram, B., Pruhs, K.R.: Minimizing flow time nonclairvoyantly. Journal of the ACM 50(4), 551–567 (2003)
9. Becchetti, L., Leonardi, S.: Nonclairvoyant scheduling to minimize the total flow time on single and parallel machines. Journal of the ACM 51(4), 517–539 (2004)
10. Lipton, R.J., Tomkins, A.: Online Interval Scheduling. In: Proc. Of the 5th Annual ACM-SIAM Symposium on Discrete Algorithm (SODA 1994), pp. 302–311. New York (1994)
11. Goldwasser, M.H.: Patience is a Virtue: The effect of slack on competitiveness for admission control. Journal of Scheduling 6(2), 183–211 (2003)

Large-Scale Parallel Collaborative Filtering for the Netflix Prize

Yunhong Zhou, Dennis Wilkinson, Robert Schreiber, and Rong Pan

HP Labs, 1501 Page Mill Rd, Palo Alto, CA, 94304
{yunhong.zhou,dennis.wilkinson,rob.schreiber,rong.pan}@hp.com

Abstract. Many recommendation systems suggest items to users by utilizing the techniques of *collaborative filtering* (CF) based on historical records of items that the users have viewed, purchased, or rated. Two major problems that most CF approaches have to contend with are scalability and sparseness of the user profiles. To tackle these issues, in this paper, we describe a CF algorithm *alternating-least-squares with weighted-λ-regularization* (ALS-WR), which is implemented on a parallel Matlab platform. We show empirically that the performance of ALS-WR (in terms of *root mean squared error* (RMSE)) monotonically improves with both the number of features and the number of ALS iterations. We applied the ALS-WR algorithm on a large-scale CF problem, the Netflix Challenge, with 1000 hidden features and obtained a RMSE score of 0.8985, which is one of the best results based on a pure method. In addition, combining with the parallel version of other known methods, we achieved a performance improvement of 5.91% over Netflix's own CineMatch recommendation system. Our method is simple and scales well to very large datasets.

1 Introduction

Recommendation systems try to recommend items (movies, music, webpages, products, etc) to interested potential customers, based on the information available. A successful recommendation system can significantly improve the revenue of e-commerce companies or facilitate the interaction of users in online communities. Among recommendation systems, *content-based* approaches analyze the content (e.g., texts, meta-data, features) of the items to identify related items, while *collaborative filtering* uses the aggregated behavior/taste of a large number of users to suggest relevant items to specific users. Collaborative filtering is popular and widely deployed in Internet companies like Amazon [16], Netflix [2], Google News [7], and others.

The Netflix Prize is a large-scale data mining competition held by Netflix for the best recommendation system algorithm for predicting user ratings on movies, based on a training set of more than 100 million ratings given by over 480,000 users to 17,700 movies. Each training data point consists of a quadruple (user, movie, date, rating) where rating is an integer from 1 to 5. The test dataset consists of 2.8 million data points with the ratings hidden. The goal is

R. Fleischer and J. Xu (Eds.): AAIM 2008, LNCS 5034, pp. 337–348, 2008.

to minimize the RMSE (root mean squared error) when predicting the ratings on the test dataset. Netflix's own recommendation system (CineMatch) scores 0.9514 on the test dataset, and the grand challenge is to improve it by 10%.

The Netflix problem presents a number of practical challenges. (Which is perhaps why, as yet, the prize has not been won.) First, the size of the dataset is 100 times larger than previous benchmark datasets, resulting in much longer model training time and much larger system memory requirements. Second, only about 1% of the user-movie matrix has been observed, with the majority of (potential) ratings missing. This is, of course, an essential aspect of collaborative filetering in general. Third, there is noise in both the training and test dataset, due to human behavior – we cannot expect people to be completely predictable, at least where their feelings about ephemera like movies is concerned. Fourth, the distribution of ratings per user in the training and test datasets are different, as the training dataset spans many years (1995-2005) while the testing dataset was drawn from recent ratings (year 2006). In particular, users with few ratings in the training set are well represented in the test set. Intuitively, it is hard to predict the ratings of a user who is sparsely represented in the training set.

In this paper, we introduce the problem in detail. Then we describe a parallel algorithm, alternating-least-squares with weighted-λ-regularization. We use parallel Matlab on a Linux cluster as the experimental platform, and our core algorithm is parallelized and optimized to scale up well with large, sparse data. When we apply the proposed method to the Netflix Prize problem, we achieve a performance improvement of 5.91% over Netflix's own CineMatch system.

The rest of the paper is organized as follows: in Section 2 we introduce the problem formulation. In Section 3 we describe our novel parallel Alternative-Least-Squares algorithm. Section 4 describes experiments that show the effectiveness of our approach. Section 5 discusses related work and Section 6 concludes with some future directions.

2   Problem Formulation

Let $R = \{r_{ij}\}_{n_u \times n_m}$ denote the user-movie matrix, where each element r_{ij} represents the rating score of movie j rated by user i with its value either being a real number or missing, n_u designates the number of users, and n_m indicates the number of movies. As in most recommendation systems our task is to estimate some of the missing values in R based on the known values. (The Netflix dataset consists of $n_m = 17770$ movies, $n_u = 488000$ users, and $n_r \approx 100$ million known ratings.)

We start with a low-rank approximation of the ratings matrix R. This approach models both users and movies by giving them coordinates in a low dimensional feature space. Each user and each movie has a feature vector, and each rating (known or unknown) of a movie by a user is modeled as the inner product of the corresponding user and movie feature vectors. More specifically, let $U = [\mathbf{u_i}]$ be the user feature matrix, where $\mathbf{u_i} \in \mathbb{R}^{n_f}$, $i = 1 \ldots n_u$, denotes the ith column of U, and let $M = [\mathbf{m_j}]$ be the movie feature matrix, where

$\mathbf{m_j} \in \mathbb{R}^{n_f}$, $j = 1 \ldots n_m$, is the jth column of M. Here n_f is the dimension of the feature space, that is, the number of hidden variables in the model. It is a system parameter that can be determined by a hold-out dataset or cross-validation. If user ratings were fully predictable and n_f sufficiently large, we could expect that $r_{ij} = <\mathbf{u_i}, \mathbf{m_j}>$, $\forall\ i, j$. In practice, however, we minimize a loss function (of U and M) to obtain the matrices U and M. In this paper, we study the mean-square loss function. The loss due to a single rating is defined as the squared error:

$$\mathcal{L}^2(r, \mathbf{u}, \mathbf{m}) = (r - <\mathbf{u}, \mathbf{m}>)^2. \tag{1}$$

Then we can define the empirical, total loss (for a given pair U and M) as the summation of loss on all the known ratings in Eq. (2).

$$\mathcal{L}^{emp}(R, U, M) = \frac{1}{n} \sum_{(i,j) \in I} \mathcal{L}^2(r_{ij}, \mathbf{u_i}, \mathbf{m_j}), \tag{2}$$

where I is the index set of the known ratings and n is the size of I.

We can formulate the low-rank approximation problem as follows.

$$(U, M) = \arg \min_{(U, M)} \mathcal{L}^{emp}(R, U, M). \tag{3}$$

where U and M are real, have n_f rows, but are otherwise unconstrained.

In this problem, (Eq. (3)), there are $(n_u + n_m) \times n_f$ free parameters to be determined. Our results show that allowing n_f to be quite large, 1000 in our tests, improves the quality of the results. Thus, for Netflix we have over $480,000,000$ model parameters, more data than there are points in the training set I. For this reason, problem 3 is underdetermined. Indeed, very few users will have rated n_f movies when n_f is greater than a few hundred, yet we must assign each user a point in an n_f dimensional feature space.

Thus, when n_f is relatively large, solving the problem Eq. (3) overfits the data. To avoid overfitting, a common method appends a Tikhonov regularization [22] term to the empirical risk function:

$$\mathcal{L}_\lambda^{reg}(R, U, M) = \mathcal{L}^{emp}(R, U, M) + \lambda(\|U\Gamma_U\|^2 + \|M\Gamma_M\|^2), \tag{4}$$

for a certain suitably selected Tikhonov matrices Γ_U and Γ_M.[1] We will discuss the details in the next section.

3   Our Approaches

In this section, we describe an iterative algorithm, alternating-least-squares with weighted-λ-regularization (ALS-WR), to solve the low rank approximation problem. Then we develop a parallel implementation of ALS-WR based on a parallel Matlab platform.

---

[1] Throughout the paper, $\|X\|$ denotes the Frobenius norm of the matrix X.

3.1   ALS with Weighted-λ-Regularization

As the rating matrix contains both signals and noise, it is important to remove noise and use the recovered signal to predict missing ratings. *Singular Value Decomposition* (SVD) is a natural approach that approximates the original user-movie rating matrix R by the product of two rank-k matrices $\tilde{R} = U^T \times M$. The solution given by the SVD minimizes the Frobenious norm of $R - \tilde{R}$, which is equivalent to minimizing the RMSE over all elements of R. However, as there are many missing elements in the rating matrix R, standard SVD algorithms cannot find U and M.

In this paper, we use *alternating-least-squares* (ALS) to solve the low-rank matrix factorization problem as follows:

Step 1. Initialize matrix M by assigning the average rating for that movie as the first row, and small random numbers for the remaining entries.

Step 2. Fix M, Solve for U by minimizing the objective function (4);

Step 3. Fix U, solve for M by minimizing the objective function similarly;

Step 4. Repeat Steps 2 and 3 until a stopping criterion is satisfied.

Observe that when the regularization matrices $\Gamma_{(U,M)}$ are nonsingular, each of the problems of Steps 2 and 3 of the algorithm has a unique solution, which we derive below. Note that the sequence of achieved errors $\mathcal{L}_\lambda^{reg}(R, U, M)$ is monotone nonincreasing and bounded below, hence this sequence converges.

Rather than going all the way to convergence, we use a stopping criterion based on the observed RMSE on the probe dataset. The probe dataset is provided by Netflix, and it has the same distribution as the hidden test dataset. After one round of updating both U and M, if the change in RMSE on the probe dataset is less than 1 bps[2], the iteration stops and we use the obtained U, M to make final predictions on the test dataset.

As we mention in Section 2, there are many free parameters. Without regularization, ALS might lead to overfitting. A common fix is to use Tikhonov regularization, which penalizes large parameters. We tried various regularization matrices, and eventually found the following weighted-λ-regularization to work the best, as it never overfits the test data (empirically) when we increase the number n_f of features or the number of ALS iterations.

$$f(U, M) = \sum_{(i,j) \in I} (r_{ij} - \mathbf{u}_i^T \mathbf{m}_j)^2 + \lambda \left(\sum_i n_{u_i} \|\mathbf{u}_i\|^2 + \sum_j n_{m_j} \|\mathbf{m}_j\|^2 \right), \quad (5)$$

where n_{u_i} and n_{m_j} denote the number of ratings of user i and movie j respectively. This corresponds to Tikhonov regularization (4), where $\Gamma_U = \mathrm{diag}(\sqrt{n_{u_i}})$ and $\Gamma_M = \mathrm{diag}(\sqrt{n_{m_j}})$. [3]

---

[2] 1 bps equals 0.0001.

[3] The same objective function was used previously by Salakhutdinov et al. [20] and solved using gradient descent. We will discuss more on their approach in Section 5.2.

Now we demonstrate how to find the matrix U when M is given. Let I_i^U denote the set of indices of movies that user i rated, so that n_{u_i} is the cardinality of I_i^U; similarly I_j^M denotes the set of indices of users who rated movie j, and n_{m_j} is the cardinality of I_j^M. A given column of U, say u_i, is determined by solving a regularized linear least squares problem involving the known ratings of user i, and the feature vectors $\mathbf{m}_j$ of the movies $j \in I_i^U$ that user i has rated.

$$\frac{1}{2}\frac{\partial f}{\partial u_{ki}} = 0, \quad \forall i, k$$

$$\Rightarrow \sum_{j \in I_i^U} (\mathbf{u}_i^T \mathbf{m}_j - r_{ij}) m_{kj} + \lambda n_{u_i} u_{ki} = 0, \quad \forall i, k$$

$$\Rightarrow \sum_{j \in I_i^U} m_{kj} \mathbf{m}_j^T \mathbf{u}_i + \lambda n_{u_i} u_{ki} = \sum_{j \in I_i^U} m_{kj} r_{ij}, \quad \forall i, k$$

$$\Rightarrow \left(M_{I_i^U} M_{I_i^U}^T + \lambda n_{u_i} E \right) \mathbf{u}_i = M_{I_i^U} R^T(i, I_i^U), \quad \forall i$$

$$\Rightarrow \mathbf{u}_i = A_i^{-1} V_i, \quad \forall i$$

where $A_i = M_{I_i^U} M_{I_i^U}^T + \lambda n_{u_i} E$, $V_i = M_{I_i^U} R^T(i, I_i^U)$, and E is the $n_f \times n_f$ identity matrix. $M_{I_i^U}$ denotes the sub-matrix of M where columns $j \in I_i^U$ are selected, and $R(i, I_i^U)$ is the row vector where columns $j \in I_i^U$ of the i-th row of R are selected.

Similarly, when M is updated, we can compute individual m_j's via a regularized linear least squares solution, using the feature vectors of users who rated movie j, and their ratings of it, as follows:

$$\mathbf{m}_j = A_j^{-1} V_j, \quad \forall j,$$

where $A_j = U_{I_i^M} U_{I_i^M}^T + \lambda n_{m_j} E$ and $V_j = U_{I_i^M} R(I_i^M, j)$. $U_{I_i^M}$ denotes the sub-matrix of U where columns $i \in I_i^M$ are selected, and $R(I_i^M, j)$ is the column vector where rows $i \in I_i^M$ of the j-th column of R is taken.

Let n_r denote the total number of ratings, then we have $n_r \ll n_u n_m$ for a sparse rating matrix. We can easily derive the running time of the above algorithm based on standard matrix operations:

Theorem 1. *For ALS-WR, each step of updating U takes $O\left(n_f^2(n_r + n_f n_u)\right)$ time while each step of updating M takes $O\left(n_f^2(n_r + n_f n_m)\right)$. If ALS-WR takes totally n_t rounds to stop, it runs in time $O\left(n_f^2(n_r + n_f n_u + n_f n_m)n_t\right)$.*

3.2   Parallel ALS with Weighted-λ-Regularization

We parallelize ALS by parallelizing the updates of U and of M. We are using the latest verstion of Matlab, which allows parallel Matlab computation in which several separate copies of Matlab, each with its own private workspace, and each running on its own hardware platform, collaborate and communicate to solve problems. Each such running copy of Matlab is referred to as a "lab", with its own identifier (labindex) and with a static variable (numlabs) telling how many labs there are. Matrices can be private (each lab has its own copy, and their values differ), replicated (private, but with the same value on all labs) or distributed (there is one matrix, but with rows, or columns, partitioned among the labs). Distributed matrices are a convenient way to store and use very large datasets, too large to be stored on one processor and its associated local memory. In our case, we use two distributed copies of the ratings matrix R, one distributed by rows (i.e., by users) and the other by columns (i.e., by movies). We will compute distributed, updated matrices U and M. In computing U we will require a replicated version of M, and *vice versa*. Thus, our labs communicate to make replicated versions of U and of M from the distributed versions that are first computed. Matlab's "gather" function performs the inter-lab communication needed for this.

To update M, we require a replicated copy of U, local to each lab. We use the ratings data distributed by columns (movies). The distribution is by blocks of equal numbers of movies. The lab that stores the ratings of movie j will, naturally, be the one that updates the corresponding column of M, which is movie j's feature vector. Each lab computes $\mathbf{m}_j$ for all movies in the corresponding movie group, in parallel. These values are then "gathered" so that every node has all of M, in a replicated array. To update U, users are partitioned into equal-size user groups and each lab just update user vectors in the corresponding user group, using the ratings data partitioned by rows. The following Matlab snippet implements the procedure of updating M given U:

```
function M = updateM(lAcols, U)
    lamI = lambda * eye(Nf);
    lM = zeros(Nf,Nlm); lM = single(lM);
    for m = 1:Nlm
        users = find(lAcols(:,m));
        Um = U(:, users);
        vector = Um * full(lAcols(users, m));
        matrix = Um * Um' + locWtM(m) * lamI;
        X = matrix \ vector;
        lM(:, m) = X;
    end
    M = gather(darray(lM));
end
```

For the above Matlab code, lAcols is the local copy of R distributed by columns (movies), locWtM is the vector of n_{m_j} for all movies in the partitioned movie group, and Nlm is the number of movies in the movie group. Nf and lambda correspond to n_f and λ, and they are the only tunable parameters of ALS-WR.

The gather operation is the only place where we incur communication cost due to using a distributed (as opposed to a shared-memory) algorithm. For our method, it takes less than 5% of the total run time. Therefore, *the parallel algorithm achieves a nearly linear speedup.* As an example, for $n_f = 100$, it takes about 2.5 hours to update M and U once with a single processor; with 30 processors, it takes about 5 minutes for one iteration, and the converged solution (with 30 ALS iterations) can be computed in 2.5 hours.

4   Performance for the Netflix Prize Problem

We ran on a 30-processor Linux cluster of HP ProLiant DL380 G4 machines. All processors are Xeon 2.8GHz and every four processors share 6GB of RAM. For each fixed n_f, we run between 10 to 25 rounds of ALS-WF and stop when one round of U, M update improves by less than 1 bps the RMSE score on the probe dataset. The optimal value of λ is determined by trial and error.[4] The test RMSE is obtained by submission to the Netflix prize website[5].

The probe dataset is a subset of the training dataset. It consists of 1,408,395 ratings from the year 2006. In it, users are selected at random with uniform probability, and at most 9 ratings are included for each selected user. The test dataset is hidden by Netflix but the distribution of the test dataset is the same as the distribution of the probe dataset. For model building and parameter tuning, we exclude the probe dataset from the training dataset (this is how we determine the set I, above.) We use the probe for convergence criterion in ALS-WR and for testing.

4.1   Postprocessing

For postprocessing of the prediction results, we first apply a global bias correction technique to the prediction solution. Given a prediction P, if the mean of P is not equal to the mean of the test dataset, we can shift all predicted values by a fixed constant $\tau = \text{mean}(\text{test}) - \text{mean}(P)$, thus improving the RMSE. We also use convex combinations of several predictors to obtain a better predictor. For example, given two predictors P_0 and P_1, we can obtain a family of predictors $P_x = (1-x)P_0 + xP_1$, $x \in [0,1]$, and use linear regression to find x^* minimizing $\text{RMSE}(P_x)$. Therefore we obtain P_{x^*} which is at least as good as P_0 or P_1.

4.2   Experimental Results for ALS

The most important discovery we made is that ALS-WR never overfits the data if we either increase the number of iterations or the number of hidden features. As Figure 1 shows, with fixed n_f and λ, each iteration improves the RMSE score of the probe dataset, and it converges after about 20 rounds. Different λ values give different final score, and we normally need to try a small number of

---

[4] Empirically we found that for fixed n_f, the convergence RMSE score is a convex function of λ, and the optimal value of λ is monotone decreasing with respect to n_f. Based on these observations, we are able to find the best value of λ for each n_f with only 2-3 experiments.

[5] See http://www.netflixprize.com/rules for the detailed rules of the competition.

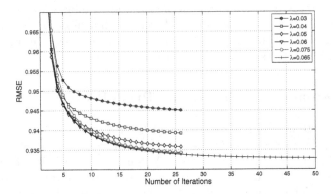

Fig. 1. Comparisons of different λ values for ALS-WR with $n_f = 8$. The best performer with 25 rounds is $\lambda = 0.065$. For this fixed λ, after 50 rounds, the RMSE score still improves but only less than 0.1 bps for each iteration afterwards.

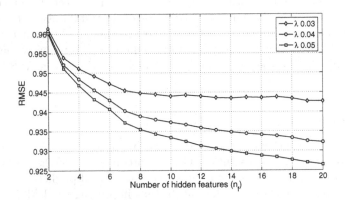

Fig. 2. Performance of ALS-WR with fixed λ and varying n_f

λ values to get a good RMSE score. Figure 2 shows the performance of ALS-WR with fixed λ value and varying number of hidden features (n_f ranges from 2 to 20). For each experiment, ALS-WR iterations continue until the RMSE over the probe dataset improves less than 1 bps. From the figure we can tell that the RMSE monotonically decreases with larger n_f, even though the improvement diminishes gradually.

Next we conduct experiments with real submissions using large values of n_f. For ALS with simple λ regularization ($\Gamma_u = \Gamma_m = E$), we obtain a RMSE of 0.9184. For ALS with weighted-λ-regularization, we obtained a RMSE of 0.9114 with $n_f = 50$, 0.9066 with $n_f = 150$. With $n_f = 300$ and global bias correction, we obtain a RMSE of 0.9017; with $n_f = 400$ and global bias correction, a score of 0.9006 was obtained; with $n_f = 500$ and global bias shift, a score of 0.9000 was obtained. Ultimately, we experimented with $n_f = 1000$ and obtained a RMSE score of 0.8985.[6] Given that 6 bps improvement is obtained from 400 to 500 features, and assuming diminishing (equal-decrease) return with increasing number of features, moving from 500 to 1000 features improves approximately $5 + 4 + 3 + 2 + 1 = 15$ bps. Therefore, 0.8985 is likely the limit we can achieve using ALS with Weighted-λ-Regularization. A RMSE score of 0.8985 translates into a 5.56% improvement over Netflix's CineMatch, and it represents one of the top *single-method* performance according to our knowledge.

4.3 Other Methods and Linear Blending

We also implement parallel versions of two other popular collaborative filtering techniques as described in this section. In each case, the speedup as compared to a single-processor version is roughly a factor of n on a cluster of n processors.

The Restricted Boltzmann Machine (RBM) is a kind of neural network where there are visible states and hidden states, and undirected edges connecting each visible state to each hidden state. There are no connections among visible states or among hidden states, thus the name "restricted." RBM was previously demonstrated to work well for the Netflix challenge [20]. We implemented RBM using Matlab, and converted it to Pmode. For a model with 100 hidden units, it takes about 1 hour for one iteration without Pmode; using Pmode with 30 labs, it takes 3 minutes for one iteration.

The k-nearest neighbor (kNN) method is also popular for prediction. With a properly defined distance metric, for each data point needed for prediction, the weighted average of the ratings of its k closest neighbors is used to predict the rating of this point. Since there are so many user-user pairs for us to handle in reasonable time and space, we use a simplified approach with only movie-movie similarities. Again, we parallelize kNN by partitioning users into user groups so that each lab processes one user group.

For RBM itself, a score of 0.9181 is obtained. For kNN with $k = 21$ and a good similarity function, a RMSE of 0.9270 is obtained. Linear blending of ALS with kNN and RBM yields a RMSE of 0.8952 (ALS + kNN + RBM), and it represents a 5.91% improvement over Netflix's CineMatch system.

5 Related Work

There is a lot of academic and industrial work on recommendation systems, low-rank matrix approximation, and the Netflix prize. In the following we briefly discuss related work most relevant to ours.

---

[6] The experiment with $n_f = 1000$ is technically challenging as U takes $2G$ memory with single precision entries for each processor. We managed to run the procedure of updateM in two batches, while in each batch only two processors for each server are active and U is only replicated in these processors. This avoids memory thrashing using the 6G shared memory for each server. And ALS-WR stops in 10 rounds while each rounds takes 1.5 hours.

5.1   Recommendation Systems

Recommendation systems can be mainly categorized into content-based and collaborative filtering, and are well-studied in both academia and industry [16,2,7]. *Content-based* recommendation systems analyze the content (e.g., texts, metadata, features) of the item to identify related items, with exemplary systems InfoFinder [12], NewsWeeder [14]. *Collaborative Filtering* uses aggregated behavior/taste of a large number of users to suggest relevant items to specific users. Recommendations generated by CF are based solely on the user-user and/or item-item similarities, with exemplary systems GroupLens [19] and Bellcore Video Recommender [11]. Efforts to combine both content-based approach and collaborative filtering include the Fab system [3] and unified probabilistic framework [18].

5.2   The Netflix Prize Approaches

For the Netflix prize, Salakhutdinov et al. [20] used Restricted Boltzmann Machines (RBM), obtaining an RMSE score of slightly below 0.91. They also presented a low-rank approximation approach using gradient descent. Their low-rank approximation obtained an RMSE score slightly above 0.91 using between 20-60 hidden features.[7] The objective function of their SVD approach is the same as our ALS-WR, however we use alternating least squares instead of gradient descent to solve the optimization problem, and we are able to use a much large number of features (1000 rather than 40) to obtain significant improvement in RMSE score.

Among many other approaches to the Netflix problem, Bell et al. [5] proposed a neighborhood-based technique which combines k-nearest-neighbor (kNN) and low-rank approximation to obtain significantly better results compared to either technique alone. Their team won the progress prize in October 2007, obtaining an RMSE score on the qualifying dataset of 0.8712, improving the CineMatch score by 8.5%. However, their solution [4] is a linear combination of 107 individual solutions, while multiple solutions are derived by variants of three classes of solutions (ALS, RBM, and kNN). For ALS alone, their best result was obtained using 128 hidden features with an RMSE score above 0.9000. For a comprehensive treatment of various approaches for the Netflix prize, see the individual papers presented in KDD Cup & Workshop 2007 [21,17,15,13,23].

5.3   Low-Rank Approximation

When a fully specified matrix is to be approximated by a low-rank matrix factorization, variants of singular value decomposition are used, for example in information retrieval (where SVD techniques are known as latent semantic indexing [9]). Other matrix factoring methods, for example nonnegative matrix factorization and maximum margin matrix factorization have also been proposed for the Netflix prize [23].

---

[7] We obtained similar results and got 0.9114 with $n_f = 50$.

For a partially specified matrix, the SVD is not applicable. To minimize the sum of squared differences between the known elements and the corresponding elements of the factored low rank matrix, ALS has proven to be an effective approach. It provides non-orthogonal factors, unlike SVD. The SVD can be computed one column at a time, whereas for the partially specified case, no such recursive formulation holds. An advantage of ALS is its easy parallelization. Like Lanczos for the sparse, fully specified case, ALS preserves the sparse structure of the known matrix elements and is therefore storage-efficient.

6 Concluding Remarks

We introduced a simple parallel algorithm for large-scale collaborative filtering which, in the case of the Netflix prize, performed as well as any single method reported in the literature. Our algorithm is designed to be scalable to very large datasets. Moderately better scores can be obtained by refining the RBM and kNN implementation or using more complicated blending schemes. ALS-WR in particular is able to achieve good results without using date or movie title information. The fast runtime achieved through parallelization is a competitive advantage for model building and parameter tuning in general. It will be interesting to develop a theory to explain why ALS-WR never overfits the data.

As the world shifts into Internet computing and web applications, large-scale data intensive computing becomes pervasive. Traditional single-machine, single-thread computing is no longer viable, and there is a paradigm shift in computing models. Parallel and/or distributed computing becomes an essential component for any computing environment. Google, the leading Internet company, is building its own proprietary parallel/distributed computing infrastructure, based on MapReduce [8], Google File System [10], Bigtable [6], etc. Most technology companies do not have the capital and expertise to develop an in-house large-scale parallel/distributed computing infrastructure, and prefer instead to use readily available solutions to solve computing infrastructure problems. Hadoop [1] is an open-source project sponsored by Yahoo!, which tries to replicate the Google computing infrastructure with open-source development. We have found parallel Matlab to be flexible and efficient, and very straightforward to program. Thus, from our experience, it seems to be a strong candidate for widespread, easily scalable parallel/distributed computing.

References

1. The Hadoop Project, http://lucene.apache.org/hadoop/
2. Netflix CineMatch, http://www.netflix.com
3. Balabanovi, M., Shoham, Y.: Fab: content-based, collaborative recommendation. Communications of the ACM 40(3), 66–72 (1997)
4. Bell, R., Koren, Y., Volinsky, C.: The bellkor solution to the netflix prize. Netflix Prize Progress Award (October 2007),
   http://www.netflixprize.com/assets/ProgressPrize2007_KorBell.pdf

5. Bell, R., Koren, Y., Volinsky, C.: Modeling relationships at multiple scales to improve accuracy of large recommender systems. In: Proc. KDD 2007, pp. 95–104 (2007)
6. Chang, F., et al.: Bigtable: A distributed storage system for structured data. In: Proc. of OSDI 2006, pp. 205–218 (2006)
7. Das, A., Datar, M., Garg, A., Rajaram, S.: Google news personalization: Scalable online collaborative filtering. In: Proc. of WWW 2007, pp. 271–280 (2007)
8. Dean, J., Ghemawat, S.: Mapreduce: Simplified data processing on large clusters. In: Proc. OSDI 2004, San Francisco, pp. 137–150 (2004)
9. Deerwester, S., Dumais, S.T., Furnas, G.W., Landauer, T.K., Harshman, R.: Indexing by latent semantic analysis. J. Amer. Soc. Info. Sci. 41(6), 391–407 (1999)
10. Ghemawat, S., Gobioff, H., Leung, S.-T.: The Google File System. In: Proc. of SOSP 2003, pp. 29–43 (2003)
11. Hill, W., Stead, L., Rosenstein, M., Furnas, G.: Recommending and evaluating choices in a virtual community of use. In: Proc. of CHI 1995, Denver (1995)
12. Krulwich, B., Burkey, C.: Learning user information interests through extraction of semantically significant phrases. In: Proc. AAAI Spring Symposium on Machine Learning in Information Access, Stanford, CA (March 1996)
13. Kurucz, M., Benczur, A.A., Csalogany, K.: Methods for large scale SVD with missing values. In: Proc. KDD Cup and Workshop (2007)
14. Lang, K.: NewsWeeder: Learning to filter Netnews. In: Proc. ICML 1995, pp. 331–339 (1995)
15. Lim, Y.J., Teh, Y.W.: Variational bayesian approach to movie rating prediction. In: Proc. KDD Cup and Workshop (2007)
16. Linden, G., Smith, B., York, J.: Amazon.com recommendations: Item-to-item collaborative filtering. IEEE Internet Computing 7, 76–80 (2003)
17. Paterek, A.: Improving regularized singular value decomposition for collaborative filtering. In: Proc. KDD Cup and Workshop (2007)
18. Popescul, A., Ungar, L., Pennock, D., Lawrence, S.: Probabilistic models for unified collaborative and content-based recommendation in Sparse-Data Environments. In: Proc. UAI, pp. 437–44 (2001)
19. Resnick, P., Iacovou, N., Suchak, M., Bergstrom, P., Riedl, J.: GroupLens: an open architecture for collaborative filtering of Netnews. In: Proc. the ACM Conference on Computer-Supported Cooperative Work, Chapel Hill, NC (1994)
20. Salakhutdinov, R., Mnih, A., Hinton, G.E.: Restricted boltzmann machines for collaborative filtering. In: Proc. ICML, pp. 791–798 (2007)
21. Takacs, G., Pilaszy, I., Nemeth, B., Tikk, D.: On the gravity recommendation system. In: Proc. KDD Cup and Workshop (2007)
22. Tikhonov, A.N., Arsenin, V.Y.: Solutions of Ill-posed Problems. John Wiley, New York (1977)
23. Wu, M.: Collaborative filtering via ensembles of matrix factorizations. In: Proc. KDD Cup and Workshop (2007)

Author Index

Lecture Notes in Computer Science

Sublibrary 3: Information Systems and Application, incl. Internet/Web and HCI

For information about Vols. 1– 4606
please contact your bookseller or Springer

Vol. 4816: B. Falcidieno, M. Spagnuolo, Y. Avrithis, I. Kompatsiaris, P. Buitelaar (Eds.), Semantic Multimedia. XII, 306 pages. 2007.

Vol. 4813: I. Oakley, S.A. Brewster (Eds.), Haptic and Audio Interaction Design. XIV, 145 pages. 2007.

Vol. 4810: H.H.-S. Ip, O.C. Au, H. Leung, M.-T. Sun, W.-Y. Ma, S.-M. Hu (Eds.), Advances in Multimedia Information Processing – PCM 2007. XXI, 834 pages. 2007.

Vol. 4809: M.K. Denko, C.-s. Shih, K.-C. Li, S.-L. Tsao, Q.-A. Zeng, S.H. Park, Y.-B. Ko, S.-H. Hung, J.-H. Park (Eds.), Emerging Directions in Embedded and Ubiquitous Computing. XXXV, 823 pages. 2007.

Vol. 4808: T.-W. Kuo, E. Sha, M. Guo, L.T. Yang, Z. Shao (Eds.), Embedded and Ubiquitous Computing. XXI, 769 pages. 2007.

Vol. 4806: R. Meersman, Z. Tari, P. Herrero (Eds.), On the Move to Meaningful Internet Systems 2007: OTM 2007 Workshops, Part II. XXXIV, 611 pages. 2007.

Vol. 4805: R. Meersman, Z. Tari, P. Herrero (Eds.), On the Move to Meaningful Internet Systems 2007: OTM 2007 Workshops, Part I. XXXIV, 757 pages. 2007.

Vol. 4804: R. Meersman, Z. Tari (Eds.), On the Move to Meaningful Internet Systems 2007: CoopIS, DOA, ODBASE, GADA, and IS, Part II. XXIX, 683 pages. 2007.

Vol. 4803: R. Meersman, Z. Tari (Eds.), On the Move to Meaningful Internet Systems 2007: CoopIS, DOA, ODBASE, GADA, and IS, Part I. XXIX, 1173 pages. 2007.

Vol. 4802: J.-L. Hainaut, E.A. Rundensteiner, M. Kirchberg, M. Bertolotto, M. Brochhausen, Y.-P.P. Chen, S.S.-S. Cherfi, M. Doerr, H. Han, S. Hartmann, J. Parsons, G. Poels, C. Rolland, J. Trujillo, E. Yu, E. Zimányie (Eds.), Advances in Conceptual Modeling – Foundations and Applications. XIX, 420 pages. 2007.

Vol. 4801: C. Parent, K.-D. Schewe, V.C. Storey, B. Thalheim (Eds.), Conceptual Modeling - ER 2007. XVI, 616 pages. 2007.

Vol. 4797: M. Arenas, M.I. Schwartzbach (Eds.), Database Programming Languages. VIII, 261 pages. 2007.

Vol. 4796: M. Lew, N. Sebe, T.S. Huang, E.M. Bakker (Eds.), Human–Computer Interaction. X, 157 pages. 2007.

Vol. 4794: B. Schiele, A.K. Dey, H. Gellersen, B. de Ruyter, M. Tscheligi, R. Wichert, E. Aarts, A. Buchmann (Eds.), Ambient Intelligence. XV, 375 pages. 2007.

Vol. 4777: S. Bhalla (Ed.), Databases in Networked Information Systems. X, 329 pages. 2007.

Vol. 4761: R. Obermaisser, Y. Nah, P. Puschner, F.J. Rammig (Eds.), Software Technologies for Embedded and Ubiquitous Systems. XIV, 563 pages. 2007.

Vol. 4747: S. Džeroski, J. Struyf (Eds.), Knowledge Discovery in Inductive Databases. X, 301 pages. 2007.

Vol. 4744: Y. de Kort, W. IJsselsteijn, C. Midden, B. Eggen, B.J. Fogg (Eds.), Persuasive Technology. XIV, 316 pages. 2007.

Vol. 4740: L. Ma, M. Rauterberg, R. Nakatsu (Eds.), Entertainment Computing – ICEC 2007. XXX, 480 pages. 2007.

Vol. 4730: C. Peters, P. Clough, F.C. Gey, J. Karlgren, B. Magnini, D.W. Oard, M. de Rijke, M. Stempfhuber (Eds.), Evaluation of Multilingual and Multi-modal Information Retrieval. XXIV, 998 pages. 2007.

Vol. 4723: M. R. Berthold, J. Shawe-Taylor, N. Lavrač (Eds.), Advances in Intelligent Data Analysis VII. XIV, 380 pages. 2007.

Vol. 4721: W. Jonker, M. Petković (Eds.), Secure Data Management. X, 213 pages. 2007.

Vol. 4718: J. Hightower, B. Schiele, T. Strang (Eds.), Location- and Context-Awareness. X, 297 pages. 2007.

Vol. 4717: J. Krumm, G.D. Abowd, A. Seneviratne, T. Strang (Eds.), UbiComp 2007: Ubiquitous Computing. XIX, 520 pages. 2007.

Vol. 4715: J.M. Haake, S.F. Ochoa, A. Cechich (Eds.), Groupware: Design, Implementation, and Use. XIII, 355 pages. 2007.

Vol. 4714: G. Alonso, P. Dadam, M. Rosemann (Eds.), Business Process Management. XIII, 418 pages. 2007.

Vol. 4704: D. Barbosa, A. Bonifati, Z. Bellahsène, E. Hunt, R. Unland (Eds.), Database and XML Technologies. X, 141 pages. 2007.

Vol. 4690: Y. Ioannidis, B. Novikov, B. Rachev (Eds.), Advances in Databases and Information Systems. XIII, 377 pages. 2007.

Vol. 4675: L. Kovács, N. Fuhr, C. Meghini (Eds.), Research and Advanced Technology for Digital Libraries. XVII, 585 pages. 2007.

Vol. 4674: Y. Luo (Ed.), Cooperative Design, Visualization, and Engineering. XIII, 431 pages. 2007.

Vol. 4663: C. Baranauskas, P. Palanque, J. Abascal, S.D.J. Barbosa (Eds.), Human-Computer Interaction – INTERACT 2007, Part II. XXXIII, 735 pages. 2007.

Vol. 4662: C. Baranauskas, P. Palanque, J. Abascal, S.D.J. Barbosa (Eds.), Human-Computer Interaction – INTERACT 2007, Part I. XXXIII, 637 pages. 2007.

Vol. 4658: T. Enokido, L. Barolli, M. Takizawa (Eds.), Network-Based Information Systems. XIII, 544 pages. 2007.

Vol. 4656: M.A. Wimmer, J. Scholl, Å. Grönlund (Eds.), Electronic Government. XIV, 450 pages. 2007.

Vol. 4655: G. Psaila, R. Wagner (Eds.), E-Commerce and Web Technologies. VII, 229 pages. 2007.

Vol. 4654: I.-Y. Song, J. Eder, T.M. Nguyen (Eds.), Data Warehousing and Knowledge Discovery. XVI, 482 pages. 2007.

Vol. 4653: R. Wagner, N. Revell, G. Pernul (Eds.), Database and Expert Systems Applications. XXII, 907 pages. 2007.

Vol. 4636: G. Antoniou, U. Aßmann, C. Baroglio, S. Decker, N. Henze, P.-L. Patranjan, R. Tolksdorf (Eds.), Reasoning Web. IX, 345 pages. 2007.

Vol. 4611: J. Indulska, J. Ma, L.T. Yang, T. Ungerer, J. Cao (Eds.), Ubiquitous Intelligence and Computing. XXIII, 1257 pages. 2007.

Vol. 4607: L. Baresi, P. Fraternali, G.-J. Houben (Eds.), Web Engineering. XVI, 576 pages. 2007.